Information Technology Project Management

Second Edition

Kathy Schwalbe, Ph.D., PMP

Augsburg College

Australia • Canada • Mexico • Singapore • Spain • United Kingdom • United States

Information Technology Project Management, Second Edition, is published by Course Technology.

Senior Vice President, Publisher
Kristen Duerr

Managing Editor
Jennifer Locke

Senior Product Manager
Margarita Donovan

Development Editor
Marilyn R. Freedman

Production Editor
Daphne E. Barbas

Associate Product Manager
Matthew Van Kirk

Marketing Manager
Toby Shelton

Editorial Assistant
Janet Aras

Text Designer
GEX Publishing Services

Cover Designer
Betsy Young

Printed in Canada

1 2 3 4 5 6 7 8 9 WC 02 01

For more information, contact Course Technology, 25 Thomson Place, Boston, Massachusetts, 02210.

Or find us on the World Wide Web at: www.course.com

ISBN 0-619-03528-5

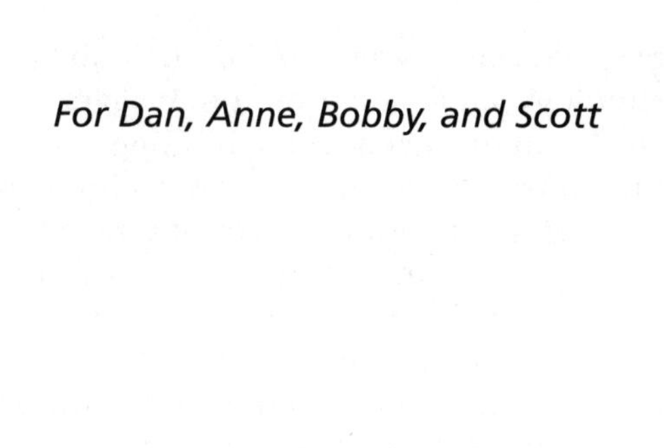

For Dan, Anne, Bobby, and Scott

Preface

The future of many organizations depends on their ability to harness the power of information technology, and good project managers are in high demand. Colleges are responding to this need by establishing courses in project management and making them part of the information technology or management curriculum. Corporations are investing in continuing education to help develop information technology project managers and effective project teams. This book provides a much-needed framework for teaching courses in information technology project management. The first edition of this book was extremely well received by people in academia and the workplace. The second edition builds on the strong points of the first edition and adds even more important information and features.

It's impossible to read a newspaper, magazine, or Web page without hearing about the impact of information technology on our society. Information is traveling faster and being shared by more individuals than ever before. Now you can do your banking or order your groceries on the World Wide Web; companies have linked their many systems together to help them fill orders on time and better serve their customers; and software companies are coming out with new products every day to help us streamline our work and get better results. Technology is changing almost everything about the way we live and work today. When this technology works well, it is almost invisible. But did it ever occur to you to ask, "Who makes these complex technologies and systems happen?"

Because you're reading this book, you must have an interest in the "behind-the-scenes" aspects of technology. If I've done my job correctly, as you read you'll begin to see the many innovations we are currently experiencing as the result of hundreds of successful information technology projects. In this book you'll read about projects that went well, like Northwest Airlines ResNet reservation system, Bank Of America's interstate banking project, Lucent Technology's fiber optic cable project, and Kodak's Advantix Advanced Photo System project. Of course, not all projects are successful; factors such as time, money, and unrealistic expectations, among many others, can sabotage a promising effort if it is not properly managed. In this book, you'll also learn from the mistakes people made on many projects that were not successful. I have written this book in an effort to educate you, tomorrow's information technology project managers, about what will help to make a project succeed—and what can make it fail.

Although project management has been an established field for many years, managing information technology projects requires ideas and information that go beyond standard project management. For example, many information technology projects fail because of a lack of user input, incomplete and changing requirements, and a lack of executive support. This book includes suggestions on dealing with these issues. New technologies can also aid in managing information technology projects, and examples of using software to assist in project management are included throughout the text.

Information Technology Project Management, Second Edition, is still the only textbook to apply all nine project management knowledge areas—project integration, scope, time, cost, quality, human resource, communications, risk, and procurement management—and all five process groups—initiating, planning, executing, controlling, and closing—to information technology projects. This text builds on the PMBOK Guide 2000, an American National Standard, to provide a solid framework and context for managing information technology projects. It also includes an appendix, *A Guide to Microsoft Project 2000,* and a 120-day trial version of Microsoft Project 2000 software. A second appendix provides advice on earning and maintaining Project Management Professional (PMP) certification from the Project Management Institute (PMI) as well as information on other certification programs, such as CompTIA's IT Project+ certification.

Information Technology Project Management, Second Edition, was written to provide practical lessons in project management for students and practitioners alike. By weaving together theory and practice, this text presents an understandable, integrated view of the many concepts, skills, tools, and techniques involved in information technology project management. The comprehensive design of the text provides a strong foundation for students and practitioners in information technology project management.

NEW TO THE SECOND EDITION

Building on the success of the first edition, *Information Technology Project Management, Second Edition,* introduces a uniquely effective combination of features. The main changes made to the second edition include the following:

- This edition reflects information from the PMBOK Guide 2000. The PMBOK Guide 2000 is a key document in the project management profession and is ANSI approved. Several major changes were made from the PMBOK Guide 1996 version, especially in the project risk management knowledge area. Other changes include new terminology for earned value management and new and revised definitions for several other project management terms. For example, the PMBOK Guide 2000 now includes a brief definition of critical chain scheduling. *Information Technology Project Management, Second Edition,* helps you understand critical chain scheduling by providing several pages and figures explaining this important new concept and examples of how companies use this technique to improve productivity.

- Microsoft Project 2000 examples are used throughout the text, and Appendix A provides an excellent guide for using this popular project management software tool. Students can work through the appendix using their own 120-day copy of the software and really learn how to use Project 2000 effectively. The example project used in Appendix A has been simplified, and there are more screen shots and subheadings to help students learn how to use this powerful software.
- Minicases are now included at the end of chapters. Students learn best when they can apply what they are learning. These new minicases provide great examples of how various project management concepts, tools, and techniques can be applied to real-world problems. Several minicases involve use of Project 2000, Excel, web authoring tools, and other software to improve students' hands-on skills.
- Updated examples are provided throughout the text. You'll notice several new examples in the second edition to reflect recent events in managing real information technology projects. Several of the What Went Right and What Went Wrong examples have been updated, and many new suggested readings have been added to keep you up-to-date. Results of new studies are also included throughout the text.
- User feedback is incorporated. Based on student and faculty feedback on the first edition of this text, you'll see several changes, mostly additions, to help clarify information. For example, Chapter 1, *Introduction to Project Management,* includes a new section describing recent developments in project management software products. Chapter 2, *The Project Management Context and Processes,* includes a new section on developing an information technology project management methodology. Chapter 4, *Project Scope Management,* provides even more suggestions for creating a good WBS. Chapter 9, *Project Communications Management,* includes a new table describing how different types of media are suited to different communication needs as well as additional suggestions for managing conflict.

APPROACH

Many people have been practicing some form of project management with little or no formal study in this area. New books and articles are being written each year as we discover more about the field of project management, and project management software continues to advance. Because the project management field and the technology industry are changing rapidly, you cannot assume that what worked twenty years ago is still the best approach today. This text provides up-to-date information on how good project management and effective use of project management software can help you manage information technology projects. Three distinct features of this book include its relationship to the Project Management Body of Knowledge, its bundling with

Microsoft Project 2000, and its value in preparing for Project Management Professional and other certification exams.

Based on the PMBOK Guide 2000

The Project Management Institute created the Guide to the Project Management Body of Knowledge (the PMBOK Guide[1]) as a framework and starting point for understanding project management. It includes an introduction to project management, brief descriptions of all nine project management knowledge areas, and a glossary of terms. The PMBOK Guide is, however, just that, a guide. This text uses the PMBOK Guide 2000 as a foundation. It goes beyond the Guide by providing more details, highlighting additional topics, and providing a real-world context for project management. *Information Technology Project Management, Second Edition,* explains project management specifically as it applies to managing information technology projects in the 21st century. It includes several unique features to bring to its readers the excitement of this dynamic field (for more information on features, see the section entitled "Pedagogical Features," below).

Bundled with Microsoft Project 2000

Software has advanced tremendously in recent years, and it is important for project managers and their teams to use software to assist them in managing information technology projects. Each copy of *Information Technology Project Management, Second Edition*, includes a 120-day trial version of the leading project management software on the market—Microsoft Project 2000. Examples using Project 2000 and other software tools are provided throughout the text. Several chapters include sections describing how software can enhance project management. Appendix A, *A Guide to Microsoft Project 2000*, explains how to use this powerful software to help in project scope, time, cost, human resource, and communications management.

Resource for PMP and Other Certification Exams

Professional certification is an important factor in recognizing and ensuring quality in a profession. PMI provides certification as a Project Management Professional (PMP), and this text is an excellent resource for studying for the certification exam. This text will also help you pass other certification exams,

[1] The Project Management Body of Knowledge (PMBOK Guide) is a key document in the project management profession. Excerpts of the 2000 edition are available free of charge from the Project Management Institute's Web site, www.pmi.org

such as PMI's Certificate of Added Qualification (CAQ) in systems development and CompTIA's IT+ Project+ exam. Having experience working on projects does not mean you can easily pass the PMP or other certification exams.

I like to tell my students a story about taking a driver's license test after moving to Minnesota. I had been driving very safely and without accidents for over sixteen years, so I thought I could just walk in and take the test. I was impressed by the sophisticated computer system used to administer the test. The questions were displayed on a large touch-screen monitor, often along with an image or video to illustrate different traffic signs or driving situations. I became concerned when I found I had no idea how to answer several questions, and I was perplexed when the test seemed to stop and a message displayed saying, "Please see the person at the service counter." This was a polite way of saying I had failed the test! After controlling my embarrassment, I picked up one of the Minnesota driving test brochures, studied it for an hour or two that night, and successfully passed the test the next day.

The point of this story is that it is important to study information from the organization that creates the test and not be over-confident that your experience is enough. Because this text is based on PMI's PMBOK Guide, it provides a valuable reference for studying for PMP certification. Appendix B provides specific advice on PMP and other certification exams. Although the current PMP exam tests for all types of project management, PMI is also currently working on an extension to the exam specifically for information technology professionals. *Information Technology Project Management, Second Edition,* is an excellent resource for PMI's new exams for IT project managers.

ORGANIZATION AND CONTENT

Information Technology Project Management, Second Edition, is organized into three main sections to provide a framework for project management, a detailed description of each project management knowledge area, and an application of the project management process groups to a successful information technology project. The first two chapters form a first section that provides an introduction to the project management framework and sets the stage for the remaining chapters. Chapter 1 introduces the critical need for better project management in the information technology field. It provides an overview of the field of project management, including definitions of fundamental terms, the relationship between project management and other disciplines, a brief history of the field and how it has changed, and an introduction to the project management profession, including information on careers, certification, ethics, and project management software. Chapter 2 provides a context for project management. It focuses on the need to understand how projects fit into an entire organizational system and applies common project management terms to concepts used in the information technology field. Chapter 2 also introduces the project management process groups—initiating, planning, executing, controlling, and closing. It briefly describes these processes and how they differ from product process groups and provides a matrix relating the process groups to the knowledge areas. A new section in this chapter describes how to develop an information technology project management methodology based on the PMBOK Guide and unique organizational needs.

Chapters 3 through 11 form a second section of the book that describes each of the project management knowledge areas—project integration, scope, time, cost, quality, human resource, communications, risk, and procurement management—in the context of information technology projects. An entire chapter is dedicated to each knowledge area. Each knowledge area chapter includes sections that map to their major processes as described in the PMBOK Guide 2000. For example, the chapter on project integration management includes sections on project plan development, project plan execution, and integrated change control. Additional sections highlight other important concepts related to each knowledge area. Each chapter also includes detailed examples of key project management tools and techniques as applied to information technology projects. For example, the chapter on project scope management includes samples of a project charter, net present value analysis, and work breakdown structures for several information technology projects.

Chapters 12 through 16 form a third section that documents a highly successful information technology project from start to finish. These chapters apply the project management process groups—initiating, planning, executing, controlling, and closing—to a real information technology project. Northwest Airlines' ResNet projects in the mid-1990s and early 2000s resulted in a new reservation system and increased profits for the company. In its first year of operation, ResNet saved Northwest Airlines over $15 million, and savings were over $33 million in the second year. The new reservation system allowed the company to effectively manage a fourfold growth in revenues in a five-year period. Techniques discussed in earlier chapters (project justification, work breakdown structures, Gantt charts, cost estimates, status reports, audit reports, and so on) are illustrated using Northwest's real project documents. This running case, based on a real, large-scale information technology project, will give you a better understanding of how many of the topics covered in earlier chapters are used in the real world. Several faculty reviewers of this text commented that this section went far beyond other texts by turning a spotlight on an actual major project for an extended period of time and from many standpoints.

PEDAGOGICAL FEATURES

Several pedagogical features are included in this text to enhance presentation of the materials so that readers can more easily understand the concepts and apply them. Throughout the book, emphasis is placed on applying concepts to up-to-date, real-world information technology project management.

Learning Objectives, Chapter Summaries, Discussion Questions, and Exercises

Learning Objectives, Chapter Summaries, Discussion Questions, and Exercises are designed to function as integrated study tools. Learning Objectives reflect what readers should be able to accomplish after completing each chapter. Chapter Summaries highlight key concepts readers should master. The Discussion Questions help guide critical thinking about those key concepts. Exercises provide opportunities to practice important techniques.

Chapter-Opening Case and Case Wrap Up

To set the stage, each chapter begins with a case related to the materials in that chapter. These scenarios (most based on the author's experiences) spark student interest and introduce important concepts in a real-world context. As project management concepts and techniques are discussed, they are applied to the opening case and other similar scenarios. Each chapter then closes the case—with some ending successfully and some, realistically, failing—to further illustrate the real world of project management.

What Went Right and What Went Wrong

Failures, as much as successes, can be valuable learning experiences. Each chapter of the text includes one or more examples of real information technology projects that went right as well as examples of projects that went wrong. These examples further illustrate the importance of mastering key concepts in each chapter.

Minicases

Minicases, a new feature in the second edition, provide more challenging assignments to help you apply some of the concepts, tools, and techniques discussed in each chapter. Each minicase has two parts based on the same business scenario. Instructors can assign minicases for homework or use them as in-class exercises to help students apply what they are learning.

Suggested Readings

Every chapter ends with a list of suggested readings, annotated by the author, that offer opportunities for further research. Several exercises require readers to review these readings or research related topics. The World Wide Web is widely referenced throughout the text as a research tool, and many of the Suggested Readings can be found on the Web.

Key Terms

The fields of information technology and project management both include many unique terms that are key to creating a workable language when the two fields are combined. Key terms are displayed in bold face and are defined the first time they appear. Definitions of key terms are provided in alphabetical order at the end of each chapter and in a glossary at the end of the book.

Application Software

Learning becomes much more dynamic with hands-on practice using the top project management software tool in the industry, Microsoft Project 2000, as well as other tools, such as spreadsheet software and Internet browsers. Each chapter offers you many opportunities to get hands-on and build new software skills. This text is written from the point of view that reading about something only gets you so far; in order to understand project management, you have to do it for yourself. In addition to the exercises and minicases found at the end of each chapter, several challenging projects are provided at the end of Appendix A, *A Guide to Using Microsoft Project 2000*.

SUPPLEMENTS

When this text is used in an academic setting, an Instructor's Resource Kit accompanies it. The CD-ROM-based kit includes an Instructor's Manual with Solutions, PowerPoint-based Course Presenter, and Course Test Manager, Course Technology's cutting-edge Windows-based testing software. The Instructor's Manual includes sample syllabi, discussion topics, essay topics, and solutions to all of the end-of-chapter material in the text. Course Presenter is a presentation tool developed in Microsoft PowerPoint that utilizes all of the art from the text to generate impressive screen shows for use in your lectures. Course Test Manager includes up to 100 questions per chapter, and allows you to create printed and online pretests, practice tests, and actual examinations.

ACKNOWLEDGEMENTS

I never would have taken on this project—writing this book, the first or second edition—without the help of many people. I would like to thank the staff at Course Technology for their dedication and hard work in helping me produce this book and in doing such an excellent job of marketing it. Jennifer Locke was a great project manager and editor in coordinating all of the people involved in producing this book. I owe tremendous gratitude to Marilyn Freedman, my development editor for both the first and second edition. It was valuable having someone read every word that I wrote and offer helpful suggestions on how to make the book even better. I learned a lot about collaborative writing from Marilyn. Daphne Barbas did an excellent job as the production editor. It's amazing how many little details are involved in writing a good book, and Daphne and the other editors and artists supporting Course Technology did a great job in producing a high quality text.

I thank my many colleagues who contributed information to this book. I again used materials in the project risk management chapter provided by David Jones of Lockheed Martin and an adjunct professor at Augsburg College and the University of St. Thomas; cost management information provided by Jodi Curtis, a project manager at SafeNet Consulting and member of my PMP study group; and procurement management information provided by Rita Mulcahy, who ran my PMP review course. I also thank Karen Boucher of the Standish Group, Bill Munroe of Blue Cross Blue Shield, and Tess Galati of Practical Communications, Inc. for providing excellent materials for the second edition of this book.

Special thanks go to Rachel Hollstadt, CEO of Hollstadt & Associates, for providing materials on testing for Chapter 7, *Project Quality Management,* and also for suggesting that I contact people at Northwest Airlines to find a large information technology project to include in this book. Rachel and I met with Bob Borlik, Vice President of Information Services at Northwest Airlines, to see if Northwest Airlines would be willing to share information on one of their information technology projects. Bob immediately suggested the ResNet project as an example of one of the most successful information technology projects of which he was aware. He then recommended I talk to Lori Manacke, Director of Business Results in his department. Lori then introduced me to Arvid Lee and Kathy Christenson.

Arvid and Kathy were wonderful collaborators. Their enthusiasm for the ResNet projects was obvious, as were their intelligence, diligence, and fun-loving natures. We had several marathon sessions discussing and documenting the initiating, planning, executing, controlling, and closing of the ResNet projects. Many thanks also go to Peeter Kivestu, the project manager for the ResNet projects. Peeter used many creative techniques to make the ResNet projects so successful. I owe special thanks to Arvid Lee, project manager for the ResNet+ project, for providing new information in 2001 related to NWA's reservation systems and project management practices.

I also want to thank my students at Augsburg College and the University of Minnesota, for providing feedback on the first edition of this book. I received many valuable comments from them on ways to improve the text and structure of my courses. I am also grateful for the articles students gave me, which provided additional examples of information technology projects, material for the What Went Right and What Went Wrong features, and additional references. Student papers and discussions on critical chain scheduling also helped me to further understand that topic and write about it in this edition.

Faculty and staff at Augsburg College were very supportive of my work. I am grateful for summer research money provided by the faculty development office, which encouraged me to begin writing this book. I also thank Lee Clarke at Augsburg College for agreeing to adjust his teaching schedule so I could have more time to write this second edition.

Three faculty reviewers provided excellent feedback for me in writing this second edition. Sherry Thatcher of the University of Arizona, Ann Digilio of The University of Findlay, and Steven White of Anne Arundel Community College, provided outstanding suggestions for improving the text in this second edition. I also wish to thank the reviewers of the first edition, Barbara Denison, Wright State University; Barbara Grabowski, Benedictine University; Bruce Hartman, Lucent Corporation; and Tom Logan, National American University.

I thank the people involved in Minnesota's PMI chapter. Cliff Sprague and Michael Branch gave me the opportunity to solicit information for the book at a dinner meeting and I made several valuable contacts there. I continue to serve on the board of the Minnesota PMI chapter and enjoy working with all of the people involved.

Most of all, I am grateful to my family. Without their support, I never could have written this book. My wonderful husband, Dan, was very patient and supportive of me working like crazy to write the first edition of this book. The second edition wasn't quite as hectic, but I still needed support from Dan, especially when we experienced several computer problems. Dan has written textbooks himself, so he provided some great advice. He even referred to parts of my book to help him in his new job as a lead architect for software development with Com Squared Systems, Inc. Our three children, Anne, Bobby, and Scott, continue to be very supportive of their mom's work. Our oldest child, Anne, starts college in the fall of 2001, and she actually thinks it's cool that her mom wrote a textbook. Our children all understand the main reason why I write—I have a passion for educating future leaders of the world, including them.

As always, I am eager to receive your feedback on this book. Please send all feedback to Course Technology at mis@course.com. My editors will make sure it gets to me!

Kathy Schwalbe, Ph.D., PMP
Augsburg College

ABOUT THE AUTHOR

Kathy Schwalbe is an Associate Professor in the Department of Business Administration at Augsburg College in Minneapolis, where she teaches courses in project management, computing for business, systems analysis and design, information systems projects, and electronic commerce. She is the area coordinator for the management information systems major and program manager for Augsburg's Information Technology Certificate program. Kathy is also an adjunct faculty member at the University of Minnesota, where she teaches a graduate level course in project management. Kathy worked for ten years in industry before entering academia in 1991. She was an Air Force officer, systems analyst, project manager, senior engineer, and information technology consultant. Kathy is an active member of PMI, serving as the Student Chapter Liaison for PMI-Minnesota, Editor of the ISSIGreview, and member of PMI's test writing team. Kathy earned her Ph.D. in Higher Education atthe University of Minnesota, her MBA at Northeastern University's High Technology MBA program, and her B.S. in mathematics at the University of Notre Dame.

Brief Contents

Chapter 1 Introduction to Project Management 1
Chapter 2 The Project Management Context and Processes 23
Chapter 3 Project Integration Management 58
Chapter 4 Project Scope Management 83
Chapter 5 Project Time Management 119
Chapter 6 Project Cost Management 157
Chapter 7 Project Quality Management 191
Chapter 8 Project Human Resource Management 227
Chapter 9 Project Communications Management 267
Chapter 10 Project Risk Management 301
Chapter 11 Project Procurement Management 335
Chapter 12 Initiating 361
Chapter 13 Planning 379
Chapter 14 Executing 401
Chapter 15 Controlling 417
Chapter 16 Closing 433
Appendix A Guide to Using Microsoft Project 2000 447
Appendix B Advice for the PMP Exam and Related Certifications 525
Glossary 535
Index 547

Contents

Chapter 1 Introduction to Project Management

Introduction 2

What is a Project? 4

What is Project Management? 7

How Project Management Relates to Other Disciplines 10

History of Project Management 11

The Project Management Profession 13

Project Management Careers 14
Project Management Certification 15
Code of Ethics 16
Project Management Software 16

Chapter 2 The Project Management Context and Processes

A Systems View of Project Management 25

Project Phases and the Project Life Cycle 27

Product Life Cycles 29
The Importance of Project Phases and Management Reviews 31

Understanding Organizations 32

Organizational Structures 34
Stakeholder Management 37
Top Management Commitment 38

Suggested Skills for a Project Manager 41

Project Management Process Groups 43

Developing an Information Technology Project Management Methodology 48

Chapter 3 Project Integration Management

What is Project Integration Management? 59

Project Plan Development 62

Project Plan Execution 68

Integrated Change Control 71

Change Control on Information Technology Projects 73
Change Control System 74

Chapter 4 Project Scope Management

What Is Project Scope Management? 84

Project Initiation: Strategic Planning and Project Selection 85
Identifying Potential Projects 85
Methods for Selecting Projects 88
Project Charters 96
Scope Planning and the Scope Statement 98
The Scope Statement 98
Scope Definition and the Work Breakdown Structure 99
The Work Breakdown Structure 100
Approaches to Developing Work Breakdown Structures 104
Advice for Creating a WBS 106
Scope Verification and Scope Change Control 107
Suggestions for Improving User Input 109
Suggestions for Reducing Incomplete and Changing Requirements 109

Chapter 5 Project Time Management

Importance of Project Schedules 120
Where Do Schedules Come From? Defining Activities 122
Activity Sequencing 123
Project Network Diagrams 124
Activity Duration Estimating 127
Schedule Development 128
Gantt Charts 129
Critical Path Method 132
Critical Chain Scheduling 138
Program Evaluation and Review Technique (PERT) 141
Controlling Changes to the Project Schedule 142
Reality Checks on Scheduling 142
Working with People Issues 143
Using Software to Assist in Time Management 145
Words of Caution on Using Project Management Software 146

Chapter 6 Project Cost Management

The Importance of Project Cost Management 158
Basic Principles of Cost Management 160
Resource Planning 164
Cost Estimating 165
Types of Cost Estimates 165
Cost Estimation Techniques and Tools 167
Typical Problems with Information Technology Cost Estimates 169

Sample Cost Estimates and Supporting Detail for an Information Technology Project 170

Cost Budgeting 173

Cost Control 175

Earned Value Management 175

Using Software to Assist in Cost Management 183

Chapter 7 Project Quality Management

Quality of Information Technology Projects 192

What Is Project Quality Management? 195

Modern Quality Management 196

Deming 196

Juran 197

Crosby 198

Ishikawa 198

Taguchi 199

Feigenbaum 200

Malcolm Baldrige Award and ISO 9000 200

Quality Planning 201

Quality Assurance 203

Quality Control 204

Tools and Techniques for Quality Control 204

Pareto Analysis 204

Statistical Sampling and Standard Deviation 206

Quality Control Charts, Six Sigma, and the Seven Run Rule 208

Testing 211

Improving Information Technology Project Quality 213

Leadership 213

The Cost of Quality 215

Organizational Influences, Workplace Factors, and Quality 216

Maturity Models 217

Chapter 8 Project Human Resource Management

The Importance of Human Resource Management 228

Current State of Human Resource Management 228

Implications for the Future of Human Resource Management 231

What is Project Human Resource Management? 232

Keys to Managing People 233

Motivation Theories 233

Influence and Power 236

Improving Effectiveness 238

Organizational Planning 241
Issues in Project Staff Acquisition and Team Development 247
Staff Acquisition 247
Resource Loading and Leveling 249
Team Development 252
Using Software to Assist in Human Resource Management 257

Chapter 9 Project Communications Management

The Importance of Project Communications Management 268
Communications Planning 270
Information Distribution 272
Using Technology to Enhance Information Distribution 273
Formal and Informal Methods for Distributing Information 273
Determining the Complexity of Communications 275
Performance Reporting 278
Administrative Closure 279
Suggestions for Improving Project Communications 280
Using Communication Skills to Manage Conflict 281
Developing Better Communication Skills 282
Running Effective Meetings 283
Using Templates for Project Communications 285
Developing a Communications Infrastructure 291
Using Software to Assist in Project Communications 292

Chapter 10 Project Risk Management

The Importance of Project Risk Management 302
Risk Management Planning 305
Common Sources of Risk on Information Technology Projects 307
Risk Identification 310
Qualitative Risk Analysis 313
Calculating Risk Factors Using Probability/Impact Matrixes 313
Top 10 Risk Item Tracking 316
Expert Judgment 317
Quantitative Risk Analysis 318
Decisions Trees and Expected Monetary Value 318
Simulation 320
Risk Response Planning 321
Risk Monitoring and Control 323
Using Software to Assist in Project Risk Management 323
Results of Good Project Risk Management 326

Chapter 11 Project Procurement Management

Importance of Project Procurement Management 336

Procurement Planning 339

Procurement Planning Tools and Techniques 340

Types of Contracts 341

Statement of Work 344

Solicitation Planning 345

Solicitation 348

Source Selection 348

Contract Administration 351

Contract Close-out 352

Using Software to Assist in Project Procurement Managment 352

Chapter 12 Initiating

What Is Involved in Project Initiation? 362

Background on Northwest Airlines 363

Background on ResNet 364

Selecting the Project Manager 368

Preparing Business Justification for the Projects 368

ResNet Beta Project 368

ResNet 1995 and ResNet 1996 369

Developing the Project Charter 373

Actions of the Project Manager and Senior Management in Project Initiation 374

Chapter 13 Planning

What Is Involved in Project Planning? 380

Developing the Project Plans 383

ResNet Beta Project Planning 383

ResNet 1995 Planning 384

ResNet 1996 Planning 384

Determining Project Scope and Schedules 386

ResNet Cost Estimates 391

Human Resource and Communications Planning 394

Quality, Risk, and Procurement Planning 396

Chapter 14 Executing

What Is Involved in Executing Projects? 402

Providing Project Leadership 403

Developing the Core Team 404
Verifying Project Scope 405
Assuring Quality 405
Disseminating Information 407
Procuring Necessary Resources 408
Training Users to Develop Code 410

Chapter 15 Controlling

What Is Involved in Controlling Projects? 418
Schedule Control 419
Scope Change Control 421
Quality Control 423
Performance and Status Reporting 425
Managing Resistance to Change 426

Chapter 16 Closing

What Is Involved in Closing Projects? 434
Administrative Closure 435
ResNet Audit 436
ResNet Final Recognition Party and Personnel Transition 438
Lessons Learned 439
ResNet+ 440
ResNet+ and Project Management at nwa in the 21st Century 442

Appendix A Guide to Using Microsoft Project 2000

Introduction 448
Overview of Microsoft Project 2000 449
- Starting Project 2000 and the Project Help Window 449
- Main Screen Elements 452
- Project 2000 Views 454
- Project 2000 Filters 457

Project Scope Management 459
- Creating a New Project File 460
- Developing a Work Breakdown Structure 462
- Saving Project Files with or Without a Baseline 465

Project Time Management 466
- Entering Task Durations 466
- Establishing Task Dependencies 472
- Changing Task Dependency Types and Adding Lead or Lag Time 475
- Gantt Charts 478

Network Diagrams 480
Critical Path Analysis 482
Project Cost Management 484
Fixed and Variable Cost Estimates 484
Assigning Resources to Tasks 488
Baseline Plan, Actual Costs, and Actual Times 493
Earned Value Management 498
Project Human Resource Management 499
Resource Calendars 499
Resource Histograms 501
Resource Leveling 505
Project Communications Management 507
Common Reports and Views 507
Using Templates and Inserting Hyperlinks and Comments 509
Saving Files as Web Pages 512
Using Project 2000 in Workgroups 515
Project Central 516
Exercises 518
Exercise A-1: Web Site Development 518
Exercise A-2: Software Training Program 520
Exercise A-3: Project Tracking Database 522
Exercise A-4: Real Project Application 524

Appendix B Advice for the PMP Exam and Related Certifications
What is PMP Certification? 525
What Are the Requirements for Earning and Maintaining PMP Certification? 526
What Other Certification Exams Related to Information Technology Project Management Are Available? 527
What Is the Structure and Content of the PMP Certification Exam? 528
How Should You Prepare for the PMP Exam? 529
Ten Tips for Taking the PMP Exam 530
Sample Questions 531
Final Advice on PMP Certification and Project Management in General 533

Glossary 535

Index 547

1

Introduction to Project Management

Objectives

After reading this chapter, you will be able to:

1. *Explain what a project is and provide examples of information technology projects*
2. *Describe what project management is and discuss key elements of the project management framework*
3. *Describe how project management relates to other disciplines*
4. *Understand the history of project management*
5. *Explain the growing need for better project management, especially for information technology projects*
6. *Describe the project management profession, including recent trends in project management certification and software products*

Five hundred people, all members of the information technology department, gathered in the large corporate auditorium at the request of Anne Roberts, the new Vice President (VP) of Operations. Although the company, a large retail chain, was doing fairly well, several changes were occurring in the industry. Many of its competitors had implemented new information systems to improve inventory control and streamline the sales and distribution processes. Several start-up companies were taking away traditional business by selling products over the Internet. Anne's company was best known for its quality products and brand names, but its information systems were nothing to be proud of. The company's president hired Anne to focus on improving operations by taking advantage of new technologies.

Anne began to address the audience, "I have some good news and some bad news. The good news is that, compared to last year's industry average, our dismal information technology project success rate of 30 percent was above average. The bad news is that it is totally unacceptable. Most companies have improved information technology project management processes, and our direct competitors are at least two years ahead of us in their use of information technology. I stand before you today to extend an invitation and a challenge. I invite you all to work with me and your business colleagues in developing new systems and processes that will ensure our success in the 21st century. I have been authorized to spend whatever it takes to turn things around. Our challenge is to prove to the executive board of directors that our internal information technology department can improve our operations by delivering quality information technology products and services on time and on budget. To meet this challenge, we must work together to focus on finding solutions to complex problems. We must decide what information technology projects will most benefit the company, and then we must plan and successfully execute those projects. If we succeed, the board will provide profit-sharing bonuses for everyone. If we fail, the board is considering outsourcing the majority of our information technology functions."

INTRODUCTION

Many people and companies today seem to have a new or renewed interest in project management. Until the 1980s, project management primarily focused on providing schedule and resource data to senior management. This tracking of a few key project parameters is still an important element, but today's project management involves much more. Beginning in the late 1990s, business environments became more complex than those of earlier decades. Today, new technologies have become a significant factor in many businesses. Computer hardware, software, and networks, as well as the use of interdisciplinary and global work teams, have radically changed the work environment. According to the Standish Group, there has been an information technology "project gold rush." In 1998, corporate America issued 200,000 new-start application software development projects. During the year 2000, there were 300,000 new starts—up by 100,000 projects—and more than half a million new projects will be initiated during 2001. These changes have fueled the need for more sophisticated and better project management. Today's corporations are recognizing that, to be successful, they need to be conversant with and use modern project management techniques.

What Went Wrong?

In 1995, the Standish Group published an often-quoted study entitled "CHAOS".[1] This prestigious consulting firm surveyed 365 information technology executive managers in the U.S. who managed over 8,380 different information technology applications. As the title suggests, information technology projects were in a state of chaos. United States companies spent more than $250 billion each year in the early 1990s on information technology application development of approximately 175,000 projects. Examples of these projects included creating a new database for a state department of motor vehicles, developing a new system for car rental and hotel reservations, and implementing a client-server architecture for the banking industry. The survey found that the average cost of an information technology development project for a large company was over $2.3 million; for a medium company, it was over $1.3 million; and for a small company, it was over $434,000. Their study reported that the overall success rate of information technology projects was *only* 16.2 percent. They defined success as meeting project goals on time and on budget. The study also found that in 1995 over 31 percent of information technology projects were canceled before completion, costing U.S. companies and government agencies over $81 billion. The authors of this study were adamant about the need for better project management in the information technology industry. They said, "Software development projects are in chaos, and we can no longer imitate the three monkeys—hear no failures, see no failures, speak no failures."

Many organizations claim that using project management grants them advantages, such as:

- Better control of financial, physical, and human resources
- Improved customer relations
- Shorter development times
- Lower costs
- Higher quality and increased reliability
- Higher profit margins
- Improved productivity
- Better internal coordination
- Higher worker morale

This chapter provides an introduction to projects and project management, compares project management to other disciplines, gives a brief history of project management, and provides information on this growing profession. Although project management can be applied to many different industries and types of projects, this textbook focuses on applying project management to information technology projects.

[1] The Standish Group, "CHAOS" (**www.standishgroup.com/chaos.html**) (1995). Another reference is Johnson, Jim, "CHAOS: The Dollar Drain of IT Project Failures," *Application Development Trends* (January 1995).

WHAT IS A PROJECT?

To discuss project management, it is important to first understand the concept of a project. A **project** is a temporary endeavor undertaken to accomplish a unique purpose. Projects normally involve several people performing interrelated activities, and the main sponsor for the project is often interested in the effective use of resources to complete the project in an efficient and timely manner. The following attributes help to further define a project:

- *A project has a unique purpose.* Every project should have a well-defined objective. For example, Anne Roberts, the VP of Operations in the opening case, might sponsor an information technology collaboration project to develop a list and initial analysis of potential information technology projects that might improve operations for the company. The unique purpose of this project would be to create a collaborative report with ideas from people throughout the company. The results would provide the basis for further discussions and projects. As in this example, projects provide a unique product, service, or result.
- *A project is temporary.* A project has a definite beginning and a definite end. In the information technology collaboration example, Anne might form a team of people to immediately work on the information technology collaboration project, and then expect a report and an executive presentation of the results in one month.
- *A project requires resources, often from various areas.* Resources include people, hardware, software, or other assets. Many projects cross departmental or other boundaries in order to achieve their unique purposes. For the information technology collaboration project, people from information technology, marketing, sales, distribution, and other areas of the company would need to work together to develop ideas. They might also hire outside consultants to provide input. Once the project team finishes selecting key projects, they will probably require additional hardware, software, and network resources. People from other companies—product vendors and consulting companies—will become resources for meeting new project objectives. Resources, however, are not unlimited. They must be used effectively in order to meet project and other corporate goals.
- *A project should have a primary sponsor or customer.* Most projects have many interested parties or stakeholders, but someone must take the primary role of sponsorship. The project sponsor usually provides the direction and funding for the project. In this case, Anne Roberts would be the sponsor for the information technology collaboration project. Once further information technology projects are selected, however, the sponsors for those projects would be senior managers in charge of the main parts of the company affected by the projects. For example, if a project were initiated to provide direct product sales via the Internet, the head of sales

might be the project sponsor. If several projects related to Internet technologies were undertaken, a program might be formed. A **program** is a group of projects managed in a coordinated way. A program director would provide leadership for these projects, and the sponsors might come from several difficult business areas.

- *A project involves uncertainty.* Because every project is unique, it is sometimes difficult to clearly define the project's objectives, estimate how long it will take to complete, or how much it will cost. This uncertainty is one of the main reasons project management is so challenging, especially on projects involving new technologies.

A good project manager is the key to a project's success. Project managers work with the project sponsors, the project team, and the other people involved in a project to try to meet goals.

Every project is constrained in different ways by its scope, time goals, and cost goals. These limitations are sometimes referred to in project management as the **triple constraint**. To create a successful project, scope, time, and cost must all be taken into consideration, and it is the project manager's duty to balance these three often competing goals. He or she must consider the following:

- *Scope:* What is the project trying to accomplish? What unique product or service does the customer or sponsor expect from the project?
- *Time:* How long should it take to complete the project? What is the project's schedule?
- *Cost:* What should it cost to complete the project?

Figure 1-1 illustrates the three dimensions of the triple constraint. Note that each area—scope, time, and cost—has a target at the beginning of the project. For example, the information technology collaboration project might have an initial scope of producing a fifty-page report and a one-hour presentation on about thirty different potential information technology projects. The project scope might be defined further by providing a description of each potential project, an investigation of what other companies have implemented for similar projects, a rough time and cost estimate, and assessments of the risk and potential payoff as high, medium, or low. The initial time estimate for this project might be one month, and the cost estimate might be $50,000. These expectations would provide the targets for the scope, time, and cost dimensions of the project.

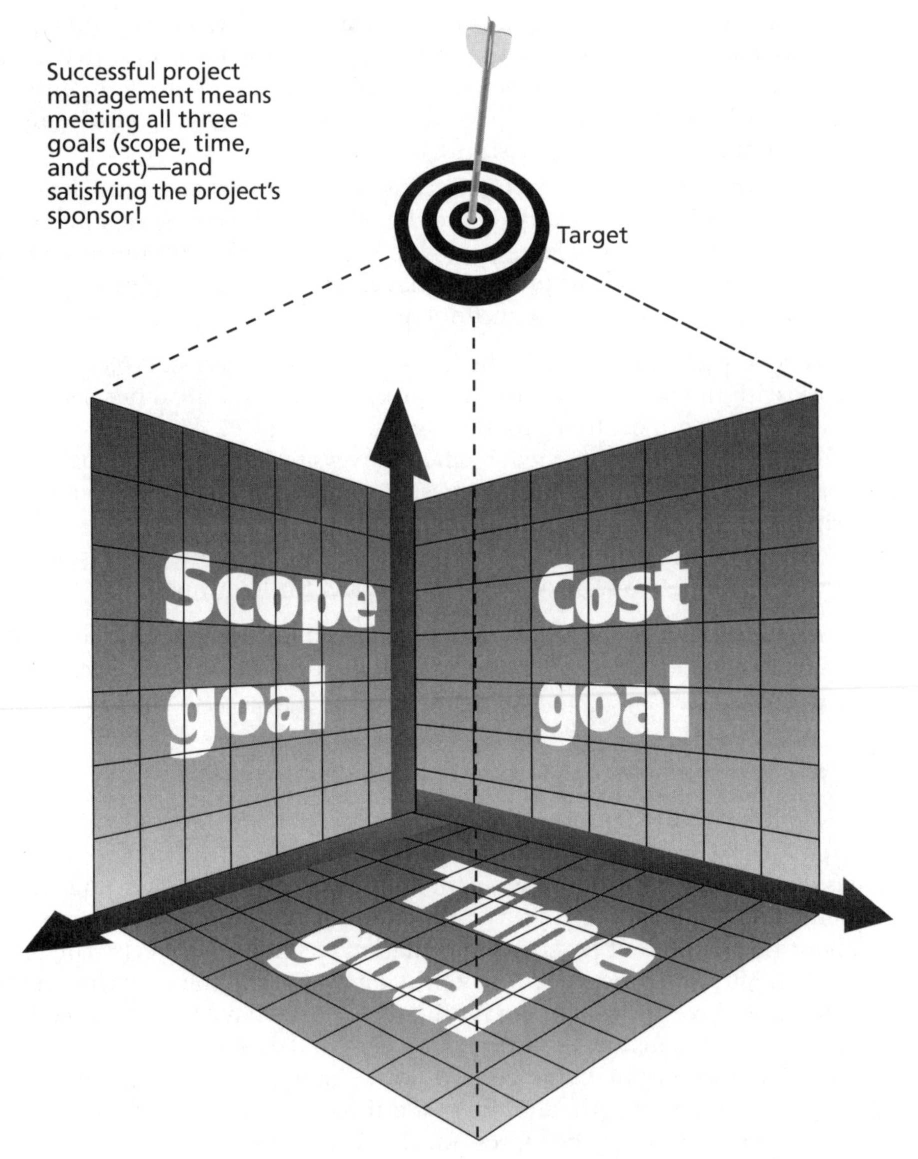

Figure 1-1. The Triple Constraint of Project Management

Managing the triple constraint involves making trade-offs between scope, time, and cost goals for a project. Because of the uncertain nature of projects and competition for resources, it is rare to complete many projects according to the exact scope, time, and cost plans originally predicted. The project's sponsor, team members, or other stakeholders might have different views of the project as time progresses. For example, to generate project ideas, suppose the project manager for the information technology collaboration project began by sending an e-mail survey to all employees. In addition, to coordinate survey responses, suppose the project manager assigned one key contact in each department of the company to the project for one week. Now, suppose the e-mail survey generated only a few good project ideas, and several department contacts were uncooperative or unqualified to work on the project. What if the project team did not have the in-house expertise to analyze the survey data? Should the project manager ask for more money to hire an outside consultant? Should he or she ask for more time? Or should the project manager renegotiate the scope of the project? As another example, suppose the CEO heard about the project and thought the project team should come up with at least forty potential projects instead of thirty. Should the project team try to accomplish this increase in scope without changing the cost and schedule goals? It is the project manager's job to negotiate with the project team and sponsor in order to make qualified decisions about scope, time, and cost goals.

Although the triple constraint describes how the basic elements of a project—scope, time, and cost—interrelate, other elements can also play significant roles. Quality is often a key factor in projects, as is customer or sponsor satisfaction. Some people, in fact, refer to the "quadruple" constraint of project management, including quality along with scope, time, and cost. Others believe that quality considerations, including customer satisfaction, must be inherent in setting the scope, time, and cost goals of a project. A project team may meet scope, time, and cost goals but fail to meet quality standards or satisfy their sponsor, if they have not adequately addressed these concerns. For example, Anne Roberts may receive a fifty-page report describing thirty potential information technology projects and hear a presentation on the findings of the report. The work may have been completed on time and within the cost constraint, but the quality may have been unacceptable. Anne's view of a project description may be very different from the project team's view.

How can you avoid the problems that occur when you meet scope, time, and cost goals, but lose sight of quality or customer satisfaction? The answer is *good project management, which includes even more than meeting the triple constraint.*

WHAT IS PROJECT MANAGEMENT?

Project management is "the application of knowledge, skills, tools, and techniques to project activities in order to meet project requirements."[2] Project managers must not only strive to meet specific scope, time, cost, and quality goals of projects, they must also facilitate the entire process to meet the needs and expectations of the people involved in or affected by project activities.

Figure 1-2 provides a framework for beginning to understand project management. Key elements of this framework include the project stakeholders, project management knowledge areas, and project management tools and techniques.

[2] Project Management Institute (PMI) Standards Committee, *A Guide to the Project Management Body of Knowledge (PMBOK Guide)* (2000). The PMBOK Guide is a key document in the project management profession and is ANSI approved. Excerpts are available free of charge from PMI's Web site, **www.pmi.org.**

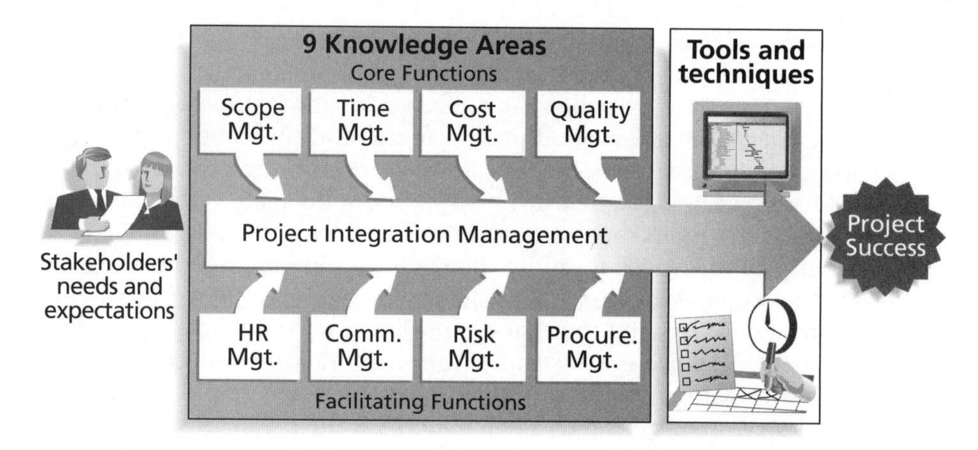

Figure 1-2. Project Management Framework

Stakeholders are the people involved in or affected by project activities and include the project sponsor, project team, support staff, customers, users, suppliers, and even opponents to the project. People's needs and expectations are important in the beginning and throughout the life of a project. Successful project managers work on developing good relationships with project stakeholders to ensure their needs and expectations are understood and met.

Knowledge areas describe the key competencies that project managers must develop. The center of Figure 1-2 shows the nine knowledge areas of project management. The four core knowledge areas of project management include project scope, time, cost, and quality management. These are considered to be core knowledge areas because they lead to specific project objectives. Brief descriptions of each are provided below:

- Project scope management involves defining and managing all the work required to successfully complete the project.
- Project time management includes estimating how long it will take to complete the work, developing an acceptable project schedule, and ensuring timely completion of the project.
- Project cost management consists of preparing and managing the budget for the project.
- Project quality management ensures that the project will satisfy the stated or implied needs for which it was undertaken.

The four facilitating knowledge areas of project management are human resource, communications, risk, and procurement management. These are called facilitating areas because they are the means through which the project objectives are achieved. Brief descriptions of each are provided below:

- Project human resource management is concerned with making effective use of the people involved with the project.

- Project communications management involves generating, collecting, disseminating, and storing project information.
- Project risk management includes identifying, analyzing, and responding to risks related to the project.
- Project procurement management involves acquiring or procuring goods and services that are needed for a project from outside the performing organization.

Project integration management, the ninth knowledge area, is an overarching function that affects and is affected by all of the other knowledge areas. Project managers must have knowledge and skills in all nine of these areas.

Project management **tools and techniques** assist project managers and their teams in carrying out scope, time, cost, and quality management. Additional tools help project managers and teams carry out human resource, communications, risk, procurement, and integration management. For example, some popular time management tools and techniques include Gantt charts, network diagrams (sometimes referred to as PERT charts), and critical path analysis. Project management software is a tool that can facilitate management processes in all the knowledge areas. These and many other project management tools and techniques are described throughout this text.

What Went Right?

A follow-up study done by the Standish Group in 1998 showed some improvement in the statistics for information technology projects. The cost of failed projects went down from $81 billion in 1995 to an estimated $75 billion in 1998. There was a major decrease in cost overruns from the $59 billion spent in 1995 to the estimated $22 billion in 1998. The 1998 study also showed that 26 percent of information technology projects succeeded in meeting scope, time, and cost goals. In spite of these improvements, 46 percent of information technology projects were challenged—completed over budget and past the original deadline—and 28 percent failed.[3]

The 2001 Standish Group report showed decided improvement in information technology project management compared to the 1995 study.

- Time overruns significantly decreased to 63%, compared to 222%.
- Cost overruns were down to 45%, compared to 189%.
- Required features and functions were up to 67%, compared to 61%.
- 78,000 U.S. projects were successful, compared to 28,000.
- 28% of information technology projects succeeded, compared to 16%.[4]

Even though there have been significant improvements in managing information technology projects, there is still much room for improvement. The best news is that project managers are

[3] The Standish Group, "1998 CHAOS Report" (1998). Cabanis, Jeanette, "Standish Research Indicates IT Project Success," *PM Network* (September 1998) 7.

[4] The Standish Group, "CHAOS 2001: A Recipe for Success" (2001).

learning how to succeed more often. "The reasons for the increase in successful projects vary. First, the average cost of a project has been more than cut in half. Better tools have been created to monitor and control progress and better skilled project managers with better management processes are being used. The fact that there are processes is significant in itself."[5]

In spite of the advantages that project management offers, it is not a silver bullet that guarantees success on all projects. Project management is a very broad, often complex discipline. What works on one project may not work on another, so it is essential for project managers to continue to develop their knowledge and skills in managing projects. The unique nature of projects and the challenges involved in managing them are what excites many people about working in project management.

HOW PROJECT MANAGEMENT RELATES TO OTHER DISCIPLINES

Much of the knowledge needed to manage projects is unique to the discipline of project management. However, project managers must also have knowledge and experience in general management and must understand the application area of the project in order to work effectively with specific industry groups and technologies. For example, project managers must understand general management areas such as organizational behavior, financial analysis, and planning techniques, to name a few. If a project involves sales force automation, the project manager needs to understand the sales process, sales automation software, and mobile computing. Figure 1-3 shows the relationships between project management, general management, and application areas.

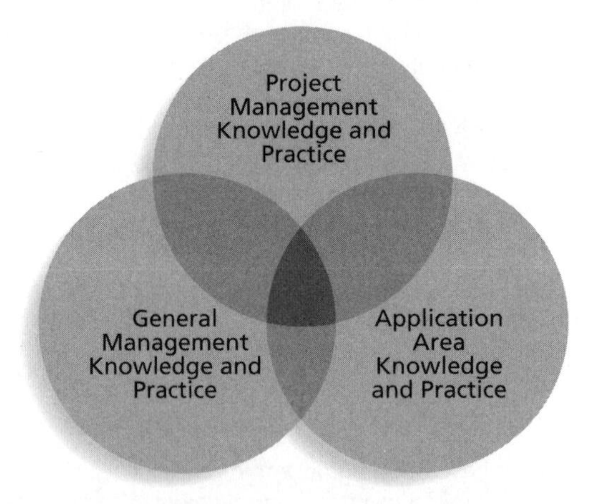

Figure 1-3. Project Management and Other Disciplines

[5] Ibid.

Although being a project manager requires some knowledge of and practice in general management areas, the role of a project manager is different from the role of a corporate manager or executive. What distinguishes project management from general or operations management is the nature of the projects. Since projects are unique, temporary, and involve various resources, project managers must focus on integrating all the various activities required to successfully complete the project. In contrast, most of the tasks performed by a general manager or operations manager are repetitive, ongoing, and done as day-to-day activities.

General or operations managers also focus more on a particular discipline or functional area. For example, a manager of an accounting department focuses on the discipline of accounting. If a project manager were selected to manage an information technology project for the accounting department, then he or she would need to know some things about accounting as well as information technology. However, his or her main job would be to manage the project, not to perform accounting or information technology functions.

Project management also requires some knowledge of the particular industry or knowledge area of the project itself. For example, this textbook focuses on information technology projects—projects that involve computer hardware, computer software, and telecommunications technology. It would be very difficult for someone with little or no background in information technology to become the project manager for a large information technology project. Not only would it be difficult to work with other managers and suppliers, it would also be difficult to earn the respect of the project team. Although new information technology project managers need to draw on their information technology expertise, they must spend more time becoming better project managers and less time being information technology experts in order to successfully lead their project teams.

HISTORY OF PROJECT MANAGEMENT

Some people might argue that the building of the Egyptian pyramids could be considered a project. Others might argue that building the Great Wall of China was also a project. Most people agree, however, that the more modern concept of project management began with the Manhattan Project, which the U.S. military led to develop the atomic bomb.

In fact, the military was the key industry behind the development of several project management techniques. In 1917 Henry Gantt developed the famous Gantt chart as a tool for scheduling work in job shops. Managers drew charts by hand to show tasks to be performed against a calendar timeline. This tool provided a standard format for planning and reviewing all the work that needed to be done on early military projects.

Today's project managers still use the Gantt chart as the primary tool to communicate project schedule information, but with the aid of computers, it is no longer necessary to draw the charts by hand. Figure 1-4 displays a Gantt chart for developing an intranet. This version of the chart was created with Microsoft Project, the most widely used project management software today. Note that a Gantt chart lists tasks that need to be done and the time needed to perform these tasks in a common calendar format. A Gantt chart can also display actual times it took to complete tasks, which helps project managers measure performance.

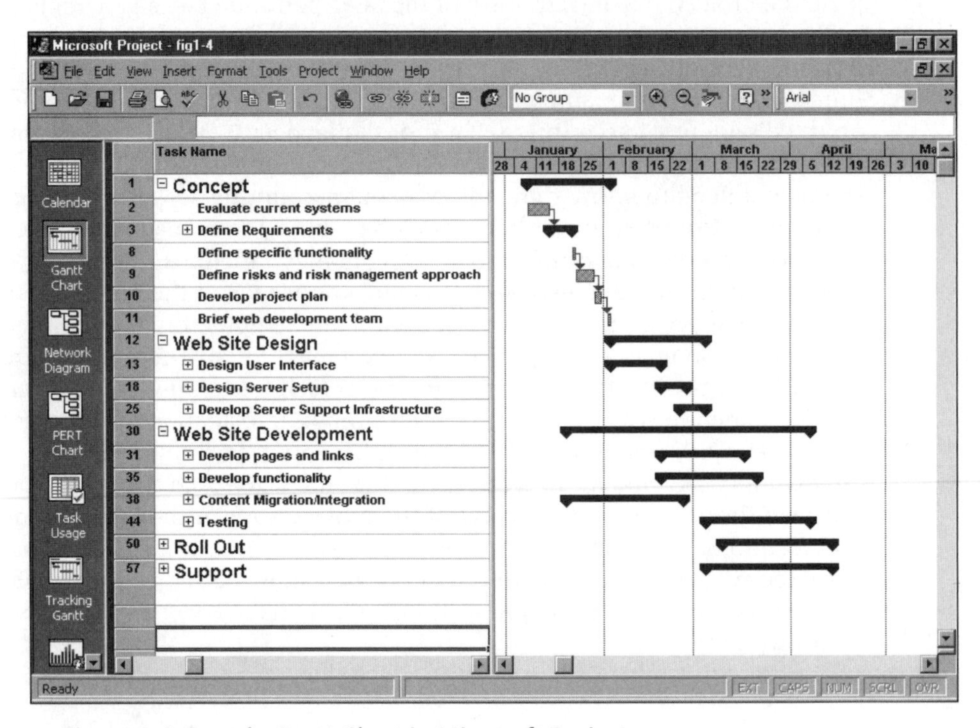

Figure 1-4. Sample Gantt Chart in Microsoft Project

Network diagrams were first used in 1958 for the Navy Polaris missile/submarine project. These diagrams helped managers model the relationships among project tasks, which allowed them to create even more realistic schedules. Figure 1-5 displays a network diagram created using Microsoft Project. Note that the diagram includes arrows that show which tasks are related and the sequence in which tasks must be performed. The concept of determining relationships among tasks is key in helping to improve project scheduling. This concept allows you to find and monitor the critical path—the longest path through a project network diagram that determines the earliest completion of a project. Chapter 5 on project time management provides more details about Gantt charts, network diagrams, and critical path analysis, and Appendix A explains how to use Microsoft Project 2000 to generate Gantt charts and network diagrams.

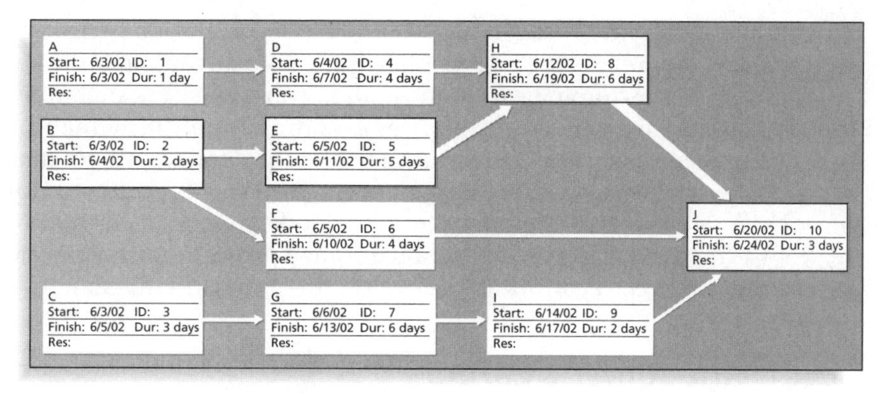

Figure 1-5. Sample Network Diagram

By the 1970s, the military had begun to use software to help manage large projects. Early project management software products were very expensive and ran on mainframe computers. For example, Artemis was an early project management software product that helped managers analyze complex schedules for designing aircraft. A full-time person was often required to run the complicated software and expensive pen plotters were used to draw network diagrams and Gantt charts. As computer hardware became smaller and more affordable, and software became more graphical and easy to use, project management software became less expensive, easier to use, and more popular. Today, many different industries use project management software on all types and sizes of projects. New software makes basic tools, such as Gantt charts and network diagrams, inexpensive, easy to create, and available for anyone to update.

In the latter part of the 20th century, people in virtually every industry began to investigate and apply different aspects of project management to their projects. The sophistication and effectiveness with which project management tools are being applied and used today is influencing the way companies do business, use resources, and respond to market requirements with speed and accuracy. The job title of project manager describes someone leading the construction of a new sports arena, planning a fund-raiser for a charitable organization, or managing development of an electronic commerce application. All of these examples illustrate one key point: No matter what industry, you need to understand the problems at hand if you are to manage a project successfully. The real challenge of the manager is to understand the concepts of project management and determine what tools and techniques should be applied on specific projects.

THE PROJECT MANAGEMENT PROFESSION

Like it or not, change is a constant in this world, especially in information technology. Year 2000 issues dominated the work of many information technology departments in the late 1990s. The Year 2000 focus produced unmet demand for other types of information technology projects—projects meant to exploit Web-based applications, take advantage of faster telecommunications technologies, establish data warehousing capabilities, and so on. Having experienced the backlog resulting from this demand, many companies need to adopt a far more rigorous approach to project management to make up for lost time while improving the success rate of information technology projects. To meet these demands, people need to know more about the need for information technology project managers, and project managers need to continue to develop the project management profession.

Project Management Careers

A 1999 *ComputerWorld* article listed "project manager" as the number one position that information technology managers say they will need for contract help.[6] Senior programmer/analyst, network administrator, and Web developer followed. Many people know about the need to fill these other information technology career fields, but most do not know about the need for project managers or really understand what project management is all about.

Most people in information technology work on projects in some capacity. Some people become project managers, at least on a part-time basis, very early in their careers. Many people think you need many years of experience and vast technical knowledge to be a project manager. Although this need for experience and knowledge might be the case for very large, complex, and expensive projects, many information technology projects can be, and are, led by project managers who are just starting their careers. Project managers need some general management and technical knowledge, but they primarily need the skills and desire to be project managers.

The project management profession is growing at a very rapid pace. *Fortune* magazine called project management the "number one career choice" in its article "Planning a Career in a World Without Managers."[7] A 2000 report by the **Project Management Institute (PMI)**, an international professional society for project managers, estimated that the average salary for a project manager was over $81,000[8]. Although many professional societies are suffering from declining memberships, PMI membership continues its fast-paced growth, and the organization included over 70,000 members worldwide at the beginning of 2001. A large percentage of PMI members work in the information technology field, and over 15,000 pay additional dues to join the Information Systems Specific Interest Group.[9]

[6] Fryer, Bronwyn, "Jobs Forecast '99," *ComputerWorld Web site* (**www.computerworld.com**) (January 4, 1999).

[7] Stewart, Thomas A. and McGowan, Joe, "Planning a Career in a World Without Managers," *Fortune Magazine* (March 20, 1996).

[8] Project Management Institute, PMI's 2000 Salary Survey (2000).

[9] Project Management Institute (PMI), PMI Web site (**www.pmi.org**) (2001).

In the 1990s, many companies began creating project offices to develop project management expertise in their organizations and to create a more formal career path for project managers. Many colleges, universities, and firms now offer courses related to various aspects of project management. The problems in managing projects, the publicity about project management, and the belief that it really can make a difference continue to contribute to the growth of this field.

Project Management Certification

Professional certification is an important factor in recognizing and ensuring quality in a profession. PMI provides certification as a **Project Management Professional (PMP)**—someone who has documented sufficient project experience, agreed to follow the PMI code of ethics, and demonstrated knowledge of the field of project management by passing a comprehensive examination. Appendix B provides more information on PMP certification.

The number of people earning certification in project management continues to increase. In 1993, there were about 1,000 certified project management professionals. By the end of 2000, there were approximately 28,000 certified project management professionals. Figure 1-6 shows the rapid growth in the number of people earning project management professional certification from 1993 to 2000.

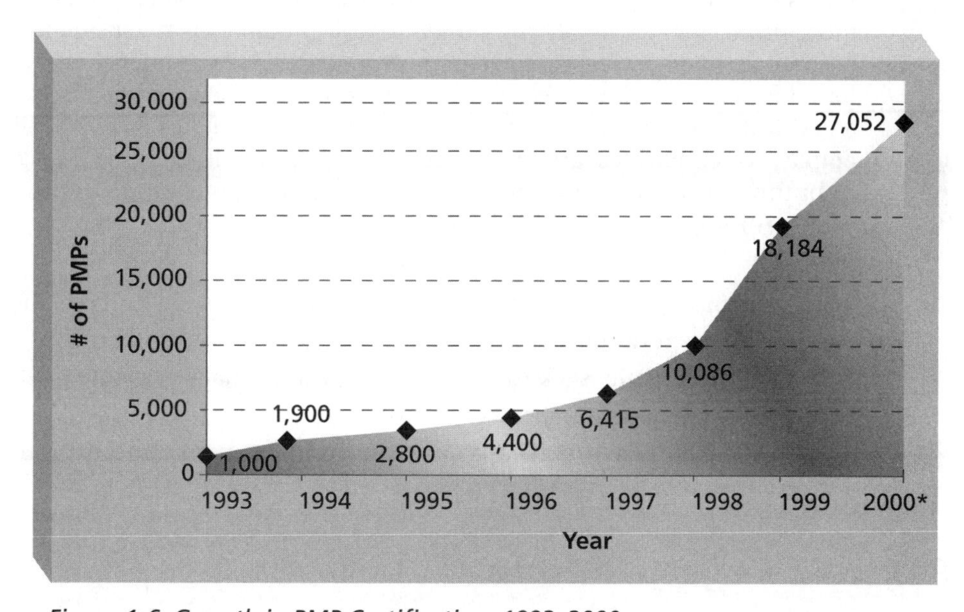

Figure 1-6. Growth in PMP Certification, 1993–2000

International Data Corporation conducted a study in 1995 (on behalf of Drake Prometric, the IBM Corporation, Lotus Development Corporation, and Microsoft) to learn whether a company's information technology support function is more productive when the staff includes a greater percentage of professionally certified employees. Certification could include technical certification in products such as NT or Lotus Notes, or broader certification in areas such as data processing or project management. The study found that companies supporting certification tend to operate in more complex information technology environments and are more efficient than companies that do not support certification.

As information technology projects become more complex and global in nature, the need for people with demonstrated knowledge and skills in project management will continue. Just as passing the CPA exam is a standard for accountants, passing the PMP exam is becoming a standard for project managers. Some companies are requiring that all project managers be PMP certified. Project management certification is also enabling professionals in the field to share a common base of knowledge. For example, any PMP can list, describe, and use the nine project management knowledge areas. Sharing a common base of knowledge is important because it helps advance the theory and practice of project management.

Code of Ethics

Ethics is an important part of all professions. PMI developed a project management professional code of ethics that all project managers must sign in order to become certified project management professionals. PMI states that it is vital for all PMPs to conduct their work in an ethical manner. Conducting work in an ethical manner helps the profession earn the confidence of the public, employers, employees, and project team members. Following is an excerpt from the code of ethics for the project management profession.

As professionals in the field of project management, PMI members pledge to uphold and abide by the following:

- I will maintain high standards of integrity and professional conduct.
- I will accept responsibility for my actions.
- I will continually seek to enhance my professional capabilities.
- I will practice with fairness and honesty.
- I will encourage others in the profession to act in an ethical and professional manner.[10]

Project Management Software

Unlike the cobbler neglecting to make shoes for his own children, the project management and software development communities have definitely responded to the need to provide more software to assist in managing projects.

[10] Project Management Institute (PMI), "Member Code of Ethics" (**www.pmi.org/membership/standards/ethical.htm#CodeofEthics**) (revised October 2000).

PMI published a Project Management Software Survey in 1999 that describes, compares, and contrasts more than 200 project management software tools. ALLPM.COM, an Internet site for information technology project management information and resources, provides an alphabetical listing of and links to over one hundred products that help in managing projects.[11] Deciding which project management software to use has become a project in itself. This section provides a brief summary of the basic types of project management software and references for finding more information. Appendix A provides a brief guide to using Microsoft Project 2000, the most widely used project management software tool today.

Many people still use basic productivity software, like Microsoft Word or Excel, to perform many project management functions, such as determining project scope, time, and cost, assigning resources, preparing project documentation, and so on. People often use productivity software instead of specialized project management software because they already have it and know how to use it. However, there are hundreds of different project management software tools that provide specific functionality for managing projects. These project management tools can be divided into three general categories based on functionality and price:

- *Low-end tools:* These tools provide basic project management features and generally cost under $200 per user. They are often recommended for small projects and single users. Most of these tools allow users to create Gantt charts, which cannot be done using current productivity software. For example, Milestones Simplicity by KIDASA Software, Inc., has a Schedule Setup Wizard interface that walks users through simple steps to produce a Gantt chart. For only $49 per user, this tool also includes a large assortment of symbols, flexible formatting, an outlining utility, and an Internet Publishing Wizard.[12] Another product, called "How's it going?", was written in Microsoft Access 97 by LogicAbility. For $120 per user, this tool includes an online guide and templates for many project management deliverables; reports for project tracking, status reporting, and budgeting; time reporting and resource management features; and scheduling features for creating Gantt charts and performing critical path analysis.[13]
- *Midrange tools*: A step up from low-end tools, midrange tools are designed to handle larger projects, multiple users, and multiple projects. All of these tools can produce Gantt charts and network diagrams, and can assist in critical path analysis, resource allocation, project tracking, status reporting, and so on. Prices range from about $200 to $500 per user, and several tools require additional server software for using workgroup features. Microsoft Project is still the most popular project management software used today, and Project 2000 won the "Readers' Choice" award at the Web2000 conference. Microsoft's Project Central, a companion product,

[11] ALLPM.com, PM Products (**www.allpm.com/links/Products**) (2001).

[12] Howe, Charlie, "A Fast Schedule Is Mere Simplicity," PMNetwork (April 2000).

[13] Howe, Charlie, "Good Surprises Do Come in Small Packages," PMNetwork (January 2000).

facilitates collaboration and communication of project information over a corporate intranet. Several other companies provide companion products to Project 2000, such as Project Assistants, Inc.'s Project Commander ($199 per user), which provides additional functionality and ease of use to Project 2000. Other companies that sell midrange project management tools include Artemis, PlanView, Primavera, and Welcom, to name just a few.

- *High-end tools:* Another category of project management software includes high-end tools, sometimes referred to as enterprise project management software. These tools provide robust capabilities to handle very large projects, dispersed workgroups, and enterprise functions that summarize and combine individual project information to provide an enterprise view of all projects. These products are generally licensed on a per-user basis. For example Advanced Management Solutions (AMS) sells a $40,000 fifty-seat license of its AMS REALTIME suite of products to a number of large clients. This particular product requires users to have Oracle installed on their computers, even though it runs on multiple operating systems. AMS REALTIME interfaces with midrange tools such as Microsoft Project and Primavera Project Planner, and it also offers a complete suite of application programming interfaces (APIs) to enable integration with other business information systems.[14] Several companies, including Microsoft, that provide midrange tools are also starting to offer enterprise versions of their software.

As mentioned earlier, there are many reasons to study project management, particularly as it relates to information technology projects. The number of information technology projects continues to grow, the complexity of these projects continues to increase, and the profession of project managers continues to expand and mature. As more people study and work in this important field, the success rate of information technology projects should continue to improve.

CASE WRAP-UP

The information technology collaboration project was a huge success. Anne Roberts hired an experienced facilitator to help run several brainstorming meetings. She also authorized the VP of Information Technology to hire other consultants to provide necessary expertise and share extensive reports on new technologies that were proven to be effective at similar companies. Anne also found some opinion leaders in the company who were instrumental in getting people

[14] Howe, Charlie, "Get to Your Project Information in REALTIME," PMNetwork (January 2001).

excited about working on the project. One of the project ideas generated was to form a project office or project center of excellence. There were a few certified project management professionals in the company with experience working in project offices, and they collected compelling data to support the formation of such a group for their own company. At this point, Anne and the other members of the information technology department were very confident that they could improve operations.

CHAPTER SUMMARY

A project is a temporary endeavor undertaken to accomplish a unique purpose. Projects are unique, temporary, require resources, have a sponsor, and involve uncertainty.

The "triple constraint" of project management refers to managing the scope, time, and cost dimensions of a project.

Project management is the application of knowledge, skills, tools, and techniques to project activities in order to meet project requirements. Stakeholders are the people involved in or affected by project activities.

A framework for project management includes the project stakeholders, project management knowledge areas, and project management tools and techniques. The knowledge areas include project integration management, scope, time, cost, quality, human resources, communications, risk, and procurement management.

The discipline of information technology (IT) is focusing on improving project management methods to help improve the low success rate of information technology projects. A 1995 study found that only 16 percent of information technology projects met scope, time, and cost goals. This performance improved to 28 percent in a 2001 follow-up study, but there is still great room for improvement.

Project management overlaps somewhat with general management and application area knowledge. Project managers, however, use unique project management skills to focus on integrating all of the knowledge areas required to successfully complete a project.

The Manhattan Project was the first project that used modern project management techniques. The Gantt chart, network diagram, and critical path analysis are project scheduling tools that were developed in the early 1900s. Today, project management is used in some form in virtually all organizations and disciplines, and project management software is making specialized project management techniques easier to use.

The need for more project managers, especially in information technology, is growing. The Project Management Institute (PMI) is an international professional society that provides certification as a Project Management Professional (PMP) and a code of ethics for project managers. PMI publishes many documents, including comparisons of the hundreds of different project management software products available on the market. These products fall into three basic categories: low-end tools, midrange tools, and high-end or enterprise tools. Microsoft Project is the most popular project management software tool in use today.

Discussion Questions

1. What is a project? How is it different from what most people do in their day-to-day jobs?
2. Give three examples of activities that are projects and three examples of activities that are not projects.
3. How is project management different from general management?
4. Explain in your own words what the triple constraint means. Give an example of it on a real project with which you are familiar.
5. Give an example of an information technology project that went well. Give an example of one that did not go so well.
6. Why do you think so many information technology projects are unsuccessful?

Exercises

1. Read the 1995 "CHAOS" article (the fourth suggested reading). Write a one-page summary of the article, its key conclusions, and your opinion of the article. If possible, find materials from the 1998 and 2001 reports and include findings from them in your analysis.
2. Read one of the other suggested readings. Write a one-page summary of the article, its key conclusions, and your opinion.
3. Using three different search engines, do an Internet search for the terms *project management, project management careers, project offices,* and *information technology project management*. Write down the number of hits that you received for each of these phrases. Find at least three good Web sites that provide interesting information on one of the topics. In a one- to two-page paper, summarize key information about these three Web sites as well as the Project Management Institute's Web site at **www.pmi.org.**
4. Scan information technology industry journals or Web sites such as *Information Week, Computer World,* and *Information World,* and find one good article about information technology project management. Write a one- to two-page paper describing the article.

5. Skim through Appendix A on Microsoft Project 2000. Write a one- to two-page paper answering the following questions: What functions does this software provide that cannot be done easily using other tools such as a spreadsheet or database? Does the software appear easy to learn? What are key inputs and outputs of the software?

SUGGESTED READINGS

1. Ibbs, C. William and Young, H. Kwak. "Calculating Project Management's Return on Investment." *Project Management Journal* (June 2000).

   Many companies want proof that project management is a good investment. This paper describes a procedure to measure the return on investment for project management. Information is based on thirty-eight different companies and government agencies in four different industries.

2. Project Management Institute. "Information Source Guide 2001." (2001).

   This booklet describes hundreds of books related to the field of project management. You can also search for books from PMI's online bookstore at **www.pmibookstore.org** *or at other popular online book sites such as* **amazon.com**.

3. Project Management Institute. *Project Management Research at the Turn of the Millennium*. (2000).

   PMI sponsored its first international research conference in Paris, France in June 2000. This document contains over forty invited papers written by some of the best researchers in the field of project management.

4. The Standish Group. "CHAOS." (1995) (**www.standishgroup.com/visitor/chaos.html**).

   This article is often quoted for its survey results showing the poor track record for information technology projects. It provides statistics and good examples of successful and unsuccessful information technology projects. The Standish Group has also done two updates to this study, published in 1998 and 2001.

5. Thamhain, Hans J. "Best Practices for Controlling Technology-Based Projects." *Project Management Journal* (December 1996).

   Hans Thamhain is well known for his research on technology-based project management and team building. This paper supports the position that modern project management tools and techniques can significantly enhance overall project performance and be integrated into the business process. It summarizes many analytical, process-oriented, and people-oriented techniques for project control and offers suggestions on using them effectively.

6. Yourdon, Ed. "Surviving a Death March Project." *Software Development* (July 1997).

Ed Yourdon is famous for his books on software development, including a 1997 text titled Death March. *Yourdon offers practical advice on how to handle software projects that are plagued with problems from the start. He describes the importance of a project manager being able to negotiate schedules, budgets, and other aspects of a project with users, managers, and other stakeholders.*

KEY TERMS

- **Program** — a group of projects managed in a coordinated way
- **Project** — a temporary endeavor undertaken to accomplish a unique purpose
- **Project management** — the application of knowledge, skills, tools, and techniques to project activities in order to meet project requirements
- **Project Management Institute (PMI)** — international professional society for project managers
- **Project management knowledge areas** — project integration management, scope, time, cost, quality, human resource, communications, risk, and procurement management
- **Project Management Professional (PMP)** — certification provided by PMI that requires documenting project experience, agreeing to follow the PMI code of ethics, and passing a comprehensive examination
- **Project management tools and techniques** — methods available to assist project managers and their teams. Some popular tools in the time management knowledge area include Gantt charts, network diagrams, critical path analysis, and project management software
- **Stakeholders** — people involved in or affected by project activities
- **Triple constraint** — balancing scope, time, and cost goals

2

The Project Management Context and Processes

Objectives

After reading this chapter, you will be able to:

1. *Understand the systems view of project management and how it applies to information technology projects*
2. *Explain the four general phases in the project life cycle*
3. *Distinguish between project development and product development*
4. *Analyze a formal organization using the four frames of organizations*
5. *Explain the differences among functional, matrix, and project organizational structures*
6. *Explain why top management commitment to project management is critical for a project's success*
7. *List important skills and attributes of a good project manager*
8. *Briefly describe the five process groups of project management, the typical level of activity for each, and the interactions among them*
9. *See how the project process groups relate to the project management knowledge areas and how organizations can develop an information technology project management methodology*

Tom Walters recently accepted a new position at his college as the Director of Information Technology. Tom had been a faculty member at the college for the past fifteen years and was respected by his colleagues. The college—a small, private college in the Southwest—offered a variety of programs in the liberal arts and professional areas. Enrollment included 1,500 full-time traditional students and about 1,000 working-adult students attending an evening program. Like most colleges, its use of information technology had grown tremendously in the past five years. There were a few classrooms on campus with computers for the instructors and students, and a few more with just instructor stations and projection systems. Tom knew that several colleges throughout the country had begun to require all students to lease laptops and that these colleges incorporated technology components into most courses. This idea fascinated him. He and two other members of the Information Technology Department visited a local college that had required all students to lease laptops for the past three years, and they were very impressed with what they saw and heard. Tom and his staff developed plans to start requiring students to lease laptops at their college the next year.

Tom sent an e-mail to all faculty and staff in September, which briefly described this and other plans. He did not get much response, however, until the February faculty meeting when, as he described some of the details of his plan, the chairs of the History, English, Philosophy, and Economics Departments all voiced their opposition to the idea. They eloquently stated that the college was not a technical training school, and that they thought the idea was ludicrous. Members of the Computer Science Department voiced their concern that all of their students already had state-of-the art desktop computers and would not want to pay a mandatory fee to lease less powerful laptops. The director of the adult education program expressed her concern that many adult-education students would balk at an increase in fees. Tom was in shock to hear his colleagues' responses. Now what should he do?

Many of the theories and concepts of project management are not difficult to understand. What *is* difficult is implementing them as part of good project management. Project managers must consider many different components when managing projects. This chapter discusses some of these components, such as using a systems approach, following a project life cycle, understanding organizations, gaining top management commitment for project management, developing important skills for project management, and integrating the project management process groups.

A SYSTEMS VIEW OF PROJECT MANAGEMENT

Even though projects are temporary and intended to provide a unique product or service, organizations cannot run projects in isolation. If projects are run in isolation, it is unlikely that those projects will ever truly serve the needs of the organization. Therefore, projects must operate in a broad organizational environment, and project managers need to consider projects within the greater organizational context. To effectively handle complex situations, project managers need to take a holistic view of a project and understand how it is situated within the larger organization. Taking this type of holistic view of projects and the organizations in which they are carried out is called **systems thinking**.

The term **systems approach** emerged in the 1950s to describe a holistic and analytical approach to solving complex problems that includes using a **systems philosophy**, systems analysis, and systems management. A systems philosophy is an overall model for thinking about things as systems. **Systems** are sets of interacting components working within an environment to fulfill some purpose. For example, the human body is a system composed of many subsystems—the brain, the skeletal system, the circulatory system, the digestive system, and so on. **Systems analysis** is a problem-solving approach that requires defining the scope of the system to be studied, dividing it into its component parts, and then identifying and evaluating its problems, opportunities, constraints, and needs. The analyst then examines alternative solutions for improving the current situation, identifies an optimum, or at least satisfactory, solution or action plan, and examines that plan against the entire system. **Systems management** addresses the business, technological, and organizational issues associated with making a change to a system.

Using a systems approach is critical to successful project management. Senior managers and project managers must identify key business, technological, and organizational issues related to each project in order to identify and satisfy key stakeholders and do what is best for the entire organization.

In the opening case, when Tom Walters planned the laptop project, he did not use a systems management approach. Members of the Information Technology Department did all of the planning. Even though Tom sent an e-mail describing the laptop project to all faculty and staff, he did not address many of the organizational issues involved in such a complex project. Most faculty and staff are very busy at the beginning of fall term and many may not have read the entire message. Others may have had concerns, but were too busy to communicate those concerns to the Information Technology Department. Tom was unaware of the effects the laptop project would have on other parts of the college. He did not clearly define the business, technological, and organizational issues associated with the project. Tom and the Information Technology Department began work on the laptop project in isolation. If they had taken a systems management approach, considered other dimensions of the project, and involved key stakeholders, they could have identified many of the issues raised at the February faculty meeting and addressed them *before* the meeting.

Figure 2-1 provides a sample of some of the business, organizational, and technological issues that could be factors in the laptop project. In this case, technological issues, though not simple by any means, are probably the least difficult to identify and resolve. However, projects must address issues in all three spheres of the systems management model—business, organizational, and technological. Although it is easier to focus on the immediate and sometimes narrow concerns of a particular project, project managers and other staff must keep in mind the effects of any project on the health and needs of the entire system or organization.

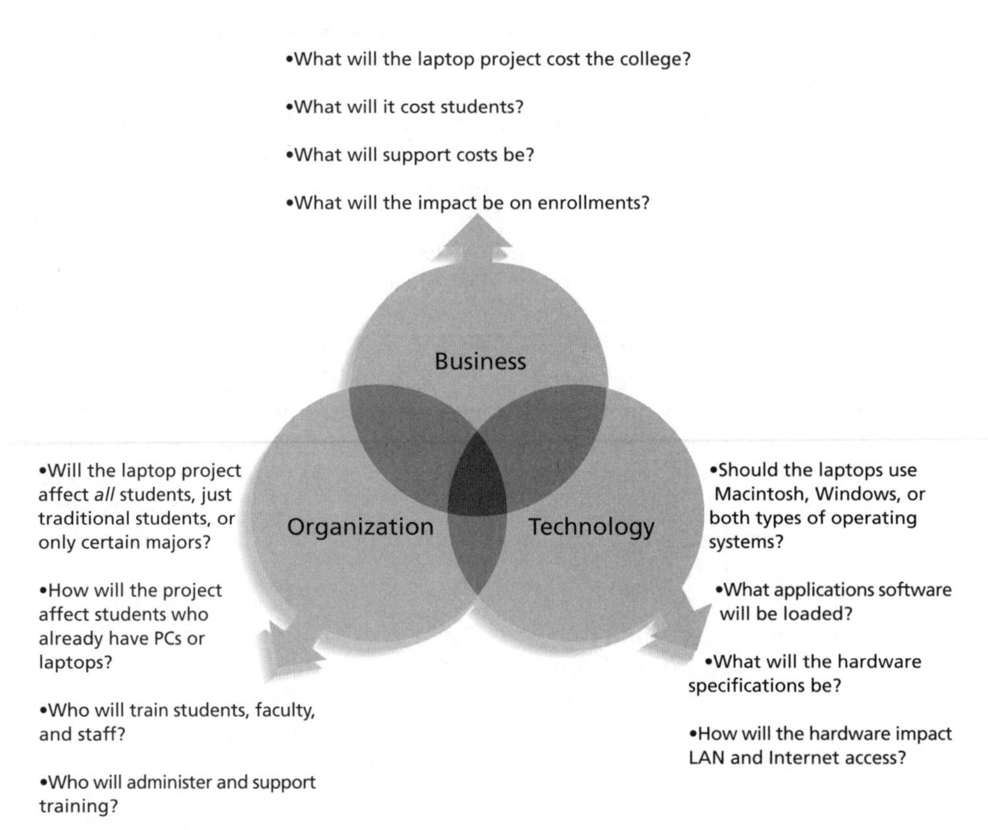

Figure 2-1. Three-Sphere Model for Systems Management

Many information technology professionals become enamored with the technology and day-to-day problem solving involved in working on information systems. They tend to become frustrated with many of the people problems or politics involved in most organizations. In addition, many information technology professionals ignore important business issues. Does it make financial sense to pursue this new technology? Should this software be developed in-house or purchased off-the-shelf? Using a more holistic approach helps project managers integrate business and organizational issues into their planning. It also helps them look at projects as a series of interrelated phases. When you integrate business and organizational issues into project planning and look at projects as a series of interrelated phases, you do a better job of ensuring project success.

PROJECT PHASES AND THE PROJECT LIFE CYCLE

Because projects operate as part of a system and involve uncertainty, it is good practice to divide projects into several phases. A **project life cycle** is a collection of project phases. Project phases vary by project or industry, but some general phases include concept, development, implementation, and close-out. The first two phases (concept and development) focus on planning and are often referred to as **project feasibility**. The last two phases (implementation and close-out) focus on delivering the actual work and are often referred to as **project acquisition**. A project must successfully complete each phase before moving on to the next. This project life cycle approach provides better management control and appropriate links to the ongoing operations of the organization.

Figure 2-2 provides a summary framework for the general phases of the project life cycle. In the concept phase of a project, managers usually briefly describe the project—they develop a very high-level or summary plan for the project, which describes the need for the project and basic underlying concepts. A preliminary or very rough cost estimate is developed in this first phase, and an overview of the work involved is created. Project work is usually defined in a **work breakdown structure (WBS)**—an outcome-oriented document that defines the total scope of the project. (The work breakdown structure is explained further in Chapter 4, *Project Scope Management*.) For example, if Tom Walters (from the opening case) had followed the project life cycle instead of moving full-steam ahead with the laptop project, he could have created a committee of faculty and staff to study the concept of increasing the use of technology on campus. This committee might have developed a management plan that included a first, smaller project to investigate alternative ways of increasing the use of technology. They might have estimated that it would take six months and $20,000 to do a detailed technology study. The WBS at this phase of the study might have three levels and partition the work to include a competitive analysis of what five similar campuses were doing, a survey of local students, staff, and faculty, and a rough assessment of how using more technology would affect costs and enrollments. At the end of the concept phase, the committee would be able to deliver a report and presentation on its findings. The report and presentation would be an example of a **deliverable**—a product produced as part of a project.

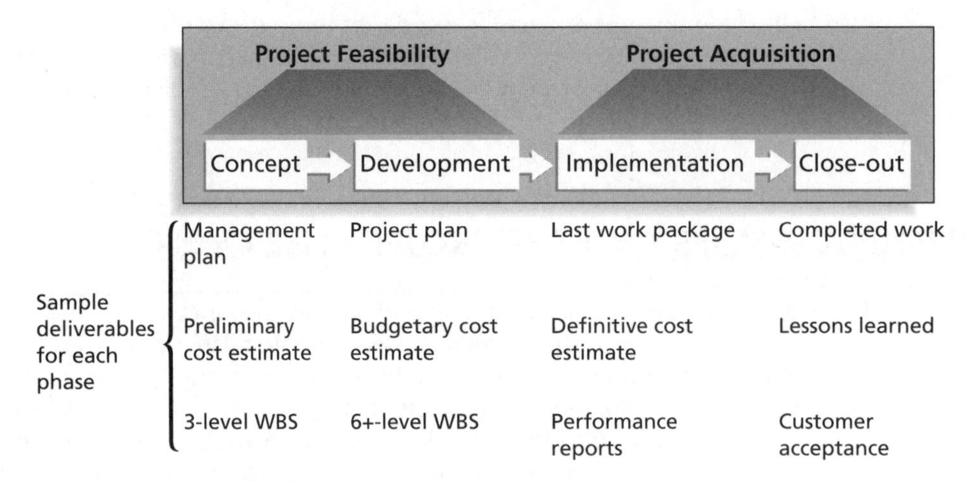

Figure 2-2. Phases of the Project Life Cycle

After the concept phase is completed, the next project phase of development begins. In the development phase, the project team creates a more detailed project plan, a more accurate cost estimate, and a more thorough WBS. In the example under discussion, suppose the concept phase report suggested that requiring students to have laptops was one means of increasing the use of technology on campus. The project team could then further expand this idea in the development phase. They would have to decide if students would purchase or lease the laptops, what type of hardware and software the laptops would require, how much to charge students, how to handle training and maintenance, how to integrate the use of the new technology with current courses, and so on. If, however, the concept phase report showed that the laptop idea was not a good idea for the college, then increasing the use of technology by requiring laptops would no longer be considered in the development phase. This phased approach minimizes the time and money spent developing inappropriate projects. A project idea must pass the concept phase before evolving during the development phase.

The third phase of the project life cycle is called implementation. In this phase, the project team delivers the required work, creates a definitive or very accurate cost estimate, and provides performance reports to stakeholders. Suppose Tom Walter's college took the idea of requiring students to have laptops through the development phase. During the implementation phase, the project team would need to obtain the required hardware and software, install the necessary network equipment, deliver the laptops to the students, create a process for collecting fees, provide training to students, faculty, and staff, and so on. Other people on campus would also be involved in the implementation phase. Faculty would need to consider how best to take advantage of the new technology. The recruiting staff would have to update their materials to reflect this new feature of the college. Security would need to address new problems that might result from having students carry around expensive equipment. The bulk of a project team's efforts and the most money are usually spent during the implementation phase of projects.

The last phase of the project life cycle is called close-out. In the close-out phase, all of the work is completed, and there should be some sort of customer acceptance of the entire project. The project team should document their experiences on the project in a lessons-learned report. If the laptop idea made it all the way through the implementation phase and all students received laptops, the project team would then complete the project by closing out any related activities. They might administer a survey to students, faculty, and staff in order to gather opinions on how the project fared. They would ensure that any contracts with suppliers were completed and appropriate payments had been made. They would transition future work related to the laptop project to other parts of the organization. The project team could also share their lessons learned with other college campuses who are considering doing the same thing.

Just as a *project* has a life cycle, so does a *product*. Information technology projects help to produce products such as new software, hardware, networks, research reports, and training on new systems. Understanding the product life cycle is just as important to good project management as understanding the phases of the project life cycle.

Product Life Cycles

Most information technology professionals are familiar with the concept of a **systems development life cycle (SDLC)**, which is a framework for describing the phases involved in developing and maintaining information systems. Common names for these general phases are information systems planning, analysis, design, implementation, and support. Some popular models of a systems development life cycle include the waterfall model, the spiral model, the incremental release model, the Rapid Application Development (RAD) model, and the prototyping model.

- The waterfall model has well-defined, linear stages of systems development and support.
- The spiral model was developed based on experience with various refinements of the waterfall model as applied to large government software projects. It recognizes the fact that most software is developed using an iterative or spiral approach rather than a linear approach.
- The incremental release model provides for progressive development of operational software, with each release providing added capabilities.
- The RAD model, used to produce systems quickly without sacrificing quality, includes four phases—requirements planning, user design, construction, and cutover. RAD tools are available to facilitate rapid prototyping and code generation.
- The prototyping model is used for developing software prototypes to clarify user requirements for operational software.

These models are all examples of product life cycles, and most introductory management information systems texts describe each of them in detail. The type of software and complexity of the information system being developed determines which model is used. Figure 2-3 shows Boehm's famous spiral model of software development.[1] The spiral model illustrates how complex the process of developing an information system can be.

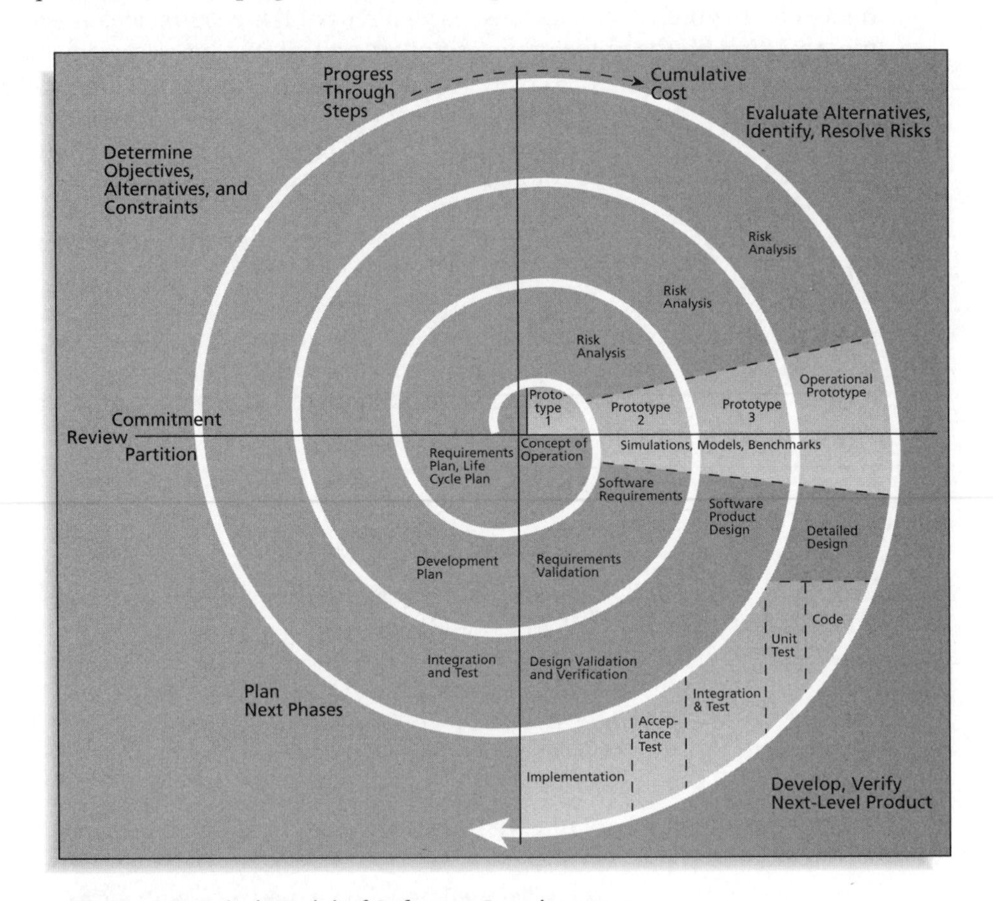

Figure 2-3. Spiral Model of Software Development

It is important not to confuse the project life cycle with the product life cycle. The project life cycle applies to all projects, regardless of the products being produced. On the other hand, product life cycle models vary considerably based on the nature of the product.

Most large information technology products are developed as a series of projects. For example, the systems planning phase for a new information system can include a project to hire an outside consulting firm to help identify and evaluate potential strategies for developing a particular business application,

[1] Boehm, Barry,"A Spiral Model of Software Development and Enhancement," *IEEE Computer* (May 1988) 5, 61–72.

such as a new order processing system or general ledger system. It can also include a project to develop, administer, and evaluate a survey of users to get their opinions on the current information systems used for performing that business function in the organization. The systems analysis phase might include a project to create process models for certain business functions in the organization. It can also include a project to create data models of existing databases in the company related to the business function and application. The implementation phase might include a project to hire contract programmers to code a part of the system. The support phase might include a project to develop and run several training sessions for users of the new application. All of these examples show that large information technology products are usually composed of several smaller projects.

When developing information systems, project management is a cross-life cycle activity; it is done in all of the product phases of developing information systems. Because project management needs to occur during all phases of the systems development life cycle, it is critical for information technology professionals to understand and practice good project management.

The Importance of Project Phases and Management Reviews

Due to the complexity and importance of many information technology projects and their resulting products, it is important to take time to review how the projects are going. A project should successfully pass through each of the project phases before continuing to the next. Because the organization usually commits more money as a project continues, a management review should occur after each phase to evaluate progress, likely success, and continued compatibility with organizational goals. These management reviews, called **phase exits** or **kill points**, are very important for keeping projects on track and determining if they should be continued, redirected, or terminated. Recall that projects are just one part of the entire system of an organization. Changes in other parts of the organization might affect what is happening on projects, and what happens on projects might likewise affect what is happening in other parts of the organization. By breaking projects into phases, senior management can make sure that the projects are still compatible with the needs of the rest of the company.

Let's take another look at the opening case. Suppose Tom Walters' college did a study on increasing the use of technology. At the end of the concept phase, the project team could have presented information to the faculty and administration that described different options for increasing the use of technology, an analysis of what competing colleges were doing, and results of a survey of local stakeholders' opinions on the subject. This presentation at the end of the concept phase represents a management review. Suppose the study reported that 90 percent of students, faculty, and staff surveyed strongly opposed the idea of requiring all students to have laptops and that many adult students said they would attend other colleges if they were required to pay for the additional technology. The college would probably decide not to pursue this idea any further. Had Tom taken a phased approach, he and his staff would not have wasted the time and money it took to develop more detailed plans.

What Went Right?

Having specific deliverables and kill points at the end of project phases helps managers make better decisions about whether to proceed, redefine, or kill a project. Improvement in information technology project success rates reported by the Standish Group has been due, in part, to an increased ability to know when to cancel failing projects. Standish Group Chairman Jim Johnson made the following observation:

"The real improvement that I saw was in our ability to—in the words of Thomas Edison—know when to stop beating a dead horse . . . Edison's key to success was that he failed fairly often; but as he said, he could recognize a dead horse before it started to smell . . . as a result he had 14,000 patents and was very successful. . . . In information technology we ride dead horses—failing projects—a long time before we give up. But what we are seeing now is that we are able to get off them; able to reduce cost overrun and time overrun. That's where the major impact came on the success rate."[2]

UNDERSTANDING ORGANIZATIONS

The systems approach to project management requires that project managers always view their projects in the context of the larger organization. Organizational issues are often the most difficult part of working on and managing projects. For example, many people believe that most projects fail for political reasons. Project managers often do not spend enough time identifying all the stakeholders involved in projects, especially the people opposed to the projects. Similarly, they often do not consider the political context of a project in the organization. In order to improve the success rate of information technology projects, it is important for project managers to develop a better understanding of people as well as organizations.

Organizations can be viewed as having four different frames: structural, human resources, political, and symbolic:[3]

- The **structural frame** deals with how the organization is structured (usually depicted in an organizational chart) and focuses on different groups' roles

[2] Cabanis, Jeannette, "A Major Impact: The Standish Group's Jim Johnson On Project Management and IT Project Success," *PM Network*, PMI (September 1998) 7.

[3] Bolman, Lee G. and Deal, Terrence E., *Reframing Organizations*, Jossey-Bass Publishers, 1991.

and responsibilities in order to meet the goals and policies set by top management. This frame is very rational and focuses on coordination and control. For example, a key issue in information technology related to the structural frame is whether information technology personnel should be centralized in one department or decentralized across several different departments. Different organizational structures are described in the next section.

- The **human resources frame** focuses on producing harmony between the needs of the organization and the needs of the people. It recognizes that there are often mismatches between the needs of the organization and the needs of individuals and groups and works to resolve any potential problems. For example, many projects might be more efficient for the organization if personnel worked 80 or more hours a week for several months. This work schedule would probably conflict with the personal lives of those people. Important issues in information technology related to the human resources frame are the shortage of information technology workers and unrealistic schedules imposed on many projects.
- The **political frame** addresses organizational and personal politics. **Politics** in organizations take the form of competition among groups or individuals for power and leadership. The political frame assumes that organizations are coalitions composed of varied individuals and interest groups. Often, important decisions need to be made based on the allocation of scarce resources. Competition for scarce resources makes conflict a central issue in organizations, and power improves the ability to obtain scarce resources. Project managers must pay attention to politics and power if they are to be effective. It is important to know who opposes your projects as well as who supports them. Important issues in information technology related to the political frame are the power shifts from central functions to operating units or from functional managers to project managers.

What Went Wrong?

Many data warehousing projects are sidetracked or derailed completely by politics. Data warehousing projects are always potentially political because they cross departmental boundaries, change both the terms of data ownership and data access, and affect the work practices of highly autonomous and powerful user communities. Many organizations fail to admit that data warehousing projects often fail primarily because management and project teams do not understand and manage politics. Marc Demarest found over 1,200 articles on the topic of data warehousing through a journal search he did from July 1995 to July 1996. Many of those articles offer advice on how to run successful data warehousing projects and focus on the importance of design, technical, and procedural factors, when, in fact, political factors are often the most important in helping these projects succeed. As Demarest states: "If we were honest with ourselves, as professionals, we would admit what Rosabeth Moss Kanter suggested in 1979 in a famous Harvard Business Review article:"

Power is America's last dirty word. It is easier to talk about money—and much easier to talk about sex—than it is to talk about power. People who have it deny it; people who want it do not want to appear to hunger for it; and people who engage in its machinations do so secretly.
Power is also the last dirty word in data warehousing. And it's crippling the discipline, in my view.[4]

- The **symbolic frame** focuses on symbols and meanings. What is most important about any event in an organization is not what actually happened, but what it means. Was it a good sign that the CEO came to a kick-off meeting for a project, or was it a threat? The company's culture is also related to this frame. How do people dress? How many hours do they work? How do they run meetings? Important issues in information technology related to the symbolic frame are the meaning of work in high-technology environments and the image of information technology workers as being either key partners in the business or a necessary cost.

Project managers must learn to use all four of these frames in order to function well in organizations. Chapter 8, *Project Human Resource Management*, and Chapter 9, *Communications Management*, further develop some of the organizational issues involved in the human resources and symbolic frames. The following sections on organizational structures, stakeholder management, and the need for top management commitment provide additional information related to the structural and political frames.

Organizational Structures

Many discussions of organizations focus on organizational structure. Three general classifications of organizational structures are: **functional**, **project**, and **matrix**. Figure 2-4 portrays these three organizational structures. A functional organization is the hierarchy most people think of when picturing an organizational chart. Functional managers or vice presidents in specialties such as engineering, manufacturing, information technology (IT), and human resources (HR) report to the chief executive officer (CEO). Their staffs have specialized skills in their respective disciplines. For example, most colleges and universities have very strong functional organizations. Only faculty in the Business Department teach business courses; faculty in the History Department teach history; faculty in the Art Department teach art, and so on.

A project organization also has a hierarchical structure, but instead of functional managers or vice presidents reporting to the CEO, project managers report to the CEO. Their staffs have a variety of skills needed to complete their particular projects. Many large defense organizations use project structures. For example, major aircraft corporations usually have vice presidents in charge of

[4] Demarest, Marc, "The Politics of Data Warehousing" **(www.hevanet.com/demarest/marc/dwpol.html)** (June 1997).

each aircraft they produce. Many consulting firms also follow a project organization and hire people specifically to work on particular projects.

A matrix organization represents the middle ground between functional and project structures. Personnel often report to both a functional manager and one or more project managers. For example, information technology personnel at 3M, and many other companies, often split their time between two or more projects, but they report to their manager in the Information Technology Department. Project managers in matrix organizations have staff from various functional areas working on their projects, as shown in Figure 2-4. Matrix structures can be strong, weak, or balanced, based on the amount of control exerted by the project managers.

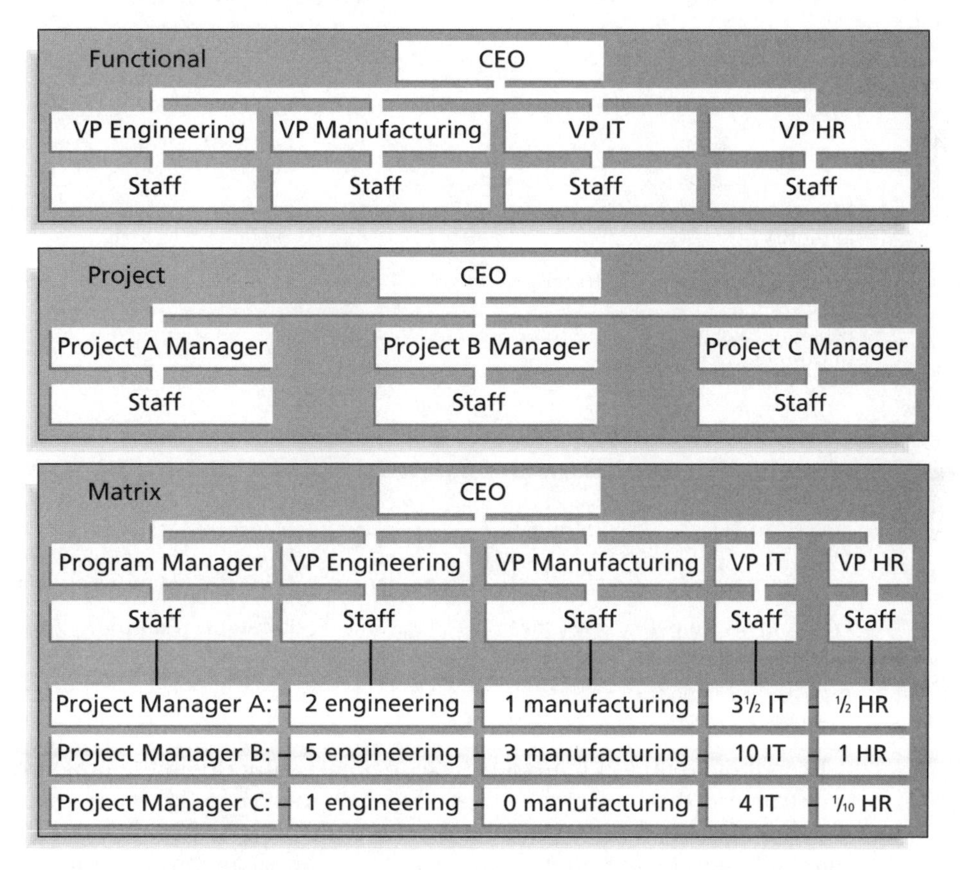

Figure 2-4. Functional, Project, and Matrix Organizational Structures

Table 2-1 summarizes how organizational structure influences projects and project managers.[5] Project managers have the most authority in a pure project

[5] PMI Standards Committee, *A Guide to the Project Management Body of Knowledge (PMBOK Guide)*, PMI, 2000, 19.

organization and the least amount of authority in a pure functional organization. It is important that project managers understand the current structure under which they are working. For example, if someone in a functional organization is asked to lead a project that requires strong support from several different functional areas, he or she should ask for senior management sponsorship. This sponsor should solicit support from all relevant functional managers to make sure that they cooperate on the project and that qualified people are available to work as needed. The project manager might also ask for a separate budget to be able to pay for project-related trips, meetings, and training or to provide financial incentives to the people supporting the project.

Table 2-1: Organizational Structure Influences on Projects

PROJECT CHARACTERISTICS	ORGANIZATION TYPE				
	FUNCTIONAL	MATRIX			PROJECT
		Weak Matrix	Balanced Matrix	Strong Matrix	
Project manager's authority	Little or none	Limited	Low to moderate	Moderate to high	High to almost total
Percent of performing organization's personnel assigned full-time to project work	Virtually none	0-25%	15-60%	50-95%	85-100%
Project manager's role	Part-time	Part-time	Full-time	Full-time	Full-time
Common title for project manager's role	Project Coordinator/ Project Leader	Project Coordinator/ Project Leader	Project Manager/ Project Officer	Project Manager/ Program Manager	Project Manager/ Program Manager
Project management administrative staff	Part-time	Part-time	Part-time	Full-time	Full-time

PMBOK Guide, 2000, 19.

Even though project managers have the most authority in the project organization structure, this type of organization is often inefficient for the company as a whole. Since people are assigned full-time to the project, they may not always be utilized. Project organizations may also miss out on economies of scale available through the pooling of requests for materials with other projects. Disadvantages such as these illustrate the benefit of using a systems approach to managing projects. When project managers use a systems approach, they are better able to make decisions that address the needs of the whole organization.

Stakeholder Management

Recall from Chapter 1 that project stakeholders are the people involved in or affected by project activities. Stakeholders can be internal to the organization, external to the organization, directly involved in the project, or just affected by the project. Internal project stakeholders generally include the project sponsor, project team, support staff, and internal customers for the project. Other internal stakeholders include top management, other functional managers, and other project managers. Since organizations do not have unlimited resources, projects affect top management, other functional managers, and other project managers by using some of the organization's limited resources. Thus, whereas additional internal stakeholders may not be directly involved in the project, they are still stakeholders because the project affects them in some way. External project stakeholders include the project's customers (if they are external to the organization), competitors, suppliers, and other external groups who may be involved in or affected by the project, such as government officials or concerned citizens. Since the purpose of project management is to meet project requirements and satisfy stakeholders, it is critical that project managers take adequate time to identify, understand, and manage relationships with all project stakeholders. Using the four frames of organizations to think about project stakeholders can help you meet their expectations.

Consider again the laptop project from the opening case. Tom Walters seemed to focus on just a few internal project stakeholders. He viewed only part of the structural frame of the college. Since his department would do most of the work in administering the laptop project, he focused on those stakeholders. Tom did not even involve the main customers for this project—the students at the college. Even though Tom sent an e-mail to faculty and staff, he did not hold meetings with senior administration or faculty at the college. Tom's view of who the stakeholders were for the laptop project was very limited.

During the faculty meeting, it became evident that the laptop project had many stakeholders in addition to just the Information Technology Department and students. If Tom had expanded his view of the structural frame of his organization by reviewing an organizational chart for the entire college, he could have identified other key stakeholders. He would have been able to see that academic department heads and members of different administrative areas would be affected by the laptop project. If Tom had focused on the human resources frame, he would have been able to tap his knowledge of the college and identify individuals who would most support or oppose requiring laptops. By using the political frame, Tom could have considered the main interest groups that would be most affected by this project's outcome. Had he used the symbolic frame, Tom could have tried to address what moving to a laptop environment would really mean for the college. He then could have anticipated some of the opposition from people who were not in favor of increasing the use of technology on campus. He also could have solicited a strong endorsement from the college president or dean before talking at the faculty meeting.

Tom Walters, like many new project managers, learned the hard way that his technical and analytical skills were not enough to guarantee success in project management. To be more effective, he had to identify and address the needs of different stakeholders and understand how his project related to the entire organization.

Top Management Commitment

A very important factor in helping project managers successfully lead projects is the level of commitment and support they receive from top management. In fact, without top management commitment, many projects will fail. As described earlier, projects are part of the larger organizational environment, and many factors that might affect a project are out of the project manager's control. Several studies cite executive support as one of the key factors associated with project success for virtually all projects.

Table 2-2 summarizes the results of the 2001 Standish Group study describing, in order of importance, what factors contribute most to the success of information technology projects. Note that executive support was listed as the most important factor, overtaking user involvement, which was most important in the 1995 CHAOS study. Executive support can help projects to achieve several of the other success factors, too, when top managers make it a priority to encourage users to actively participate in projects, assign experienced project managers, provide clear business objectives, and so on.

Table 2-2: What Helps Projects Succeed?

• Executive support
• User involvement
• Experienced project manager
• Clear business objectives
• Minimized scope
• Standard software infrastructure
• Firm basic requirements
• Formal methodology
• Reliable estimates

The Standish Group, "CHAOS 2001: A Recipe for Success" (2001).

Top management commitment is crucial to project managers for the following reasons:

- Project managers need adequate resources. The best way to kill a project is to withhold the required money, people, resources, and visibility for the project. If project managers have top management commitment, they will also have adequate resources and not be distracted by events that do not affect their specific projects.
- Project managers often require approval for unique project needs in a timely manner. For example, on large information technology projects, top management must understand that unexpected problems may result from the nature of the products being produced and the specific skills of the people on the project team. The project might need additional hardware and software halfway through, for proper testing, for example; or the project manager might need to offer special pay and benefits to attract and retain key project personnel. With top management commitment, project managers can meet these specific needs in a timely manner.
- Project managers must have cooperation from people in other parts of the organization. Since most information technology projects cut across functional areas, top management must help project managers deal with the political issues that often arise in these types of situations. If certain functional managers are not responding to project managers' requests for necessary information, top management must step in to encourage functional managers to cooperate.
- Project managers often need someone to mentor and coach them on leadership issues. Many information technology project managers come from technical positions and are inexperienced as managers. Senior managers should take the time to pass on advice on how to be good leaders. They should encourage new project managers to take classes to develop leadership skills and allocate the time and funds for them to do so.

Project managers for information technology projects work best in an environment in which top management values information technology. Working in an organization which values good project management and sets standards for its use also helps project managers succeed.

The Need for Organizational Commitment to Information Technology

Another factor affecting top management commitment to information technology projects is the organization's commitment to information technology in general. It is very difficult for a large information technology project (or a small one, for that matter) to be successful if the organization itself does not value information technology. Many companies have realized that information technology is integral to their business and have created a vice president or equivalent level position for the head of information technology, often called the Chief Information Officer (CIO). Some companies assign people from non-information technology areas to work on large projects full-time to increase involvement from end users of the systems. Some CEOs even take a strong leadership role in promoting the use of information technology in their organizations.

The Gartner Group, Inc., a well-respected information technology consulting firm, awarded Boston's State Street Bank and Trust Company's CEO, Marshall Carter, the 1998 Excellence in Technology Award. Carter provided the vision and leadership for his organization to successfully implement new information technology that expanded the bank's business. They had to gather, coordinate, and analyze vast amounts of data from around the globe to provide new asset management services to their customers. It took six years to transform State Street into a company providing state-of-the-art tools and services to its customers. The bank's revenues, profits, and earnings per share more than doubled during Carter's first five years. One key to Carter's success was his vision that technology was an integral part of the business and not just a means of automating old banking services. Carter used a highly personal style to keep his people motivated, and he often showed up at project review meetings to support his managers on information technology projects.[6]

The Need for Organizational Standards

Another problem in most organizations is not having standards or guidelines to follow that could help in performing project management. These standards or guidelines might be as simple as providing outlines and examples of what should be in a project plan, and guidelines on how the project manager should provide status information to senior management. The content of a project plan and how to provide status information might seem like common sense to senior managers, but many new information technology project managers have never before created plans or given a nontechnical status report.

Some organizations invest heavily in project management by creating a project management office or center of excellence. A **project management office** or **center of excellence** is an organizational entity created to assist project managers in achieving project goals. Rachel Hollstadt, founder and CEO of a project management consulting firm, suggests that organizations consider adding a new position, a Chief Project Officer (CPO), to elevate the importance of project management even more. Some organizations develop career paths for project managers. Some require that all project managers have project management professional (PMP) certification (*see* Chapter 1). All of these examples of setting standards demonstrate an organization's commitment to project management.

[6] Melymuka, Kathleen, "Old Bank, New Ideas," *ComputerWorld* (February 15, 1999).

SUGGESTED SKILLS FOR A PROJECT MANAGER

As you can imagine, a good project manager needs many skills. Achieving high performance on information technology projects requires strong management skills, particularly strong communication, leadership, and political skills. Project managers also need skills in organization, teamwork, coping, and making effective use of technology.

Why do project managers need strong management skills? One reason is that to understand, navigate, and meet stakeholders' needs and expectations, project managers need to lead, communicate, negotiate, problem solve, and influence the organization at large. They need to be able to actively listen to what others are saying, help develop new approaches for solving problems, and then persuade others to work toward achieving project goals. Project managers must lead their project teams by providing vision, delegating work, creating an energetic and positive environment, and setting an example of appropriate and effective behavior.

Project managers must also have strong organizational skills to be able to plan, analyze, set, and achieve project goals. Project managers must focus on teamwork skills in order to use their people effectively. They need to be able to motivate different types of people and develop *esprit de corps* within the project team and with other project stakeholders. Since most projects involve changes and trade-offs between competing goals, it is important for project managers to also have strong coping skills. Project managers must be flexible, creative, and sometimes patient in working toward project goals; they must also be persistent in making project needs known.

Lastly, project managers must be able to make effective use of technology as it relates to the specific project. Making effective use of technology often includes special product knowledge or experience with a particular industry. Project managers must make many decisions and deal with people in a wide variety of disciples, so it helps tremendously to have a project manager who is confident in using the special tools or technologies that are the most effective in particular settings.

Several studies have been done to determine what project managers actually do in their jobs and what characteristics are associated with success. The National Science Foundation's 1999 study, titled "Building a Foundation for Tomorrow: Skills Standards for Information Technology, Millennium Edition" documents results of a collaborative effort between industry and academia to improve the education of the information technology workforce. This study found that project management is one of the job skills needed in every major information technology field, from database administrator to network specialist to technical writer. Table 2-3 lists fifteen project management job functions that are essential for good project management.

Table 2-3: Fifteen Project Management Job Functions

1. Define scope of project
2. Identify stakeholders, decision-makers, and escalation procedures
3. Develop detailed task list (work breakdown structures)
4. Estimate time requirements
5. Develop initial project management flow chart
6. Identify required resources and budget
7. Evaluate project requirements
8. Identify and evaluate risks
9. Prepare contingency plan
10. Identify interdependencies
11. Identify and track critical milestones
12. Participate in project phase review
13. Secure needed resources
14. Manage the change control process
15. Report project status

"Building a Foundation for Tomorrow: Skills Standards for Information Technology, Millenium Edition," Northwest Center for Emerging Technologies, Belleview, WA, 1999.

Each of the job functions listed in Table 2-3 requires different performance criteria, technical knowledge, foundation skills, and personal qualities. In a recent study, 100 practicing project managers were asked what they believed were the critical characteristics necessary to be effective and what made project managers ineffective.[7] Table 2-4 lists the results. Their study found that effective project managers provide leadership by example, are visionary, technically competent, decisive, good communicators, good motivators, stand up to upper management when necessary, support team members, and encourage new ideas. They also found that their respondents believed that positive leadership contributes the most to project success. The most important characteristics and behaviors of positive leaders include being a team builder and communicator, having high self-esteem, focusing on results, demonstrating trust and respect, and setting goals.

[7] Zimmerer, Thomas W. and Mahmoud M. Yasin, "A Leadership Profile of American Project Managers," *Project Management Journal* (March 1998), 31–38.

Table 2-4: Most Significant Characteristics of Effective and Ineffective Project Managers

EFFECTIVE PROJECT MANAGERS	INEFFECTIVE PROJECT MANAGERS
Lead by example	Set bad examples
Are visionaries	Are not self-assured
Are technically competent	Lack technical expertise
Are decisive	Are poor communicators
Are good communicators	Are poor motivators
Are good motivators	
Stand up to upper management when necessary	
Support team members	
Encourage new ideas	

Zimmerman and Yasin, 1998.

Even if organizations could find people with all of the skills and characteristics identified in these and other studies, it does not mean that every project would be a roaring success. Success is more likely, however, when project managers work to develop these skills and organizations promote the use of good project management.

PROJECT MANAGEMENT PROCESS GROUPS

Recall from Chapter 1 that project management consists of nine knowledge areas: integration, scope, time, cost, quality, human resource, communications, risk, and procurement. Chapters 3 through 11 of this book provide detailed information about each of these knowledge areas and how they apply to managing information technology projects. Another important concept related to project management are the five project management process groups: initiating, planning, executing, controlling, and closing. Chapters 12 through 16 of this book describe each process group in detail through a real case study of a large information technology project. It is important to understand what is involved in each project management process group and how they relate to the nine knowledge areas.

Project management is an integrative endeavor; decisions and actions taken in one knowledge area at a certain time will usually affect other knowledge areas. Managing these interactions often requires making trade-offs among the project objectives of scope, time, and cost—the triple constraint of project management described in Chapter 1. A project manager may also need to make trade-offs between knowledge areas, such as between risk and human resources. As a consequence, you can view project management as a number of interlinked processes. A **process** is a series of actions directed toward a particular result. **Project management process groups** progress from initiation activities to planning activities, executing activities, controlling activities, and closing activities.

Initiating processes include actions to commit to begin or end projects and project phases. Several things must be done to initiate a project in the concept phase. Someone must define the business need for the project, someone must sponsor the project, and someone must take on the role of project manager. Initiating processes take place during each phase of a project. For example, project managers and teams should reexamine the business need for the project during every phase of the project life cycle to determine if the project is worth continuing. Initiating processes are even required to end a project. Someone must initiate activities to ensure that all the work will be completed, that the customer will accept the work, that the project team will document lessons learned on the project, and that all project resources will be reassigned.

Planning processes include devising and maintaining a workable scheme to accomplish the business need that the project was undertaken to address. Project plans are created to define each knowledge area as it relates to the project at that point in time. For example, plans must be developed to define the scope of the project, to define the schedule as to when various project activities should be done and who will do them, to estimate costs, to decide what resources will need to be procured, and so on. To account for changing conditions on the project and in the organization, plans are often revised during each phase of the project life cycle.

Executing processes include coordinating people and other resources to carry out the project plans and produce the products or deliverables of the project or phase. Examples of executing processes include developing the project team, providing leadership, assuring project quality, disseminating information, procuring necessary resources, and delivering the actual work.

Controlling processes ensure that project objectives are met. The project manager and staff monitor and measure progress against plans and take corrective action when necessary. A common controlling process is having performance and status reviews. If changes are necessary, someone must identify, analyze, and manage those changes.

Closing processes include formalizing acceptance of the phase or project and bringing it to an orderly end. Administrative activities are often involved, such as archiving project files, documenting lessons learned, and receiving formal acceptance of work delivered as part of the phase or project.

Figure 2-5 shows the project management process groups and how they relate to each other in terms of typical level of activity, time frame, and overlap. Notice that the process groups are not discrete, one-time events. They occur at varying levels of intensity throughout each phase of a project. The level of activity and time, length of each process group vary for every project. Normally the executing processes take the most resources and the most amount of time, followed by planning. The initiating and closing processes are usually the shortest and require the least amount of resources and time. However, every project is unique, so there can be exceptions.

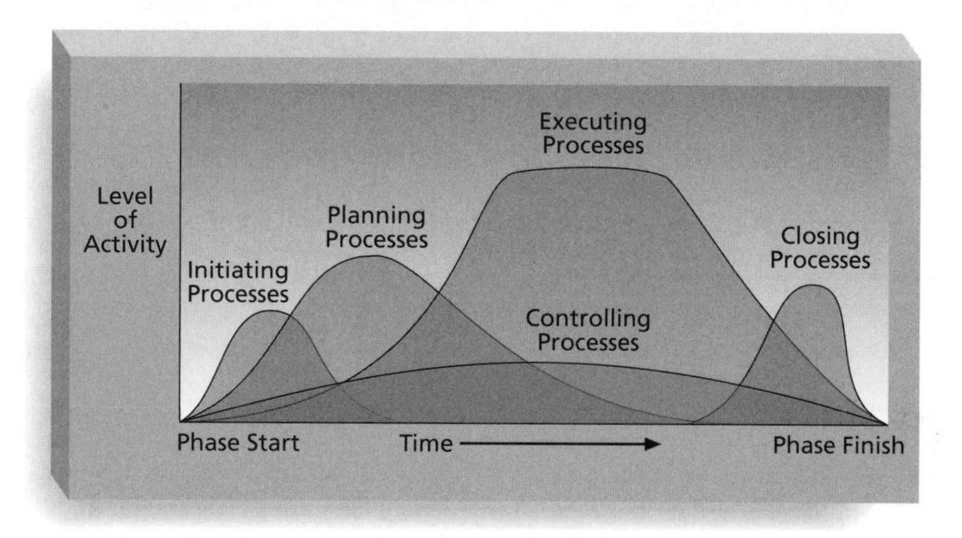

Figure 2-5. Overlap of Process Groups in a Phase
PMBOK Guide, 2000, 31

Each of the five project management process groups is characterized by the completion of certain tasks. During initiating processes for a new project, the organization recognizes that a new project exists. Often, this recognition is accomplished by completing a stakeholder analysis, requirements document, and feasibility study. These reports outline the potential supporters and opponents of a project, a definition of the project, and the high-level goals, scope, deliverables, deadlines, and resources of the project. The main outcomes of the initiating process at the beginning phase of a project are the creation of a project charter (described in Chapter 4) and selection of a project manager.

Key outcomes of the planning process include completion of a work breakdown structure, project schedule, and project budget (*see* Chapters 5, 6, and 7, respectively). Planning is especially important for information technology projects. Everyone who has ever worked on a large information technology project involving new technology knows the saying "A dollar spent up front in planning is worth one hundred dollars spent after the system is implemented." Why is planning so crucial in information technology projects? It takes a huge amount of effort to change a system once it has been implemented.

The executing process involves taking the actions necessary to ensure that the work described in the planning activities will be completed. The main outcome of this process is the delivery of the actual work of the project. For example, if an information technology project involves providing new hardware, software, and training, the executing processes would include leading the project team and other stakeholders to purchase the hardware, develop and test the software, and deliver and participate in the training. This process group should overlap with all of the other process groups and take the most resources to accomplish.

Controlling is the process of measuring progress towards the project objectives, monitoring deviation from the plan, and taking corrective action to match progress with the plan. The ideal outcome of controlling is to successfully complete a project by delivering the agreed upon project scope within time, cost, and quality constraints. If changes to project objectives or plans are required, controlling processes ensure that they are made in an efficient and effective manner to meet stakeholder needs and expectations. Controlling processes overlap all of the other process groups because changes can be made at any time.

During the closing processes, the project team works to gain acceptance of the end product and bring the phase or project to an orderly end. Key outcomes are formal acceptance of the work and creation of closing documents such as a project audit and lessons-learned report.

You can map the main activities of each project management process group, which apply to an entire project or a phase of a project, into the nine project management knowledge areas. Table 2-5 provides a big picture of the relationships among the thirty-nine project management activities, the time in which they are typically completed, and the knowledge areas into which they fit. The activities listed in the table are the main processes for each knowledge area listed in the *PMBOK Guide* 2000. Notice that the majority of project management activities occur as part of the planning process group. Since each project is unique, project teams are always trying to do something that has not been done before. To succeed at doing unique and new activities, project teams must do a fair amount of planning.

Table 2.5: Relationships Among Project Process Groups, Activities, and Knowledge Areas

KNOWLEDGE AREA	PROJECT PROCESS GROUPS				
	INITIATING	PLANNING	EXECUTING	CONTROLLING	CLOSING
Integration		Project plan development	Project plan execution	Integrated change control	
Scope	Initiation	Scope planning		Scope verification	
		Scope definition		Scope change control	
Time		Activity definition		Schedule control	
		Activity sequencing			
		Activity duration estimating			

Table 2.5: Relationships Among Project Process Groups, Activities, and Knowledge Areas (continued)

KNOWLEDGE AREA	PROJECT PROCESS GROUPS				
	INITIATING	PLANNING	EXECUTING	CONTROLLING	CLOSING
		Schedule development			
Cost		Resource planning		Cost control	
		Cost estimating			
		Cost budgeting			
Quality		Quality planning	Quality assurance	Quality control	
Human resources		Organizational planning	Team development		
		Staff acquisition			
Communications		Communica-tions planning	Information distribution	Performance reporting	Administrative closure
Risk		Risk management planning		Risk monitoring and control	
		Risk identification			
		Qualitative risk analysis			
		Quantitative risk analysis			
		Risk response planning			
Procurement		Procurement planning	Solicitation		Contract close-out
		Solicitation planning	Source selection		

DEVELOPING AN INFORMATION TECHNOLOGY PROJECT MANAGEMENT METHODOLOGY

This chapter has discussed various topics, including systems thinking and systems analysis, the project life cycle, product life cycles, management reviews, the need for understanding organizations, and project management process groups. It is easy to get confused trying to understand the difference between life cycles and process groups, or between product development methodologies and a general project management approach. Many organizations talk about project management and spend much time and expense in training efforts, but after they have completed the training, people still do not know how to apply good project management that is tailored for their particular needs. Because of this problem, some organizations develop their own internal information technology project management methodologies.

For example, after implementing a systems development life cycle (SDLC) at Blue Cross Blue Shield of Michigan, the methods area became aware that developers and managers were not always doing the same things on every information technology project. Deliverables were often missing or looked different from project to project. There was a general feeling of a lack of consistency and a need for a standard to guide both new and experienced project managers. Senior management decided to authorize funds to develop a methodology for project managers that could also become the basis for information technology project management training within the organization. It was also viewed as part of an overall effort to help raise the company's Software Capability Maturity Model level. (See Chapter 7, *Project Quality Management*, for more information on maturity models.)

Blue Cross Blue Shield launched a three-month project to develop its own project management methodology. Some of the project team members had already passed the PMP certification examination, so they decided to base their methodology on the *PMBOK Guide*, making adjustments as needed to best describe how their organization managed information technology projects. Figure 2-6 shows a one-page picture of the resulting information technology project management processes and the information flow among them.[8]

[8] Munroe, William, "Developing and Implementing an IT Project Management Process," *ISSIGreview* (First Quarter 2001). This article was also published in *Trends in Software Engineering Process Management* (www.tsepm.com) (Fall 2000).

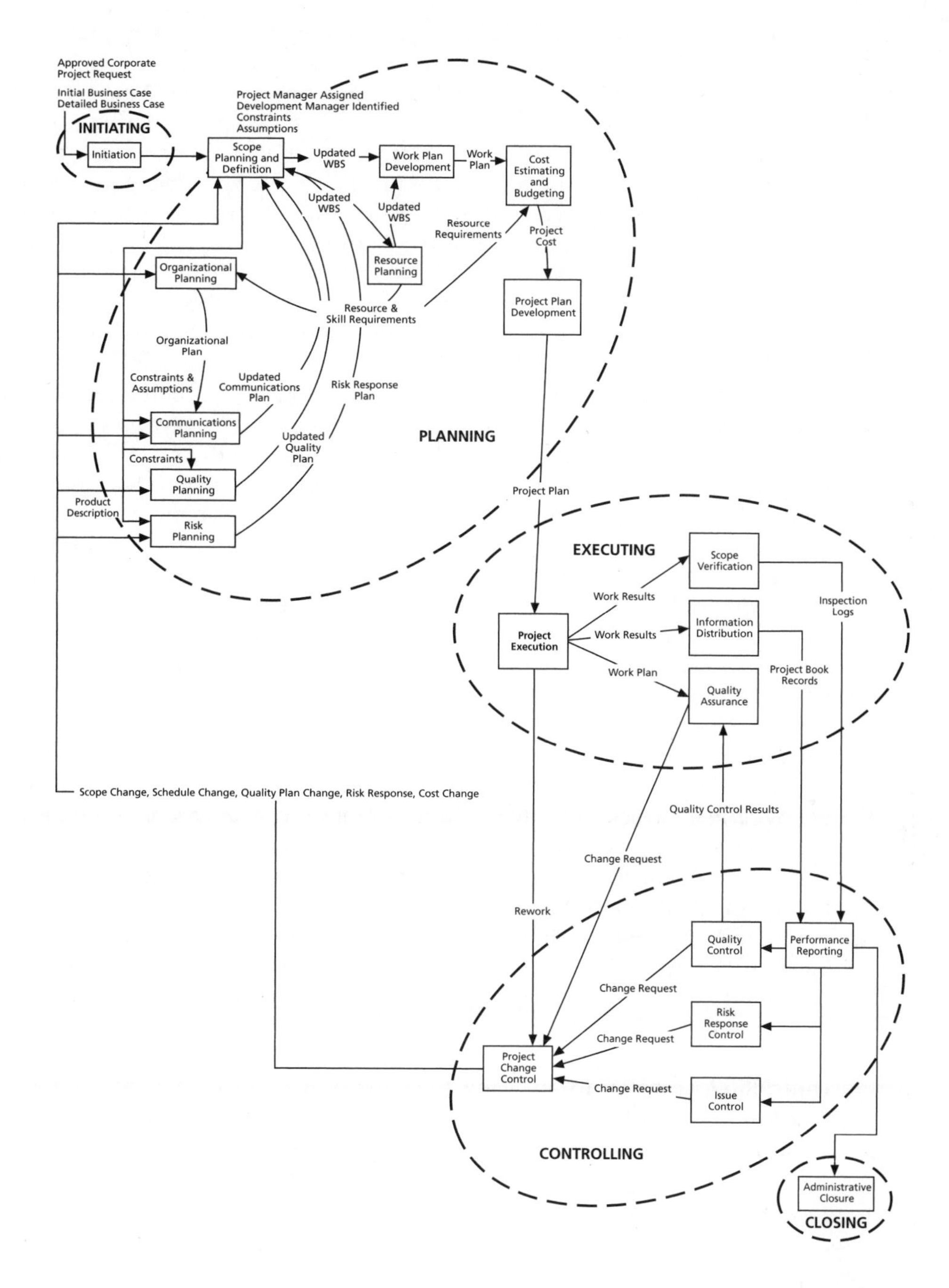

Figure 2-6. Information Technology Project Management Methodology at Blue Cross Blue Shield

The Blue Cross Blue Shield team realized that some processes in the *PMBOK Guide* would have to be dropped or deemphasized to fit their organization's needs. For example, in contrast to some industries, the overriding financial investment in software development was made in people's salaries, not in materials. In addition, at Blue Cross Blue Shield negotiations with contracting firms were not done in the information technology area. Therefore, most of the procurement functions were absorbed into other processes, such as Scope Planning and Definition, and Resource Planning.

Additional processes were also added. For example, to keep track of the large amount of documentation necessary on an information technology project, the team decided to develop a process for maintaining and updating a project workbook that would serve as an information resource for team members and a hardcopy record of project activities. Project Book Records was therefore added as a separate process under Information Distribution. Another new process, Issue Control, was added because of its importance to information technology projects. Issues are prevalent in information technology projects, in part because of the inherent complexity of information systems and rapidly changing technology.

The team also decided to combine the *PMBOK Guide's* processes of Activity Sequencing, Activity Definition, Activity Duration Estimating, and Schedule Development into one process that would be called Work Plan Development. Also, to enhance usability and simplify the overall process, the PMBOK Integrated Change Control, Scope Change Control, Schedule Control, and Cost Control processes were absorbed into a process called Project Change Control.

Blue Cross Blue Shield wanted their information technology project management methodology to work with any of the several systems development life cycle (SDLC) models. This forced team members who may have been involved in all aspects of software development to strictly separate their thinking of product creation from managing that activity. Understanding the difference between the SDLC and the information technology project management methodology amounted to a paradigm shift for some members of the team who had been used to filling a dual role of developer and project manager. Today Blue Cross Blue Shield's information technology project management methodology is the basis for their training programs and is used to develop and implement their information technology projects.

The project management process groups—initiating, planning, executing, controlling, and closing—provide a useful framework for understanding project management. They apply to most projects (information technology and non-information technology) most of the time, and, along with the project management knowledge areas, they help project managers look at the big picture of managing a project in their particular organizations.

CASE WRAP-UP

After several people voiced concerns about the laptop idea at the faculty meeting, the president of the college directed that a committee be formed to formally review the concept of requiring students to have laptops in the near future. Because the college was dealing with several other important enrollment-related issues, the president named the vice president of enrollment to head the committee. Other people soon volunteered or were assigned to the committee, including Tom Walters as head of Information Technology, the director of the adult education program, the chair of the Computer Science Department, and the chair of the History Department. The president also insisted that the committee include at least two members of the student body. The president knew everyone was busy, and he did not think that the laptop idea was a high-priority issue for the college. Therefore, he directed the committee to present a proposal at the next month's faculty meeting, either to recommend the creation of a formal project team (of which these committee members would commit to be a part) to fully investigate requiring laptops, or to recommend terminating the concept. At the next facility meeting, few people were surprised to hear the recommendation to terminate the concept. Tom Walters learned that he had to pay much more attention to the needs of the entire college before proceeding with detailed information technology plans.

CHAPTER SUMMARY

Projects operate in an environment broader than the project itself. Project managers need to take a systems approach when working on projects; they need to consider projects within the greater organizational context.

A project life cycle is a collection of project phases. Most projects include concept, development, implementation, and close-out phases. Examples of product life cycles include the waterfall model, spiral model, incremental release model, RAD model, and prototyping model. Most product life cycle models, such as these, comprise many distinct projects. Project management is a cross life cycle activity in information technology projects and should be done during all phases of information technology product development.

A project must successfully pass through each of the project phases in order to continue. A management review called a phase exit or kill point should occur at the end of each project phase. These reviews are important for keeping projects on track and determining if projects should be continued, redirected, or terminated.

Organizations have four different frames: structural, human resources, political, and symbolic. Project managers need to understand all of these aspects of organizations in order to be successful. The structural frame focuses on different groups' roles and responsibilities to meet the goals and policies set by top management. The human resources frame focuses on producing harmony between the needs of the organization and the needs of people. The political frame addresses organizational and personal politics. The symbolic frame focuses on symbols and meanings.

The structure of an organization has strong implications for project managers, especially in terms of the amount of authority the project manager has. The three basic organizational structures include functional, matrix, and project. Project managers have the most authority in a pure project organization, an intermediate amount of authority in the matrix organization, and the least amount of authority in a pure functional organization.

Project stakeholders are individuals and organizations who are actively involved in the project or whose interests may be positively or negatively affected as a result of project execution or successful project completion. Project managers must identify and understand the different needs of all stakeholders on their projects.

Top management commitment is a key factor associated with project success. Since projects often cut across many areas in an organization, top management must assist project managers if they are to do a good job of project integration.

Project managers perform a variety of job functions, and they need to have many different skills and personal characteristics to do their jobs well. Project managers need strong management skills, as well as particularly strong communication, leadership, and political skills. Project managers also need skills in organization, teamwork, coping, and making effective use of technology.

Project management can be viewed as a number of interlinked processes. The five process groups of project management are initiating, planning, executing, controlling, and closing. These processes occur at varying levels of intensity throughout each phase of a project, and specific outcomes are produced as a result of each process. Normally the executing processes take the most resources and the most amount of time, followed by planning.

Mapping the main activities of each project management process group into the nine project management knowledge areas provides a big picture of what activities are involved in project management. Some organizations develop their own information technology project management methodologies, using the *PMBOK Guide* as a foundation. It is important to distinguish between the work involved in creating products and the project management processes involved in their creation.

DISCUSSION QUESTIONS

1. What does it mean to take a systems view? How does taking a systems view apply to project management?
2. How does a project life cycle differ from the systems development life cycle (SDLC)? Describe several projects that could be done in each stage of the SDLC.
3. Briefly explain the difference between functional, matrix, and project organizations. Describe how each structure affects the management of the project.
4. Discuss the importance of top management commitment and the development of standards for successful project management. Give examples of projects that failed due to a lack of top management commitment and a lack of organizational standards.
5. Which skills do you think are most important for an information technology project manager? Can they all be learned, or do you think some are innate?
6. Briefly describe what happens in each of the five process groups (initiating, planning, executing, controlling, and closing). On which processes should the most team members spend the most time? What are some of the deliverables of each process?
7. How do the project management process groups differ from the processes with which most information technology professionals are familiar? How are they similar? How are they related?

EXERCISES

1. Apply the information on the four frames of organizations to an information technology project you are familiar with. If you cannot think of a good information technology project, use your personal experience in deciding where to attend college to apply this framework. Write a one- to two-page paper describing key issues related to the structural, human resources, political, and symbolic frames. Which frame seemed to be the most important and why? For example, did you decide where to attend college primarily because of the curriculum and structure of the program? Did you follow your friends? Did your parents have a lot of influence in your decision? Did you like the culture of the campus?
2. Search the Internet for information on software development life cycles. Find two good Web sites related to this subject. Do these sources mention project management at all? Write a one- to two-page summary of your findings.

3. Read the suggested readings by Crawford, Posner, or Zimmerer and Yasin, or find another reference about the skills required for a good project manager. Write a one- to two-page paper describing the article.
4. Search the Internet and scan information technology industry magazines or Web sites to find an example of an information technology project that had problems. Write a one- to two-page paper summarizing who the key stakeholders were for the project and how they influenced the outcome.
5. Write a one- to two-page summary of an article about the importance of top management support for project success. You may use the Wiegers article from the suggested reading list.
6. Read the article by William Munro regarding Blue Cross Blue Shield's information technology project management methodology. Write a one- to two-page summary of the article, its key conclusions, and your opinion of it. Do you think this methodology could be applied in many other organizations, or does each organization need to create its own methodology?

MINICASE

You have been part of your company's information technology department for three years. You have learned a lot about the company and about many new technologies in your latest assignment—developing applications for your corporate intranet. Since you are an avid recreational athlete, you have spent a fair amount of time thinking about how you would write a sophisticated application to help people learn about the many corporate athletic teams, register on-line, determine team schedules, post team statistics, and so on. You have heard some rumors about profits not being as high as expected in the past year, and you know that mostly the junior employees participate on the athletic teams.

Part 1. Using Figure 2-1 as a guide, use the three-sphere model of systems management to identify potential issues that could be factors in deciding whether you should proceed with your idea to develop an application for recreational athletics at your company. Include at least three questions for each sphere.

Part 2: Your immediate boss likes your idea of developing a recreational sports application on the intranet, but he has to convince his boss that this project is valuable to the company. Prepare a short presentation with five to ten slides and speaker notes to convince senior management to approve the recreational athletics application project. Be sure to list benefits of the project and suggest a phased approach. For example, the first phase might involve just posting information about various sports teams on the intranet. The second phase might include on-line registration, and so on.

SUGGESTED READINGS

1. Boehm, Barry. "Anchoring the Software Process." *IEEE Software* (1996); "A Spiral Model of Software Development and Enhancement." *IEEE Computer* (May 1988): 5, 61-72.

    Boehm is well known for his work on the software development process. In the 1996 article he proposes using three common milestones to serve as a basis for managing software projects. The 1988 article summarizes various software development models and explains the spiral model in detail.

2. Bolman, Lee G. and Deal, Terrence E. *Reframing Organizations*. Jossey-Bass Publishers (1991).

    This book describes the four frames of organizations: structural, human resources, political, and symbolic. It explains how to look at situations from more than one vantage point to help bring order out of confusion.

3. Crawford, Lynn. "Profiling the Competent Project Manager." *Project Management Research at the Turn of the Millennium*. Project Management Institute (2000).

    The competence of the project manager is a key factor in project success. In this article, Crawford presents a review and analysis of research-based literature concerning the knowledge, skills, and personal attributes of project managers.

4. Northwest Center for Emerging Technologies. *Building a Foundation for Tomorrow: Skills Standards for Information Technology, Millennium Edition*. Belleview, WA (1999). (Copies may be ordered by calling the Belleview Community College bookstore at (206) 641-2066.)

    The National Science Foundation provided the vision and major funding for this collaborative study done by industry and academia. The report documents suggested skill standards for information technology career clusters. It provides detailed information on the performance criteria, technical knowledge, foundation skills, and personal qualities for each of fifteen major job functions of project management.

5. Posner, Barry Z. "What It Takes to Be a Good Project Manager." *Project Management Institute* (March 1987).

    This article summarizes key attributes of successful project managers and lists common causes for problems in managing projects based on a survey of 287 project managers.

6. Robert, Daniel W. "Creating an Environment for Project Success." *Information Systems Management* (Winter 1997).

    This article states that IS project teams are set up for failure because the project environment is established in a way that precludes success. The author presents specific strategies for taking control of the project environment to increase the likelihood of success.

7. Wiegers, Karl. "Software Process Improvement: 10 Traps to Avoid." *Software Development* (May 1996).

This article lists ten common traps that can undermine software process improvement programs. Lack of management commitment and unrealistic expectations are the top two.

8. Zimmerer, Thomas W. and Mahmoud M. Yasin. "A Leadership Profile of American Project Managers." *Project Management Journal* (March 1998).

This article summarizes recent literature on project managers and provides detailed results of a recent survey of 100 senior project managers on what makes project managers successful, what project management tools are most useful, and what leadership factors have the most positive influence on organizations.

KEY TERMS

- **closing processes** — formalizing acceptance of the project or phase and bringing it to an orderly end
- **controlling processes** — actions to ensure that project objectives are met
- **deliverable** — a product, such as a report or segment of software code, produced as part of a project
- **executing processes** — coordinating people and other resources to carry out the project plans and produce the products or deliverables of the project
- **functional organizational structure** — an organizational structure that groups people by functional areas such as information technology, manufacturing, engineering, and accounting.
- **human resources frame** — focuses on producing harmony between the needs of the organization and the needs of people
- **initiating processes** — actions to commit to begin or end projects and project phases
- **matrix organizational structure** — an organizational structure in which employees are assigned to both functional and project managers
- **phase exit** or **kill point** — management review that should occur after each project phase to determine if projects should be continued, redirected, or terminated
- **planning processes** — devising and maintaining a workable scheme to accomplish the business need that the project was undertaken to address
- **political frame** — addresses organizational and personal politics
- **politics** — competition between groups or individuals for power and leadership
- **process** — a series of actions directed toward a particular result
- **project acquisition** — a common reference to the last two project phases—implementation and close-out
- **project feasibility** — a common reference to the first two project phases—concept and development

- **project life cycle** — the collection of project phases—concept, development, implementation, and close-out
- **project management process groups** — the progression of project activities from initiation to planning, executing, controlling, and closing
- **project organizational structure** — an organizational structure that groups people by major projects, such as specific aircraft programs
- **structural frame** — deals with how the organization is structured (usually depicted in an organizational chart) and focuses on different groups' roles and responsibilities to meet the goals and policies set by top management
- **symbolic frame** — focuses on the symbols, meanings, and culture of an organization
- **systems** — sets of interacting components working within an environment to fulfill some purpose
- **systems analysis** — a problem-solving approach that requires defining the scope of the system to be studied, and then dividing it into its component parts for identifying and evaluating its problems, opportunities, constraints, and needs
- **systems approach** — a holistic and analytical approach to solving complex problems that includes using a systems philosophy, systems analysis, and systems management
- **systems development life cycle (SDLC)** — a framework for describing the phases involved in developing and maintaining information systems
- **systems management** — addressing the business, technological, and organizational issues associated with making a change to a system
- **systems philosophy** — an overall model for thinking about things as systems
- **systems thinking** — taking a holistic view of an organization to effectively handle complex situations
- **work breakdown structure (WBS)** — an outcome-oriented analysis of the work involved in a project that defines the total scope of the project

3

Project Integration Management

Objectives

After reading this chapter, you will be able to:

1. *Understand the importance of project integration management*
2. *Describe an overall framework for project integration management as it relates to the other project management knowledge areas and the project life cycle*
3. *Describe project plan development and the major components of a good project plan*
4. *Explain project plan execution and key aspects of getting work results*
5. *Describe the integrated change control process, including project plan updates, corrective actions, and lessons learned*

Nick Carson was recently made project manager of a critical biotech project for his company in Silicon Valley. This project involved creating the hardware and software for a DNA-sequencing instrument used in assembling and analyzing the human genome. Each instrument sold for approximately $200,000, and various clients would purchase several instruments. One hundred instruments running 24 hours per day could decipher the entire human genome in less than two years. The biotech project was the company's largest project, and it had tremendous potential for future growth and revenues. Unfortunately, there were problems managing this large project. It had been underway for three years and had already gone through three different project managers. Nick had been the lead software developer on the project before senior management made him project manager. Senior management told him to do whatever it took to deliver the first version of the software for the DNA sequencing instrument in four months and a production version in nine months.

Their urgency in getting the project out was influenced by negotiations for a potential corporate buyout with a larger company.

Nick was highly energetic and intelligent and had the technical background to make the project a success. He delved into the technical problems and found some critical flaws that were keeping the DNA sequencing instrument from working. He was having difficulty, however, in his new role as project manager. Although Nick and his team got the product out on time, senior management was upset because Nick did not focus on managing. He never provided management with accurate schedules or detailed plans of what was happening on the project. Instead of performing the work of project manager, Nick had taken on the role of software integrator and troubleshooter. Nick, however, did not understand senior management's problem—he delivered the product, didn't he? Didn't they realize how valuable he was?

WHAT IS PROJECT INTEGRATION MANAGEMENT?

Project integration management involves coordinating all of the other project management knowledge areas throughout a project's life cycle. It ensures that all the elements of a project come together at the right times to complete a project successfully. The main processes involved in project integration management include:

1. **Project plan development**, which involves putting the results of other planning processes into a consistent, coherent document—the project plan.
2. **Project plan execution**, which involves carrying out the project plan by performing the activities included in it.
3. **Integrated change control**, which involves coordinating changes across the entire project.

Figure 3-1 provides an overview of the processes, inputs, tools and techniques, and outputs of project integration management. For example, inputs to project plan development include other planning outputs, historical information, organizational policies, constraints, and assumptions. Tools and techniques to assist in project plan development include a project planning methodology, stakeholder skills and knowledge, a project management information system or project management software, and earned value management. Outputs of project plan development are the project plan and supporting detail.

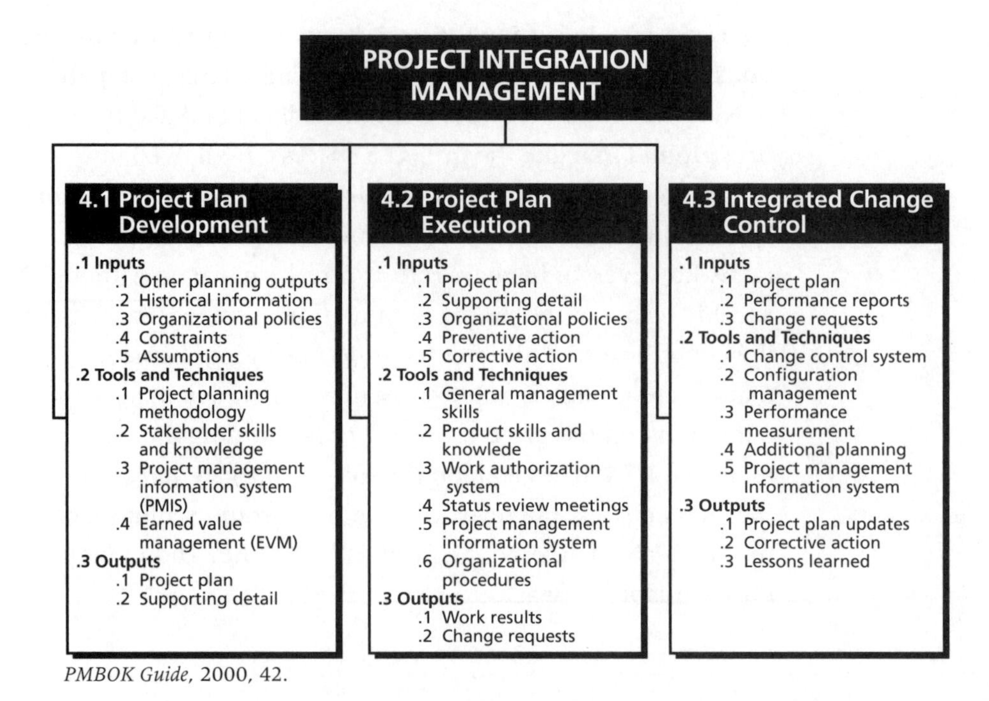

PMBOK Guide, 2000, 42.

Figure 3-1. Project Integration Management Overview

To accomplish project integration management, you must engage in project scope, quality, time, and cost management and human resource, communications, risk, and procurement management. Because it ties together all the knowledge areas, project integration management depends on activities from all eight of the other knowledge areas. It also requires commitment from top management in the organization sponsoring the project throughout the life of the project.

Figure 3-2 provides a framework for understanding how project integration management serves as the guiding force in managing a project. Recall from Chapter 2 that projects pass through the basic phases of concept, development, implementation, and close-out. The x-axis of this figure represents the project life cycle phases, and the y-axis represents the eight other project management knowledge areas. The project integration management knowledge area is represented as an arrow that becomes more focused as the project progresses through its life cycle. Project integration management pulls together all of these elements to guide the project toward successful completion. By performing good project planning, execution, and change control, project managers and their teams can meet or exceed stakeholder needs and expectations.

Many people consider integration management to be the key to overall project success. Someone must take responsibility for coordinating all of the people, plans, and work required to complete a project. Someone must focus on the big picture of the project and steer the project team toward successful completion. Someone must make the final decisions when there are conflicts among project goals or people involved. Someone must communicate key project information to senior management. This someone is the project manager, and the project manager's chief tool for accomplishing all these tasks is project integration management.

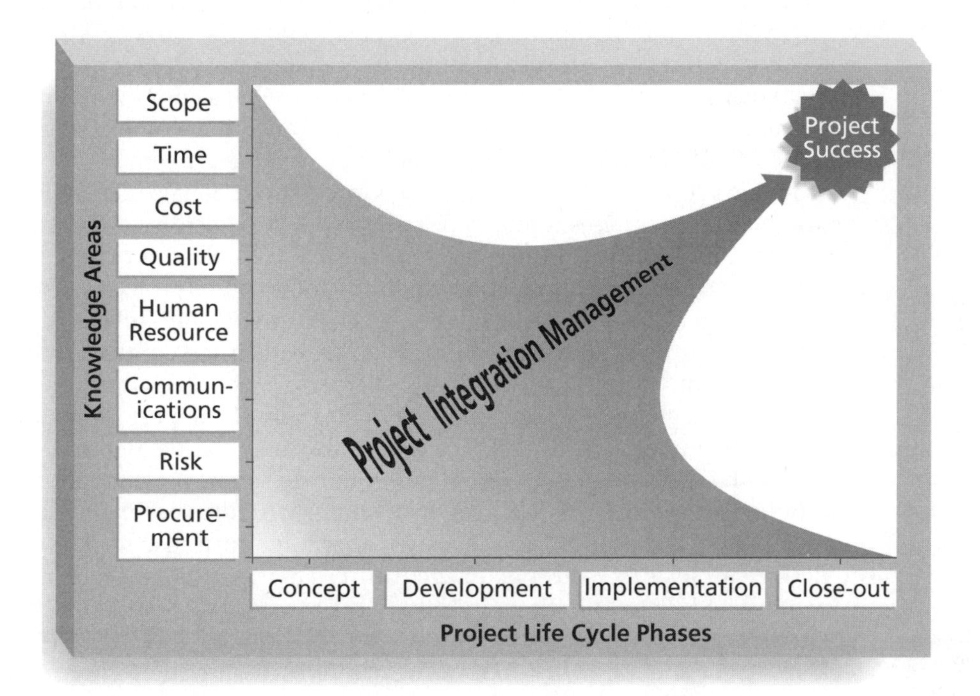

Figure 3-2. Framework for Project Integration Management

Good project integration management is critical to providing stakeholder satisfaction. Integration management includes interface management. **Interface management** involves identifying and managing the points of interaction between various elements of the project. The number of interfaces can increase exponentially as the number of people involved in a project increases. Thus, one of the most important jobs of a project manager is that of establishing and maintaining good communication and relationships across organizational interfaces. The project manager must communicate well with all project stakeholders, including customers, the project team, senior management, other project managers, and opponents of the project.

What happens when a project manager does not communicate well with all stakeholders? In the opening case, Nick Carson seemed to ignore a key stakeholder for the DNA sequencing instrument project—his senior management. Nick was comfortable working with other members of the project team, but he was not familiar with his new job as project manager or the needs of the company's top management. Nick continued to do his old job of software developer and took on the added role of software integrator. He mistakenly thought integration management meant software integration management and focused on the project's technical problems. He totally ignored what project integration is really about—integrating the work of all of the people involved in the project by focusing on good communication and relationship management. Recall that project management is applying knowledge, skills, tools, and techniques in order to meet project requirements, while also meeting or exceeding stakeholder needs and expectations. Nick did not take the time to find out what senior management expected from him as the project manager; he assumed that completing the project on time was sufficient to make senior management happy.

In addition to not understanding project integration management, Nick did not use holistic or systems thinking (*see* Chapter 2). He burrowed into the technical details of his particular project. He did not stop to think about what it meant to be the project manager, how this project related to other projects in the organization, or what senior management's expectations were of him and his team.

Project integration management must occur within the context of the whole organization, not just within a particular project. The work of the project must be integrated with the ongoing operations of the performing organization. In the opening case, Nick's company was negotiating a potential buyout with a larger company. As a consequence, senior management needed to know when the DNA sequencing instrument would be ready, how big the market was for the product, and if they had enough in-house staff to continue to manage projects like this one in the future. They wanted to see a project plan and schedule to help them monitor the project's progress and show their potential buyer what was happening. When senior managers tried to talk to Nick about these issues, Nick soon returned to discussing the technical details of the project. Even though Nick was very bright, he had no experience or real interest in many of the business aspects of how his company operated. Project managers must always view their projects in the context of the changing needs of their organizations and respond to requests from senior management.

Project integration management, therefore, involves integrating the other knowledge areas within the project as well as integrating different areas outside the project. In order to integrate across knowledge areas and across the organization, there must be a good project plan. Therefore, the first process of integration management is project plan development.

PROJECT PLAN DEVELOPMENT

A **project plan** is a document used to coordinate all project planning documents and help guide a project's execution and control. Project plans also document project planning assumptions and decisions regarding choices, facilitate communication among stakeholders, define the content, extent, and timing of key management reviews, and provide a baseline for progress measurement and project control. Plans should be dynamic, flexible, and subject to change when the environment or project changes. Plans should greatly assist the project manager in leading the project team and assessing project status.

To create and assemble a good project plan, the project manager must practice the art of integration since information is required from all of the knowledge areas. Working with the project team and other stakeholders to create a project plan will help the project manager have a good understanding of the overall project and how to guide its execution.

Just as projects are unique, so are project plans. A small project involving a few people over a couple of months might have a project plan consisting of a two-page description of the project along with a work breakdown structure and Gantt chart. A large project involving a hundred people over three years would have much more detailed project plans. It is important to tailor project plans to fit the needs of specific projects. The plans should guide the work, so they should be only as detailed as needed for each project.

There are, however, common elements to most project plans. Parts of a project plan include an introduction or overview of the project, a description of how the project is organized, the management and technical processes used on the project, and sections describing the work to be done, the schedule, and the budget.

The introduction or overview of the project should include, as a minimum, the following information:

- The project name: Every project should have a unique name. Unique names help distinguish each project and avoid confusion among related projects.
- A brief description of the project and the need it addresses: This description should clearly outline the goals of the project and why it is being done. It should be written in layman's terms, avoid technical jargon, and include a rough time and cost estimate.
- The sponsor's name: Every project needs a sponsor. Include the name, title, and contact information of the sponsor in the introduction.
- The names of the project manager and key team members: The project manager should always be the contact for project information. Depending on the size and nature of the project, names of key team members may also be included.
- Deliverables of the project: This section should briefly list and describe the products that will be produced as part of the project. Software packages, pieces of hardware, technical reports, and training materials are examples of deliverables.
- A list of important reference materials: Many projects have a history preceding them. Listing important documents or meetings related to a project helps project stakeholders understand that history. This section should reference

the plans produced for other knowledge areas. (Recall from Chapter 2 that every single knowledge area includes some planning processes.) For example, the overall project plan should reference and summarize important parts of the scope management, schedule management, cost management, quality management, staffing management, communications management, risk management, and procurement management plans.

- A list of definitions and acronyms, if appropriate: Many projects, especially information technology projects, involve terminology unique to a particular industry or technology. Providing a list of definitions and acronyms helps to avoid confusion.

The description of how the project is organized should include the following information:

- Organizational charts: In addition to an organizational chart for the company sponsoring the project and for the customer's company (if it is an external customer), there should be a project organizational chart to show the lines of authority, responsibilities, and communication for the project. You can find an example of a project's organizational chart in Northwest Airline's reservation system project in Chapter 13.
- Project responsibilities: This section of the project plan should describe the major project functions and activities and identify those individuals who are responsible for them. A responsibility assignment matrix (described in Chapter 8) is a tool often used for displaying this information.
- Other organizational or process related information: Depending on the nature of the project, there may be a need to document major processes followed on the project. For example, if the project involves releasing a major software upgrade, it might help everyone involved in the project to see a diagram or timeline of the major steps involved in this process.

The section of the project plan describing management and technical approaches should include the following information:

- Management objectives: It is important to understand senior management's view of the project, what the priorities are for the project, and any major assumptions or constraints.
- Project controls: This section describes how to monitor project progress and handle changes. Will there be monthly status reviews and quarterly progress reviews? Will specific forms or charts be used to monitor progress? Will the project use earned value management (described in Chapter 6) to assess and track performance? What is the process for change control? What level of management is required to approve different types of changes? (Change control is discussed later in this chapter.)
- Risk management: This section briefly addresses how risks will be identified, managed, and controlled. It should refer to the risk management plan, if one is required for the project.

- Project staffing: This section describes the number and types of people required for the project. It should refer to the staffing management plan, if one is required for the project.
- Technical processes: This section describes specific methodologies a project might use and how information is to be documented. For example, many information technology projects follow specific software development methodologies or use particular Computer Aided Software Engineering (CASE) tools. Many companies or customers also have specific formats for technical documentation. It is important to clarify these technical processes in the project plan.

The section of the overall project plan describing the work to be done should reference the scope management plan and summarize the following:

- Major work packages: Project work is usually organized into several work packages using a work breakdown structure (WBS), and a statement of work (SOW) is often produced to describe the work in more detail (*see* Chapter 4). This section should briefly summarize the main work packages for the project and refer to appropriate sections of the scope management plan.
- Key deliverables: This section lists and describes the key products produced as part of the project. It should also describe the quality expectations for the deliverables.
- Other work related information: This section highlights key information related to the work done on the project. For example, it might list specific hardware or software that must be used on the project or certain specifications that must be followed. It should document major assumptions made in defining the project work.

The project schedule information section should include the following:

- Summary schedule: It is helpful to see a one-page summary of the overall project schedule. Depending on the size and complexity of the project, the summary schedule might list only key deliverables and their planned completion dates. For smaller projects, it might include all of the work and associated dates for the entire project in a Gantt chart.
- Detailed schedule: This section provides more detailed information on the project schedule. It should reference the schedule management plan and discuss dependencies among project activities that could affect the project schedule. For example, it might explain that a major part of the work cannot be started until funding is provided by an external agency. A project network diagram can be used to show these dependencies (*see* Chapter 5, *Project Time Management*).
- Other schedule related information: Many assumptions are often made in preparing project schedules. This section should document major assumptions and highlight other important information related to the project schedule.

The budget section of the overall project plan should include the following:

- Summary budget: The summary budget includes the total estimate of the overall project's budget. It could also include the budget estimate for each month or year by certain budget categories. It is important to provide some explanation of what these numbers mean. For example, is the total budget estimate a firm number than cannot change, or is it a rough estimate based on projected costs over the next three years?
- Detailed budget: This section summarizes what is in the cost management plan and includes more detailed budget information. For example, what are the fixed and recurring cost estimates for the project each year? What are the projected financial benefits of the project? What types of people are needed to do the work, and how are the labor costs calculated?
- Other budget related information: This section documents major assumptions and highlights other important information related to financial aspects of the project.

Many organizations have guidelines for creating project plans. For example, Department of Defense (DOD) Standard 2167, Software Development Plan, describes the format for contractors to use in creating a plan for software development for DOD projects. IEEE Standard 1058.1 describes the contents of a Software Project Management Plan. Table 3-1 provides the format for the IEEE Standard Software Project Management Plan. Note that this format includes five main sections—introduction, project organization, managerial process, technical process, and work packages, schedule, and budget. Companies working on software development projects for the Department of Defense must follow this standard. In most private organizations, specific documentation standards are not as rigorous; however, there are usually guidelines for developing project plans. It is good practice to follow standards or guidelines for developing project plans in an organization to facilitate the development and execution of those plans. The organization can work more efficiently if all project plans follow a similar format.

Table 3-1: Sample Outline for a Software Project Management Plan (SPMP)

	PROJECT MANAGEMENT PLAN SECTIONS				
	Introduction	**Project Organization**	**Managerial Process**	**Technical Process**	**Work Packages, Schedule, and Budget**
Section Topics	Project overview; project deliverables; evolution of the SPMP; reference materials; and definitions and acronyms	Process model; organizational structure; organizational boundaries and interfaces; and project responsibilities	Management objectives and priorities; assumptions, dependencies, and constraints; risk management; monitoring and controlling mechanisms; and staffing plan	Methods, tools, and techniques; software documentation; and project support functions	Work packages; dependencies; resource requirements; budget and resource allocation; and schedule

IEEE Std 1058.1-1987

Because the ultimate goal of project management is to meet or exceed stakeholder needs and expectations from a project, it is important to include a stakeholder analysis as part of project planning. A **stakeholder analysis** documents information such as key stakeholders' names and organizations, their roles on the project, unique facts about each stakeholder, their level of interest in the project, their influence on the project, and suggestions for managing relationships with each stakeholder. Since a stakeholder analysis often includes sensitive information, it should not be part of the overall project plan, which would be available for all stakeholders to review. In many cases, only project managers and other key team members should see the stakeholder analysis. Table 3-2 provides an example of a stakeholder analysis that Nick Carson could have used to help him manage the DNA sequencing instrument project described in the opening case. It is important for project managers to take the time to perform a stakeholder analysis of some sort in order to help understand and meet stakeholder needs and expectations. A stakeholder analysis will also help the project manager lead the execution of the project plan.

Table 3-2: Sample Stakeholder Analysis

	KEY STAKEHOLDERS				
	AHMED	SUSAN	ERIK	MARK	DAVID
Organization	Internal senior management	Project team	Project team	Hardware vendor	Project manager for other internal projects
Role on project	Sponsor of project and one of the company's founders	DNA sequencing expert	Lead programmer	Supplies some instrument hardware	Competing for company resources
Unique facts	Demanding, likes details, business focus, Stanford MBA	Very smart, Ph.D. in biology, easy to work with, has a toddler	Best programmer I know, weird sense of humor	Start-up company, he knows we can make him rich if this works	Nice guy, one of oldest people at company, has three kids in college
Level of interest	Very high	Very high	High	Very high	Low to medium
Level of influence	Very high; can call the shots	Subject matter expert; critical to success	High; hard to replace	Low; other vendors available	Low to medium
Suggestions on managing relationship	Keep informed, let him lead conversations, do as he says, and quickly	Make sure she reviews specifications and leads testing; can do some work from home	Keep him happy so he stays; emphasize stock options; likes Mexican food	Give him enough lead time to deliver hardware	He knows his project takes a back seat to this one, but I can learn from him

PROJECT PLAN EXECUTION

The second process of integration management is project plan execution. Project plan execution involves managing and performing the work described in the project plans. The majority of time on a project is usually spent on execution, as is most of the project's budget. The application area of the project directly affects project execution because the products of the project are produced during project execution. For example, the DNA sequencing instrument from the opening case and all associated software and documentation would be produced during project execution. The project team would need to use their expertise in biology, hardware and software development, and testing to successfully produce the product. The project manager would need to focus on leading the project team and managing stakeholder relationships to successfully execute the project plan.

Project integration management views project planning and execution as intertwined and inseparable activities. The main function of creating project plans is to guide project execution. A good plan should help produce good products or work results. What good work results consist of should be documented in a plan, and knowledge gained from doing work early in the project should be reflected in updates to the project plans. Anyone who has tried to write a computer program from poor specifications appreciates the importance of a good plan. Anyone who has had to document a poorly programmed system appreciates the importance of good execution.

A common sense approach to improving the coordination between project plan development and execution is to follow this simple rule: Those who will do the work should plan the work. All project personnel need to experience and develop both planning and executing skills. In information technology projects, programmers who have had to write detailed specifications and then create the code from their own specifications become better at writing specifications. Likewise, most systems analysts begin their careers as programmers so they understand what type of analysis and documentation is really needed to write good code. Although project managers are responsible for developing the overall project plan, they must solicit inputs from the project team members who are developing plans in each knowledge area.

Project managers must lead by example to demonstrate the importance of creating good plans and then following them in project execution. Project managers often create plans for things they need to do themselves. If project managers follow through on their own plans, their team members are more likely to do the same.

What Went Wrong?

Many people have a poor view of plans based on past experiences. Senior managers often require a plan, but then no one follows up on whether the plan was followed. For example, one project manager said he would meet with each project team leader within two months to review their plans. The project manager created a detailed schedule for these reviews. He cancelled the first meeting due to another business commitment. He rescheduled the next meeting for unexplained personal reasons. Two months later, the project manager had still not met with over half of the project team leaders. Why should project members feel obligated to follow their own plans when the project manager obviously did not follow his?

Although project planning is different from strategic planning, there are many similarities. Henry Mintzberg, former president of the Strategic Management Society, describes the so-called "pitfalls" of planning and shows how the planning process itself can destroy commitment, narrow a company's vision, discourage change, and breed an atmosphere of politics. He also states that planning must improve. "People called planners can sometimes do strange things . . . . We need to delineate the word carefully if it is not to be eventually dropped from the management literature as hopelessly contaminated."[1]

Good project plan execution requires many skills. General management skills—leadership, communication, and political skills—are essential. For example, organizational procedures can help or hinder project plan execution. If an organization has guidelines for creating overall project plans and plans for each project management knowledge area that everyone in the organization follows, it will be easier to create the plans. Likewise, if the organization uses the project plans as the basis for performing and monitoring progress during execution, the culture will promote the relationship between good planning and execution. Project managers must provide leadership on their specific projects to interpret these planning and executing guidelines. Project managers must also communicate well with the project team and other project stakeholders to create and execute good project plans.

Project managers may sometimes find it necessary to break organizational rules to produce project results in a timely manner. When project managers break the rules, politics will play a role in the results. For example, if a particular project requires use of nonstandard software, the project manager must use his or her political skills to convince concerned stakeholders of the need to break the rules on using only standard software. Breaking organizational rules—and getting away with it—requires excellent leadership, communication, and political skills.

Product skills and knowledge are also critical for successful project execution. Project managers and their staffs must have the required expertise to produce the products of the project. If they do not, it is the project manager's job to help develop the necessary skills, find someone else to do some of the work, or alert senior management to the problem. The information technology labor shortage is a huge concern because there are not enough skilled workers to meet the demand for information technology projects. Organizations must be selective in deciding what information technology projects to work on so that they can

[1] Mintzberg, Henry. "The Rise and Fall of Strategic Planning." *Free Press* (1993).

ensure they have adequate resources to do the work. It is also important for information technology project managers to have product knowledge so they can help plan and lead projects that take advantage of new technologies.

What Went Right?

Michael Michalski, IT Manager for BELMARK Inc., provides an excellent example of how a project manager can use product skills and knowledge to help plan and deliver information technology projects more effectively. BELMARK is a small company that produces all types of labels. Michalski knew that they could use the Internet to reduce the time it took to create and deliver labels to their customers. He worked with a programmer for about three months to develop an application called "Web-to-Web." This application is an "extranet" that ties together several different companies via the Web. Web-to-Web uses basic Hypertext Markup Language (HTML) codes and other Internet technologies to link information between BELMARK, their customers, and shipping services. For example, customers using Web-to-Web can view and edit their BELMARK label designs using Portable Document Format (PDF) files on the Web, specifying colors, fonts, sizes, and so on, and can then send them back to BELMARK via the Web with an order for a specific quantity. Customers can also select their shipping service (such as Federal Express) and track when they will receive their shipment of labels. Michalski said this project reduced the time it took for customers to have their final label designs from four days down to as little as fifteen minutes.[2]

Project plan execution also requires specialized tools and techniques, some of which are unique to project management. Tools and techniques that help project managers perform activities that are part of execution processes include:

Work Authorization System. A work authorization system is a method for ensuring that qualified people do the work at the right time and in the proper sequence. The work authorization system can be a manual process in which written forms and signatures are used to authorize work to begin on specific project activities or work packages. Automated work authorization systems are also available to streamline this process. On smaller projects, verbal authorizations are usually adequate.

Status Review Meetings. Status review meetings are regularly scheduled meetings used to exchange project information. Status review meetings are also excellent tools for motivating people to make progress on the project. If project members know they have to give formal reports every month to key project stakeholders, they will make sure to accomplish some work. Sending a written weekly or monthly status report is normally not as effective as having a formal verbal presentation. Status review meetings will be discussed further in Chapter 9, *Project Communications Management*.

[2] Michalski, Mike. "Using Internet Technologies for Business-to-Business Communications—the Extranet," *Keynote Address at the Information Systems Education Conference* (October, 1998).

Project Management Software. Project management software can greatly assist in creating and executing the project plan. For example, project managers or other team members can create Gantt charts for overall project plans using software such as Microsoft Project 2000. These Gantt charts can include hyperlinks to other planning documents. For example, an overall project plan might include a deliverable for creating software test plans. This item on the Gantt chart could include a hyperlink to a Microsoft Word file of that particular software test plan. If a project team member updates the Word file containing the test plan, the hyperlink feature on the Gantt chart automatically links to the updated file. Once a baseline plan is set for the project, project team members can enter information on when activities actually started and ended, how many hours were spent completing them, and so on. The project manager can then use project management software to compare the baseline and actual information by viewing reports on project progress—running a report to show what tasks are past due or what the actual costs are to date, versus what was planned. Appendix A provides detailed information on using Microsoft Project 2000 to assist in various aspects of project management.

Although these tools and techniques can aid in project execution, the project manager must remember that positive leadership is the real key to successful project management. To enable project managers to focus on providing leadership for the whole project, they should delegate the detailed work involved in using these tools to other team members.

INTEGRATED CHANGE CONTROL

Integrated change control involves identifying, evaluating, and managing changes throughout the project life cycle. The three main objectives of integrated change control are:

1. Influencing the factors that create changes to ensure that changes are beneficial: To ensure that changes are beneficial and that a project is successful, project managers and their teams must make trade-offs among key project dimensions such as scope, time, cost, and quality.
2. Determining that a change has occurred: To determine that a change has occurred, the project manager must know the status of key project areas at all times. In addition, the project manager must communicate significant changes to top management and key stakeholders. Top management and other key stakeholders do not like surprises, especially ones that mean the project might produce less, take longer to complete, cost more than planned, or be of lower quality than desired.
3. Managing actual changes as they occur: Managing change is a key job of project managers and their staffs. It is important that project managers exercise discipline in managing the project to help minimize the number of changes that occur.

Important inputs to the integrated change control process include project plans, performance reports, and change requests. Important outputs include project plan updates, corrective actions, and documentation of lessons learned. Figure 3-3 provides a schematic of the integrated change control process.

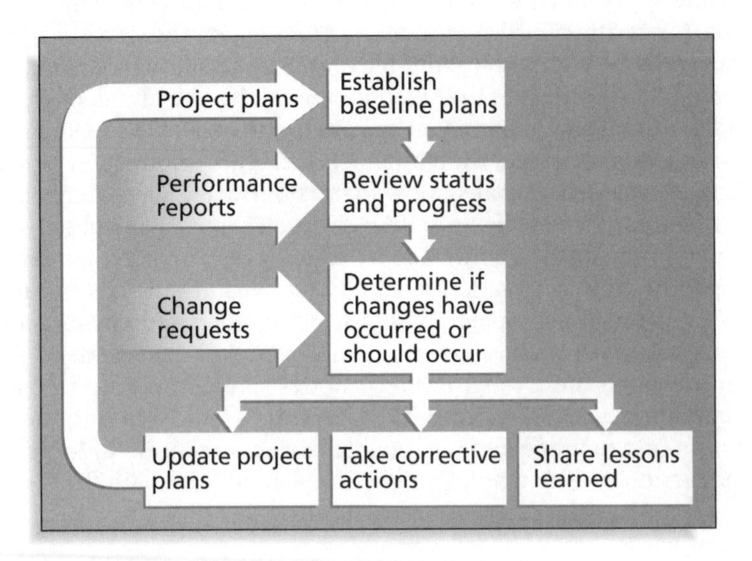

Figure 3-3. Integrated Change Control Process

The project plan provides the baseline for identifying and controlling project changes. For example, the project plan includes a section describing the work to be done on a project. This section of the plan describes the key deliverables for the project, the products of the project, and quality requirements. The schedule section of the project plan lists the planned dates for completing key deliverables, and the budget section of the project plan provides the planned cost for these deliverables. The project team must focus on delivering the work as planned. If changes are made during project execution, the project plans must be revised.

Performance reports provide status information on how project execution is going. The main purpose of these reports is to alert the project manager and team of issues that might cause problems in the future. The project manager and team must decide if corrective actions are needed, what the best course of action is, and when to act. For example, suppose one of the key deliverables in a project plan is installation of a new Web server for the project. If one of the project team members reports that he or she is having problems coordinating the purchase and installation of this Web server, the project manager should assess what will happen to the project if this deliverable is a little late. If late installation will cause problems in other areas of the project, the project manager should take necessary actions to help the team member complete this task on time. Perhaps the purchase request is held up because one of the purchasing clerks is on vacation. The project manager could talk to the head of the purchasing department to make sure someone processes the order. If nothing can be done to meet the planned installation date, the project manager should alert the other people who would be affected by this schedule change. The project manager should also look at the big picture of how the project is going. If there is a consistent trend that schedules are slipping, the project manager should alert key stakeholders and negotiate a later completion date for the project. Performance reports are described more in Chapter 9, *Project Communications Management*.

Change requests are common on projects and occur in many different forms. They can be oral or written, formal or informal. For example, the team member responsible for installing the Web server might ask the project manager at a progress review meeting if it is all right to order a server with a faster processor than planned, from the same manufacturer for the same approximate cost. Since this change is positive and should have no negative effects on the project, the project manager might give a verbal approval at the progress review meeting. Nevertheless, it is still important that this change be documented to avoid any potential problems in the future. The appropriate team member should update the section of the scope management plan with the new specifications for the server. Still, keep in mind that many change requests can have a major impact on a project. For example, customers changing their minds about the number of Web servers they want as part of a project will have a definite impact on the scope and cost of the project. Such a change might also affect the project's schedule. More significant changes must be presented in written form, and there should be a formal review process for analyzing and deciding whether these changes should be approved.

In addition to updating project plans and taking corrective actions, documenting lessons learned is an important output from the integrated change control process. The project manager and team should share what they have learned with each other and other people in the organization. There should be some documentation of these lessons learned on change control, and discussing them at an open meeting is an effective way to share this information. Discussing lessons learned on change control should help manage the change control process in future projects.

Change is unavoidable and often expected on many information technology projects. Technologies change, personnel change, organizational priorities change, and so on. Careful change control on information technology projects is a critical success factor. A good change control system is also important for project success. These topics are discussed next.

Change Control on Information Technology Projects

A widely held view of information technology project management from the 1950s to the 1980s was that the project team should strive to do exactly what was planned on time and within budget. The problem with this view was that project teams could rarely meet original project goals, especially on projects involving new technologies. Stakeholders rarely agreed up front on what the scope of the project really was or what the end product should really look like. Time and cost estimates created early in a project were rarely accurate.

Starting in the 1990s, most project managers and top management realized that project management is a process of constant communication and negotiation about project objectives and stakeholder expectations. This view assumes that changes will be made throughout the project life cycle and recognizes the fact that changes are often beneficial to some projects. For example, if a project team member discovers a new hardware or software technology that could satisfy the customers' needs for less time and money, the project team and key stakeholders should be open to making major changes in the project.

All projects will have some changes, and managing them is a key issue in project management, especially for information technology projects. Many information technology projects involve the use of hardware and software that is updated frequently. For example, the initial plan for the specifications of the Web server described earlier may have been the state of the art at that time. If the actual ordering of the Web server occurred six months later, it is quite possible that a more powerful server could be ordered at the same cost. This example illustrates a positive change. On the other hand, the manufacturer of the server specified in the project plan could go out of business, which would result in a negative change. Information technology project managers should be accustomed to changes such as these and build some flexibility into their project plans and execution. Customers for information technology projects should also be open to meeting project objectives in different ways.

Even if project managers, teams, and customers are flexible, it is important that projects have a formal change control system. To plan for managing change, a project must have a good change control system.

Change Control System

A **change control system** is a formal, documented process that describes when and how official project documents may be changed. It also describes the people authorized to make changes, the paperwork required, and any automated or manual tracking systems the project will use. A change control system often includes a change control board (CCB), configuration management, and a process for communicating changes.

A **change control board (CCB)** is a formal group of people responsible for approving or rejecting changes on a project. The primary functions of a change control board are to provide guidelines for preparing change requests, evaluate change requests, and manage the implementation of approved changes. An organization could have key stakeholders for the entire organization on this board, and a few members could rotate based on the unique needs of each project. By creating a formal board and process for managing changes, better overall change control should result.

CCBs can have some drawbacks, however. One drawback is the time it takes for decisions to be made on proposed changes. CCBs often meet only once a week or once a month and may not make decisions in one meeting. Some organizations have streamlined processes for making quick decisions on smaller project changes. One company created a "48 hour policy." In this policy, task leaders on a large information technology project would reach agreements on key decisions or changes within their expertise and authority. The person in the area most affected by this decision or change then had 48 hours to go to his or her senior management to seek approval. If for some reason the project team's decision could not be implemented, the senior manager consulted would have 48 hours to reverse a decision; otherwise, the project team's decision was approved. This type of process is a great way to deal with the many time-sensitive decisions or changes that must be made on many information technology projects.

Configuration management is another important technique for integrated change control. **Configuration management** ensures that the descriptions of the project's products are correct and complete, and concentrates on the management of technology by identifying and controlling the functional and physical design characteristics of products and their support documentation. Members of the project team, often called configuration management specialists, are often assigned to perform configuration management for large projects. Their job is to identify and document the functional and physical characteristics of the project's products, control any changes to such characteristics, record and report the changes, and audit the products to verify conformance to requirements.

Another factor in change control is communication. Project managers should use written and oral performance reports to help identify and manage project changes. In addition to more formal reports, some project managers have stand-up meetings once a week or even every morning, depending on the nature of the project. The goal of a stand-up meeting is to quickly communicate what is most important on the project. For example, the project manager might have an early morning stand-up meeting every day with all of his or her team leaders. There might be a weekly stand-up meeting every Monday morning with all interested stakeholders. Standing keeps meetings short and forces everyone to focus on the most important things going on in a project.

Why is good communication so critical to success? One of the most frustrating aspects of project change is not having everyone in sync and informed about the latest project information. Again, it is the project manager's responsibility to integrate all project changes so that the project stays on track. The project manager and his or her staff must develop a system for notifying everyone affected by a change in a timely manner. E-mail and the World Wide Web make it easier to disseminate the most current project information. Using special project management software also helps project managers track and communicate project changes. (You can find more information about good communication in Chapter 9, *Project Communications Management*.)

Table 3-3 lists suggestions for managing integrated change control. As described earlier, project management is a process of constant communication and negotiation. Project managers should plan for changes and use appropriate tools and techniques such as a change control board, configuration management, and good communication. It is helpful to define procedures for making timely decisions on small changes, use written and oral performance reports to help identify and manage changes, and use software to assist in planning, updating, and controlling projects.

Table 3-3: Suggestions for Managing Integrated Change Control

■	View project management as a process of constant communication and negotiation
■	Plan for change
■	Establish a formal change control system, including a Change Control Board (CCB)
■	Use good configuration management
■	Define procedures for making timely decisions on smaller changes
■	Use written and oral performance reports to help identify and manage change
■	Use project management and other software to help manage and communicate changes

Project managers must also provide strong leadership to steer the project to successful completion. They must not get too involved in managing project changes. Project managers should delegate much of the detailed work to project team members and focus on providing overall leadership for the project in general. Remem-ber, project managers must focus on the big picture and perform project integration management well to lead their team and organization to success.

CASE WRAP-UP

Without talking to Nick Carson or his team, Nick's CEO hired a new person, Jim, to act as a middle manager between himself and the people in Nick's department. The CEO and other senior managers really liked Jim, the new middle manager. He met with them often, shared ideas, and had a great sense of humor. He started developing standards the company could use to help manage projects in the future. For example, he developed templates for creating project plans and progress reports and put them on the company's intranet. However, Jim and Nick did not get along. Jim accidentally sent an e-mail to Nick that was supposed to go to the CEO. In this e-mail, Jim said that Nick was hard to work with and preoccupied with the birth of his son.

Nick was furious when he read the e-mail and stormed into the CEO's office. The CEO suggested that Nick move to another department, but Nick did not like that option. Without considering the repercussions, the CEO offered Nick a severance package to leave the company. Because of the planned corporate buyout, the CEO knew they might have to let some people go anyway. Nick talked the CEO into giving him a two-month sabbatical he had not yet taken plus a higher percentage on his stock options. After discussing the situation with his wife and realizing that he would get over $70,000 if he resigned, Nick took the severance package. He had such a bad experience as a project manager that he decided to stick with being a software developer, and there was no shortage of those jobs where he lived. After Nick resigned, the CEO found several other bright, young, technical people at his door asking for their severance packages. Today Nick is very happy with his position as a senior research fellow at a competing biotech firm.

CHAPTER SUMMARY

Project integration management is often viewed as the most important project management knowledge area, since it ties together all the others. A project manager's primary focus should be on project integration.

Key processes of integration management include project plan development, project plan execution, and integrated change control.

The main purpose of a project plan is to facilitate action. Plans should be dynamic, flexible, and subject to change. A project plan includes an introduction or overview of the project, a description of how the project is organized, the management and technical processes used on the project, and sections describing the work to be done, schedule information, and budget information.

The project manager should create a stakeholder analysis as part of project planning to understand who all the stakeholders are on a project and how to work with them to satisfy their needs.

The majority of a project's budget should be spent on project plan execution. Project planning and execution are intertwined. Those who do the work should plan the work.

Integrated project change control involves making sure changes are beneficial to the project, determining when changes should occur, and managing those changes. A change control system often includes a change control board (CCB), configuration management, and a process for communicating changes.

Discussion Questions

1. Describe project integration management in your own words. Give an example of a project that succeeded because of good project plan development, project execution, or integrated change control.
2. How does integration management relate to the project life cycle, stakeholders, and the other project management knowledge areas?
3. What are some crucial elements of a good project plan? Describe what might be in a plan for a project to develop a Web-based information system for providing transfer credit information for all colleges and universities in the world.
4. What are crucial elements of successful project plan execution? Describe a well-executed project you are familiar with. Describe a disaster. What were some of the main differences between these projects?
5. Discuss the importance of following a good integrated change control process on information technology projects. What do you think of the suggestions made in this chapter? Think of three additional suggestions for change control on information technology projects.

Exercises

1. Develop an outline (major headings and subheadings only) for a project plan to create a Web site for your class, and then fill in the details for the introduction or overview section. Assume that this Web site would include a home page with links to a syllabus for the class, lecture notes or other instructional information, links to the Web site for this textbook, links to other Web sites with project management information, and links to personal pages for each member of your class and future classes. Also include a bulletin board and chat room feature where students and the instructor can exchange information. Assume your instructor is the project's sponsor, you are the project manager, your classmates are your project team, and you have up to one year to complete the project.
2. Read the *IEEE Standard for Software Project Management Plans*, one of the suggested readings. Write a one-page summary of this standard and your opinion of its contents.
3. Write a one- to two-page paper based on the opening case. Answer the following questions:
   - What do you think the real problem was in this case?
   - Does the case present a realistic scenario? Why or why not?
   - Was Nick Carson a good project manager? Why or why not?
   - What could Nick have done to be a better project manager?
   - What should senior management have done to help Nick?
4. Write a one- to two-page summary of an article about project control. You may use the Ward or Wideman articles from the suggested reading list.

5. Interview two to three different people to get their views on the differences between staying in a more technical career path and moving into project management. Ask them if they agree with the findings in Johnson's article stating that the best information technology people make the worst project managers. Write a one- to two-page summary of what you find, including your opinions on this subject.

MINICASE

Robin, a senior consultant for a large consulting firm, was just assigned to be the project manager for a $100 million project to upgrade the information systems for the state Department of Corrections. Robin's boss and mentor, Jill, knew that Robin had a lot of potential and wanted to give her a challenging assignment. Robin had a lot of experience with the technology involved in the project, but she had never worked with the state government and knew little about the state legal system. Her firm assigned its top legal person to the project to assist Robin. Fred had worked on several projects with the state and understood the corrections process, but his expertise was in law, not technology. Fred also had never worked for a woman before and was surprised that Robin was ten years younger than he.

At the kickoff meeting for the project, Robin could see that understanding and meeting stakeholder needs and expectations for the project would be a huge challenge. The new governor of the state attended the first few minutes of the meeting, an indicator of the high profile of the project. The governor, who was known for not trusting consultants, questioned the value of spending more money on information systems for the Department of Corrections, but the decision to fund this project had been made before he was elected. The federal government would be funding half of the project, since the new system would have to interact with the new federal system. The head of the state Department of Corrections, Donna, had worked in the state legal system for twenty-five years. Donna knew the corrections process and the problems they faced, but she knew very little about information systems and preferred doing work the old-fashioned way. She had heard horror stories about inmates being released as a result of computer errors. Donna's new assistant, Jim, was very computer savvy and seemed most supportive of the project. No federal representatives were invited to the meeting. The meeting did not go well, as it seemed disorganized and highly political.

Part 1: Prepare a stakeholder analysis for this project, using Table 3-2 as a guide. Then write a memo to Robin's boss, Jill, making suggestions on what senior management could do to help Robin manage this challenging project.

Part 2: As part of the project planning, Robin and her team decide to emphasize the importance of communications. They decide to document a process for communicating with stakeholders, especially for status review meetings and change control. They plan to discuss these draft documents with their key stakeholders to reach agreement and get off to a good start. Write a draft document describing the process for status review meetings and a process for change control. Include templates of key documents that could be used in both processes, such as an agenda for a monthly status review meeting and a request for change form. (*See* the section "Using Templates for Project Communications," Chapter 9, *Project Communications Management*, for sample templates.)

Suggested Readings

1. IEEE. "Std 1058.1-1987." *IEEE Standard for Software Project Management Plans* (1987).

   This standard describes the format and content of software project management plans. It identifies the minimum set of elements that should appear in all software project management plans.

2. Johnson, Jim. "Turning CHAOS into SUCCESS." *Software Magazine* (www.softwaremag.com/archive/1999dec/Success.html) (December 1999).

   This article provides suggestions on how to improve the likelihood of success for information technology projects. Johnson states that many Chief Information Officers say that the best information technology people make the worst project managers. Project managers must have strong business skills and people skills, such as diplomacy and negotiation.

3. Stuckenbruck, Linn C. "The Job of the Project Manager: Systems Integration." *The Implementation of Project Management: The Professional's Handbook*. Reading, MA: Addison-Wesley Publishing Company Inc. (1996): 141-155.

   This chapter of Stuckenbruck's book provides more details on how project managers perform the integration management function. She stresses the need for maintaining communication links across the organizational interfaces.

4. Ward, James. "Productivity Through Project Management: Controlling the Project Variables." *Information Systems Management* (Winter 1994).

   Ward suggests that effectively planning and controlling three project variables—work, resources, and time—is what ensures a project's success. This article describes the typical problems and opportunities associated with project control and the dynamics and interrelationships of work, resources, and time.

5. Wideman, R. M. "Total Project Management of Complex Projects, Improving Performance with Modern Techniques." New Delhi, India: *Consultancy Development Centre* (January 1990).

   This article provides more details on the framework for project integration management presented in Figure 3-2. It emphasizes that project integration management, particularly project control, must be applied to each and every other knowledge area throughout the entire project life cycle.

KEY TERMS

- **Change control board (CCB)** — a formal group of people responsible for approving or rejecting changes on a project
- **Change control system** — a formal, documented process that describes when and how official project documents may be changed
- **Configuration management** — a process that ensures that the descriptions of the project's products are correct and complete
- **Integrated change control** — coordinating changes across the entire project
- **Interface management** — identifying and managing the points of interaction between various elements of a project
- **Project integration management** — includes the processes involved in coordinating all of the other project management knowledge areas throughout a project's life cycle. The main processes involved in project integration management include project plan development, project plan execution, and integrated change control.
- **Project management office** or **center of excellence** — an organizational entity created to assist project managers in achieving project goals
- **Project management software** — software specifically designed for project management
- **Project plan** — a document used to coordinate all project planning documents and guide project execution and control. Key parts of a project plan include an introduction or overview of the project, a description of how the project is organized, the management and technical processes used on the project, and sections describing the work to be done, schedule information, and budget information.
- **Project plan development** — taking the results of other planning processes and putting them into a consistent, coherent document—the project plan
- **Project plan execution** — carrying out the project plan by performing the activities it includes
- **Stakeholder analysis** — an analysis of information such as key stakeholders' names and organizations, their roles on the project, unique facts about each stakeholder, their level of interest in the project, their influence on the project, and suggestions for managing relationships with each stakeholder
- **Status review meetings** — regularly scheduled meetings used to exchange project information
- **Work authorization system** — a method for ensuring that qualified people do the work at the right time and in the proper sequence

4

Project Scope Management

Objectives

After reading this chapter, you will be able to:

1. *Understand the importance of good project scope management*
2. *List the reasons why most firms invest in information technology projects*
3. *Describe the strategic planning process and how it relates to information technology project selection*
4. *Apply different project selection methods such as a weighted scoring model and net present value analysis*
5. *Explain the purpose of a project charter, scope statement, and work breakdown structure*
6. *Construct a work breakdown structure*
7. *Describe tools and techniques to assist in scope verification and change control on information technology projects*

Kim Nguyen was leading a project team meeting to help define the scope of the information technology upgrade project assigned to her last week by upper management. This project was necessary for implementing several high priority Internet-based applications the company was developing. The upgrade project involved creating and implementing a plan to get all employees' information technology assets to meet new corporate standards within nine months. These standards specified the minimum equipment required for each desktop computer: type of processor, amount of memory, hard disk size, type of network connection, and software. Kim knew that in order to perform the upgrades, they would first have to create a detailed inventory of all of the current hardware, networks, and software in the entire company of 2,000 employees.

Kim had a project charter explaining the key objectives of the project and the stakeholders' roles and responsibilities. The charter also included a rough cost and schedule estimate. Kim called a meeting with her project team and other stakeholders to further plan and define the scope of the project. She wanted to get everyone's ideas on what was involved in doing the project, who would do what, and how they could avoid potential scope creep. Kim's boss suggested that the team start creating a work breakdown structure (WBS) to clearly define all of the work involved in doing the upgrade project. Kim was not quite sure where to start in creating a WBS. She had used them before, but this was the first time she was the project manager leading a team to create one.

WHAT IS PROJECT SCOPE MANAGEMENT?

Recall from Chapter 2 that several factors are associated with project success. Many of these factors, such as user involvement, clear business objectives, minimized scope, and firm basic requirements, are elements of project scope management. William V. Leban, program manager at Keller Graduate School of Management, cites lack of proper project definition and scope as the main reasons projects fail.[1]

One of the most important and most difficult aspects of project management, therefore, is defining the scope of a project. **Scope** refers to all the work involved in creating the products of the project and the processes used to create them. Project stakeholders must come to an agreement on what the products of the project are and, to some extent, how they should be produced.

Project scope management includes the processes involved in defining and controlling what is or is not included in a project. It ensures that the project team and stakeholders have the same understanding of what products will be produced as a result of the project and what processes will be used in producing them. The main processes involved in project scope management include:

- **Initiation**, which involves committing the organization to begin a project or continue to the next phase of a project. An output of initiation processes is a **project charter**, which is a key document for formally recognizing the existence and providing a broad overview of a project.

[1] Chalfin, Natalie, "Four Reasons Why Projects Fail," *PM Network*, PMI (June 1998) 7.

- **Scope planning**, which involves developing documents to provide the basis for future project decisions, including the criteria for determining if a project or phase has been completed successfully. The project team creates a scope statement and scope management plan as a result of the scope planning process.
- **Scope definition**, which involves subdividing the major project deliverables into smaller, more manageable components. The project team creates a work breakdown structure (WBS) during this process.
- **Scope verification**, which involves formalizing acceptance of the project scope. Key project stakeholders, such as the customer and sponsor for the project, formally accept the deliverables of the project during this process.
- **Scope change control**, which involves controlling changes to project scope. Scope changes, corrective action, and lessons learned are outputs of this process.

PROJECT INITIATION: STRATEGIC PLANNING AND PROJECT SELECTION

Successful managers look at the big picture or strategic plan of the organization to determine what types of projects will provide the most value. Therefore, the project initiation process involves identifying potential projects, using realistic methods to select which projects to work on, and then formalizing their initiation by issuing some sort of project charter.

Identifying Potential Projects

The first step in scope management is deciding what projects to do in the first place. Figure 4-1 shows a four-stage planning process for selecting information technology projects. Note the hierarchical structure of this model and the results produced from each stage. The first step in information technology planning, starting at the top of the hierarchy, is to develop an information technology strategic plan based on the organization's overall strategic plan. **Strategic planning** involves determining long-term objectives by analyzing the strengths and weaknesses of an organization, studying opportunities and threats in the business environment, predicting future trends, and projecting the need for new products and services. Many people are familiar with the term "SWOT" analysis—analyzing **S**trengths, **W**eaknesses, **O**pportunities, and **T**hreats—which is used to aid in strategic planning. It is very important to have managers from outside the information technology department assist in information technology strategic planning process, as they can help information technology personnel understand organizational strategies and identify the business areas that support them.

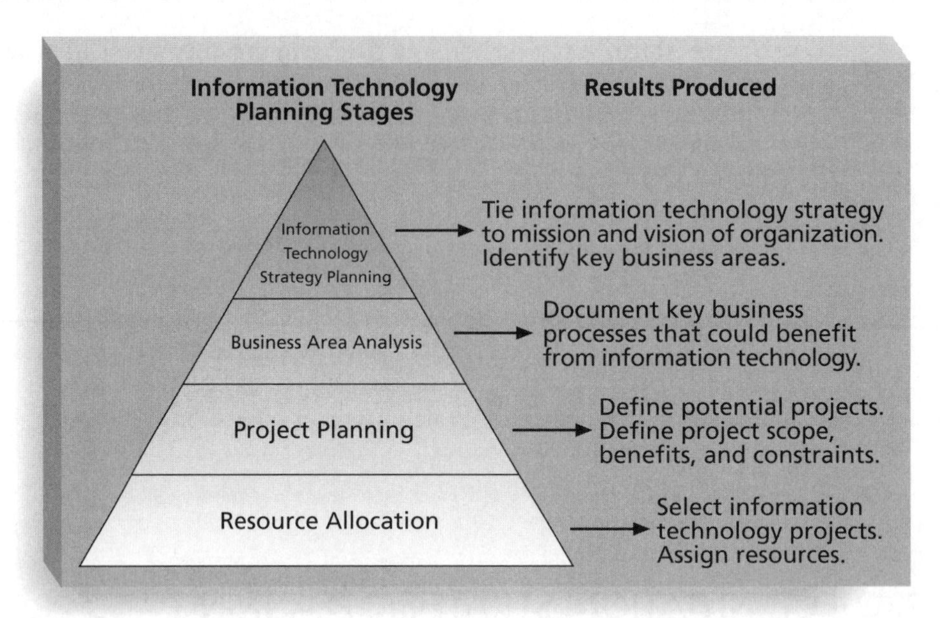

Figure 4-1. Information Technology Planning Process

After identifying business areas to focus on, the next step in the information technology planning process is to perform a business area analysis. This analysis documents business processes that are central to achieving strategic goals, and aids in discovering which ones could most benefit from information technology. Then, the next step is to start defining potential information technology projects, their scope, benefits, and constraints. The last step in the information technology planning process is selecting which projects to do and assigning resources for working on them.

Information systems can be and often are central to business strategy. Michael Porter, who developed the concept of the strategic value of competitive advantage, has written several books and articles on strategic planning and competition. He and many other experts have emphasized the importance of using information technology to support strategic plans and provide a competitive advantage. Many information systems are classified as being "strategic information systems" because they directly support key business strategies. For example, information systems can help an organization support a strategy of being a low-cost producer. Wal-Mart's inventory control system is a classic example of such a system. Information systems can support a strategy of providing specialized products or services that set a company apart from others in the industry. Consider, for example, Federal Express's introduction of online package tracking systems. Information systems can also support a strategy of selling to a particular market or occupying a specific project niche. Owens-Corning developed a strategic information system that boosted the sales of its home-insulation products by providing their customers with a system for evaluating the energy efficiency of building designs.

Even though many information technology projects do not produce "strategic information systems" or receive great publicity, it is critical that the information technology project planning process start by analyzing the overall strategy of the organization. Organizations must develop a strategy for using information technology to define how it will support the organization's objectives. Information technology plans and strategy must be aligned with the organization's strategic plans and strategy. Most organizations face thousands of problems and opportunities for improvement. Therefore, an organization's strategic plan should guide the project selection process. In fact, research shows that supporting explicit business objectives is the number one reason cited for why firms invest in information technology projects. Other top criteria for investing in information technology projects include supporting implicit business objectives and providing financial incentives such as a good internal rate of return (IRR) or net present value (NPV). These financial criteria are discussed later in this chapter. Table 4-1 summarizes the main reasons why firms invest in information technology projects.

Table 4-1: Why Firms Invest in Information Technology Projects

Reason for Investing in Information Technology Projects	Rank Based on Overall Value of Projects
Supports explicit business objectives	1
Has good internal rate of return (IRR)	2
Supports implicit business objectives	3
Has good net present value (NPV)	4
Has reasonable payback period	5
Used in response to competitive systems	6
Supports management decision making	7
Meets budgetary constraints	8
High probability of achieving benefits	9
Good accounting rate of return	10
High probability of completing project	11
Meets technical/system requirements	12
Supports legal/government requirement	13
Good profitability index	14
Introduces new technology	15

Bacon, James. "The Use of Decision Criteria in Selecting Information Systems/Technology Investments," *MIS Quarterly*, Vol. 16, No. 3 (September 1992).

Methods for Selecting Projects

Organizations identify many potential projects as part of their strategic planning processes, but the list of potential projects needs to be narrowed down to those projects that will be of most benefit. Selecting projects is not an exact science, but it is a critical part of project management. Many methods exist for selecting from among possible projects. Four common techniques are:

- Focusing on broad organizational needs
- Categorizing information technology projects
- Performing net present value or other financial analyses
- Using a weighted scoring model

In practice, organizations usually use a combination of these approaches to select projects. Each approach has advantages and disadvantages, and it is up to management to decide the best approach for selecting projects based on their particular organization.

Focusing on Broad Organizational Needs

Senior level managers must focus on meeting their organization's many different needs when deciding what projects to undertake, when to undertake them, and to what level. Projects that address broad organizational needs are much more likely to be successful because they will be important to the organization. However, it is often difficult to provide a strong justification for many information technology projects related to these broad organizational needs. For example, it is often impossible to estimate the financial value of such projects, but everyone agrees that they do have a high value. As it is said, "It is better to measure gold roughly than to count pennies precisely."

One method for selecting projects based on broad organizational needs is to determine whether they first meet three important criteria: need, funding, and will. Do people in the organization agree that the project needs to be done? Is the organization willing to provide adequate funds to do the project? Is there a strong will to make the project succeed? For example, many visionary CEOs can describe a broad need to improve certain aspects of their organizations, such as communications. Although they cannot specifically describe how to improve communications, they might allocate funds to projects that address this need. One project might be started to develop a strong information technology infrastructure, providing all employees, customers, and suppliers with the hardware and software they need to access information. The information technology upgrade project described in the opening case might fit into this category. As projects progress, the organization must reevaluate the need, funding, and will for each project to determine if the projects should be continued, redefined, or terminated.

Categorizing Information Technology Projects

Another method for selecting projects is based on various categorizations. One type of categorization assesses whether projects provide a response to a problem, an opportunity, or a directive.

- **Problems** are undesirable situations that prevent an organization from achieving its goals. These problems can be current or anticipated. For example, users of an information system may be having trouble logging onto a system or getting information in a timely manner because the system has reached its capacity. In response, the company could initiate a project to enhance the current system by adding more access lines or upgrading the hardware with a faster processor, more memory, or more storage space.
- **Opportunities** are chances to improve the organization. For example, a company might believe it could enhance sales by selling products directly to customers over the Internet. The company could initiate a project to provide direct sales of products from its Web site.
- **Directives** are new requirements imposed by management, government, or some external influence. For example, an important customer might require that all of its vendors use a certain form of electronic data interchange (EDI) in order to do business with them. Management would then initiate a project to implement this form of EDI to maintain business with that customer.

Organizations select projects for any of these reasons. It is often easier to get approval and funding for projects that address problems or directives because the organization must respond to these categories of projects to avoid hurting their business. Many problems and directives must be resolved quickly, but managers must also make sure to take a holistic view and seek opportunities for improving the organization through information technology projects.

Another categorization for information technology projects is based on the time it will take to complete them or the date by which they must be done. For example, some potential projects must be finished within a specific time window. If they cannot be finished by this set date, they are no longer valid projects. Some projects can be completed very quickly—within a few weeks, days, or even minutes. Many organizations have a help desk function to handle very small projects that can be completed quickly. Even though many information technology projects can be completed quickly, it is still important to prioritize them.

A third categorization for project selection is the overall priority of the project. Many organizations prioritize information technology projects as being high, medium, or low priority. The high priority projects should always be completed first, even if a low or medium priority project could be finished in less time. Usually there are many more potential information technology projects than an organization can undertake at any one time, so it is very important to work on the most important ones first.

Net Present Value Analysis, ROI, and Payback Analysis

Financial considerations are often an important aspect of the project selection process. Three primary methods for determining the projected financial value of projects include net present value analysis, return on investment, and payback analysis.

Net Present Value Analysis

Net present value analysis (NPV) is a method of calculating the expected net monetary gain or loss from a project by discounting all expected future cash inflows and outflows to the present point in time. Only projects with a positive net present value should be considered if financial value is a key criterion for project selection. Why? Because a positive NPV means the return from a project exceeds the cost of capital—the return available by investing the capital elsewhere. Projects with higher NPVs are preferred to projects with lower NPVs, if all other things are equal. Figure 4-2 illustrates this concept for two different projects.

	A	B	C	D	E	F	G
1							
2							
3	AN. INT. RATE.>	10%					
4							
5	PROJECT 1	YEAR 1	YEAR 2	YEAR 3	YEAR 4	YEAR 5	TOTAL
6	BENEFITS	$0	$2,000	$3,000	$4,000	$5,000	$14,000
7	COSTS	$5,000	$1,000	$1,000	$1,000	$1,000	$9,000
8	CASH FLOW	($5,000)	$1,000	$2,000	$3,000	$4,000	$5,000
9	NPV	$2,316					
10		Formula=npv(b3,b8:f8)					
11							
12	PROJECT 2	YEAR 1	YEAR 2	YEAR 3	YEAR 4	YEAR 5	TOTAL
13	BENEFITS	$1,000	$2,000	$4,000	$4,000	$4,000	$15,000
14	COSTS	$2,000	$2,000	$2,000	$2,000	$2,000	$10,000
15	CASH FLOW	($1,000)	$0	$2,000	$2,000	$2,000	$5,000
16	NPV	$3,201					
17		Formula=npv(b3,b15:f15)					
18							
19	RECOMMEND PROJECT 2 BECAUSE IT HAS THE HIGHER NPV.						
20							
21	IF STATEMENT..>	=IF(B9>B16,A5,A12)					
22	RESULT..>	PROJECT 2					
23							
24							

Figure 4-2. Net Present Value Example

Note that the sum of the cash flows is the same for both projects—$5,000. The net present values are different because they account for the time value of money. Money earned today is worth more than money earned in the future. Project 1 had a negative cash flow of $5,000 in the first year, while Project 2 had a negative cash flow of only $1,000 in the first year. Although both projects had the same total cash flow without discounting, these cash flows are not of comparable financial value. Net present value analysis, therefore, is a method for making equal comparisons between cash flows for multi-year projects.

To determine NPV, follow these steps:

1. Determine the cash inflows and outflows for the project. Figure 4-2 provides an example. Notice that the cash inflows are listed as projected benefits and the cash outflows are listed as projected costs for the project. The cash flow each year is calculated by subtracting the costs from the benefits for each year.
2. Determine the discount rate. A **discount rate** is the minimum acceptable rate of return on an investment. It is also called the required rate of return, hurdle rate, or opportunity cost of capital. Most companies use a discount rate based on the return that the organization could expect to receive elsewhere for an investment of comparable risk. In Figure 4-2, the discount rate used is 10 percent per year.
3. Calculate the net present value. There are several ways to calculate net present value. Most spreadsheet software has a built-in function to calculate NPV. For example, Figure 4-2 shows the formula that Microsoft Excel uses: =npv(discount rate, range of cash flows), where the discount rate is in cell B3 and the range of cash flows for Project 1 are in cells B8 through F8. The result of the formula yields an NPV of $2,316 for Project 1 and $3,201 for Project 2. Since both projects have positive NPVs, they are both good candidates for selection. However, since Project 2 has a higher NPV than Project 1, it would be the better choice between the two.

The mathematical formula for calculating NPV is:

$$NPV = \Sigma_{t=1...n} A/(1+r)^t$$

where t equals the year of the cash flows, A is the amount of cash flow each year, and r is the discount rate. A simpler way to use this formula is first to determine the annual discount rate and then apply it to the costs and benefits for each year. Calculate the net present value by summing the discounted benefits plus the discounted costs, assuming costs are entered as a negative number. Figures 4-3 and 4-4 illustrate this method of calculating NPV. Recall that the discount rate in this example is 10 percent or 0.10. You can calculate a **discount factor**—a multiplier for each year based on the discount rate and year—for each year as follows:

Year 1: discount factor = $1/(1+0.10)^1 = .91$

Year 2: discount factor = $1/(1+0.10)^2 = .83$

Year 3: discount factor = $1/(1+0.10)^3 = .75$

Year 4: discount factor = $1/(1+0.10)^4 = .68$

Year 5: discount factor = $1/(1+.010)^5 = .62$

You can then calculate the discounted costs each year by multiplying the discount factor by the costs for each year. You calculate the discounted benefits in the same way. To find the net present value, sum the discounted benefits plus the discounted costs, entering costs as a negative number. Notice that the NPV for Project 1 is 2,316 in Figure 4-3, just as it is in Figure 4-2. Likewise, the NPV for Project 2 is 3,201 in Figure 4-4, just as it is in Figure 4-2.

	A	B	C	D	E	F	G	H
1								
2	DISCOUNT RATE →	**10%**		**Years**				
3		**1**	**2**	**3**	**4**	**5**	**TOTAL**	
4	COSTS	($5,000)	($1,000)	($1,000)	($1,000)	($1,000)	-9,000	
5	DISCOUNT FACTOR	0.91	0.83	0.75	0.68	0.62		
6	**DISCOUNTED COSTS**	**-4,545**	**-826**	**-751**	**-683**	**-621**	**-7,427**	
7								
8	BENEFITS	$0	$2,000	$3,000	$4,000	$5,000	$14,000	
9	DISCOUNT FACTOR	0.91	0.83	0.75	0.68	0.62		
10	**DISCOUNTED BENEFITS**	**0**	**1,653**	**2,254**	**2,732**	**3,105**	**9,743**	
11								
12	DISCOUNTED BENEFITS + COSTS	-4,545	826	1,503	2,049	2,484	**2,316**	← **NPV**
13	CUMULATIVE BENEFITS + COSTS	-4,545	-3,719	-2,216	-167	2,316	4,633	
14						↑		
15	ROI	31%				Payback in this year		
16								

Figure 4-3. NPV, ROI, and Payback Analysis for Project 1

	A	B	C	D	E	F	G	H	I
1									
2	DISCOUNT RATE →	**10%**		**Years**					
3		**1**	**2**	**3**	**4**	**5**	**TOTAL**		
4	COSTS	($2,000)	($2,000)	($2,000)	($2,000)	($2,000)	-10,000		
5	DISCOUNT FACTOR	0.91	0.83	0.75	0.68	0.62			
6	**DISCOUNTED COSTS**	**-1,818**	**-1,653**	**-1,503**	**-1,366**	**-1,242**	**-7,582**		
7									
8	BENEFITS	$1,000	$2,000	$4,000	$4,000	$4,000	$15,000		
9	DISCOUNT FACTOR	0.91	0.83	0.75	0.68	0.62			
10	**DISCOUNTED BENEFITS**	**909**	**1,653**	**3,005**	**2,732**	**2,484**	**10,783**		
11									
12	DISCOUNTED BENEFITS + COSTS	-909	0	1,503	1,366	1,242	**3,201**	← **NPV**	
13	CUMULATIVE BENEFITS + COSTS	-909	-909	594	-1,960	3,201	6,403		
14				↑					
15	ROI	42%		Payback in this year					
16									

Figure 4-4. NPV, ROI, and Payback Analysis for Project 2

Return on Investment

Another important financial consideration is return on investment. **Return on investment (ROI)** is income divided by investment. For example, if you invest $100 today and next year it is worth $110, your ROI is $110/100 or 0.10 or 10 percent. It is best to consider discounted income and investment for multi-year projects when calculating ROI. Figures 4-3 and 4-4 show the calculations. For example, you calculate the ROI for Project 1 as follows:

ROI = (total discounted benefits - total discounted costs)/discounted costs

ROI = (9,747 – 7,427) / 7,427 = 31%

The higher the ROI, the better. Project 2's ROI is 42 percent, so it would be a better choice than Project 1, based on ROI.

Many organizations have a required rate of return for projects. The **required rate of return** is the minimum acceptable rate of return on an investment, and it is based on the return that the organization could expect to receive elsewhere for an investment of comparable risk.

Payback Analysis

Payback analysis is another important financial tool to use when selecting projects. **Payback period** is the amount of time it will take to recoup, in the form of net cash inflows, the net dollars invested in a project. In other words, payback analysis determines how much time will lapse before accrued benefits overtake accrued and continuing costs. Payback occurs when the cumulative discounted benefits and costs are greater than zero. Figures 4-3 and 4-4 show how to calculate the payback period, as well as NPV and ROI. For Project 1, payback occurs early in year 5 (*see* Figure 4-3), and for Project 2, it occurs in year 3 (*see* Figure 4-4). Project 2, therefore, has a better payback period because the payback period is shorter.

Many organizations have certain recommendations for the length of the payback period of an investment. They might require all information technology projects to have a payback period of less than three or even two years, regardless of the estimated NPV or ROI.

To aid in project selection, it is important for project managers to understand the organization's financial expectations for projects. It is also important for senior management to understand the limitations of financial estimates, particularly for information technology projects. For example, it is very difficult to develop good estimates of projected costs and benefits for information technology projects. This issue is discussed further in Chapter 6, *Project Cost Management*.

Weighted Scoring Model

A **weighted scoring model** is a tool that provides a systematic process for selecting projects based on many criteria. These criteria can include factors such as meeting broad organizational needs; addressing problems, opportunities, or directives; the amount of time it will take to complete the project; the overall priority of the project; and projected financial performance of the project.

The first step in creating a weighted scoring model is to identify criteria important to the project selection process. It often takes time to develop and reach agreement on these criteria. Holding facilitated brainstorming sessions or using groupware to exchange ideas can aid in developing these criteria. Some possible criteria for information technology projects include:

- Supports key business objectives
- Has strong internal sponsor
- Has strong customer support
- Uses realistic level of technology
- Can be implemented in one year or less
- Provides positive net present value
- Has low risk in meeting scope, time, and cost goals

Next, you assign a weight to each criterion. These weights indicate how much you value each criterion or how important each criterion is. You can assign weights based on percentages, and the sum of all of the criteria's weights must total 100 percent. You then assign numerical scores to each criterion (for example, 0 to 100) for each project. The scores indicate how much each project meets each criterion. At this point, you can use a spreadsheet application to create a matrix of projects, criteria, weights, and scores. Figure 4-5 provides an example of a weighted scoring model to evaluate four different projects. After assigning weights for the criteria and scores for each project, you calculate a weighted score for each project by multiplying the weight for each criterion by its score and adding the resulting values.

For example, you calculate the weighted score for Project 1 in Figure 4-5 as:

$$25\%*90 + 15\%*70 + 15\%*50 + 10\%*25 + 5\%*20 + 20\% * 50 + 10\%*20 = 56$$

Note that in this example, Project 2 would be the obvious choice for selection because it has the highest weighted score. Creating a bar chart to graph the weighted scores for each project allows you to see the results at a glance. If you create the weighted scoring model in a spreadsheet, you can enter the data, create and copy formulas, and perform "what-if" analysis. For example, suppose you change the weights for the criteria. By having the weighted scoring model in a spreadsheet, you can easily change the weights, and the weighted scores and charts will be updated automatically.

Many teachers use a weighted scoring model to determine grades. Suppose grades for a class are based on two homework assignments and two exams. To calculate final grades, the teacher would assign a weight to each of these items. Suppose Homework One is worth 10 percent of the grade, Homework Two is worth 20 percent of the grade, Test One is worth 20 percent of the grade, and Test Two is worth 50 percent of the grade. Students would want to do well on each of these items, but they should focus on performing well on Test Two since it is 50 percent of the grade.

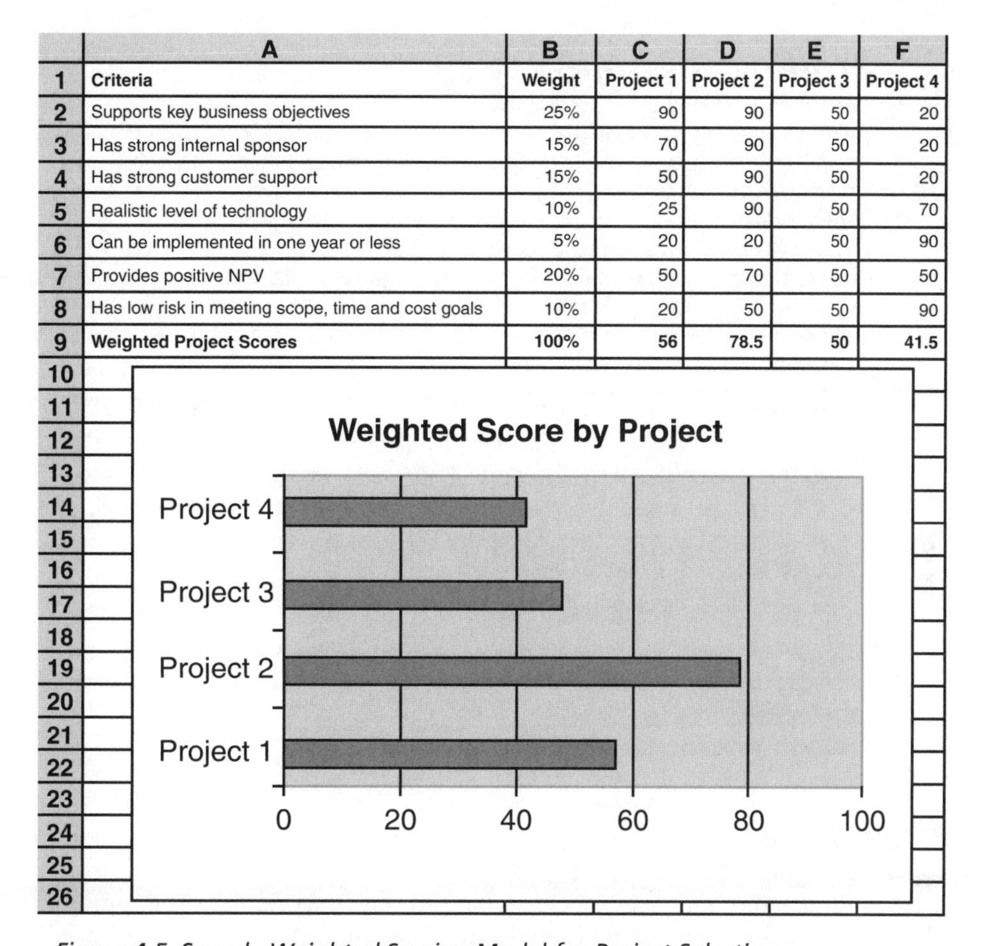

	A	B	C	D	E	F
1	**Criteria**	**Weight**	**Project 1**	**Project 2**	**Project 3**	**Project 4**
2	Supports key business objectives	25%	90	90	50	20
3	Has strong internal sponsor	15%	70	90	50	20
4	Has strong customer support	15%	50	90	50	20
5	Realistic level of technology	10%	25	90	50	70
6	Can be implemented in one year or less	5%	20	20	50	90
7	Provides positive NPV	20%	50	70	50	50
8	Has low risk in meeting scope, time and cost goals	10%	20	50	50	90
9	**Weighted Project Scores**	**100%**	**56**	**78.5**	**50**	**41.5**

Figure 4-5. Sample Weighted Scoring Model for Project Selection

You can also establish weights by assigning points. For example, a project might receive 10 points if it definitely supports key business objectives, 5 points if it somewhat supports them, and 0 points if it is totally unrelated to key business objectives. With a point model, you can simply add all the points to determine the best projects for selection, without having to multiply weights and scores and sum the results.

You can also determine minimum scores or thresholds for specific criteria in a weighted scoring model. For example, suppose a project really should not be considered if it does not score at least 50 out of 100 on every criterion. You can build this type of threshold into the weighted scoring model to reject projects that do not meet these minimum standards. As you can see, weighted scoring models can aid in project selection decisions.

What Went Right?

Virginia Housing Development Authority (VHDA) is a public mortgage finance agency that helps Virginians obtain safe, sound, and decent housing that is otherwise unaffordable to them. VHDA is entirely self-supporting and does not receive any funding from the state. Therefore, it is very careful with its information technology budget.

VHDA developed a planning process for establishing an information technology project list and related budget information. The process began with planning meetings involving all business user groups to identify potential information technology projects. They then developed cost and benefit statements for proposed projects and scored them using a weighted scoring model. The seven criteria in their scoring model were:

- The project's net present value
- Whether the project was mandated and by whom
- The relationship of the project to VHDA's strategic plan
- The relationship of the project to the information technology strategic plan
- Whether the project would replace legacy applications
- Whether there was an interim solution
- How the project would affect customer service

The new planning and budgeting process and weighted scoring model for information technology projects is very successful, and people involved in the process agree that it was a worthwhile effort.[2]

Project Charters

After senior managers decide what projects to work on, it is important to let the rest of the organization know about these projects. Management needs to create and distribute documentation to authorize beginning work on projects. This documentation can take many different forms, but one common form is a project charter. A **project charter** is a document that formally recognizes the existence of a project and provides direction on the project's objectives and management. Instead of charters, some organizations initiate projects using a simple letter of agreement, while others use formal contracts. Key project stakeholders should sign a project charter to acknowledge agreement on the need and intent of the project. A project charter is a key output of the initiation process.

Table 4-2 provides a sample of a project charter for the information technology upgrade project described in the opening case. Notice that the key parts of this project charter are:

- The project's title and date of authorization
- The project manager's name and contact information
- A brief scope statement for the project
- A summary of the planned approach for managing the project

[2] Woods, Theodore W, "The Justification, Ranking, and Budgeting Process for IT at the Virginia Housing Development Authority," *Journal of Systems Management* (May/June 1996).

- A roles and responsibilities matrix
- A sign-off section for signatures of key project stakeholders
- A comments section in which stakeholders can provide important comments related to the project

Table 4-2: Sample Project Charter

Project Title: Information Technology (IT) Upgrade Project

Project Start Date: March 4, 2002 **Projected Finish Date:** December 4, 2002

Project Manager: Kim Nguyen, 691-2784, *knguyen@abc.com*

Project Objectives: Upgrade hardware and software for all employees (approximately 2,000) within 9 months based on new corporate standards. See attached sheet describing the new standards. Upgrades may affect servers and midrange computers, as well as network hardware and software. Budgeted $1,000,000 for hardware and software costs and $500,000 for labor costs.

Approach:

- Update the information technology inventory database to determine upgrade needs
- Develop detailed cost estimate for project and report to CIO
- Issue a request for quotes to obtain hardware and software
- Use internal staff as much as possible to do the planning, analysis, and installation

ROLES AND RESPONSIBILITIES

NAME	ROLE	RESPONSIBILITY
Walter Schmidt, CEO	Project Sponsor	Monitor project
Mike Zwack	CIO	Monitor project, provide staff
Kim Nguyen	Project Manager	Plan and execute project
Jeff Johnson	Director of Information Technology Operations	Mentor Kim
Nancy Reynolds	VP, Human Resources	Provide staff, issue memo to all employees about project
Steve McCann	Director of Purchasing	Assist in purchasing hardware and software

Sign-off: (Signatures of all the above stakeholders)

Comments: (Handwritten comments from above stakeholders, if applicable)

This project must be done within ten months at the absolute latest. *Mike Zwack, CIO*

We are assuming that adequate staff will be available and committed to supporting this project. Some work must be done after hours to avoid work disruptions, and overtime will be provided. *Jeff Johnson and Kim Nguyen, Information Technology Department*

Notice that this sample charter fits on one page. A project charter can be as simple as a one-page form or a memo from a senior manager briefly describing the project and listing the responsibilities and authority of the new project manager and stakeholders. Charters may be much longer, however, depending on the nature of the project. In some cases, a contract serves as a project charter. Contracts are discussed in Chapter 11, *Project Procurement Management*.

Project charters are usually not difficult to write. What is difficult is getting people with the proper knowledge and authority to write and sign the project charters. Even though senior managers verbally approve a project, it is still important to have a formal, written charter to clarify roles and expectations. Many people assume, therefore, that senior managers must write the charter, but that does not need to be the case. As Paula Martin and Karen Tate, writing in *PM Network*, note: "What's that you said? It would be a cold day in where before your sponsor provides you with a charter? Not a problem. You can create a charter with your project team and then review it with your sponsor. That way you'll clarify the sponsor's expectations for the project."[3]

Since many projects fail because of unclear requirements and expectations, starting with a simple project charter makes a lot of sense. If project managers are having difficulty obtaining support from project stakeholders, for example, they can refer to what everyone agreed to in the project charter.

After formally recognizing the existence of a project, the next step in project scope management is more detailed scope planning.

SCOPE PLANNING AND THE SCOPE STATEMENT

Project scope planning involves developing documents to provide the basis for future project decisions, including the criteria for determining if a project or phase has been completed successfully. The project charter, descriptions of the products involved in the project, project constraints, and project assumptions are inputs to the scope planning process. The main output of this process is the creation of a written scope statement, including supporting detail, and a scope management plan.

The Scope Statement

A **scope statement** is a document used to develop and confirm a common understanding of the project scope. The scope statement should include a project justification, a brief description of the project's products, a summary of all project deliverables, and a statement of what determines project success.

- The project justification describes the business need that sparked creation of the project. For example, in the opening case, Kim Nguyen's company

[3] Martin, Paula K. and Tate, Karen, "Ready, Set, . . . Whoa," *PM Network*, PMI (June 1998) 22.

initiated the information technology upgrade project in order to support several high-priority Internet-based applications they were developing.

- The brief description of the project's products summarizes the characteristics of the products or services that the project will produce. For example, this section of the scope statement of the information technology upgrade project would describe the minimum hardware and software requirements for all of the company's computers.
- The summary of project deliverables would list the deliverables of the project. Deliverables in the information technology upgrade project might include documentation such as a project plan, a Work Breakdown Structure, a detailed cost estimate, a communications management plan, performance reports, and the like. Other deliverables would include an updated inventory of all hardware and software, upgraded hardware and software, status presentations, and so on.
- The section of the scope management plan describing what determines project success lists the quantifiable criteria that must be met for project success, such as cost, schedule, and quality measures. For example, the information technology upgrade project might be considered a success if 90 percent of all employees have computers that are capable of using the new Internet systems under development within nine months and for no more than $1.5 million. Even though the goal is to have all employees' hardware and software upgraded, stakeholders might be happy with 90 percent. This section would also stress the need to be done within ten months. It might also mention that the budget could go up to $1.7 million, if that were the case.

Scope statements also vary by type of project. Very large, complex projects have very long scope statements. Government projects often include a scope statement known as a Statement of Work (SOW). Some SOWs are hundreds of pages long, particularly if they include detailed product specifications. As with many other project management documents, the scope statement should be tailored to meet the needs of the particular project.

SCOPE DEFINITION AND THE WORK BREAKDOWN STRUCTURE

After completing scope planning, the next step in project scope management is to further define the work required for the project and to break it into manageable pieces. Breaking work into manageable pieces is called scope definition. Good scope definition is very important to project success because it helps improve the accuracy of time, cost, and resource estimates, it defines a baseline for performance measurement and project control, and it aids in communicating clear work responsibilities. The output of the scope definition process is the Work Breakdown Structure for the project.

The Work Breakdown Structure

A **work breakdown structure (WBS)** is an outcome-oriented analysis of the work involved in a project that defines the total scope of the project. It is a foundation document in project management because it provides the basis for planning and managing project schedules, costs, and changes. Some project management experts believe that work should not be done on a project if it is not included in the WBS.

A WBS is often depicted as a task-oriented family tree of activities. It is usually organized around project products or by phases. It can also be organized using the project management process groups. It can look like an organizational chart, which can help people visualize the whole project and all of its main parts. A WBS can also be shown in tabular form as an indented list of tasks that shows the same groupings of the work. Figure 4-6a shows a WBS for an intranet project. Notice that product areas provide the basis for its organization. In this case, there are main boxes or items on the WBS for developing the Web site design, the home page for the intranet, the marketing department's pages, and the sales department's pages.

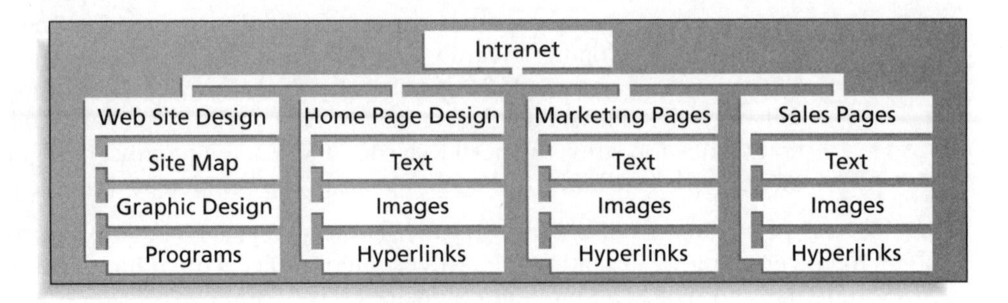

Figure 4-6a. Sample Intranet WBS Organized by Product

In contrast, a WBS for the same intranet project can be organized around project phases, as shown in Figure 4-6b.[4] Notice that project phases of concept, Web site design, Web site development, roll out, and support provide the basis for its organization.

This same WBS is shown in tabular form in Table 4-3. The items on the WBS are the same, but the numbering scheme and indentation of tasks show the structure. This tabular form is used in many documents, such as contracts. It is also used in project management software.

[4] This particular structure is based on a sample Project 98 file.

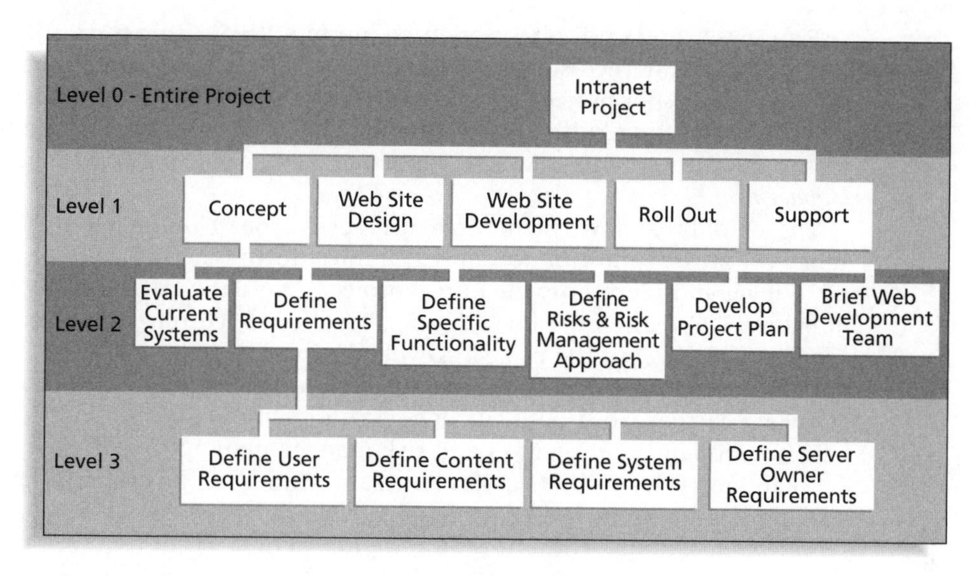

Figure 4-6b. Sample Intranet WBS Organized by Phase

Table 4-3: Intranet WBS in Tabular Form

1.0 Concept

    1.1 Evaluate current systems

    1.2 Define Requirements

        1.2.1 Define user requirements

        1.2.2 Define content requirements

        1.2.3 Define system requirements

        1.2.4 Define server owner requirements

    1.3 Define specific functionality

    1.4 Define risks and risk management approach

    1.5 Develop project plan

    1.6 Brief Web development team

2.0 Web Site Design

3.0 Web Site Development

4.0 Roll Out

5.0 Support

Figure 4-7 shows this phase-oriented intranet WBS, using the same numbering scheme, in Microsoft Project 2000. You can see from this figure that the WBS is the basis for project schedules. Notice that the WBS is in the left part of the figure, and the resulting schedule in the form of a Gantt chart is in the right part of the figure. Gantt charts are discussed further in Chapter 5, *Project Time Management.*

The work breakdown structures in Figures 4-6a, 4-6b, and 4-7 and in Table 4-3 present information in hierarchical form. The top level of a WBS is the 0 level and represents the entire project. (Note the labels on the left side of Figure 4-6b). The next level down is level 1, which represents the major products or phases of the project. Level 2 includes the major subsets of level 1. For example, in Figure 4-6b the level 2 items under the level 1 item called "Concept" include: evaluate current system, define requirements, define specific functionality, define risks and risk management approach, develop project plan, and brief Web development team. Under the level 2 item called "Define Requirements" are four level 3 items on the WBS: define user requirements, define content requirements, define server requirements, and define server owner requirements.

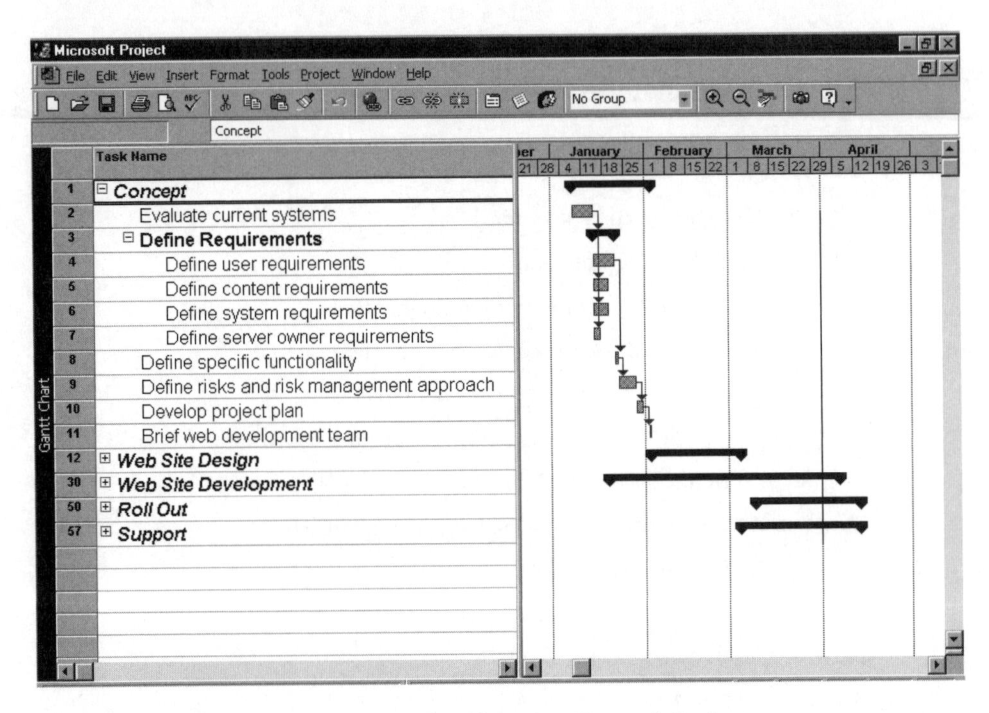

Figure 4-7. Intranet WBS and Gantt Chart in Microsoft Project

In Figure 4-6b, the lowest level provided is level 3. The lowest level of the WBS represents work packages. A **work package** is a deliverable or product at the lowest level of the WBS. As a general rule, each work package in a WBS should represent roughly eighty hours of effort. You can also think of work packages in terms of accountability and reporting. If a project has a relatively short time frame and requires weekly progress reports, a work package might represent 40 hours of work. On the other hand, if a project has a very long time frame and requires quarterly progress reports, a work package might represent over a hundred hours of work.

The sample WBSs shown here seem fairly easy to construct and understand. *Nevertheless, it is very difficult to create a good WBS.* To create a good WBS, you must understand both the project and its scope and incorporate the needs and knowledge of stakeholders. You and your project team must decide as a group how to organize the work and how many levels to include in your WBS. Many project managers have found that it is better to focus on getting the top three levels done well before getting too bogged down in more detail.

Another concern when creating a WBS is how to organize it so that it provides the basis for the project schedule. Some managers suggest creating a WBS using the project management process groups of initiating, planning, executing, controlling, and closing as level 1 in the WBS. By doing this, not only does the project team follow good project management practice, but the WBS tasks can also be mapped more easily against time. For example, Figure 4-8 shows a WBS and Gantt chart for the intranet project, organized by the product phases described earlier. Tasks under initiating include selecting a project manager, forming the project team, and developing the project charter. Tasks under planning include developing a scope statement, creating a WBS, and developing and refining other plans, which would be broken down in more detail for a real project. The tasks of concept, Web site design, Web site development, and roll out, which were WBS level 1 items in Figure 4-7, now become WBS level 2 items under executing. The executing tasks would vary the most from project to project, but many of the tasks under the other project management process groups would be similar for all projects.

It is important to involve the entire project team and customer in creating and reviewing the WBS. People who will do the work should help to plan the work by creating the WBS. Having group meetings to develop a WBS helps everyone understand what work must be done for the entire project and how it should be done, given the people involved. It also helps to identify where coordination between different work packages will be required.

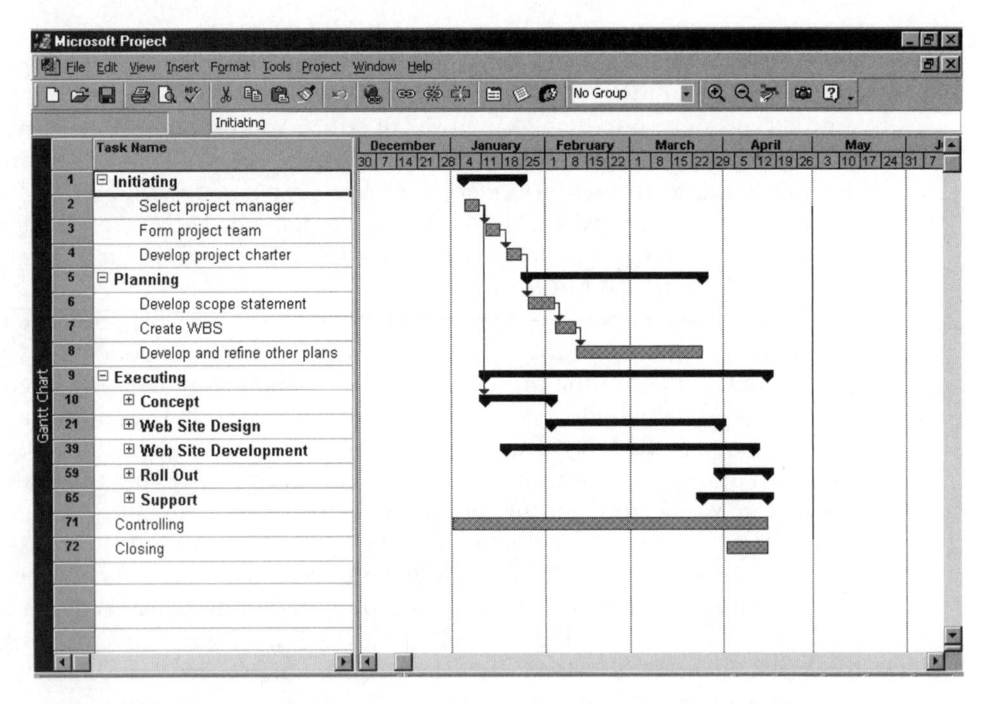

Figure 4-8. Intranet WBS and Gantt Chart Organized by Project Management Process Groups

Approaches to Developing Work Breakdown Structures

There are several approaches you can use to develop work breakdown structures. These approaches include:

- Using guidelines
- The analogy approach
- The top-down approach
- The bottom-up approach

Using Guidelines

If guidelines for developing a WBS exist, it is very important to follow them. Some organizations, for example, the U.S. Department of Defense (DOD), prescribe the form and content for WBSs for particular projects. Many DOD projects require contractors to prepare their proposals based on the DOD-provided WBS. These proposals must include cost estimates for each task in the WBS at a detailed and summary level. The cost for the entire project must be calculated by summing the costs of all of the lower level WBS tasks. When DOD personnel evaluate cost proposals, they must compare the contractors' costs with the DOD's estimates. A large variation in costs for a certain WBS task often indicates confusion as to what work must be done.

Consider a large automation project for the U.S. Air Force. In the mid-1980s, the Air Force developed a request for proposals for the Local On-Line Network System (LONS) to automate fifteen Air Force Systems Command bases. This $250 million project involved providing the hardware and developing software for sharing documents such as contracts, specifications, requests for proposals, and so on. The Air Force proposal guidelines included a WBS that contractors were required to follow in preparing their cost proposals. Level 1 WBS items included hardware, software development, training, project management, and the like. The hardware item was composed of several level 2 items, such as servers, workstations, printers, network hardware, and so on. Air Force personnel reviewed the contractors' cost proposals against their internal cost estimate, which was also based on this WBS. Having a prescribed WBS helped contractors to prepare their cost proposals and the Air Force to evaluate them.

The Analogy Approach

Another approach for constructing a WBS is the analogy approach. In the **analogy approach**, you use a similar project's WBS as a starting point. For example, Kim Nguyen from the opening case might learn that one of her organization's suppliers did a similar information technology upgrade project last year. She could ask them to share their WBS for that project to provide a starting point for her own project.

McDonnell Aircraft Company, now part of Boeing, provides an example of using an analogy approach when creating WBSs. McDonnell Aircraft Company designed and manufactured several different fighter aircraft. When creating a WBS for a new aircraft design, they started by using 74 predefined subsystems for building a fighter aircraft based on past experience. There was a level 1 WBS item for the airframe that was composed of level 2 items such as a forward fuselage, center fuselage, aft fuselage, and wings. This generic product-oriented WBS provided a starting point for defining the scope of new aircraft projects and developing cost estimates for new aircraft designs.

Some organizations keep a repository of WBSs and other project documentation on file to assist people working on projects. Project 2000 and many other software tools include sample files to assist users in creating a WBS and Gantt chart. Viewing examples of other similar projects' WBSs allows you to understand different ways to create a WBS.

The Top-Down and Bottom-Up Approaches

Two other approaches for creating WBSs are the top-down and bottom-up approaches. Most project managers consider the top-down approach of WBS construction to be conventional. To use the **top-down approach**, start with the largest items of the project and break them into their subordinate items. This process involves refining the work into greater and greater levels of detail. For example, Figure 4-6b shows how work was broken down to level 3 for part of the intranet project. After finishing the process, all resources should be assigned at the work package level. The top-down approach is best suited to project managers who have vast technical insight and a big-picture perspective.

In the **bottom-up approach**, team members first identify as many specific tasks related to the project as possible. They then aggregate the specific tasks and organize them into summary activities, or higher levels in the WBS. For example, a group of people might be responsible for creating a WBS to create an intranet. Instead of looking for guidelines on how to create a WBS or viewing similar projects' WBSs, they could begin by listing detailed tasks they think they would need to do in order to create an intranet. After listing detailed tasks, they would group the tasks into categories. Then, they would group these categories into higher-level categories. Some people have found that writing all possible tasks down on notes and then placing them on a wall helps them to see all the work required for the project and to develop logical groupings for performing the work. For example, a business analyst on the project team might know that they had to define user requirements and content requirements for the intranet. A hardware specialist might know they had to define system requirements and server requirements. As a group, they might decide to put all four of these tasks under a higher-level item called "define requirements." Later they might realize that defining requirements should fall under a broader category of concept design for the intranet. The bottom-up approach can be very time-consuming, but it can also be a very effective way to create a WBS. Project managers often use the bottom-up approach for projects that represent entirely new systems or approaches to doing a job, or to help create buy-in and synergy with a project team.

Advice for Creating a WBS

As stated previously, creating a good WBS is no easy task and usually requires several iterations. Often, it is best to use a combination of approaches to create a project WBS. There are some basic principles, however, that apply to creating any good WBS.

1. A unit of work should appear at only one place in the WBS.
2. The work content of a WBS item is the sum of the WBS items below it.
3. A WBS item is the responsibility of only one individual, even though many people may be working on it.
4. The WBS must be consistent with the way in which work is actually going to be performed; it should serve the project team first, and other purposes only if practical.
5. Project team members should be involved in developing the WBS to ensure consistency and buy-in.

6. Each WBS item must be documented to ensure accurate understanding of the scope of work included and not included in that item.
7. The WBS must be a flexible tool to accommodate inevitable changes while properly maintaining control of the work content in the project according to the scope statement.[5]

At the request of many of its members, the Project Management Institute recently developed a WBS Practice Standard to provide guidance for developing and applying the WBS to project management (*see* Suggested Readings).

SCOPE VERIFICATION AND SCOPE CHANGE CONTROL

It is difficult to create a good scope statement and WBS for a project. It is even more difficult, especially on information technology projects, to verify the project scope and minimize scope changes. Many technical projects suffer from **scope creep**—the tendency for project scope to keep getting bigger and bigger. There are many horror stories about information technology projects failing due to scope problems such as scope creep. For this reason, it is very important to verify the project scope and develop a process for controlling scope changes.

What Went Wrong?

A project scope that is too broad and grandiose can cause severe problems. Scope creep and an overemphasis on technology for technology's sake resulted in the bankruptcy of a large pharmaceutical firm, Texas-based FoxMeyer Drug. In 1994, the CIO was pushing for a $65 million system to manage the company's critical operations. He did not believe in keeping things simple, however. The company spent nearly $10 million on state-of-the-art hardware and software and contracted the management of the project to a prestigious (and expensive) consulting firm. The project included building an $18 million robotic warehouse, which looked like something out of a science fiction movie, according to insiders. The scope of the project kept getting bigger and more impractical. The elaborate warehouse was not ready on time, and the new system generated erroneous orders that cost FoxMeyer Drug over $15 million in unrecovered excess shipments. In July of 1996, the company took a $34 million charge for its fourth fiscal quarter, and by August of that year, FoxMeyer Drug filed for bankruptcy.[6]

Another major source of information technology project scope problems is a lack of user involvement. At Grumman in the late 1980s, for example, an information technology project team became convinced that it could, and should, automate the review and approval process of government proposals. They implemented a powerful workflow system to manage the whole process. Unfortunately, the end users for the system were aerospace engineers who preferred to work in a more casual, ad hoc fashion. They dubbed the system "Naziware" and

[5] Cleland, David I, *Project Management: Strategic Design and Implementation, 2nd ed.* New York: McGraw-Hill, 1994.

[6] James, Geoffrey, "Information Technology fiascoes . . . and how to avoid them," *Datamation* (November 1997) **www.datamation.com/PlugIn/issues/1997/november/11disas.html**.

refused to use it. This example illustrates an information technology project that wasted millions of dollars developing a system that was not in touch with the way end users did their work.[7]

Scope verification involves formal acceptance of the project scope by the stakeholders. To receive formal acceptance of the project scope, the project team must develop clear documentation of the project's products and procedures for evaluating if they were completed correctly and satisfactorily. In contrast, **scope change control** involves controlling changes to the project scope. In order to minimize scope change control, it is crucial to do a good job of verifying project scope.

The 1995 Standish Group CHAOS study found that key factors associated with information technology project success include user involvement and a clear statement of requirements. Likewise, the lack of either of these factors is strongly associated with information technology project failure. Table 4-4 lists factors reported to cause the most problems on information technology projects, in order of importance. Notice that the top three factors are all directly related to scope verification and change control.

Table 4-4: Factors Causing Information Technology Project Problems

FACTOR	RANK
Lack of user input	1
Incomplete requirements and specifications	2
Changing requirements and specifications	3
Lack of executive support	4
Technology incompetence	5
Lack of resources	6
Unrealistic expectations	7
Unclear objectives	8
Unrealistic time frames	9
New technology	10

Johnson, Jim. "CHAOS: The Dollar Drain of Information Technology Project Failures," *Application Development Trends* (January 1995) **www.standishgroup.com/chaos.html**.

Research and practice, therefore, indicate that in order to verify scope and control scope change, you need to improve user input and reduce incomplete and changing requirements and specifications. Suggestions for accomplishing these two tasks are provided in the next section.

[7] Ibid.

Suggestions for Improving User Input

Lack of user input is among the more important factors contributing to project failure and can lead to problems with managing scope creep and controlling change. How can you manage this important issue? Following are suggestions for improving user input.

- Develop a good project selection process for information technology projects. Insist that all projects have a sponsor from the user organization. The sponsor should not be someone in the information technology department, nor should the sponsor be the project manager. Make information technology project information, including the project charter, scope statement, and WBS, easily available in the organization. Making basic project information available will help avoid duplication of effort and ensure that the most important projects are the ones being worked on.
- Have users on the project team. Some organizations assign co-project managers to information technology projects, one from information technology and one from the main user group. Users should be assigned full-time to large information technology projects and part-time to smaller projects. A key success factor in Northwest Airline's ResNet project (*see* Chapters 12-16) was training reservation agents—the users—in how to write programming code for their new reservation system. Because the sales agents had intimate knowledge of the business, they provided excellent input and actually created most of the software.
- Have regular meetings. Meeting regularly sounds obvious, but many information technology projects fail because the information technology team members do not have regular interaction with users. They assume they understand what users need without getting direct user feedback. To encourage this interaction, users should sign off on key deliverables presented at meetings.
- Deliver something to project users and sponsors on a regular basis. If it is some sort of hardware or software, make sure it works first.
- Co-locate users with the developers. People often get to know each other better by being in close proximity. If the users cannot be physically moved to be near developers during the entire project, they could set aside certain days for co-location.

Suggestions for Reducing Incomplete and Changing Requirements

Some requirement changes are expected on information technology projects, but many projects have too many changes to their requirements, especially during later stages of the project life cycle when it is more difficult to implement them. Following are suggestions for improving the requirements process.

- Develop and follow a requirements management process that includes procedures for initial requirements determination.
- Employ techniques such as prototyping, use case modeling, and Joint Application Design to thoroughly understand user requirements. **Prototyping** involves developing a working replica of the system or some aspect of the system. These working replicas may be throwaways or an incremental component of the deliverable system. Prototyping is an effective tool for gaining an understanding of requirements, determining the feasibility of requirements, and resolving user interface uncertainties. **Use case modeling** is a process for identifying and modeling business events, who initiated them, and how the system should respond to them. It is an effective tool for understanding requirements for object-oriented systems. **Joint Application Design (JAD)** uses highly organized and intensive workshops to bring together project stakeholders—the sponsor, users, business analysts, programmers, and so on—to jointly define and design information systems. The techniques also help users become more active in defining system requirements.
- Put all requirements in writing and keep them current and readily available. Several tools are available to automate this function. For example, a type of software, called a requirements management tool, aids in capturing and maintaining requirements information, provides immediate access to the information, and assists in establishing necessary relationships between requirements and information created by other tools.
- Create a requirements management database for documenting and controlling requirements. Computer Aided Software Engineering (CASE) tools or other technologies can assist in maintaining a repository for project data.
- Provide adequate testing to verify that the project's products perform as expected. Conduct testing throughout the project life cycle.
- Use a process for reviewing requested requirements changes from a systems perspective. For example, ensure that scope changes include associated cost and schedule changes. Require approval by appropriate stakeholders. For example, at PanEnergy Corp. in Houston, Bruce Woodland completed a $10 million electronic commerce project on time and under budget. He says, "Whenever someone wanted to add or change something, we'd tell them how long it would take and then ask them how they wanted to deal with it . . . . The first few times, the users almost took us out behind the dumpster and hanged us. But when we could match specific functionalities they wanted with the time and dollars they required, we didn't have a problem."[8]

[8] King, Julia, "IS reins in runaway projects," *ComputerWorld* (September 24, 1997).

- Emphasize completion dates. For example, a project manager at Farmland Industries Inc. in Kansas City, Missouri, kept her 15-month, $7 million integrated supply-chain project on track by setting the project deadline. She says, "May 1 was the drop-dead date, and everything else was backed into it. Users would come to us and say they wanted something, and we'd ask them what they wanted to give up to get it. Sticking to the date is how we managed scope creep."[9]

Project scope management is very important, especially on information technology projects. Organizations must first select important projects, plan what is involved in performing the work of the project, break down the work into manageable pieces, verify the scope with project stakeholders, and manage changes to project scope. Using the basic project management concepts and techniques discussed in this chapter can help you successfully perform project scope management.

CASE WRAP-UP

Kim Nguyen started the Information Technology upgrade project meeting by reviewing the project charter with everyone. There were twelve people in the room representing the main stakeholders for the project. After reviewing the charter, Kim opened the floor for questions, which she answered confidently. Kim then explained that her goal for the meeting was to start planning the scope and defining the work involved in doing the project.

She asked attendees if they had any experience writing scope statements or developing work breakdown structures. A couple of people raised their hands. Yvonne, a member of the Marketing Department, one of the largest user groups in the company, had worked in the government for six years and was quite familiar with scope statements and WBSs. Kim asked Yvonne to describe what these documents were and how they were used on government projects. Kim summarized Yvonne's inputs on a flip chart, and one of her teammates wrote down key points for the meeting minutes. Then Bill, a senior manager in the information technology Operations Group, described how he had worked on a similar information technology upgrade project for another company several years ago. The meeting continued to progress, and everyone gained a better understanding of what was involved in project scope management.

[9] Ibid.

CHAPTER SUMMARY

Project scope management includes the processes required to ensure that the project includes all the work required, and only the work required, to complete the project successfully. The main processes involve initiation, scope planning, scope definition, scope verification, and scope change control.

The first step in project scope management is deciding on which projects to work. To decide on which projects to work, organizations should review the organization's strategic plan and then develop an information technology strategic plan that supports the organization's overall strategy.

Firms initiate information technology projects for many different reasons. Common techniques for selecting projects include focusing on broad organization needs, categorizing projects, performing financial analyses, and developing weighted scoring models.

Net present value analysis is the preferred financial measure for selecting projects. A positive net present value (NPV) is good, and the higher the NPV, the better. Other important financial considerations are the projected return on investment (ROI) of a project and the payback period.

A project charter is a document that formally recognizes the existence of a project. It must be signed by key project stakeholders to acknowledge agreement on the need for and intent of the project.

A scope statement is created in the scope planning process. This document includes a project justification, a brief description of the project's products, a summary of all project deliverables, and a statement of what determines project success.

A work breakdown structure (WBS) is the main output of the scope definition process. It is an outcome-oriented analysis of the work involved in a project that defines the total scope of the project. The WBS forms the basis for planning and managing project schedules, costs, and changes. Good WBSs are difficult to create. Approaches for developing a WBS include using guidelines, the analogy approach, the top-down approach, and the bottom-up approach.

Scope verification involves formal acceptance of the project scope by the stakeholders. Scope change control involves controlling changes to the project scope.

Poor scope management is one of the key reasons projects fail. For information technology projects, it is important for good scope management to have strong user involvement, a clear statement of requirements, and a process for managing scope changes.

DISCUSSION QUESTIONS

1. What are the main reasons for investing in information technology projects? Do you believe your organization uses similar reasons?
2. Give examples of information technology projects and why they were or were not selected for implementation. What type of project selection process did the organization use?

3. Develop a list of criteria for categorizing information technology projects. Apply these criteria to a few systems you are familiar with.
4. Discuss the theory and practice behind using project charters, scope statements, and WBSs.
5. Describe a project that suffered from scope creep. Could it have been avoided? How? Can scope creep be a good thing? When?

EXERCISES

1. Use spread sheet software to create the weighted scoring model presented in this chapter. Make sure your formulas work correctly. Then create your own criteria and project scores for a project selection scenario of your choice.
2. Create a WBS in chart form (similar to an Organizational chart—*see* intranet sample in Figure 4-6a) for the information technology upgrade project described in the opening case of this chapter. Break down the work to at least the third level for one of the WBS items. Make notes of questions you had while completing this exercise.
3. Create a WBS for one of the following projects:
   - Building your dream house
   - Planning a traditional wedding
   - Creating a new information system for your school or company

   Break down the work to at least the third level for one of the items on the WBS. Make notes of questions you had while completing this exercise.
4. Review the files in the MS Project 2000 templates folder. What do you think about the WBSs for them? Try drawing a WBS in chart form for one of the sample templates.
5. Perform a financial analysis for a project similar to those found in Figures 4-3 and 4-4. Assume the projected costs and benefits for this project are spread over four years as follows: Estimated costs are $1,500,000 in year 1 and $300,000 each year in years 2, 3, and 4. Estimated benefits are $0 in year 1 and $1,000,000 each year in years 2, 3, and 4. Use a 10 percent discount rate. Create a spreadsheet to calculate and clearly display the NPV, ROI, and year in which payback occurs. Also write a paragraph explaining whether or not you would recommend investing in this project, on the basis of your financial analysis.
6. Read one of the suggested readings. Write a one-page summary of the article, its key conclusions, and your opinion.

MINICASE

A financial services company has a long list of potential projects to consider this year. Managers at this company must decide which projects to pursue and how to define the scope of the projects selected for approval. The company has decided to use a weighted decision matrix to help in project selection, using criteria that map to corporate objectives. All projects selected must develop a WBS using corporate guidelines.

Part 1: You are part of a team that will analyze proposals and recommend which projects to pursue. Your team has decided to create a weighted decision matrix using the following criteria and weights:

Criteria	Weight
1. Enhances new product development	20%
2. Streamlines operations	20%
3. Increases cross-selling	25%
4. Has good NPV	35%

To determine the score for the last criterion, your team has developed the following scoring system:

- NPV is less than 0, the score is 0
- NPV is between 0 and $100,000, the score is 25
- NPV is between $100,000 and $200,000, the score is 50
- NPV is between $200,000 and $400,000, the score is 75
- NPV is above $400,000, the score is 100

The following is information for three potential projects:

- Project 1: Scores for criteria 1, 2, and 3 are 10, 20, and 80, respectively. Estimated costs the first year are $500,000, and costs for years 2 and 3 are $100,000 each. Estimated benefits for years 1, 2, and 3 are $200,000, $400,000, and $600,000, respectively.
- Project 2: Scores for criteria 1, 2, and 3 are all 50. Estimated costs the first year are $700,000, and costs for the second year are $200,000. Estimated benefits for years 1 and 2 are $300,000 and $700,000, respectively.
- Project 3: Scores for criteria 1, 2, and 3 are 0, 50, and 80, respectively. Estimated costs the first year are $300,000, and costs for years 2, 3, and 4 are $100,000 each. Estimated benefits for years 1, 2, 3, and 4 are $0, $600,000, $500,000, and $400,000, respectively.

Develop a spreadsheet to calculate the NPVs and weighted scores for the three projects. Use a 10 percent discount rate for the NPV calculations.

Part 2: One project selected for initiation is development of an expert system that current customers can use to give them advice on other financial products and services that would meet their needs. Corporate guidelines for creating project WBSs are to use the five project management process groups as level 1 in the WBS. Each process group must have at least one major deliverable, such as a signed project charter for initiating, an approved project plan for planning, monthly status reports for controlling, and a final report, presentation, and lessons learned for closing. The deliverables for execution vary by project.

Use the corporate guidelines to develop a WBS in tabular form (*see* Table 4-3). Leave the categories of initiating, planning, controlling, and closing at level 1. For the executing section, include level 2 categories of analysis, design, prototyping, testing, implementation, and support. Assume the support category includes level 3 categories of training, documentation, user support, and enhancements. Number and indent categories appropriately. For example, your WBS should start with 1.0 Initiating. In addition to creating your WBS, describe three potential deliverables you could include under the executing tasks for this project.

SUGGESTED READINGS

1. Abramovici, Adrian. "Controlling Scope Creep." *PM Network* (January 2000).

   This article states that scope creep is one of the most common problems that project managers face. The author illustrates the problems with scope creep in a short case study and provides practical advice on how to control it.

2. Hallows, Jolyon E. "The Fourth Dimension: Justifying Information Technology Projects." *PM Network* (November 1998).

   This article describes the importance of viewing project success as more then just meeting scope, time, and cost objectives. Information technology projects must realize financial and other organizational objectives to be successful.

3. Knutson, Joan. "From Making Sense to Making Cents: Measuring Project Management ROI–Part 1." *PM Network* (January 1999).

   This is the first part of a two-part article related to measuring the financial value of project management. It explains return on investment (ROI) calculations and provides examples of costs and benefits of using project management in organizations. Part II of this article appears in the February 1999 issue of PM Network.

4. Project Management Institute. "Project Management Institute Practice Standard for Work Breakdown Structures" (2001).

   The WBS Practice Standard is intended to provide guidance for developing and applying the WBS to project management. PMI released an exposure draft of this document in the fall of 2000 and plans to publish the final document in the fall of 2001. The charter for this project is available on PMI's Web site at www.pmi.org/standards/wbscharter.htm.

5. The Standish Group. "Unfinished Voyages" (1996) (www.standishgroup.com/voyages.html).

   The Standish Group used the project success criteria from the 1995 CHAOS research to create a weighted scoring model for estimating information technology project success potential. The most important success criteria, user involvement, was given 19 success points, while the least important, hard-working, focused staff, was given 3 success points. You can use this checklist to decide whether a project is worth starting or continuing.

6. Thompson, Jess. "What You Need to Manage Requirements." *IEEE Software* (March 1994).

    Thompson, the chief architect of a commercial requirements management tool, provides insight into the capabilities of these tools and the general issues to be addressed when managing requirements for software development projects.

KEY TERMS

- **Analogy approach** — creating a WBS by using a similar project's WBS as a starting point
- **Bottom-up approach** — creating a WBS by having team members identify as many specific tasks related to the project as possible and then grouping them into higher level items
- **Directives** — new requirements imposed by management, government, or some external influence
- **Discount factor** — a multiplier for each year based on the discount rate and year
- **Discount rate** — the minimum acceptable rate of return on an investment; also called the required rate of return, hurdle rate, or opportunity cost of capital
- **Initiation** — committing the organization to begin a project or continue to the next phase of a project
- **Joint Application Design (JAD)** — using highly organized and intensive workshops to bring together project stakeholders—the sponsor, users, business analysts, programmers, and so on—to jointly define and design information systems
- **Net present value analysis (NPV)** — a method of calculating the expected net monetary gain or loss from a project by discounting all expected future cash inflows and outflows to the present point in time
- **Opportunities** — chances to improve the organization
- **Payback period** — the amount of time it will take to recoup, in the form of net cash inflows, the net dollars invested in a project
- **Problems** — undesirable situations that prevent the organization from achieving its goals
- **Project charter** — a document that formally recognizes the existence of a project and provides direction on the project's objectives and management
- **Project scope management** — the processes involved in defining and controlling what is or is not included in a project

- **Prototyping** — developing a working replica of the system or some aspect of the system to help define user requirements
- **Required rate of return** — the minimum acceptable rate of return on an investment
- **Return on investment (ROI)** — income divided by investment
- **Scope** — all the work involved in creating the products of the project and the processes used to create them
- **Scope change control** — controlling changes to project scope
- **Scope creep** — the tendency for project scope to keep getting bigger and bigger
- **Scope definition** — subdividing the major project deliverables into smaller, more manageable components
- **Scope planning** — developing documents to provide the basis for future project decisions, including the criteria for determining if a project or phase has been completed successfully
- **Scope statement** — a document used to develop and confirm a common understanding of the project scope
- **Scope verification** — formalizing acceptance of the project scope
- **Strategic planning** — determining long-term objectives by analyzing the strengths and weaknesses of an organization, studying opportunities and threats in the business environment, predicting future trends, and projecting the need for new products and services
- **Top-down approach**—creating a WBS by starting with the largest items of the project and breaking them into their subordinate items
- **Use case modeling** — a process for identifying and modeling business events, who initiated them, and how the system should respond to them
- **Weighted scoring model** — a technique that provides a systematic process for basing project selection on numerous criteria
- **Work breakdown structure (WBS)** — an outcome-oriented analysis of the work involved in a project that defines the total scope of the project
- **Work package** — a deliverable or product at the lowest level of the WBS

5

Project Time Management

Objectives

After reading this chapter you should be able to:

1. *Understand the importance of good project time management*
2. *Explain the basic process for developing project schedules*
3. *Describe how various tools and techniques help project managers perform activity definition, activity sequencing, activity duration estimating, schedule development, and schedule control*
4. *Use a Gantt chart for schedule planning and tracking schedule information*
5. *Construct a project network diagram and understand its importance for determining overall project completion dates*
6. *Understand and use critical path analysis*
7. *Describe several techniques for shortening project schedules*
8. *Explain the basic concepts behind critical chain scheduling*
9. *Discuss reality checks and people issues involved in project schedule management and control*
10. *Describe how software can assist in project time management*

Sue Johnson was the project manager for a consulting company contracted to provide a new online registration system at a local university. This system absolutely had to be operational by May 1 so students could use it to register for the fall term. Her company's contract had a stiff penalty clause if the system was not ready by then, and Sue and her team would get nice bonuses for doing a good job on this project. Sue knew that it was her responsibility to meet the schedule and also manage scope, cost, and quality expectations. She and her team developed a detailed schedule and network diagram to help organize the project.

Developing the schedule turned out to be the easy part. Keeping the project on track turned out to be more difficult. Managing people

issues and resolving schedule conflicts were two of the bigger challenges. Many of the customers' employees took unplanned vacations and missed or rescheduled project review meetings. These changes made it difficult for her staff to follow their planned schedule for the system because they had to have customer sign-off at various stages of the systems development life cycle. One senior programmer on her project team quit, and she knew it would take extra time for a new person to get up to speed. It was still early in the project, but Sue knew they were falling behind. What could she do to meet the operational date of May 1?

IMPORTANCE OF PROJECT SCHEDULES

Many information technology projects are failures in terms of meeting scope, time, and cost projections. Managers often cite delivering projects on time as one of their biggest challenges. According to the 1995 CHAOS report, the average time overrun on unsuccessful information technology projects was 222 percent of the original estimate.[1] A 222 percent time overrun means that a project that was planned to take one year ended up taking 2.2 years to complete. The 2001 CHAOS report showed that time overruns decreased between 1995 and 2000—down to 63 percent—but most completed projects still finished significantly behind schedule.

Managers also cite schedule issues as the main reason for conflicts on projects throughout the project life cycle. Figure 5-1 shows the results of research on causes of conflict on projects. This figure shows that, overall, schedule issues cause the most conflict over the life of a project. (Note that project phases in this study were called project formation, early phases, middle phases, and end phases. You can interpret these phase names as concept, development, implementation, and close-out, as described in Chapter 2). During the project formation or concept phase, priorities and procedures cause more conflict than schedules. During the early or development phase, only priorities cause more conflict than schedules. During the middle or implementation phase and end or close-out phase, schedule issues are the predominant source of conflict.

Perhaps part of the reason schedule problems are so common is that time is easily and simply measured. You can debate scope and cost overruns and make actual numbers appear closer to estimates, but once a project schedule is set, anyone can quickly estimate schedule performance by subtracting the original time estimate from how long it really took to complete the project. People often compare planned and actual project completion times without taking into account approved changes in the project. Time is also the one variable that has the least amount of flexibility. Time passes no matter what happens on a project.

[1] Johnson, Jim, "CHAOS — The Dollar Drain of IT Project Failures," *Application Development Trends* (January 1995) **www.standishgroup.com/chaos.html**.

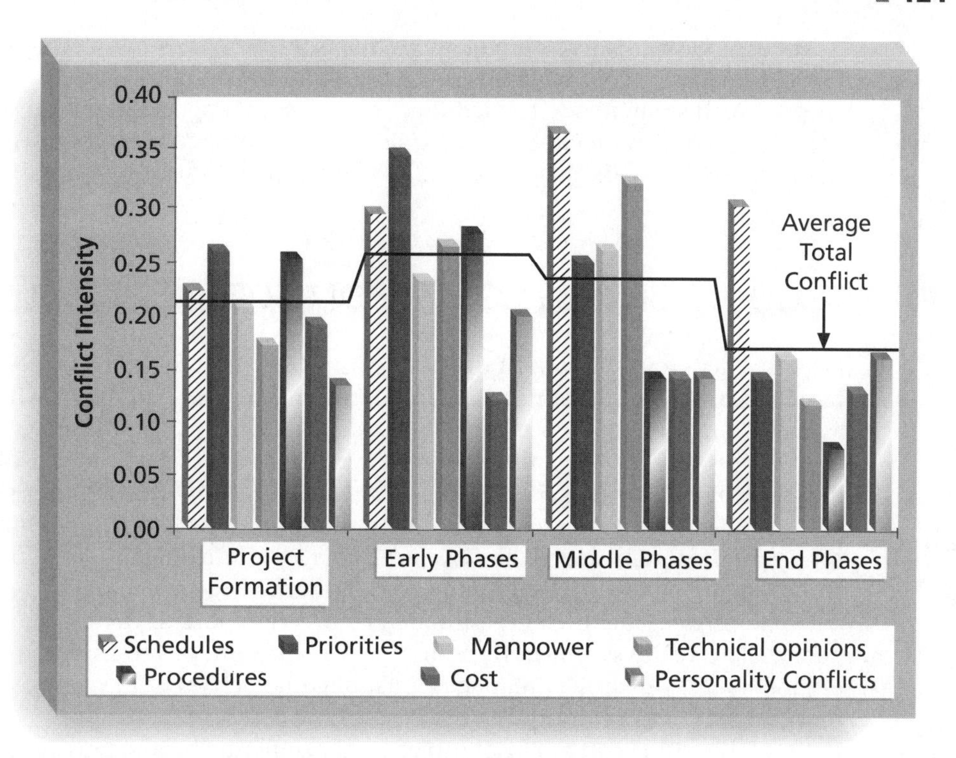

Figure 5-1. Conflict Intensity over the Life of a Project[2]

In that case, what *is* involved in project time management, and how can project managers help improve performance in this area? **Project time management**, simply defined, involves the processes required to ensure timely completion of a project. Achieving timely completion of a project, however, is by no means simple. The main processes involved in project time management include:

- **Activity definition**, which involves identifying the specific activities that the project team members and stakeholders must perform to produce the project deliverables. An **activity** or **task** is an element of work, normally found on the WBS, that has an expected duration, a cost, and resource requirements.
- **Activity sequencing**, which involves identifying and documenting the relationships between project activities.
- **Activity duration estimating**, which involves estimating the number of work periods that are needed to complete individual activities.
- **Schedule development**, which involves analyzing activity sequences, activity duration estimates, and resource requirements to create the project schedule.
- **Schedule control**, which involves controlling and managing changes to the project schedule.

[2] Thamhain, H. J., and Wilemon, D. L., "Conflict Management in Project Life Cycles," *Sloan Management Review* (Summer 1975).

You can improve time management by performing these processes and by using some basic project management tools and techniques. Every manager is familiar with some form of scheduling, but most managers have not used several of the tools and techniques unique to project time management, such as Gantt charts, network diagrams, and critical path analysis.

WHERE DO SCHEDULES COME FROM? DEFINING ACTIVITIES

Project schedules grow out of the basic documents that initiate a project. The project charter often mentions planned project start and end dates, which serve as the starting points for a more detailed schedule. The project manager starts with the project charter and develops a detailed scope statement and WBS, as mentioned in Chapter 4, *Project Scope Management*. The project charter should also include some estimate of how much money will be allocated to the project. Given this information, the project manager and his or her team use the scope statement, WBS, and budget information to begin developing a more detailed project schedule and estimated completion date. If the estimated completion date varies significantly from upper management's or the customer's original plans, the project manager must negotiate changes in scope or cost to meet schedule expectations.

Recall the triple constraint of project management—balancing scope, time, and cost goals—and note the order of these items. Ideally, the project team and key stakeholders first define the project scope, then the time or schedule for the project, and then the project's cost. The order of these three items reflects the basic order of the first three processes in time management: activity definition (further defining the scope), activity sequencing (further defining the time), and activity duration estimating (further defining the cost). These three project time management processes are the basis for creating a project schedule.

Activity definition usually results in the project team developing a more detailed WBS and supporting explanations. The goal of this process is to ensure that the project team has complete understanding of all the work they must do as part of the project scope. The WBS is often dissected further during this process as the project team members further define the activities required for performing the work. As stated earlier, activities or tasks are elements of work performed during the course of a project; they have expected durations, costs, and resource requirements. Activity definition also results in supporting detail to document important product information as well as assumptions and constraints related to specific activities. The project team should review the revised WBS and supporting detail with project stakeholders before moving on to the next step in project time management.

In the opening case, Sue Johnson and her project team had a contract and detailed specifications for the college's new online registration system. They also had to focus on meeting the May 1 date for an operational system so the college could start using the new system for fall registration. To develop a project schedule, Sue and her team had to review the contract, detailed specifications, and desired operational date, create a more detailed WBS, and highlight the most important product information, assumptions, and constraints. After developing more detailed definitions of project activities, Sue and her team would review them with their customers to ensure that they were on the right track.

ACTIVITY SEQUENCING

After defining project activities, the next step in project time management is activity sequencing. Activity sequencing involves reviewing the activities in the detailed WBS, detailed product descriptions, assumptions, and constraints to determine the relationships between activities. It also involves evaluating the reasons for dependencies and the different types of dependencies. A **dependency** or **relationship** shows the sequencing of project activities or tasks. For example, does a certain activity have to be finished before another one can start? Can several activities be done in parallel? Can some overlap? Determining these relationships or dependencies between activities has a significant impact on developing and managing a project schedule.

There are three basic reasons for creating dependencies among project activities:

- **Mandatory dependencies** are inherent in the nature of the work being done on a project. They are sometimes referred to as hard logic. For example, you cannot test code until after the code is written.
- **Discretionary dependencies** are defined by the project team. For example, a project team might follow good practice and not start detailed design of a new information system until the users sign off on all of the analysis work. Discretionary dependencies are sometimes referred to as soft logic. They should be used with care since they may limit later scheduling options.
- **External dependencies** involve relationships between project and non-project activities. The installation of a new operating system and other software may depend on delivery of new hardware from an external supplier.

As with defining activities, it is important that project stakeholders work together to discuss and define the activity dependencies that exist on their project. Some organizations have guidelines based on the activity dependencies of similar projects. Some organizations rely on the expertise of the people working on the project and their contacts with other employees and colleagues in the profession. Some people like to write each activity letter or name on a sticky note or some other moveable paper to determine dependencies or sequencing. Still others work directly in project management software to establish relationships. Just as it is easier to write a research paper by first having some thoughts on paper before typing into a word processor, it is usually easier to do some manual form of activity sequencing before entering the information into project management software.

Many organizations do not understand the importance of defining activity dependencies and do not use them *at all* in project time management. If you do not define the sequence of activities, you cannot use some of the most powerful schedule tools available to project managers: network diagrams and critical path analysis.

Project Network Diagrams

Project network diagrams are the preferred technique for showing activity sequencing. A **project network diagram** is a schematic display of the logical relationships among, or sequencing of, project activities. A sample network diagram for Project X, which uses the arrow diagramming method (ADM) or activity-on-arrow (AOA) approach, is shown in Figure 5-2.

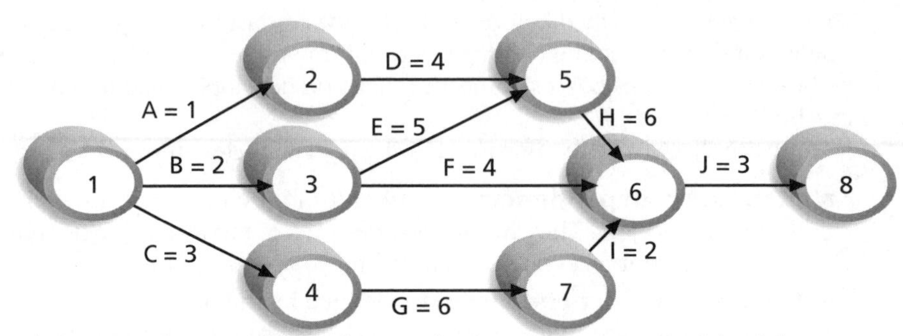

Note: Assume all durations are in days; A=1 means Activity A has a duration of 1 day.

Figure 5-2. Sample Activity-on-Arrow (AOA) Network Diagram for Project X

Note the main elements on this network diagram. The letters A, B, C, D, E, F, G, H, I, and J represent activities with dependencies that are required to complete the project. These activities come from the WBS and activity definition process described earlier. The arrows represent the activity sequencing or relationships between tasks. For example, Activity A must be done before Activity D; Activity D must be done before Activity H, and so on.

The format of this project network diagram uses the **activity-on-arrow (AOA)** or the **arrow diagramming method (ADM)**—a network diagramming technique in which activities are represented by arrows and connected at points called nodes to illustrate the sequence of activities. A **node** is simply the starting and ending point of an activity. The first node signifies the start of a project, and the last node represents the end of a project.

Keep in mind that the network diagram represents activities that must be done to complete the project. It is not a race to get from the first node to the last node. Every activity on the project network diagram must be completed in order for the project to finish.

It is also important to note that not every single item on the WBS needs to be on the project network diagram, especially on large projects. Sometimes it is enough to put summary tasks on a project network diagram or to break the project down into several smaller network diagrams. Some items that the team knows must be done and will be done regardless of other activities do not need to be included in the network diagram.

Assuming you have a list of the project activities and their start and finish nodes, follow these steps to create an AOA network diagram:

1. Find all of the activities that start at Node 1. Draw their finish nodes, and draw arrows between Node 1 and each of those finish nodes. Put the activity letter or name on the associated arrow. If you have a duration estimate, write that next to the activity letter or name, as shown in Figure 5-2. For example, A = 1 means that the duration of Activity A is one day, week, or other standard unit of time. Also be sure to put arrowheads on all arrows to signify the direction of the relationships.
2. Continue drawing the network diagram, working from left to right. Look for bursts and merges. **Bursts** occur when a single node is followed by two or more activities. A **merge** occurs when two or more nodes precede a single node. For example, in Figure 5-2, Node 1 is a burst since it goes into Nodes 2, 3, and 4. Node 5 is a merge preceded by Nodes 2 and 3.
3. Continue drawing the project network diagram until all activities are included on the diagram.
4. As a rule of thumb, all arrowheads should face toward the right, and no arrows should cross on an AOA network diagram. You may need to redraw the diagram to make it look presentable.

Even though AOA or ADM network diagrams are fairly easy to understand and create, a different method is more commonly used: the precedence diagramming method.

The **precedence diagramming method (PDM)** is a network diagramming technique in which boxes represent activities. It is particularly useful for visualizing certain types of time relationships.

Figure 5-3 illustrates the types of dependencies that can occur among project activities. After you determine the reason for a dependency between activities (mandatory, discretionary, or external), you must determine the type of dependency. Note that the terms activity and task are used interchangeably, as are relationship and dependency. The four types of dependencies or relationships between activities include:

- **Finish-to-start:** a relationship where the "from" activity must finish before the "to" activity can start. For example, you cannot provide user training until after software, or a new system, has been installed. Finish-to-start is the most common type of relationship, or dependency, and the AOA method uses only finish-to-start dependencies.
- **Start-to-start:** a relationship in which the "from" activity cannot start until the "to" activity can start. For example, on several information technology projects, a group of activities all start simultaneously, such as the many tasks that occur when a new system goes live.
- **Finish-to-finish:** a relationship where the "from" activity must finish before the "to" activity can finish. One task cannot finish before another finishes. For example, quality control efforts cannot finish before production finishes, although the two activities can be performed at the same time.
- **Start-to-finish:** a relationship where the "from" activity must start before the "to" activity can finish. This type of relationship is rarely used.

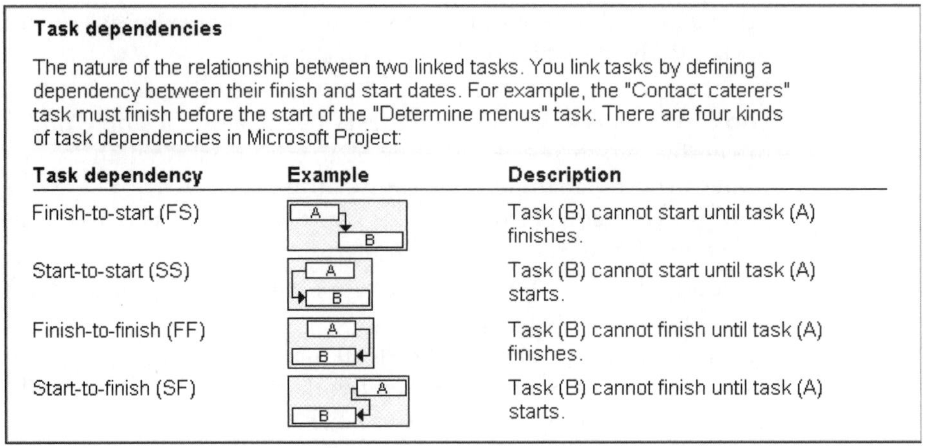

Task dependencies

The nature of the relationship between two linked tasks. You link tasks by defining a dependency between their finish and start dates. For example, the "Contact caterers" task must finish before the start of the "Determine menus" task. There are four kinds of task dependencies in Microsoft Project:

Task dependency	Example	Description
Finish-to-start (FS)	A B	Task (B) cannot start until task (A) finishes.
Start-to-start (SS)	A B	Task (B) cannot start until task (A) starts.
Finish-to-finish (FF)	A B	Task (B) cannot finish until task (A) finishes.
Start-to-finish (SF)	A B	Task (B) cannot finish until task (A) starts.

Figure 5-3. Task Dependency Types

Figure 5-4 illustrates Project X using the PDM method. Notice that the activities are placed inside boxes, which represent the nodes on this diagram. Arrows and arrowheads are again used to show the relationships between activities. This figure was created using Microsoft Project 2000, and the application automatically places additional information inside each node. Each task box includes the start and finish date, labeled Start and Finish, the task ID number, labeled ID, the task's duration, labeled Dur, and the names of resources, if any, assigned to the task, labeled Res. The outline of the boxes for tasks on the critical path appear automatically in red in the Project 2000 network diagram view. In Figure 5-4, the boxes for critical tasks have a thicker border and are not shaded.

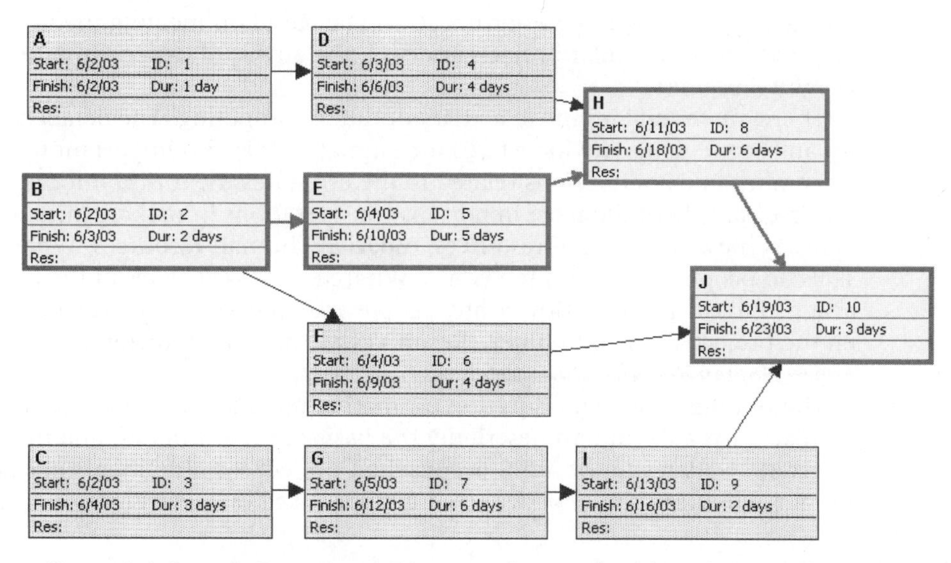

Figure 5-4. Sample Precedence Diagramming Method (PDM) Network Diagram for Project X

PDM is used more often than AOA diagrams and offers a number of advantages over the AOA technique. First, most project management software uses the PDM method. Second, the PDM method avoids the need to use dummy activities. **Dummy activities** have no duration and no resources but are occasionally needed on AOA diagrams to show logical relationships between activities. They are represented with dashed arrow lines, and you put in zero for the duration estimate. Third, the PDM method shows different dependencies among tasks, whereas the AOA method uses only finish-to-start dependencies. *See Appendix A* for more information on activity sequencing using Project 2000.

ACTIVITY DURATION ESTIMATING

After defining activities and determining their sequence, the next process in project time management is to estimate the duration of activities. It is important to note that **duration** includes the actual amount of time worked on an activity *plus* elapsed time. For example, even though it might take one workweek, or five workdays, to perform a certain activity, the duration estimate might be two weeks to allow for someone working only half-time on the activity or someone needing to wait a week to obtain outside information. As in defining activities and their sequences, it is important for the project stakeholders to discuss activity duration estimates. The people who will actually do the work, in particular, should have a lot of say in these duration estimates, since they are the ones whose performance will be based on meeting them. It is also helpful to review similar projects and seek the advice of experts in estimating activity durations.

There are several inputs to activity duration estimating. The detailed activity list and sequencing provide a basis for estimates. It is also important to review constraints and assumptions related to the estimates. Historical information is also useful. One of the most important considerations in making duration estimates is the availability of resources, especially human resources. What specific skills do people need to do the work? What are the skill levels of the people assigned to the project? How many people are expected to be available to work on the project at any one time? (Resources are discussed further in Chapter 8, *Project Human Resource Management*.)

The outputs of activity duration estimating include duration estimates for each activity, a document describing the basis of the estimates, and updates to the WBS. Updates to the WBS occur when project team members decide that certain activities should be dissected further based on their duration estimates.

SCHEDULE DEVELOPMENT

Schedule development uses the results of all the preceding project time management processes to determine the start and end date of the project. There are often several iterations of all the time management processes before a project schedule is finalized. The ultimate goal of schedule development is to create a realistic project schedule that provides a basis for monitoring project progress for the time dimension of the project.

Several tools and techniques assist in the schedule development process:

- A Gantt chart is a common tool for displaying project schedule information.
- PERT analysis is one means for evaluating schedule risk on projects.
- Critical path analysis is a very important tool for developing and controlling project schedules.
- Critical chain scheduling is a technique that accounts for resource constraints.

The following sections provide samples of each of these tools and techniques and a discussion of their advantages and disadvantages.

Gantt Charts

Gantt charts provide a standard format for displaying project schedule information by listing project activities and their corresponding start and finish dates in a calendar format. As mentioned in Chapter 1, Henry Gantt developed the first Gantt chart during World War I for scheduling work in job shops. Early versions simply listed project activities or tasks in one column to the left, calendar time units such as months to the right, and horizontal bars under the calendar units to illustrate when activities should start and end. However, Gantt charts normally do not show relationships among project activities, as network diagrams do. Today, most people use project management software to create more sophisticated versions of Gantt charts and allow for easy updates of information.

Figure 5-5 shows a simple Gantt chart for Project X. Figure 5-6 shows a more sophisticated Gantt chart based on a software launch project. Recall that the activities on the Gantt chart should coincide with the activities on the WBS. Notice that the software launch project's Gantt chart contains milestones, summary tasks, individual task durations, and arrows showing task dependencies.

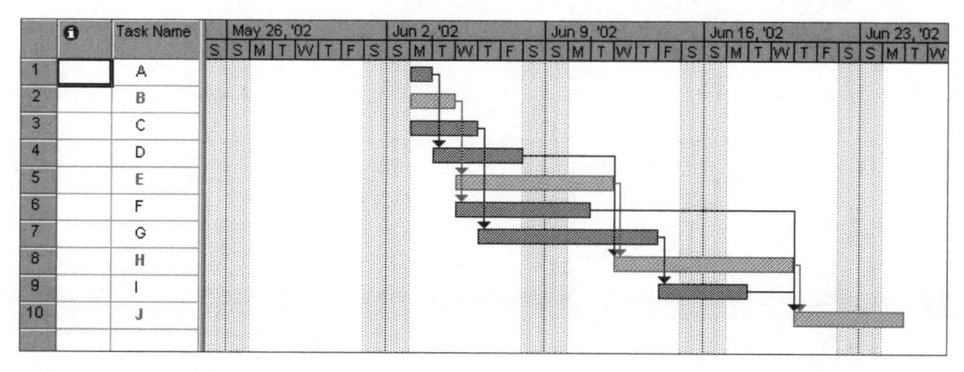

Figure 5-5. Gantt Chart for Project X

Notice the different symbols on the software launch project's Gantt chart (Figure 5-6):

- The black diamond symbol represents a **milestone**—a significant event on a project with zero duration. In Figure 5-6, Task 1, "Marketing Plan distributed," is a milestone that occurs on March 17. Tasks 3, 4, 8, 9, 14, 25, 27, 43, and 45 are also milestones. For very large projects, senior managers might want to see only milestones on a Gantt chart. Project 2000 allows you to filter information displayed on a Gantt chart so you can easily show specific tasks, such as milestones.

- The thick black bars with arrows at the beginning and end represent summary tasks. For example, Activities 12 through 15—"Develop creative briefs," "Develop concepts," "Creative concepts," and "Ad development"—are all subtasks of the summary task called Advertising, Task 11. WBS activities are referred to as tasks and subtasks in most project management software.
- The light gray horizontal bars such as those found in Figure 5-6 for tasks 5, 6, 7, 10, 12, 13, 15, 22, 24, 26 and 44, represent the duration of each individual task. For example, the light gray bar for Subtask 5, "Packaging," starts in mid-February and extends until early May.
- Arrows connecting these symbols show relationships or dependencies between tasks. Gantt charts often do not show dependencies, which is their major disadvantage. If dependencies have been established in Project 2000, they are automatically displayed on the Gantt chart.

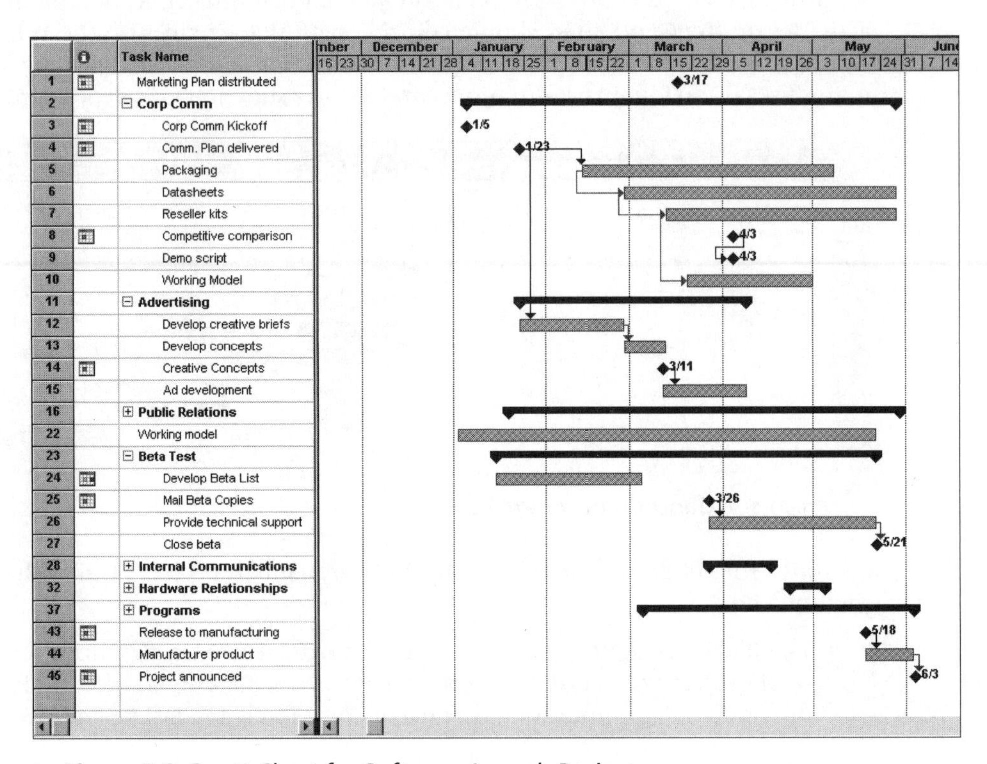

Figure 5-6. Gantt Chart for Software Launch Project

Milestones are a particularly important part of schedules, and some people use the SMART criteria to help define them. The **SMART criteria** are guidelines suggesting that milestones should be **s**pecific, **m**easurable, **a**ssignable, **r**ealistic, and **t**ime-framed. For example, distributing a marketing plan is specific, measurable, and assignable if everyone knows what should be in the marketing plan and how it should be distributed, how many copies should be distributed and to whom, and who is responsible for the actual delivery. Distributing the marketing plan is realistic and time-framed if it is an achievable event and scheduled at an appropriate time.

You can use Gantt charts to evaluate progress on a project by showing actual schedule information. Figure 5-7 shows a **tracking Gantt chart**—a Gantt chart that compares planned and actual project schedule information. The planned schedule dates for activities are called the **baseline dates**. The tracking Gantt chart includes columns (hidden in Figure 5-7) labeled "Start" and "Finish" to represent actual start and finish dates for each task, as well as columns labeled "Baseline Start" and "Baseline Finish" to represent planned start and finish dates for each task. In this example, the project is completed, but several tasks missed their planned start and finish dates.

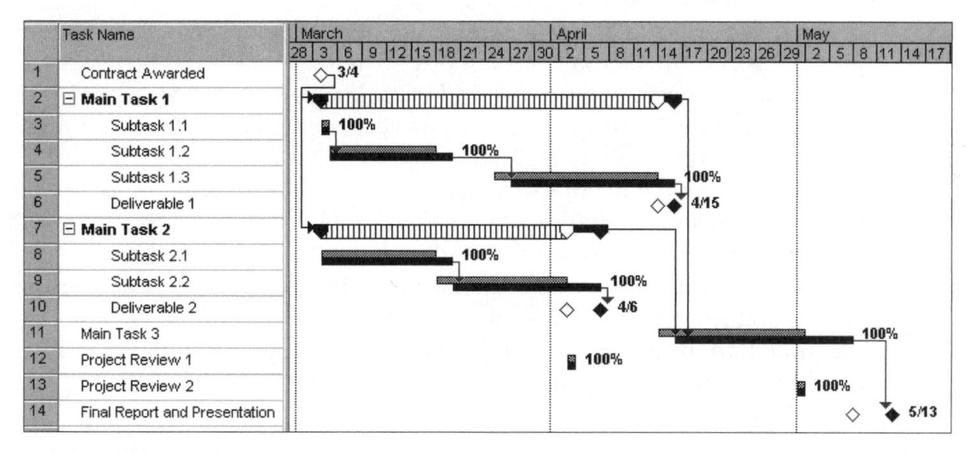

Figure 5-7. Sample Tracking Gantt Chart

To serve as a progress evaluation tool, a tracking Gantt chart uses a few additional symbols.

- Notice that the Gantt chart in Figure 5-7 often shows two horizontal bars for tasks. The top horizontal bar represents the planned or baseline duration for each task. The bar below it represents the actual duration. Subtasks 1.2 and 1.3 illustrate this type of display. If these two bars are the same length and start and end on the same date, then the actual schedule was the same as the planned schedule for that task. This scheduling occurred for Subtask 1.1, where the task started and ended as planned on 3/4/2002. If the bars do not start and end on the same date, then the actual schedule differed from the planned or baseline schedule. If the top horizontal bar is shorter than the bottom one, the task took longer than planned, as you can see for Subtask 1.2. If the top horizontal bar is longer than the bottom one, the task took less time than planned. A striped horizontal bar, illustrated by Main Task 1, represents the planned duration for summary tasks. The black bar adjoining it shows progress for summary tasks. For example, Main Task 1 shows the actual duration took three days longer than what was planned.

- A white diamond on the tracking Gantt chart represents a **slipped milestone**. A slipped milestone means the milestone activity was actually completed later than originally planned. For example, the last task provides an example of a slipped milestone since the final report and presentation were completed later than planned.
- Percentages to the right of the horizontal bars display the percentage of work completed for each task. For example, 100 percent means the task is finished. Fifty percent means the task is still in progress and is 50 percent completed.
- In the columns to the left of the Gantt chart, you can display baseline and actual start and finish dates.

A tracking Gantt chart is based on the percentage of work completed for project tasks or the actual start and finish dates. It allows the project manager to monitor schedule progress on individual tasks and the whole project. For example, Figure 5-7 shows that this project is completed. It started on time, but it finished a little late, on 5/13 versus 5/8/2002.

The main advantage of using Gantt charts is that they provide a standard format for displaying planned and actual project schedule information. In addition, they are easy to create and understand. The main disadvantage of Gantt charts is that they do not usually show relationships or dependencies between tasks. If Gantt charts are created using project management software and tasks are linked, then the dependencies will be displayed, but not as clearly as they would be displayed on project network diagrams.

Critical Path Method

Many projects fail to meet schedule expectations. **Critical path method (CPM)**—also called **critical path analysis**—is a project network analysis technique used to predict total project duration. It is an important tool that will help you combat project schedule overruns. A **critical path** for a project is the series of activities that determine the earliest time by which the project can be completed. It is the longest path through the network diagram and has the least amount of slack or float. **Slack** or **float** is the amount of time an activity may be delayed without delaying a succeeding activity or the project finish date.

To find the critical path for a project, you must first develop a good network diagram, which, in turn, requires a good activity list based on the WBS. Once you create a project network diagram, you must also estimate the duration of each activity to determine the critical path. Calculating the critical path involves adding the durations for all activities on each path through the project network diagram. The longest path is the critical path.

The AOA project network diagram for Project X is shown again in Figure 5-8. Note that you can use either the AOA or PDM network diagramming method to determine the critical path on projects. Figure 5-8 shows all of the paths—a total of four—

through the project network diagram. Note that each path starts at the first node (1) and ends at the last node (8) on the AOA diagram. This figure also shows the length or total duration of each path through the project network diagram. These lengths are computed by adding the durations of each activity on the path. Since path B-E-H-J at 16 days has the longest duration, it is the critical path for the project.

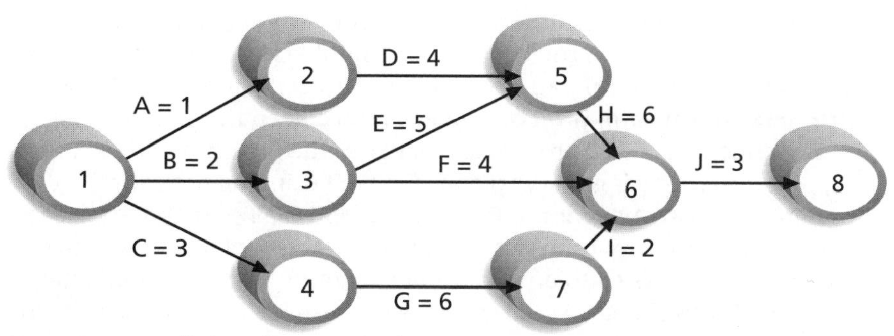

Note: Assume all durations are in days.

Path 1:	A-D-H-J	Length = 1+4+6+3 = 14 days
Path 2:	B-E-H-J	Length = 2+5+6+3 = 16 days
Path 3:	B-F-J	Length = 2+4+3 = 9 days
Path 4:	C-G-I-J	Length = 3+6+2+3 = 14 days

Since the critical path is the longest path through the network diagram, Path 2, B-E-H-J, is the critical path for Project X.

Figure 5-8. Determining the Critical Path for Project X

What does the critical path really mean? The critical path shows the shortest time in which a project can be completed. Even though the critical path is the *longest* path, it represents the *shortest* time it takes to complete a project. If one or more of the activities on the critical path takes longer than planned, the whole project schedule will slip *unless* the project manager takes corrective action.

People are often confused about what the critical path is for a project or what it really means. Some people think the critical path includes the most critical activities. However, the critical path is concerned only with the time dimension of a project. The fact that its name includes the word *critical* does not mean that it includes all critical activities. For example, Frank Addeman, Executive Project Director at Walt Disney Imagineering, explained in a keynote address at the May 2000 PMI-ISSIG Professional Development Seminar that growing grass was on the critical path for building Disney's Animal Kingdom theme park! This 500-acre park required special grass for its animal inhabitants, and some of the grass took years to grow. Another misconception is that the critical path is the shortest path through the project network diagram. In some areas, for example, transportation modeling, similar diagrams are drawn in which identifying the shortest path is the goal. For a project, however, each activity must be done in order to complete the project. It is not a matter of choosing the shortest path.

Other aspects of critical path analysis may cause confusion. Can there be more than one critical path on a project? Does the critical path ever change? In the Project X example, suppose that Activity A has a duration estimate of 3 days instead of 1 day. This new duration estimate would make the length of Path 1 equal to 16 days. Now the project has two longest paths of equal duration, so there are two critical paths. Therefore, there *can* be more than one critical path on a project. Project managers should closely monitor performance of activities on the critical path to avoid late project completion. If there is more than one critical path, project managers must keep their eyes on all of them.

The critical path on a project can change as the project progresses. For example, suppose everything is going as planned at the beginning of the project. In this example, suppose Activities A, B, C, D, E, F, and G all start and finish as planned. Then suppose Activity I runs into problems. If Activity I takes more than 4 days, it will cause path C-G-I-J to be longer than the other paths, assuming they progress as planned. This change would cause path C-G-I-J to become the new critical path. Therefore, the critical path can change on a project.

Using Critical Path Analysis to Make Schedule Trade-Offs

It is important to know what the critical path is throughout the life of a project so you *can* make trade-offs. If the project manager knows that one of the tasks on the critical path is behind schedule, he or she needs to decide what to do about it. Should the schedule be renegotiated with stakeholders? Should more resources be allocated to other items on the critical path to make up for that time? Or is it okay if the project finishes behind schedule? By keeping track of the critical path, the project manager and his or her team take a proactive role in managing the project schedule.

A technique that can help project managers make schedule trade-offs is determining the free slack and total slack for each project activity. **Free slack** or **free float** is the amount of time an activity can be delayed without delaying the early start of any immediately following activities. The **early start date** for an activity is the earliest possible time an activity can start based on the project network logic. **Total slack** or **total float** is the amount of time an activity may be delayed from its early start without delaying the planned project finish date.

You calculate free slack and total slack by doing a forward and backward pass through a network diagram. A **forward pass** determines the early start and early finish dates for each activity. The **early finish date** for an activity is the earliest possible time an activity can finish based on the project network logic. The project start date is equal to the early start date for the first network activity. Early start plus the duration of the first activity is equal to the early finish date of the first activity. It is also equal to the early start date of each subsequent activity. When an activity has multiple predecessors, its early start date is the latest of the early finish dates of those predecessors. For example, Task H in Figure 5-8 is immediately preceded by Tasks D and E. The early start date for Task H, therefore, is the early finish date of Task E, since it occurs later than the early finish date of Task D. A **backward pass** through the network diagram determines the late start and late finish dates for each activity in a similar fashion. The **late start date** for an activity is the latest possible time an activity might begin without delaying the project finish date. The **late finish date** for an activity is the latest possible time an activity can be completed without delaying the project finish date.

You can determine the early and late start dates of each activity by hand, but using project management software is much quicker. Table 5-1 shows the free and total slack for all activities on the project network diagram for Project X. This table was created by simply selecting the Schedule Table view in Project 2000. Knowing the amount of float or slack allows project managers to know whether the schedule is flexible and how flexible it might be. For example, at 7 days, Task F in this example has the most free and total slack. The most slack on any other activity is only 2 days. Understanding how to create and use slack information provides a basis for negotiating (or not negotiating) project schedules.

Table 5-1: Free and Total Float or Slack for Project X

TASK	START	FINISH	LATE START	LATE FINISH	FREE SLACK	TOTAL SLACK
A	6/2/02	6/2/02	6/4/02	6/4/02	0d	2d
B	6/2/02	6/3/02	6/2/02	6/3/02	0d	0d
C	6/2/02	6/4/02	6/4/02	6/6/02	0d	2d
D	6/3/02	6/6/02	6/5/02	6/10/02	2d	2d
E	6/4/02	6/10/02	6/4/02	6/10/02	0d	0d
F	6/4/02	6/9/02	6/13/02	6/18/02	7d	7d
G	6/5/02	6/12/02	6/9/02	6/16/02	0d	2d
H	6/11/02	6/18/02	6/11/02	6/18/02	0d	0d
I	6/13/02	6/16/02	6/17/02	6/18/02	2d	2d
J	6/19/02	6/23/02	6/19/02	6/23/02	0d	0d

Techniques for Shortening a Project Schedule

It is common for stakeholders to want to shorten a project schedule estimate. By knowing the critical path, the project manager and his or her team can use several duration compression techniques to shorten the project schedule. One technique is to reduce the duration of activities on the critical path. You can shorten the duration of critical path activities by allocating more resources to those activities or by changing their scope.

Recall that Sue Johnson in the opening case was having schedule problems on the online registration project because several users missed important project review meetings and one of the senior programmers quit. If Sue and her team created a realistic project schedule, produced accurate duration estimates, and established dependencies between tasks, they could analyze where they were in terms of meeting the May 1 deadline. If some activities on the critical path had already slipped and they did not build in extra time at the end of the project, then they would have to take corrective actions to finish the project on time. Sue could request that her company or the university provide more people to work on the project to try to make up time. She could also request that the scope of activities be reduced to complete the project on time. Sue could also use project time management techniques, such as crashing or fast tracking, to shorten the project schedule.

Crashing is a technique for making cost and schedule trade-offs to obtain the greatest amount of schedule compression for the least incremental cost. For example, suppose one of the items on the critical path for the online registration project was entering course data for the fall term into the new system. If this task is yet to be done and was originally estimated to take two weeks based on the university providing one part-time data entry clerk, Sue could suggest that the university have the clerk work full-time to finish the task in one week instead of two. This change would not cost Sue's company more money, and it could shorten the project end date by one week. If the university could not meet this request, Sue could consider hiring a temporary data entry person for one week or so to help get the task done faster. By focusing on tasks on the critical path that could be done more quickly for no extra cost or a small cost, the project schedule can be shortened.

The main advantage of crashing is shortening the time it takes to finish a project. The main disadvantage of crashing is that it often increases total project costs. *See* Chapter 6, *Project Cost Management,* for a further discussion of project costs.

Another technique for shortening a project schedule is fast tracking. **Fast tracking** involves doing activities in parallel that you would normally do in sequence. For example, Sue Johnson's project team may have planned not to start any of the coding for the online registration system until *all* of the analysis was done. Instead, they could consider starting some coding activity before the analysis is completed.

The main advantage of fast tracking, like crashing, is that it can shorten the time it takes to finish a project. The main disadvantage of fast tracking is that it can end up lengthening the project schedule since starting some tasks too soon often increases project risk and results in rework.

What Went Wrong?

The National Insurance Recording System (Nirs2) is an information technology fiasco that affected the most vulnerable members of British society, but the prestigious consulting firm that built Nirs2 claims it is working. The Contributions Agency is a 5,000-member group that helps to collect 45 billion pounds a year—one-third of the nation's annual revenue—in contributions. A central part of the agency's work is the National Insurance Recording System, which validates, processes, and stores files on who has paid what contributions. The system affects almost every adult in the country, as it determines their entitlement to pension, welfare, and other benefits.

In June 1998, the Contributions Agency claimed that the new Nirs2 system had not been delivered as promised and that contingency planning and compensation arrangements were inadequate. There were more than 1,900 system problems, three-quarters of which were still unresolved. These problems resulted in the Agency's inability to process 17 million contributions made to individual national insurance accounts in 1997-1998. There has been a clear failure to deliver services to the citizens, since the Agency cannot correctly calculate citizens' pensions and benefits. For example, over 1.2 million claims for Jobseekers Allowance were cleared without the benefit of up-to-date information; 396,000 retirement pension claims were made based on old information; and 160,000 pensioners were underpaid. The backlog of items waiting to be input to the system could take years to clear, all at the taxpayers' expense.

The primary reasons for this information technology fiasco were scheduling issues. The consulting firm suggested that the original approach of delivering the entire system by February 1997 be turned into a lower risk, phased approach, with three releases delivered between February 1997 and April 1999. Although a phased approach is normally good practice and the Contributions Agency agreed to it, the system was still ridden with problems. The hand-off of the system to end users for implementation was disastrous, and missing the original target date for the new system has resulted in serious problems for both the Contributions Agency, the consulting firm, and innocent citizens.[3]

Importance of Updating Critical Path Data

In addition to finding the critical path at the beginning of a project, it is important to update the schedule with actual data. After the project team completes activities, you should document the actual durations of those activities. You should also document revised estimates for activities in progress or yet to be started. These revisions often cause a project's critical path to change, resulting in a new estimated completion date for the project. Again, proactive project managers and their teams stay on top of changes so they can make more informed decisions and keep stakeholders informed of, and involved in, major project decisions.

[3] Collins, Tony, "Nirs2: An IT Success or Failure?" *Computer Weekly* (February 4, 1999).

Critical Chain Scheduling

Another technique that addresses the challenge of meeting or beating project finish dates is an application of the theory of constraints called critical chain scheduling. The Theory of Constraints (TOC) is a management philosophy developed by Eliyahu M. Goldratt and introduced in his book *The Goal*. The **Theory of Constraints** is based on the fact that, like a chain with its weakest link, any complex system at any point in time often has only one aspect or constraint that limits its ability to achieve more of its goal. For the system to attain any significant improvements, that constraint must be identified, and the whole system must be managed with it in mind. **Critical chain scheduling** is a method of scheduling that takes limited resources into account when creating a project schedule and includes buffers to protect the project completion date.

You can find the critical path for a project without considering resource allocation. For example, task duration estimates and dependencies can be made without considering the availability of resources. In contrast, an important concept in critical chain scheduling is the availability of resources. If a particular resource is needed full-time to complete two tasks that were originally planned to occur simultaneously, critical chain scheduling acknowledges that you must either delay one of those tasks until the resource is available or find another resource. In this case, accounting for limited resources often extends the project finish date, which is not most people's intent. Other important concepts related to critical chain scheduling include multitasking and time buffers.

Multitasking occurs when a resource works on more than one task at a time. This situation occurs frequently on projects. People are assigned to multiple tasks within the same project or different tasks on multiple projects. For example, suppose someone is working on three different tasks, Task 1, Task 2, and Task 3, for three different projects, and each task takes ten days to complete. If the person did not multitask, and instead completed each task sequentially, starting with Task 1, then Task 1 would be completed after day ten, Task 2 would be completed after day twenty, and Task 3 would be completed after day thirty, as shown in Figure 5-9a. However, because many people in this situation try to please all three people who need their tasks completed, they often work on the first task for some time, then the second, then the third, then go back to the first task, and so on, as shown in Figure 5-9b. In this example, the tasks were all half-done one at a time, then completed one at a time. Task 1 is now completed at the end of day twenty instead of day ten, Task 2 is completed at the end of day twenty-five instead of day twenty, and Task 3 is still completed on day thirty. This example illustrates how multitasking can delay task completions. Multitasking also often involves wasted setup time, which increases total durations.

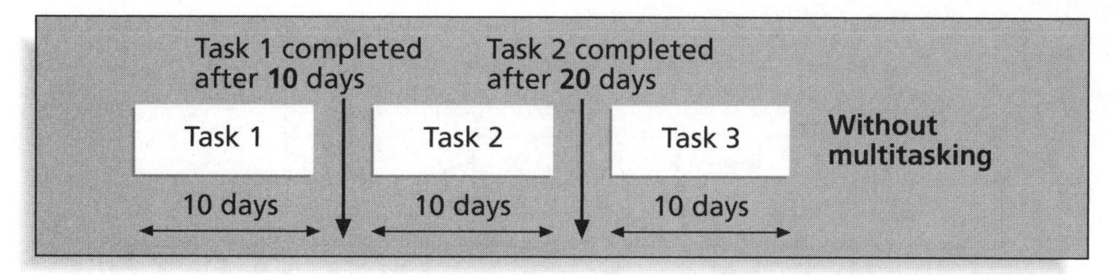

Figure 5-9a. Three Tasks Without Multitasking

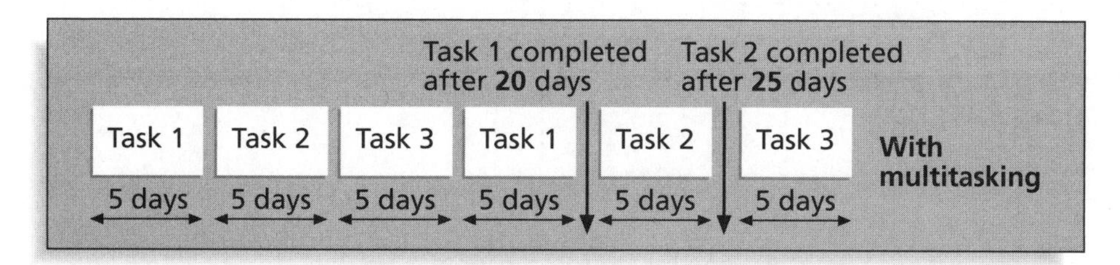

Figure 5-9b. Three Tasks with Multitasking

Critical chain scheduling assumes that resources do not multitask. Someone cannot be assigned to two tasks simultaneously on the same project, when critical chain scheduling is in effect. Likewise, critical chain theory suggests that projects be prioritized so people working on more than one project at a time know which tasks take priority. Preventing multitasking avoids resource conflicts and wasted setup time caused by shifting between multiple tasks over time.

An essential concept to improving project finish dates with critical chain scheduling is to change the way people make task estimates. Many people add a safety or **buffer**, which is additional time to complete a task, to an estimate to account for various factors. These factors include the negative effects of multitasking, distractions and interruptions, fear that estimates will be reduced, Murphy's Law, and so on. **Murphy's Law** states that if something can go wrong, it will. Critical chain scheduling removes buffers from individual tasks and instead creates a **project buffer**, which is additional time added before the project's due date. Critical chain scheduling also protects tasks on the critical path from being delayed by using **feeding buffers**, which are additional time added before tasks on the critical path that are preceded by non-critical-path tasks.

Figure 5-10 provides an example of a project network diagram constructed using critical chain scheduling. Note that the critical chain accounts for a limited resource, X, and the schedule includes use of feeding buffers and a project buffer in the network diagram. The task estimates in critical chain scheduling should be shorter than traditional estimates because they do not include their own buffers. Not having task buffers should mean less occurrence of **Parkinson's Law**, which states that work expands to fill the time allowed. The feeding buffers and project buffers protect the date that really needs to be met—the project completion date.

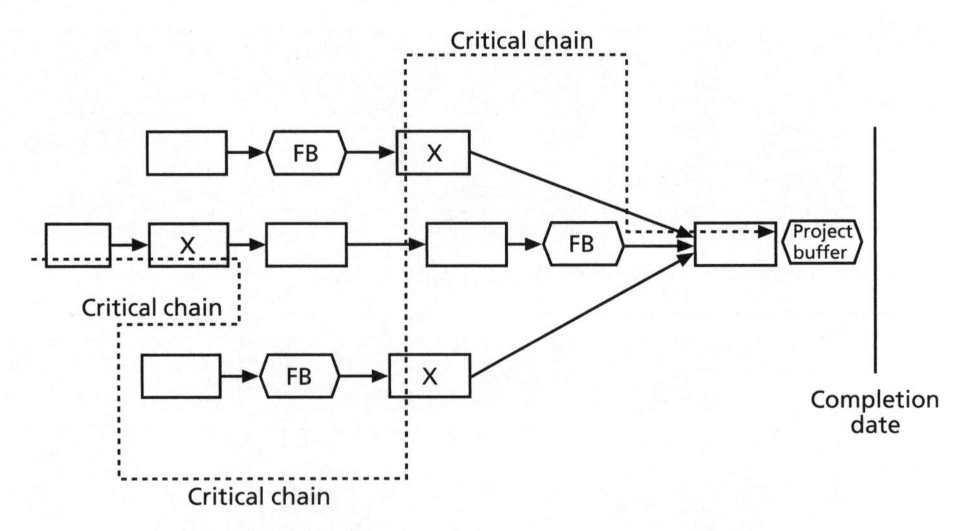

Figure 5-10. Example of Critical Chain Scheduling[4]

As you can see, critical chain scheduling is a fairly complicated yet powerful tool that involves critical path analysis, resource constraints, and changes in how task estimates are made in terms of buffers. Some people have called critical chain scheduling one of the most important new concepts in the field of project management. Others argue, however, that this concept is really the same as critical path analysis with resource leveling, a technique for resolving resource conflicts by delaying tasks (*see* Chapter 8, *Project Human Resource Management*).

Several companies have reported successes with critical chain scheduling:

- Lucent Technology's Outside Plant Fiber Optic Cable Business Unit used critical chain scheduling to reduce its product introduction interval by 50 percent, improve on-time delivery, and increase the organization's capacity to develop products.
- Synergis Technologies Group successfully implemented critical chain scheduling to manage more than 200 concurrent projects in nine locations, making on-time delivery their top priority.
- The Antarctic Support Associates project team switched to critical chain scheduling to enable the Laurence M. Gould research vessel to pass sea-trial tests and be ready to embark on its voyage to Antarctica on schedule in January 1998, rather than four months late as had been anticipated.[5]

[4] Goldratt, Eliyahu, *Critical Chain*, Great Barrington, MA, The North River Press, 1997, 218.

[5] Avraham Y. Goldratt Institute Web site, **www.goldratt.com** (January 2001).

Program Evaluation and Review Technique (PERT)

Another project time management technique is the **Program Evaluation and Review Technique (PERT)**—a network analysis technique used to estimate project duration when there is a high degree of uncertainty about the individual activity duration estimates. PERT applies the critical path method to a weighted average duration estimate.

PERT uses **probabilistic time estimates**—duration estimates based on using optimistic, most likely, and pessimistic estimates of activity durations—instead of one specific or discrete duration estimate. Like the critical path method, PERT is based on a project network diagram, normally the PDM method. To use the PERT method, you calculate a weighted average for the duration estimate of each project activity using the following formula:

$$\textbf{PERT weighted average} = \frac{\text{optimistic time} + 4\text{X most likely time} + \text{pessimistic time}}{6}$$

Returning to the opening case, suppose Sue Johnson's project team used PERT to determine the schedule for the online registration system project. They would have to collect numbers for the optimistic, most likely, and pessimistic duration estimates for each project activity. Suppose one of the activities was to design an input screen for the system. Someone might estimate that it would take about two weeks or 10 workdays to do this activity. Without using PERT, the duration estimate for that activity would be 10 workdays. Using PERT, the project team would also need to estimate the pessimistic and optimistic times for completing this activity. Suppose an optimistic estimate is that the input screen can be designed in 8 workdays, and a pessimistic time estimate is 24 workdays. Applying the PERT formula, you get the following:

$$\textbf{PERT weighted average} = \frac{8\text{ workdays} + 4\text{ X }10\text{ workdays} + 24\text{ workdays}}{6} = 12\text{ workdays}$$

Instead of using the most likely duration estimate of 10 workdays, the project team would use 12 workdays when doing critical path analysis.

The main advantage of PERT is that it attempts to address the risk associated with duration estimates. PERT has three disadvantages: it involves more work since it requires several duration estimates, there are better probabilistic methods for assessing risk (*See* Chapter 10, *Project Risk Management*), and it is rarely used in practice. In fact, many people confuse PERT with project network diagrams since the latter are often referred to as PERT charts.

CONTROLLING CHANGES TO THE PROJECT SCHEDULE

There are many issues involved in controlling changes to project schedules. It is important first to ensure that the project schedule is realistic. Many projects, especially in information technology, have very unrealistic schedule expectations. It is also important to use discipline and leadership to emphasize the importance of following and meeting project schedules. Although the various tools and techniques assist in developing and managing project schedules, project managers must manage several people-related issues to keep projects on track. "Most projects fail because of people issues, not from failure to draw a good PERT chart."[6] Project managers can perform a number of reality checks that will help them manage changes to project schedules. There are also several skills that help project managers control schedule changes related to people issues.

Reality Checks on Scheduling

One of the first reality checks a project manager should make is to review the draft schedule usually included in the project charter. Although this draft schedule might include only a project start and end date, the project charter sets some initial schedule expectations for the project. Next, the project manager and his or her team should prepare a more detailed schedule and get stakeholders' approval. To establish the schedule, it is critical to get involvement and commitment from all project team members, upper management, the customer, and other key stakeholders.

Several experts have written about the lack of realistic schedule estimates for information technology projects. Ed Yourdon, a well-known and respected software development expert, describes "death march" projects as projects doomed for failure from the start, due to unrealistic expectations, especially for meeting time constraints.[7] Many schedules for embedded software systems are "propped up by prayer and hopeful wishes."[8] Yielding to pressure from senior management or other groups such as marketing in determining project schedules is nothing short of lying.[9] Experts in chaos theory suggest that project managers must account for the complex behavior of the organization by arranging for additional resources to be committed to the project. Using an analogy of how to avoid traffic jams on highways, they suggest that project managers schedule resources on the project so that no single resource is utilized more than 75 percent.[10] It is very important, therefore, to set realistic project schedules and allow for contingencies throughout the life of a project.

[6] Bolton, Bart, "IS Leadership," *ComputerWorld* (May 19, 1997).

[7] Yourdon, Ed, "Surviving a Death March Project," *Software Development* (July 1997).

[8] Ganssle, Jack G, "Lies, Damn Lies, and Schedule," *Embedded Systems Programming* (December 1997).

[9] Ibid.

[10] Monson, Robert, *The Role of Complexity and Chaos in Project Management*, working paper (February 1999).

Another type of reality check comes from progress meetings with stakeholders. The project manager is responsible for keeping the project on track, and key stakeholders like to stay informed, often through high-level periodic reviews. Managers like to see progress made on projects each month or so. Project managers often illustrate progress with a tracking Gantt chart showing key deliverables and activities. The project manager needs to understand the schedule and why activities are or are not on track and take a proactive approach to meeting stakeholder expectations. Upper management hates surprises, so it is very important for the project manager to be clear and honest in communicating project status. By no means should project managers create the illusion that the project is going fine when, in fact, it is having serious problems. When serious conflicts arise that could affect the project schedule, the project manager must alert senior management and work with them to resolve the conflicts.

Working with People Issues

As in any endeavor involving a lot of people, it is often not the technical aspects that are most difficult. Creating a good PERT chart and detailed project schedule are significant skills for project managers, but good project managers realize that their main job is to lead the people involved in the project. Depending on the size of the project, many project managers have one or more staff members responsible for coordinating input from many other people to create and update the project schedule. Delegating the details of a project schedule allows the project manager to focus on the big picture and lead the entire project to keep it on schedule. Several leadership skills that help project managers control schedule changes include:

- Empowerment
- Incentives
- Discipline
- Negotiation

It is important for the project manager to empower project team members to take responsibility for their activities. Having team members help to create a detailed schedule and provide timely status information empowers them to take responsibility for their actions. As a result, they should feel more committed to the project.

The project manager can also use financial or other incentives to encourage people to meet schedule expectations. It is sometimes effective to use coercive power or negative incentives to stop people from missing deadlines. For example, one project manager started charging her team members twenty-five cents every time they were late in submitting their weekly time sheets for a project. She was amazed at how few people were late after implementing that simple policy.

Project managers must also use discipline to control project schedules. Several information technology project managers have discovered that setting firm dates for key project milestones helps minimize schedule changes. It is very easy for scope creep to raise its ugly head on information technology projects. Insisting that important schedule dates be met and that proper planning and analysis be completed up front helps everyone focus on doing what is most important for the project. This discipline results in meeting project schedules.

What Went Right?

Chris Higgins used the discipline he learned in the Army to transform project management into a cultural force at Bank of America. Higgins learned that taking time on the front end of a project can save significant time and money on the back end. As a quartermaster in the Army, when Higgins' people had to pack tents, he devised a contest to find the best way to fold a tent and determine the precise spots to place the pegs and equipment for the quickest possible assembly. Higgins used the same approach when he led an interstate banking initiative to integrate incompatible check processing, checking account, and savings account platforms in various states.

Law mandated that the banks solve the problem in one year or less. Higgins' project team was pushing to get to the coding phase of the project quickly, but Higgins held them back. He made the team members analyze, plan, and document requirements for the system in such detail that it took six months just to complete that phase. But the discipline up front enabled the software developers on the team to do all of the coding in only three months, and the project was completed on time.[11]

Project managers and their team members also need to practice good negotiation skills. Customers and management often pressure people to shorten project schedules. Some people blame information technology schedule overruns on poor estimating, but others state that the real problem is that software developers and other information technology professionals are terrible at defending their estimates. Unlike the general population, most information technology professionals are introverts. They have a difficult time holding their own when extroverts in marketing or sales or upper management tell them to shorten their schedule estimates. Many information technology professionals are also younger than the stakeholders pushing for shorter schedules. It is important, therefore, for the project manager and team members to defend their estimates and learn to negotiate with demanding stakeholders.

[11] Melymuke, Kathleen, "Spit and Polish," *ComputerWorld* (February 16, 1998).

USING SOFTWARE TO ASSIST IN TIME MANAGEMENT

Several types of software are available to assist in project time management. Software for facilitating communications helps project managers exchange schedule-related information with project stakeholders. Decision support models can help project managers analyze various trade-offs that can be made related to schedule issues. However, project management software, such as Microsoft Project 2000, was designed specifically for performing project management tasks. You can use Project 2000 to draw network diagrams, determine the critical path for a project, create Gantt charts, and report, view, and filter specific project time management information.

Project management software can facilitate the creation of project network diagrams and the use of the critical path method. Many projects involve hundreds of tasks with complicated dependencies. After you enter the necessary information, project management software automatically generates a network diagram and calculates the critical path(s) for the project. It also highlights the critical path in red on the network diagram. Project management software also calculates the free and total float or slack for all activities. For example, Table 5-1 was created using Project 2000 by selecting a particular view from the menu bar. Using project management software eliminates the need to perform cumbersome calculations by hand and allows for "what if" analysis as activity duration estimates or dependencies change. Recall that knowing which activities have the most slack gives the project manager an opportunity to reallocate resources or make other changes to compress the schedule or help keep it on track.

Project 2000 easily creates Gantt charts and tracking Gantt charts, which make tracking actual schedule performance versus the planned or baseline schedule much easier to do. It is important, however, to enter actual schedule information in a timely manner in order to benefit from using the Tracking Gantt chart feature. Some organizations use e-mail or other communications software to send up-to-date task and schedule information to the person responsible for updating the schedule. He or she can then quickly authorize these updates to be entered directly into the project management software. This process provides an accurate and up-to-date project schedule in Gantt chart form.

Project 2000 also includes many built-in reports, views, and filters to assist in project time management. Table 5-2 lists some of these features. For example, a project manager can quickly run a report to list all tasks that are to start soon. He or she could then send out a reminder to the people responsible for these tasks. If the project manager were presenting project schedule information to senior management, he or she could create a Gantt chart showing only summary tasks or milestones. You can also create custom reports, views, tables, and filters. *See* Appendix A for further information on these time management features of Project 2000.

Table 5-2: Project 2000 Features Related to Project Time Management

REPORTS	VIEWS AND TABLE VIEWS	FILTERS
• Overview reports: critical tasks and milestones	• Gantt chart, network diagram, Tracking Gantt, schedule, tracking, variance, constraint dates, and delay	• All tasks, completed critical tasks, incomplete tasks, and milestone tasks
• Current activities reports: unstarted tasks, tasks starting soon, tasks in progress, completed tasks, should have started tasks, and slipping tasks		
• Assignment reports: who does what when		

WORDS OF CAUTION ON USING PROJECT MANAGEMENT SOFTWARE

Many people misuse project management software because they do not understand the concepts behind creating a network diagram, determining the critical path, or setting a schedule baseline. They might also rely too heavily on sample files or templates in developing their own project schedules. Understanding the underlying concepts (even being able to work with the tools manually) is critical to successful use of project management software, as is understanding the specific needs of your own project.

Many senior managers, including software professionals, have made blatant errors using various versions of Microsoft Project and similar tools. For example, one senior manager did not know about establishing dependencies among project activities and entered every single start and end date for hundreds of activities. When asked what would happen if the project started a week or two late, she responded that she would have to reenter all of the dates.

This manager did not understand the importance of establishing relationships among the tasks. Establishing relationships among tasks allows the software to update formulas automatically when the inputs change. If the project start date slips by one week, all the other dates will be updated automatically, as long as they are not hard-coded into the software. (Hard-coding involves entering all activity dates manually instead of letting the software calculate them based on durations and relationships.) If one activity cannot start before another ends, and the first activity's actual start date is two days late, the start date for the succeeding activity will automatically be moved back two days. To achieve this type of functionality, tasks that have relationships must be linked in project management software.

Another senior manager on a large information technology project did not know about setting a baseline in Microsoft Project. He spent almost one day every week copying and pasting information from Microsoft Project into a spreadsheet and using complicated "IF" statements to figure out what activities were behind schedule. He had never received any training on Microsoft Project and did not know about a lot of its capabilities. To use any software effectively, users must have adequate training in the software and an understanding of its underlying concepts.

Many project management software programs also come with templates or sample files. It is very easy to use these files without considering unique project needs. For example, if you're the project manager for a software development project, you can use the Project 2000 Software Development template file, you can use files containing information from similar projects done in the past, or you can purchase sample files from other companies. All of these files include suggested tasks, durations, and relationships. There are benefits to using templates or sample files, such as less setup time and a reality check if you've never managed this type of project before. However, there are also drawbacks to this approach. There are many assumptions in these files that may not apply to your project, such as a design phase that takes three months to complete or the performance of certain types of testing. Project managers and their teams may overrely on templates or sample files and ignore unique concerns for their particular projects.

CASE WRAP-UP

It was now March 15, just a month and a half before the new online registration system was supposed to go live. The project was in total chaos. Sue Johnson thought she could handle all of the conflicts that kept appearing on the project, and she was too proud to admit to her senior management or the university president that things were not going well. She spent a lot of time preparing a detailed schedule for the project, and she thought she was using their project management software well enough to keep up with project status. However, the five main programmers on the project all figured out a way to generate automatic updates for their tasks every week, saying that everything was completed as planned. They paid very little attention to the actual plan

and hated filling out status information. Sue did not verify most of their work to check that it was actually completed. In addition, the head of the Registrar's Office was uninterested in the project and delegated sign-off responsibility to one of his clerks who really did not understand the entire registration process. When Sue and her team started testing the new system, she learned they were using last year's course data. Using last year's course data caused additional problems because the university was moving from quarters to semesters in the fall. How could they have missed that requirement? Sue hung her head in shame as she walked into a meeting with her manager to ask for help. She learned the hard way how difficult it was to keep a project on track. She wished she had spent more time talking face-to-face with key project stakeholders, especially her programmers and the Registrar's Office representatives, to verify that key deliverables met the customer's needs.

CHAPTER SUMMARY

Project time management is often cited as the main source of conflict on projects. Most information technology projects exceed time estimates.

The main processes involved in time management include activity definition, activity sequencing, activity duration estimating, schedule development, and schedule control.

Activity definition involves identifying the specific activities that must be done to produce the project deliverables. It usually results in a more detailed WBS and supporting detail.

Activity sequencing determines the relationships or dependencies between activities. Three reasons for creating relationships are that they are mandatory based on the nature of the work, discretionary based on the project team's experience, or external based on non-project activities. Activity sequencing must be done in order to use critical path analysis.

Project network diagrams are the preferred technique for showing activity sequencing. The two methods used for creating these diagrams are the arrow diagramming method and the precedence diagramming method. There are four types of relationships between tasks: finish-to-start, finish-to-finish, start-to-start, and start-to-finish.

Activity duration estimating creates estimates for the amount of time it will take to complete each activity. These time estimates include the actual amount of time worked plus elapsed time.

Schedule development uses results from all of the other time management processes to determine the start and end date for the project. Gantt charts are often used to display the project schedule. Tracking Gantt charts show planned and actual schedule information.

The critical path method is used to predict total project duration. The critical path for a project is the series of activities that determines the earliest completion date for the project. It is the longest path through a project network diagram. If any activity on the critical path slips, the whole project will slip unless the project manager takes corrective action.

Crashing and fast tracking are two techniques for shortening project schedules. Project managers and their team members must be careful about accepting unreasonable schedules, especially for information technology projects.

Critical chain scheduling is an application of the theory of constraints that uses critical path analysis, resource constraints, and buffers to help meet project completion dates.

Program Evaluation and Review Technique (PERT) is a network analysis technique used to estimate project duration when there is a high degree of uncertainty about the individual activity duration estimates. It uses optimistic, most likely, and pessimistic estimates of activity durations. PERT is seldom used today.

Even though scheduling techniques are very important, most projects fail because of people issues, not from failure to draw a good network diagram. Project managers must involve all stakeholders in the schedule development process. It is critical to set realistic project schedules and allow for contingencies throughout the life of a project. Several leadership skills that help project managers control schedule changes include empowerment, discipline, using incentives, and negotiation.

Software can assist in project scheduling if used properly. With project management software, you can avoid the need to perform cumbersome calculations by hand and perform "what if" analysis as activity duration estimates or dependencies change. Many people misuse project management software because they do not understand the concepts behind creating a network diagram, determining the critical path, or setting a schedule baseline. People must also avoid overrelying on sample files or templates when creating their unique project schedules.

Discussion Questions

1. Why do you think schedule issues often cause the most conflicts on projects?
2. Discuss the main processes involved in project time management.
3. Discuss diagrams you have seen that are similar to project network diagrams. How are they similar, and how are they different?
4. Explain as clearly as possible what a critical path is and why a project manager should be concerned about it.

5. Explain the symbols on a Gantt chart and how to use this tool. Find examples of Gantt charts in other texts or from your work experience.
6. Explain the difference between critical path analysis and critical chain scheduling.
7. What are some of the important people issues involved in project time management?
8. What project manager skills described earlier do you think are most important for time management?

EXERCISES

1. Using Figure 5-2, enter the activities, their durations, and their relationships in Project 2000. Use a project start date of 6/02/03. View the network diagram. Does it look like Figure 5-2? It should be very similar, but appear in PDM form, as shown in Figure 5-4. Print the network diagram on one page. Return to Gantt Chart view. Select View, Table, Schedule from the menu bar to re-create Table 5-1. Write a few paragraphs explaining what the network diagram and schedule table show concerning Project X's schedule.
2. Consider Table 5-3, Network Diagram Data for a Small Project. All times are in days; the network proceeds from node 1 to node 9.

Table 5-3. Network Diagram Data for a Small Project

ACTIVITY	INITIAL NODE	FINAL NODE	ESTIMATED TIME
A	1	2	2
B	2	3	2
C	2	4	3
D	2	5	4
E	3	6	2
F	4	6	3
G	5	7	6
H	6	8	2
I	6	7	5
J	7	8	1
K	8	9	2

a. Draw an AOA network diagram representing the project. Put the node numbers in circles and draw arrows from node to node, labeling each arrow with the activity letter and estimated time.

b. Identify all of the paths on the network diagram and note how long they are.

c. What is the critical path for this project and how long is it?

d. What is the shortest possible time it will take to complete this project?

3. Enter the information from Exercise 2 in Project 2000. View the network diagram and task schedule table to see the critical path and float or slack for each activity. Print out the Gantt chart and network diagram views and the task schedule table. Write a short paper that interprets this information for someone unfamiliar with project time management.

4. a. Using the Information Technology Upgrade Project WBS you created in Chapter 4, Exercise 2, create a rough Gantt chart for the project by hand. Make sure the project does not take more than nine months to complete. Document any assumptions you made in creating the Gantt chart.

   b. Next determine the relationships between activities listed on your WBS and the Gantt chart. Build a network diagram for this project by hand. Estimate the duration of each activity, and then enter the data into Project 2000. Make sure you can complete the project in nine months or less. Print the Gantt chart and network diagram and write a one-page paper interpreting the results.

5. Complete the project scope and time management parts of one of the exercises provided in Appendix A. Print the Gantt chart and network diagrams and write a one-page paper interpreting the results.

6. Interview someone who uses some of the techniques discussed in this chapter. How does he or she feel about network diagrams, critical path analysis, Gantt charts, critical chain, using project management software, and managing the people issues involved in project time management? Write a paper describing the responses.

7. Read one of the suggested readings. Write a one-page summary of the article or book, its conclusions, and your opinion.

MINICASE

You have been asked to determine a rough schedule for a nine-month Billing System Conversion project, as part of your job as a consultant to a Fortune 500 firm. The firm's old system was written in COBOL on a mainframe computer, and the maintenance costs are prohibitive. The new system will run on an off-the-shelf application. You have identified several high-level activities that must be done in order to initiate, plan, execute, control, and close the project. Table 5-4 shows your analysis of the project's tasks and schedule so far.

Part 1: Using the information in Table 5-4, draw horizontal bars to illustrate when you think each task would logically start and end. Then use Microsoft Project 2000 to create a Gantt chart and network diagram based on this information.

Table 5-4. Billing System Conversion Project Schedule

TASKS	MAR	APR	MAY	JUN	JUL	AUG	SEP	OCT	NOV
Initiating									
Develop project charter									
Meet with stakeholders									
Planning									
Create detailed WBS and schedule									
Estimate project costs									
Create project team									
Create communication plan									
Organize a comprehensive project plan									
Executing									
Award and manage contract for software conversion									
Install new software on servers									
Install new hardware and software on client's machines									
Test new billing system									
Train users on new system									
Controlling									
Closing									

Part 2: Identify at least two milestones that could be included under each of the process groups in Table 5-4. Then write a detailed description of each of these milestones that meets the SMART criteria.

SUGGESTED READINGS

1. Abel-Hamid, Tarek K. "Investigating the Cost/Schedule Trade-off in Software Development." *IEEE Software* (1990).

   Abel-Hamid analyzes current models for estimating the effect of schedule compression on total project cost. He then presents his own system-dynamics model to illustrate the integrated nature of projects by including subsystems related to software production, planning, control, and human resource management.

2. Bolton, Bart. "IS Leadership." *ComputerWorld* **(http://www2.computerworld.com/home/print9497.nsf/All/SL9705lead)** (May 19, 1997).

   Bolton emphasizes that project success depends more on leadership than on creating PERT charts. This article includes questionnaires to determine personal and organizational leadership scores.

3. Focused Performance, "Critical Chain Basics." (**www.focusedperformance.com/articles/cc01.html**).

This article, one of several available on the Web site of Focused Performance, states that critical chain scheduling shifts the focus from ensuring the achievement of task estimates and milestones to ensuring the only date that matters—the final project due date.

4. Ganssle, Jack G. "Lies, Damn Lies, and Schedule." *Embedded Systems Programming* (December 1997).

Ganssle writes an amusing article about the way in which many schedules for embedded software systems are "propped up by prayer and hopeful wishes." He suggests that yielding to pressure from management or other groups, such as marketing, in determining project schedules is nothing short of lying.

5. Goldratt, Eliyahu. *Critical Chain*. Great Barrington, MA, The North River Press (1997).

*This text is written as a business novel about a college professor and his project management class who together develop the concept of critical chain scheduling. Many other books and articles have been written about critical chain since then. Goldratt now runs the Avraham Y. Goldratt Institute (*www.goldratt.com*), the world's largest consulting firm specializing in the application of theory of constraints (TOC) to achieve significant bottom-line results in organizations.*

6. McConnell, Steve. "How to Defend an Unpopular Schedule." *IEEE Software* (May 1996).

This article emphasizes the fact that it is just as important, if not more so, to defend project schedule estimates as it is to create them. Most software developers must learn to defend their schedule estimates by practicing good negotiation techniques.

7. The, Lee. "IS-Friendly Project Management." *Datamation* (**http://www.datamation.com/PlugIn/issues/1996/april1/04asoft4.html**) (April 1, 1996).

This article mentions the unique nature of many information systems (IS) projects and offers suggestions on features needed in project management software for IS projects. The author also states that understanding project management in general is an important prerequisite to using project management software.

KEY TERMS

- **Activity** — an element of work, normally found on the WBS, that has an expected duration, cost, and resource requirements; also called **task**
- **Activity definition** — identifying the specific activities that the project team members and stakeholders must perform to produce the project deliverables

- **Activity duration estimating** — estimating the number of work periods that are needed to complete individual activities
- **Activity-on-arrow (AOA)** or **arrow diagramming method (ADM)** — a network diagramming technique in which activities are represented by arrows and connected at points called nodes to illustrate the sequence of activities
- **Activity sequencing** — identifying and documenting the relationships between project activities
- **Backward pass** — a project network diagramming technique that determines the late start and late finish dates for each activity in a similar fashion
- **Buffer** — additional time to complete a task, added to an estimate to account for various factors
- **Burst** — when a single node is followed by two or more activities on a project network diagram
- **Crashing** — a technique for making cost and schedule tradeoffs to obtain the greatest amount of schedule compression for the least incremental cost
- **Critical chain** — a method of scheduling that takes limited resources into account when creating a project schedule and includes buffers to protect the project completion date
- **Critical path** — the series of activities in a project network diagram that determines the earliest completion of the project. It is the longest path through the network diagram and has the least amount of slack or float
- **Critical path method (CPM)** or **critical path analysis** — a project network analysis technique used to predict total project duration
- **Dependency** — the sequencing of project activities or tasks; also called a **relationship**
- **Discretionary dependencies** — sequencing of project activities or tasks defined by the project team and used with care since they may limit later scheduling
- **Dummy activities** — activities with no duration and no resources used to show a logical relationship between two activities in the arrow diagramming method of project network diagrams
- **Duration** — the actual amount of time worked on an activity *plus* elapsed time
- **Early finish date** — the earliest possible time an activity can finish based on the project network logic
- **Early start date** — the earliest possible time an activity can start based on the project network logic
- **External dependencies** — sequencing of project activities or tasks that involve relationships between project and non-project activities
- **Fast tracking** — a schedule compression technique in which you do activities in parallel that you would normally do in sequence
- **Feeding buffers** — additional time added before tasks on the critical path that are preceded by non-critical-path tasks

- **Finish-to-finish dependency** — a relationship on a project network diagram where the "from" activity must finish before the "to" activity can finish
- **Finish-to-start dependency** — a relationship on a project network diagram where the "from" activity must finish before the "to" activity can start
- **Forward pass** — a project network diagramming technique that determines the early start and early finish dates for each activity
- **Free slack** — the amount of time an activity can be delayed without delaying the early start of any immediately following activities; also called free float
- **Gantt chart** — a standard format for displaying project schedule information by listing project activities and their corresponding start and finish dates in a calendar format
- **Late finish date** — the latest possible time an activity can be completed without delaying the project finish date.
- **Late start date** — the latest possible time an activity may begin without delaying the project finish date
- **Mandatory dependencies** — sequencing of project activities or tasks that are inherent in the nature of the work being done on the project
- **Merge** — when two or more nodes precede a single node on a project network diagram
- **Milestone** — a significant event on a project with zero duration
- **Multitasking** — when a resource works on more than one task at a time
- **Murphy's Law** — if something can go wrong, it will
- **Node** — the starting and ending point of an activity on an activity-on-arrow diagram
- **Parkinson's Law** — work expands to fill the time allowed
- **PERT weighted average** = $\frac{\text{optimistic time} + 4\text{X most likely time} + \text{pessimistic time}}{6}$
- **Precedence Diagramming Method (PDM)** — a network diagramming technique in which boxes represent activities
- **Probabilistic time estimates** — duration estimates based on using optimistic, most likely, and pessimistic estimates of activity durations instead of using one specific or discrete estimate
- **Program Evaluation and Review Technique (PERT)** — a project network analysis technique used to estimate project duration when there is a high degree of uncertainty with the individual activity duration estimates
- **Project buffer** — additional time added before the project's due date
- **Project network diagram** — a schematic display of the logical relationships or sequencing of project activities
- **Project time management** — the processes required to ensure timely completion of a project
- **Schedule control** — controlling and managing changes to the project schedule

- **Schedule development** — analyzing activity sequences, activity duration estimates, and resource requirements to create the project schedule
- **Slack** — the amount of time a project activity may be delayed without delaying a succeeding activity or the project finish date; also called **float**
- **Slipped milestone** — a milestone activity that is completed later than planned
- **SMART criteria** — guidelines to help define milestones that are **s**pecific, **m**easurable, **a**ssignable, **r**ealistic, and **t**ime-framed
- **Start-to-finish dependency** — a relationship on a project network diagram where the "from" activity cannot start before the "to" activity can finish
- **Start-to-start dependency** — a relationship in which the "from" activity cannot start until the "to" activity starts
- **Theory of constraints** (TOC) — a management philosophy that states that any complex system at any point in time often has only one aspect or constraint that is limiting its ability to achieve more of its goal
- **Total slack** — the amount of time an activity may be delayed from its early start without delaying the planned project finish date; also called **total float**
- **Tracking Gantt chart** — a Gantt chart that compares planned and actual project schedule information

6

Project Cost Management

Objectives

After reading this chapter you will be able to:

1. *Understand the importance of good project cost management*
2. *Explain basic cost management principles, concepts, and terms*
3. *Describe the resource planning, cost estimating, cost budgeting, and cost control processes*
4. *Explain the different types of cost estimates*
5. *Understand what is involved in preparing a cost estimate for an information technology project*
6. *Perform calculations for earned value management*
7. *Understand the benefits of using earned value management*
8. *Describe how software can assist in project cost management*

Juan Gonzales was a systems analyst and network specialist for a major city's waterworks department. He enjoyed helping his country, Mexico, develop its infrastructure. His next career objective was to become a project manager so he could have even more influence. One of his colleagues invited him to attend an important project review meeting for large government projects, including the surveyor's assistant project, in which Juan was most interested. The surveyor's assistant was a concept for developing a sophisticated information system that included expert systems, object-oriented databases, and wireless communications. The system would provide instant, graphical information to government surveyors to help them do their jobs. For example, after a surveyor touched a map on the screen of a hand-held device, the system would prompt him or her for the type of information needed for that area. This system would help in planning and implementing many projects, from laying fiber-optic cable to laying water lines.

He was very surprised, however, when the majority of the meeting was spent discussing cost-related issues. The government officials were reviewing many existing projects to evaluate their performance to date and the potential impact on their budgets before discussing the funding for any new projects. Juan did not understand a lot of the terms and charts the presenters were showing. What was this earned value term they kept referring to? How were they estimating what it would cost to complete projects or how long it would take? Juan thought he would learn more about the new technologies the surveyor's assistant project would use, but he discovered that the cost estimate and projected benefits were of most interest to the senior government officials at the meeting. It also seemed as if a lot of effort would be spent on detailed financial studies before any technical work could even start. Juan wished he had taken some accounting and finance courses so he could understand the acronyms and concepts people were discussing. Although Juan had a degree in electrical engineering, he had no formal education, and little experience, in finance. If Juan could understand information systems and networks, he was confident that he could understand financial issues on projects, too. He jotted down questions to discuss with his colleagues after the meeting.

THE IMPORTANCE OF PROJECT COST MANAGEMENT

Just as information technology projects have a poor track record in meeting project schedules, they also have a poor track record in project cost management. Infor-mation technology projects are expensive and known for coming in way over budget, when they are completed at all. The 1995 CHAOS report showed that the average cost overrun for information technology projects was 189 percent of their original estimates. This means a project that was estimated to cost $100,000 ended up costing $189,000. In 1995, over 31 percent of information technology projects were canceled before completion, costing United States companies and government agencies over $81 billion.[1] The 2001 CHAOS report showed that companies have made great improvements in reducing cost overruns since 1995—from 189 percent to 45 percent—but that most completed projects still went over their approved budgets. This chapter describes important concepts in project cost management, particularly the use of earned value management (EVM) to assist in cost control.

[1] Johnson, Jim, "CHAOS: The Dollar Drain of IT Project Failures," *Application Development Trends* (January 1995) **www.standishgroup.com/chaos.html**.

What Went Wrong?

According to the *San Francisco Chronicle* front-page story, "Computer Bumbling Costs the State $1 Billion," the state of California had a series of expensive information technology project failures in the late 1990s, costing taxpayers nearly $1 billion. Some of the poorly managed projects included the Department of Motor Vehicles registration and driver's license databases, a statewide child support database, the State Automated Welfare System, and a Department of Corrections system for tracking inmates. Senator John Vasconcellos thought it was ironic that the state that was leading in the creation of computers was also the state most behind in using computer technology to improve its services.[2]

Also consider the Internal Revenue Service's expensive reengineering and information technology project failures. The Internal Revenue Service (IRS) managed a series of project failures in the 1990s that cost taxpayers over $50 billion a year—roughly as much money as the annual net profit of the entire computer industry that year.[3]

What is cost? A popular textbook in cost accounting states that "Accountants usually define cost as a resource sacrificed or foregone to achieve a specific objective."[4] Webster's dictionary defines cost as "something given up in exchange." Costs are often measured in monetary amounts, such as dollars, that must be paid to acquire goods and services. Because projects cost money and take away resources that could be used elsewhere, it is very important for project managers to understand project cost management.

Many information technology professionals, however, often react to cost overrun information with a smirk. They know that many of the original cost estimates for information technology projects were low to begin with or based on very unclear project requirements, so naturally there were cost overruns. Not emphasizing the importance of realistic project cost estimates from the outset is only one part of the problem. In addition, many information technology professionals think preparing cost estimates is beneath them—a job for accountants. On the contrary, preparing good cost estimates is a very demanding, important skill that all information technology project managers need to acquire.

Another perceived reason for cost overruns is that many information technology projects involve new technology or business processes. Any new technology or business process is untested and has inherent risks. Thus, costs grow and failures are to be expected. Right? Wrong. Using good project cost management can change this false perception. In fact, the 1995 Standish Group study mentions that many of the information technology projects in their study were as mundane as a new accounting package or an order entry system. The state of California and the IRS examples also demonstrate that many information technology projects cannot blame untested technology for their cost problems. What is needed is better project cost management.

[2] Lucas, Greg, "Computer Bumbling Costs the State $1 Billion," *San Francisco Chronicle* (2/21/99) 1.

[3] James, Geoffrey, "IT Fiascoes . . . and How to Avoid Them," *Datamation* (November 1997).

[4] Horngren, Charles T., George Foster, and Srikanti M. Datar, *Cost Accounting*, 8th ed. Englewood Cliffs, NJ: Prentice-Hall, 1994.

What is project cost management? **Project cost management** includes the processes required to ensure that the project is completed within an approved budget. Notice two crucial phrases in this definition: "the project" and "approved budget." Project managers must make sure *their* projects are well defined, have accurate time and cost estimates, and have a realistic budget that *they* were involved in approving. It is the project manager's job to satisfy project stakeholders while continuously striving to reduce and control costs. The project cost management processes include:

- **Resource planning**, which involves determining what resources (people, equipment, and materials) and what quantities of each resource should be used to perform project activities. The output of the resource planning process is a list of resource requirements.
- **Cost estimating**, which involves developing an approximation or estimate of the costs of the resources needed to complete a project. The main outputs of the cost estimating process are cost estimates, supporting detail, and a cost management plan.
- **Cost budgeting**, which involves allocating the overall cost estimate to individual work items to establish a baseline for measuring performance. The main output of the cost budgeting process is a cost baseline.
- **Cost control**, which involves controlling changes to the project budget. The main outputs of the cost control process are revised cost estimates, budget updates, corrective action, estimate at completion, and lessons learned.

To understand each of the project cost management processes, you must first understand the basic principles of cost management. Many of these principles are not unique to project management; however, project managers need to understand how these principles relate to their specific projects.

BASIC PRINCIPLES OF COST MANAGEMENT

Many information technology projects are never initiated because information technology professionals do not understand the importance of knowing basic accounting and finance principles. Important concepts such as net present value analysis, return on investment, and payback analysis were discussed in Chapter 4, *Project Scope Management*. Likewise, many projects that are started never finish because of cost management problems. Most members of an executive board understand and are more interested in financial terms than information technology terms. Therefore, information technology project management professionals need to be able to present and discuss project information in financial terms as well as in technical terms. In addition to net present value analysis, return on investment, and payback analysis, project managers must understand several other cost management principles, concepts, and terms. This section describes general topics such as profits, life cycle costing, cash flow analysis, internal rate of return, tangible and intangible costs and benefits, direct costs, sunk costs, learning curve theory, and reserves. Another important topic and one of the key tools and techniques for controlling project costs, earned value management, is described in detail in the section on cost control.

Profits are revenues minus expenses. To increase profits, a company can increase revenues, decrease expenses, or try to do both. Most executives are more concerned with profits than with other issues. When justifying investments in new information systems and technology, it is important to focus on the impact on profits, not just revenues or expenses. Consider an electronic commerce application that you estimate will increase revenues for a $100 million company by ten percent. You cannot measure the potential benefits of the system without knowing the profit margin. **Profit margin** is the ratio between revenues and profits. If revenues of $100 generate $2 in profits, there is a 2 percent profit margin. If the company loses $2 for every $100 in revenue, there is a –2 percent profit margin.

Life cycle costing allows you to take a big picture view of the cost of a project over its entire life. This helps you to develop a more accurate projection of a project's financial benefits. Life cycle costing considers the total cost of ownership, or development plus support costs, for a project. For example, a company might complete a project to develop and implement a new customer service system in one or two years, but the new system could be in place for ten years. Project managers should create estimates of the costs and benefits of the project for its entire life, or ten years in the preceding example. Recall that the net present value analysis for the project would include the entire ten-year period of costs and benefits (*see* Chapter 4). Senior management and project managers need to consider the life cycle costs of projects when they make financial decisions.

Corporations have a history of not spending enough money in the early phases of information technology projects. This history has an impact on total cost of ownership. For example, it is much more cost-effective to spend money on defining user requirements and doing early testing on information technology projects than to wait for problems to appear after implementation. Table 6-1 summarizes the typical costs of correcting software defects during different stages of the system development life cycle. If you identified and corrected a software defect during the user requirements phase of a project, you would need to add $100 to $1,000 to the project's total cost of ownership. In contrast, if you waited until after implementation to correct the same software defect, you would need to add possibly millions of dollars to the project's total cost of ownership.

Table 6-1: Costs of Software Defects[5]

When Defect Is Detected	Typical Costs of Correction
User Requirements	$100 - $1,000
Coding/Unit Testing	$1,000 or more
System Testing	$7,000 - $8,000
Acceptance Testing	$1,000 - $100,000
After Implementation	Up to millions of dollars

Cash flow analysis is a method for determining the estimated annual costs and benefits for a project and the resulting *annual* cash flow. It must be done to determine net present value (*see* Chapter 4, *Project Scope Management*). Most consumers understand the basic concept of cash flow. If they do not have enough money in their wallets or checking accounts, they cannot purchase something. Managers must take cash flow concerns into account when selecting projects in which to invest. If managers select too many projects having high cash flow needs in the same year, the company will not be able to support all of the projects and maintain its profitability. It is important to define clearly the year on which the dollar amounts are based. For example, if you based all costs on 2002 estimates, you would need to account for inflation and other factors when projecting costs and benefits in future-year dollars.

Internal rate of return (IRR) is the discount rate that makes net present value equal to zero. It is also called the **time-adjusted rate of return**. Some companies prefer to estimate the internal rate of return instead of, or in addition to, net present value and set minimum values that must be achieved for projects to be selected or continued. For example, assume a three-year project had projected costs in year one of $100 and projected benefits in years two and three of $100 each year. Assume there were no benefits in year one and no costs in years two and three. Using a 10 percent discount rate, you could compute the net present value to be about $67. (*See* Chapter 4 for details on performing net present value calculations.) The internal rate of return for this project is the discount rate that makes the net present value equal to zero. In this example, the internal rate of return is 62 percent.

Tangible and intangible costs and benefits are categories for determining how definable the estimated costs and benefits are for a project. **Tangible costs** or **benefits** are those costs or benefits that can be easily measured in dollars. For example, suppose that the surveyor's assistant project described in the opening case included a preliminary feasibility study. If a firm completed this study for $100,000, the tangible cost of the study is $100,000. If Juan's government estimated that it would have cost them $150,000 to do the study

[5] Collard, Ross, *Software Testing and Quality Assurance*, working paper (1997).

themselves, the tangible benefits of the study would be $50,000. In contrast, **intangible costs** or **benefits** are costs or benefits that are difficult to measure in monetary terms. Suppose Juan and a few other people, out of personal interest, spent some time using government-owned computers, books, and other resources to research areas related to the study. Although their hours and the government-owned materials were not billed to the project, they could be considered to be intangible costs. Intangible benefits of the study might be the perceived potential benefit of having a system that could help the government lay water lines or fiber-optic cable faster or for less money. Because intangible costs and benefits are difficult to quantify, they are often harder to justify.

Direct costs are costs related to a project that can be traced back in a cost-effective way. You can attribute direct costs directly to a certain project. For example, the salaries of people working full-time on the project and the cost of hardware and software purchased specifically for the project are direct costs. Project managers should focus on direct costs, since they can control them.

Indirect costs are costs related to a project that cannot be traced back in a cost-effective way. For example, the cost of electricity, paper towels, and so on in a large building housing a thousand employees who work on many projects would be indirect costs. Indirect costs are allocated to projects, and project managers have very little control over them.

Sunk cost is money that has been spent in the past. Consider it gone, like a sunken ship that can never be returned. When deciding what projects to invest in or continue, you should not include sunk costs. For example, in the opening case, suppose Juan's office had spent $1 million on a project over the past three years to create a geographic information system, but nothing valuable was ever produced. If his government was evaluating what projects to fund next year and someone suggested that they keep funding the geographic information system project because they had already spent $1 million on it, he or she would be incorrectly making sunk cost a key factor in the project selection decision. Many people fall into the trap of considering how much money has been spent on a failing project and, therefore, hate to stop spending money on it. This trap is similar to gamblers not wanting to stop gambling because they have already lost money. Sunk costs should be forgotten.

Learning curve theory states that when many items are produced repetitively, the unit cost of those items decreases in a regular pattern as more units are produced. For example, suppose the surveyor's assistant project would potentially produce 1,000 hand-held devices that could run the new software and access information via satellite. The cost of the first hand-held device or unit would be much higher than the cost of the 1,000th unit. Learning curve theory should be used to estimate costs on projects involving the production of large quantities of items. Learning curve theory also applies to the amount of time it takes to complete some tasks. For example, the first time a new employee performs a specific task, it will probably take longer than the tenth time that employee performs a very similar task.

Reserves are dollars included in a cost estimate to mitigate cost risk by allowing for future situations that are difficult to predict. **Contingency reserves** allow for future situations that may be partially planned for (sometimes called **known unknowns**) and are included in the project cost baseline. For example, if an organization knows it has a 20 percent rate of turnover for information technology personnel, it should include contingency reserves to pay for recruiting and training costs for project personnel. **Management reserves** allow for future situations that are unpredictable (sometimes called **unknown unknowns**). For example, if a project manager gets sick for two weeks or an important supplier goes out of business, management reserve could be set aside to cover the resulting costs.

RESOURCE PLANNING

To estimate, budget, and control costs, project managers and their teams must determine what physical resources (people, equipment, and materials) and what quantities of those resources are required to complete the project. The nature of the project and the organization will affect resource planning. Expert judgment and the availability of alternatives are the only real tools available to assist in resource planning. It is important that the people who help determine what resources are necessary include people who have experience and expertise in similar projects and with the organization performing the project.

Important questions to answer in resource planning include:

- How difficult will it be to do specific tasks on this project?
- Is there anything unique in this project's scope statement that will affect resources?
- What is the organization's history in doing similar tasks? Have similar tasks been done before? What level of personnel did the work?
- Does the organization have people, equipment, and materials that are capable and available for performing the work?
- Does the organization need to acquire more resources to accomplish the work? Would it make sense to outsource some of the work?
- Are there any organizational policies that might affect the availability of resources?

A project's work breakdown structure, scope statement, historical information, resource information, and policies are all important inputs to answering these questions.

It is important to thoroughly brainstorm and evaluate alternatives related to resources, especially on projects that involve people from multiple disciplines and companies. Since most projects involve many human resources, it is often effective to solicit ideas from different people to help develop alternatives and address resource and cost-related issues early in a project.

What Went Right?

Ford Motor Company had great success with its Taurus/Sable project. Ford excels in team product development and seeks to remove barriers among design, engineering, production, marketing, sales, and purchasing. However, the Taurus team went far beyond removing barriers by creating a car that excelled in design and quality at half the typical development cost. They brought together all the relevant disciplines as a team and took the various steps in designing, producing, marketing, and selling the cars simultaneously as well as sequentially. Ford even involved people outside the company for advice—car dealerships, insurance companies, and suppliers. One supplier even offered the services of its own drafting department to prepare initial designs for Ford's approval. Bringing together all the key players provided extremely valuable ideas at the conceptual stage, the time when changes can be made without much extra cost. "Not only were there substantial savings in cost and design time, but major production contracts were being negotiated and set up some three years ahead of production, with the duration of the contract some five years. This also led to cost economies."[6]

The main output of the resource planning process is a list of resource requirements, including people, equipment, and materials. In addition to providing the basis for cost estimates, budgets, and cost controls, the resource requirements list provides vital information for project human resource management (Chapter 8) and project procurement management (Chapter 11).

COST ESTIMATING

Project managers must take cost estimates seriously if they want to complete projects within budget constraints. After developing a good resource requirements list, project managers and members of their project team must develop several estimates of the costs for these resources. This section describes the various types of cost estimates, tools and techniques for cost estimation, typical problems associated with information technology cost estimates, and a detailed example of a cost estimate for an information technology project.

Types of Cost Estimates

One of the main outputs of project cost management is a cost estimate. Project managers normally prepare several types of cost estimates for most projects. These estimates—rough order of magnitude, budgeting, and definitive—vary primarily on when they are done, how they will be used, and how accurate they are.

[6] Pinto, Jeffrey K. and O. P. Kharbanda, *Successful Project Managers*. New York: Van Nostrand Reinhold, 1995.

- A **rough order of magnitude (ROM) estimate** provides a rough idea of what a project will cost. This estimate is done very early in a project or even before a project is officially started. Project managers and upper management use this estimate to help make project selection decisions. The timeframe for this type of estimate is often three or more years prior to project completion. A ROM estimate's accuracy is typically –25 percent to +75 percent, meaning the project's actual costs could be 25 percent below the ROM estimate or 75 percent above. For information technology project estimates, this accuracy range is often much wider. Many information technology professionals automatically double estimates for software development because of the history of cost overruns on information technology projects.
- A **budgetary estimate** is used to allocate money into an organization's budget. Many organizations develop budgets at least two years into the future. Budgetary estimates are made one to two years prior to project completion. The accuracy of budgetary estimates is typically –10 percent to +25 percent, meaning the actual costs could be 10 percent less or 25 percent more than the budgetary estimate.
- A **definitive estimate** provides an accurate estimate of project costs. Definitive estimates are used for making many purchasing decisions for which accurate estimates are required and for estimating final project costs. For example, if a project involves purchasing 1,000 personal computers from an outside supplier in the next three months, a definitive estimate would be required to aid in evaluating supplier proposals and allocating the funds to pay the chosen supplier. Definitive estimates are made one year or less prior to project completion. A definitive estimate should be the most accurate of the three types of estimates. The accuracy of this type of estimate is normally –5 percent to +10 percent, meaning the actual costs could be 5 percent less or 10 percent more than the definitive estimate. Table 6-2 summarizes the three basic types of cost estimates.

Table 6-2: Types of Cost Estimates

TYPE OF ESTIMATE	WHEN DONE	WHY DONE	HOW ACCURATE
Rough Order of Magnitude (ROM)	Very early in the project life cycle, often 3–5 years before project completion	Provides estimate of cost for selection decisions	–25%, +75%
Budgetary	Early, 1–2 years out	Puts dollars in the budget plans	–10%, +25%
Definitive	Later in the project, less than 1 year out	Provides details for purchases, estimates actual costs	–5%, +10%

Two additional outputs of the cost estimating process are supporting details and a cost management plan. It is very important to include supporting detail with all cost estimates. The supporting details include the ground rules and assumptions used in creating the estimate, a description of the project (scope statement, WBS, and so on) used as a basis for the estimate, and details on the cost estimation tools and techniques used to create the estimate. This supporting detail should make it easier to prepare an updated estimate or similar estimate when one is needed.

A **cost management plan** is a document that describes how cost variances will be managed on the project. For example, if a definitive cost estimate provides the basis for evaluating supplier cost proposals for all or part of a project, the cost management plan describes how to respond to proposals that are higher or lower than the estimates. Some organizations assume that a cost proposal within 10 percent of the estimate is fine and only negotiate items that are over 10 percent higher or 20 percent lower than the estimated costs. The cost management plan is part of the overall project plan described in Chapter 3, *Project Integration Management*.

Cost Estimation Techniques and Tools

There are three basic techniques for cost estimating: analogous, bottom-up estimating, and parametric modeling. There are also many computerized tools that can help you implement these techniques.

Analogous estimates, also called **top-down estimates**, use the actual cost of a previous, similar project as the basis for estimating the cost of the current project. It is a form of expert judgment. This method is generally less costly than others, but it is also less accurate. Analogous estimates are most reliable when the previous projects are similar in fact and not just in appearance. In addition, the groups preparing estimates must have the needed expertise to determine whether certain parts of the project will be more or less expensive than analogous projects. However, if the project to be estimated involves a new programming language or working with a new type of hardware or network, the analogous approach could easily result in too low an estimate.

Bottom-up estimating involves estimating individual work items and summing them to get a project total. The size of the individual work items and the experience of the estimators drive the accuracy of the estimates. If a detailed WBS is available for a project, the project manager could have each person responsible for a work package develop his or her own cost estimate for that work package. All of the estimates would then be added to create estimates for each higher level WBS item and finally for the entire project. Using smaller work items increases the accuracy of the estimate because the people assigned to do the work develop the estimate instead of someone unfamiliar with the work. The drawback with bottom-up estimates is that they are usually time-intensive and therefore expensive to develop.

Parametric modeling uses project characteristics (parameters) in a mathematical model to estimate project costs. A parametric model might provide an estimate of $50 per line of code for a software development project based on the programming language the project is using, the level of expertise of the programmers, the size and complexity of the data involved, and so on. Parametric models are most reliable when the historical information that was used to create the model is accurate, the parameters are readily quantifiable, and the model is flexible in terms of the size of the project. For example, in the 1980s, engineers at McDonnell Douglas Corporation (now part of Boeing) developed a parametric model for estimating aircraft costs based on a large historical database. The model included the following parameters: the type of aircraft (fighter aircraft, cargo aircraft, or passenger aircraft), how fast the plane would fly, the thrust-to-weight ratio of the engine, the estimated weights of various parts of the aircraft, the number of aircraft produced, the amount of time available to produce them, and so on. In contrast to this sophisticated model, some parametric models involve very simple heuristics or rules of thumb. For example, a large office automation project might use a ballpark figure of $10,000 per workstation based on a history of similar office automation projects developed during the same time period. More complicated parametric models are usually computerized.

One popular parametric model is the **Constructive Cost Model (COCOMO)**, which is used for estimating software development costs based on parameters such as the source lines of code or function points. COCOMO was developed by Barry Boehm, a well-known expert in the field of software development and cost estimating. **Function points** are technology-independent assessments of the functions involved in developing a system. For example, the number of inputs and outputs, the number of files maintained, and the number of updates are examples of function points. **COCOMO II** is a newer, computerized version of Boehm's model that allows you to estimate the cost, effort, and schedule when planning a new software development activity. Boehm suggests that only algorithmic or parametric models do not suffer from the limits of human decision-making capability. As a consequence, Boehm and many other software experts favor using algorithmic models when estimating software project costs.

Computerized tools, such as spreadsheets and project management software, can make working with different cost estimates and cost estimation tools easier. Computerized tools, when used properly, can also help improve the accuracy of estimates. In addition to spreadsheets and project management software, more sophisticated tools are available for estimating software project costs.

Typical Problems with Information Technology Cost Estimates

Although there are many tools and techniques to assist in creating project cost estimates, many information technology project cost estimates are still very inaccurate, especially those involving new technologies or software development. Tom DeMarco, a well-known author on software development, suggests four reasons for these inaccuracies and some ways to overcome them.[7]

1. Developing an estimate for a large software project is a complex task requiring a significant amount of effort. Many estimates must be done quickly and before clear system requirements have been produced. For example, the surveyor's assistant project described in the opening case involves a lot of complex software development. Before fully understanding what information surveyors really need in the system, someone would have to create a rough order-of-magnitude estimate and budgetary estimates for this project. Rarely are the more precise, later estimates less than the earlier estimates for information technology projects. It is important to remember that estimates are done at various stages of the project, and project managers need to explain the rationale for each estimate.
2. The people who develop software cost estimates often do not have much experience with cost estimation, especially for large projects. There is also not enough accurate, reliable project data available on which to base estimates. If companies use good project management and develop a history of keeping good project information, including estimates, it should help them to improve their estimates. Enabling information technology people to receive training and mentoring on cost estimating will also improve cost estimates.
3. Human beings have a bias toward underestimation. For example, senior information technology professionals or project managers might make estimates based on their own abilities and forget that many junior people will be working on a project. Estimators might also forget to allow for extra costs needed for integration and testing on large information technology projects. It is important for project managers and senior management to review estimates and ask important questions to make sure the estimates are not biased.
4. Management might ask for an estimate, but really desire a number to help them create a bid to win a major contract or get internal funding. This problem is similar to the situation discussed in Chapter 5, *Project Time Management*, in which senior managers or other stakeholders want project schedules to be shorter than the estimates. It is important for project managers to help develop good cost and schedule estimates and to use their leadership and negotiation skills to stand by those estimates.

7 DeMarco, Tom, *Controlling Software Projects*. New York: Yourdon Press, 1982.

Sample Cost Estimates and Supporting Detail for an Information Technology Project

One of the best ways to learn how the cost estimating process works is by studying a sample cost estimate. Every cost estimate is unique, just as every project is unique. This section provides information from an actual information technology project cost estimate. The entire estimate was much longer, but parts of it are provided here to illustrate what a good cost estimate entails.

Before beginning the creation of any cost estimate, you must first know how the estimate will be used. If the estimate will be the basis for contract awards and performance reporting, it should be a definitive estimate and fairly accurate, as described earlier. In the example presented here, a budgetary estimate was done for replacing **legacy systems**, older information systems that run on old mainframe computers. Legacy systems support basic business processes such as general ledger, accounts payable and receivable, and project accounting. The name of this project was the Business Systems Replacement project. The total cost estimate for this three-year project was about $7.5 million. The life of the new system was estimated to be eight years.

The main objective of the Business Systems Replacement (BSR) project was to replace several legacy systems with a suite of packaged financial applications software (provided by Oracle). The suite of financial applications would provide more timely information for management decision making, easier access to data by end users, and reduced costs through productivity improvements throughout the company. The legacy systems were all written in-house specifically for that company. By purchasing packaged software, the company would no longer need its old mainframe computer and would benefit from having additional features and regular upgrades to the software without needing to write its own code.

Table 6-3 shows the overview for the BSR cost estimate. Notice the main categories found in the cost estimate overview: objective, scope, assumptions, and cost/benefit analysis and internal rate of return (IRR). The overview highlights information important to senior management and others interested in the cost estimate for the project.

Table 6-3: Business Systems Replacement Project Cost Estimate Overview

CATEGORY	DESCRIPTION
Objective	Install a suite of packaged financial applications software that will enable more timely information for management decision making, provide easier access to data by the ultimate end user, and allow for cost savings through productivity improvements throughout the company.
Scope	The core financial systems will be replaced by Oracle financial applications. These systems include: ■ General Ledger ■ Fixed Assets ■ Operations Report ■ Accounts Payable ■ Accounts Receivable ■ Project Accounting ■ Project Management
Assumptions	Oracle's software provides: ■ Minimal customization ■ No change in procurement systems during accounts payable implementation
Cost/Benefit Analysis & Internal Rate of Return (IRR)	BSR was broken down into a three-year cash outlay without depreciation. Costs are represented in thousands. Capital and expenses are combined in this example.

Another important part of the BSR cost estimate was the cash flow analysis. Table 6-4 shows the estimated costs, benefits or savings, and net cost or savings from fiscal year (FY) 1995 through 1997. Numbers in parentheses represent negative numbers or costs. The table also shows the total project costs and savings and projected annual savings in future years beyond 1997. Notice that this company chose to focus on the first three years of the project in this summary cash flow analysis and estimated all future annual savings in one column. Estimates are less reliable the farther out in time you go, so some estimates use this approach.

Table 6-4: Business Systems Replacement Project Cash Flow Analysis

	FY95 ($000)	FY96 ($000)	FY97 ($000)	3 YEAR TOTAL ($000)	FUTURE ANNUAL COSTS/SAVINGS ($000)
Costs					
Oracle/PM Software (List Price)	992	500	0	1492	0
60% Discount	(595)			(595)	
Oracle Credits	(397)	0		(397)	
Net Cash for Software	0	500		500	
Software Maintenance	0	90	250	340	250
Hardware & Maintenance	0	270	270	540	270
Consulting & Training	205	320	0	525	0
Tax & Acquisition	0	150	80	230	50
Total Purchased Costs	205	1330	**600**	2135	570
Information Services & Technology (IS&T)	500	1850	**1200**	3550	0
Finance/Other Staff	200	990	580	1770	
Total Costs	905	4170	2380	7455	570
Savings					
Mainframe		(101)	(483)	(584)	(597)
Finance/Asset/PM		(160)	(1160)	(1320)	(2320)
IS&T Support/Data Entry		(88)	(384)	(472)	(800)
Interest		0	(25)	(25)	(103)
Total Savings		(349)	(2052)	(2401)	(3820)
Net Cost (Savings)	905	3821	328	5054	(3250)
8 Year Internal Rate of Return	35%				

In Table 6-4, you can see that the main cost categories for this estimate include the purchase cost for Oracle's software, hardware and software maintenance related to the project, consulting and training, tax and acquisition, Information Services and Technology (IS&T) effort, and Finance/other departments' effort. These cost categories are the same as summary tasks from the project's work breakdown structure. In this example, the main supplier, Oracle, provided detailed information for determining the cost of purchasing their software. The company received discounts and credits from previous purchases that made the net cash cost for the purchased software equal to zero. Oracle also provided estimates for software maintenance, consulting, and training costs. Note that the largest item in this cost estimate is the Information Services and Technology (IS&T) effort, which involves the people in the company's IS&T department who worked with Oracle and other internal departments to plan, implement, and support the new system. Major savings or benefits items were the savings from replacing the mainframe system, savings in various departments, a reduction in Information Services and Technology staff needed to support the legacy system, and interest savings from loans related to the old system.

The cash flow analysis provided the basis for determining the internal rate of return (IRR) for the life of the project. In this example, the IRR was 35 percent over eight years, which is a very good return.

Estimates provide the basis for cost budgeting and cost control, as described in the following sections. To use an estimate as the basis for budgeting, the estimator must understand how his or her organization prepares budgets. Likewise, to use estimates as the basis for cost control, the estimator must understand how the organization performs cost control.

COST BUDGETING

Project cost budgeting involves allocating the project cost estimate to individual work items. These work items are based on the work breakdown structure for the project. The WBS, therefore, is a required input to the cost budgeting process. Likewise, the project schedule is required to allocate costs over time. The main goal of the cost budgeting process is to prepare budgetary estimates and to produce a cost baseline for measuring project performance.

Most organizations have a well-established process for preparing budgets. Continuing with the Business Systems Replacement example, the company involved required budget estimates to include the number of full-time equivalent (FTE) staff, often referred to as headcount, for each year of the project. This number provided the basis for estimating total compensation costs each year, as shown in Table 6-5. The budget also required inputs in the categories of consultant/purchased services, travel, depreciation, rents/leases, and other supplies and expenses. Notice that the total cost for the IS&T budget for FY97, $1,800,000, is based on the costs highlighted in the cash flow analysis in Table 6-4 ($600,000 for total purchased costs plus $1,200,000 for IS&T effort in FY97).

Table 6-5: Business Systems Replacement Project Budget Estimates for FY97 and Explanations

Budget Category	Estimated Costs	Explanation
Headcount (FTE)	13	Included are 9 programmer/analysts, 2 database analysts, and 2 infrastructure technicians.
Compensation	$1,008,500	Calculated by employee change notices (ECNs) and assumed a 4% pay increase in June. Overload support was planned at $10,000.
Consultant/Purchased Services	$424,500	Expected consulting needs in support of the Project Accounting and Cascade implementation efforts; maintenance expenses associated with the Hewlett-Packard (HP) computing platforms; maintenance expenses associated with the software purchased in support of the BSR project.
Travel	$25,000	Incidental travel expenses incurred in support of the BSR project.
Depreciation	$91,000	Included is the per head share of workstation depreciation, the Cascade HP platform depreciation, and the depreciation expense associated with capitalized software purchases.
Rents/Leases	$98,000	Expenses associated with the Mach1 computing platforms.
Other Supplies and Expenses	$153,000	Incidental expenses associated with things such as training, reward and recognition, long-distance phone charges, and miscellaneous office supplies.
Total Costs	**$1,800,000**	

Table 6-5 also includes explanations to support the budget estimates. Staff from the IS&T department included nine programmer/analysts, two database analysts, and two infrastructure technicians. Their compensation is based on employee change notices (ECNs), which provide actual salary and benefit information for employees when they are hired or transferred within the company. Notice that the compensation costs are the main part of the budget estimate. Having the largest part of cost estimates come from compensation costs is typical of estimates on many information technology projects. Also notice that the budget amount for compensation includes an allowance for raises and overload support or overtime pay. Other explanations include brief descriptions of the travel, depreciation, rents/leases, and other supplies and expenses for the project. It is very important to document assumptions and explanations when preparing cost estimates and cost budgets.

In addition to providing input for budgetary estimates, cost budgeting provides a cost baseline. A **cost baseline** is a time-phased budget that project managers use to measure and monitor cost performance. Estimating costs for each major project activity over time provides project managers and senior management with a foundation for project cost control.

COST CONTROL

Project cost control includes monitoring cost performance, ensuring that only appropriate project changes are included in a revised cost baseline, and informing project stakeholders of authorized changes to the project that will affect costs. The cost baseline, performance reports, change requests, and the cost management plan are inputs to the cost control process. Outputs of this process are revised cost estimates, budget updates, corrective action, revised estimates for project completion, and lessons learned.

Several tools and techniques assist in project cost control. There must be some change control system to define procedures for changing the cost baseline. This cost control change system is part of the integrated change control system described in Chapter 3, *Project Integration Management*. Since many projects do not progress exactly as planned, new or revised cost estimates are often required, as are estimates to evaluate alternate courses of action. Another very important tool for cost control is performance measurement. Although many general accounting approaches are available for measuring cost performance, there is a very powerful cost control tool that is unique to the field of project management—earned value management (EVM).

Earned Value Management

Earned value management (EVM) is a project performance measurement technique that integrates scope, time, and cost data. Earned value management is sometimes called earned value analysis (EVA), which should not be confused with economic value added, also referred to as EVA. Given a cost performance baseline, project managers and their teams can determine how well the project is meeting scope, time, and cost goals by entering actual information and then comparing it to the baseline. A **baseline** is the original project plan plus approved changes. Actual information includes whether or not a WBS item was completed or approximately how much of the work was completed, when the work actually started and ended, and how much it actually cost to do the work that was completed.

Figure 6-1 shows the Business Systems Replacement project's input form for collecting information for cost control and earned value management. This internal company form includes the following information:

- Descriptive information: The top line of this form lists the WBS number for an activity, a description of the activity, a revision number, and a revision date. This example of an input form reports information for WBS Item 6.8.1.2, a level 4 WBS activity (based on the numbering scheme) called "Design Interface Process-Customer Information."
- Assignment information: This activity is assigned to someone with the initials SMC. Other assignment information includes the responsible person's role and number of hours available per day for the activity.
- Forecast information: The BSR project used a PERT weighted average to estimate the number of hours it would take to complete activities. Recall from Chapter 5 that the PERT weighted average uses most likely, optimistic, and pessimistic estimates. In this case, the PERT calculation results in a planned effort of thirty hours. The planned duration is five days.
- Description: This section provides a detailed description of the activity.
- Assumptions: Major assumptions related to this particular WBS item are documented here.
- Results/Deliverables: This section briefly lists the main outcomes from the WBS item.
- Dependencies: This section lists the WBS number for any predecessor or successor activities.

WBS#: 6.8.1.2	Description: Design Interface Process-Customer Information	Revision:	Revision Date:

Assignments

Hours per day

Responsible: SMC Role: PA Availability: 6

Involved: Role: Availability:

Involved: Role: Availability:

Involved: Role:

Forecast

Effort (in hours) Calculated

Optimistic: 20

Most Likely: 30

Pessimistic: 40

Delay (Days):

Plan Effort: 30 Hrs

Plan Duration: 5 Days

Description

Develop an operational process design for the Customer Information interface from the Invoicing System to Oracle Receivables. This task will accept as input the business/functional requirements developed during the tactical analysis phase and produce as output a physical operational design, which provides the specifications, required for code development.

Assumptions

- All business rules and issues will be resolved prior to this task.
- The Entity Relationship Diagram (ERD) and data model for Oracle Receivables and any Oracle extension required will be completed and available prior to this task.
- The ERD for the Invoicing System will be completed and available prior to this task.
- Few iterations of the review/modify cycle will be required.

Results / Deliverables

Process Design Document - Technical
- Operation/Physical Data Flow Diagram
- Process Specifications
- Interface Data Map

Dependencies

Predecessors (WBS#): 4.7

Successors (WBS#):

Figure 6-1. Cost Control Input Form for Business Systems Replacement Project

This particular activity has not yet started, so it does not include the actual number of hours or the actual duration for the activity. Actual cost, actual duration, and percentage complete information is required in order to perform earned value management.

Earned value management involves calculating three values for each activity or summary activity from a project's WBS.

1. The **planned value (PV)**, formerly called the budgeted cost of work scheduled (BCWS), also called the budget, is that portion of the approved total cost estimate planned to be spent on an activity during a given period. Table 6-6 shows an example of earned value calculations. Suppose a project included a summary activity of purchasing and installing a new Web server. Suppose further that according to the plan, it would take one week and cost a total of $10,000 for the labor hours, hardware, and software involved. The planned value (PV) for that activity that week is, therefore, $10,000.
2. The **actual cost (AC)**, formerly called the actual cost of work performed (ACWP), is the total direct and indirect costs incurred in accomplishing work on an activity during a given period. For example, suppose it actually took two weeks and cost $20,000 to purchase and install the new Web server. Assume that $15,000 of these actual costs was incurred during week 1 and $5,000 was incurred during week 2. These amounts are the actual cost (AC) for the activity each week.

3. The **earned value (EV)**, formerly called the budgeted cost of work performed (BCWP), is the percentage of work actually completed multiplied by the planned value. Again, using the example in Table 6-6, suppose you estimated the activity of purchasing and installing a Web server to be 75 percent complete after week 1. To calculate the earned value (EV) for week 1, you would multiply the planned value of $10,000 for week 1 by 75 percent to get an earned value of $7,500 for that activity at that point in time.

Table 6-6: Earned Value Calculations for One Activity After Week One

ACTIVITY	WEEK 1	WEEK 2	TOTAL	% COMPLETE AFTER WEEK 1	EARNED VALUE AFTER WEEK 1 (EV)
Purchase Web server	10,000	0	10,000	75%	**7,500**
Planned Value (PV)	10,000	0	10,000		
Actual Cost (AC)	15,000	5,000	20,000		
Cost Variance (CV)	–7,500				
Schedule Variance (SV)	–2,500				
Cost Performance Index (CPI)	50%				
Schedule Performance Index (SPI)	75%				

Table 6-7 summarizes the formulas used in earned value management. Note that all of these formulas start with EV, the earned value. Variances are calculated by subtracting the actual cost or planned value from EV, and indexes are calculated by dividing EV by the actual cost or planned value.

The earned value calculations in Table 6-6 are carried out as follows:

$$EV = \$10,000 \times 75\% = \$7,500$$

$$CV = 7,500 - 15,000 = -7,500$$

$$SV = 7,500 - 10,000 = -2,500$$

CPI = 7,500/15,000 = 50%

SPI = 7,500/10,000 = 75%

Table 6-7: Earned Value Formulas

TERM	FORMULA
Earned Value	EV = PV to date × percent complete
Cost Variance	CV = EV – AC
Schedule Variance	SV = EV – PV
Cost Performance Index	CPI = EV/AC
Schedule Performance Index	SPI = EV/PV

Cost variance (CV) is the earned value minus the actual cost. In other words, cost variance shows the difference between the estimated cost of an activity and the actual cost of that activity. If cost variance is a negative number, it means that performing the work cost more than planned. If cost variance is a positive number, it means that performing the work cost less than planned.

Schedule variance (SV) is the earned value minus the planned value. Schedule variance shows the difference between the scheduled completion of an activity and the actual completion of that activity. A negative schedule variance means that it took longer than planned to perform the work, and a positive schedule variance means that it took less time than planned to perform the work.

The **cost performance index (CPI)** is the ratio of earned value to actual cost and can be used to estimate the projected cost of completing the project. If the cost performance index is equal to one or 100 percent, then the planned and actual costs are equal, or the costs are exactly as budgeted. If the cost performance index is less than one or less than 100 percent, the project is over budget. If the cost performance index is greater than one or more than 100 percent, the project is under budget.

The **schedule performance index (SPI)** is the ratio of earned value to planned value and can be used to estimate the projected time to complete the project. Similar to the cost performance index, a schedule performance index of one or 100 percent means the project is on schedule. If the schedule performance index is greater than one or 100 percent, then the project is ahead of schedule. If the schedule performance index is less than one or 100 percent, the project is behind schedule.

Note that in general, *negative numbers for cost and schedule variance indicate problems in those areas*. Negative numbers mean the project is costing more than planned or taking longer than planned. Likewise, *CPI and SPI less than one or less than 100 percent also indicate problems*.

Earned value calculations for all project activities (or summary level activities) are required to estimate the earned value for the entire project. Some activities may be over budget or behind schedule, but others may be under budget and ahead of schedule. By adding all of the earned values for all project activities, you can determine how the project as a whole is performing.

Figure 6-2 provides sample earned value information for a one-year project. This project had a planned total cost of $100,000. The spreadsheet shows actual cost and percentage complete information for the first five months, or through the end of May. Notice the "% Complete" column at the upper-right side, or in column O, of the spreadsheet. The earned value (EV) for each activity is calculated by multiplying the percent complete value by the planned value. The "Monthly Actual Cost" row (row 15 of the spreadsheet) shows the actual cost each month for the project's activities through May. By calculating the total planned costs, the total actual costs, and the earned value costs, you can determine the cost variance, schedule variance, cost performance index, and schedule performance index for the entire project (see cells A18 through B25 in Figure 6-2).

	A	B	C	D	E	F	G	H	I	J	K	L	M	N	O	P
1	**Activity**	**Jan**	**Feb**	**Mar**	**Apr**	**May**	**Jun**	**Jul**	**Aug**	**Sep**	**Oct**	**Nov**	**Dec**	**PV**	**% Complete**	**EV**
2	Plan and staff project	4,000	4,000											8,000	100	8,000
3	Analyze requirements		6,000	6,000										12,000	100	12,000
4	Develop ERDs			4,000	4,000									8,000	100	8,000
5	Design database tables				6,000	4,000								10,000	100	10,000
6	Design forms, reports, and queries					8,000	4,000							12,000	50	6,000
7	Construct working prototype						10,000							10,000	-	-
8	Test/evaluate prototype						2,000	6,000						8,000	-	-
9	Incorporate user feedback							4,000	6,000	4,000				14,000	-	-
10	Test system									4,000	4,000	2,000		10,000	-	-
11	Document system											3,000	1,000	4,000	-	-
12	Train users												4,000	4,000	-	-
13	Monthly Planned Value (PV)	4,000	10,000	10,000	10,000	12,000	16,000	10,000	6,000	8,000	4,000	5,000	5,000	100,000		44,000
14	Cumulative Planned Value (PV)	4,000	14,000	24,000	34,000	46,000	62,000	72,000	78,000	86,000	90,000	95,000	100,000			
15	Monthly Actual Cost (AC)	4,000	11,000	11,000	12,000	15,000										
16	Cumulative Actual Cost (AC)	4,000	15,000	26,000	38,000	53,000										
17	Monthly Earned Value (EV)	4,000	10,000	10,000	10,000	10,000										
18	Cumulative Earned Value (EV)	4,000	14,000	24,000	34,000	44,000										
19	Project EV as of May 31	44,000														
20	Project PV as of May 31	46,000														
21	Project AC as of May 31	$ 53,000														
22	**CV=EV-AC**	**$ (9,000)**														
23	**SV=EV-PV**	**$ (2,000)**														
24	**CPI=EV/AC**	**83%**														
25	**SPI=EV/PV**	**96%**														
26	**Estimate at Completion (EAC)**	**$120,455**	(original plan of $100,000 divided by CPI of 83%)													
27	**Estimated time to complete**	**12.55**	(original plan of 12 months divided by SPI of 96%)													

Figure 6-2. Earned Value Calculations for a One-Year Project After Five Months

In this example, the cost variance is –$9,000, and the schedule variance is –$2,000. These values mean that the project is both over budget and behind schedule after five months. The cost performance and schedule performance indexes are 83 percent and 96 percent, respectively. The cost performance index can be used to calculate the **estimate at completion (EAC)**—an estimate of what it will cost to complete the project based on performance to date. Similarly, the schedule performance index can be used to calculate an estimated time to complete the project. For example, in Figure 6-2 the estimated cost at completion is $120,455, or $100,000/83 percent. The estimated time to complete the project is 12.55 months, or twelve months/96 percent (see cells A26 through B27).

You can graph earned value information to track project performance. Figure 6-3 shows an earned value chart for the one-year sample project from Figure 6-2. The chart includes three lines and two points, as follows:

- Planned value (PV), the cumulative planned amounts for all activities by month. Note that the planned value line extends for the estimated length of the entire project.
- Actual cost (AC), the cumulative actual amounts for all activities by month
- Earned value (EV), the cumulative earned value amounts for all activities by month
- **Budget at Completion (BAC)**, the original total budget for the project, or $100,000 in this example. The BAC point is plotted on the chart at the original time estimate of twelve months.
- Estimate at Completion (EAC), estimated to be $120,455, as described earlier. This EAC point is plotted on the chart at the estimated time to complete of 12.55 months.

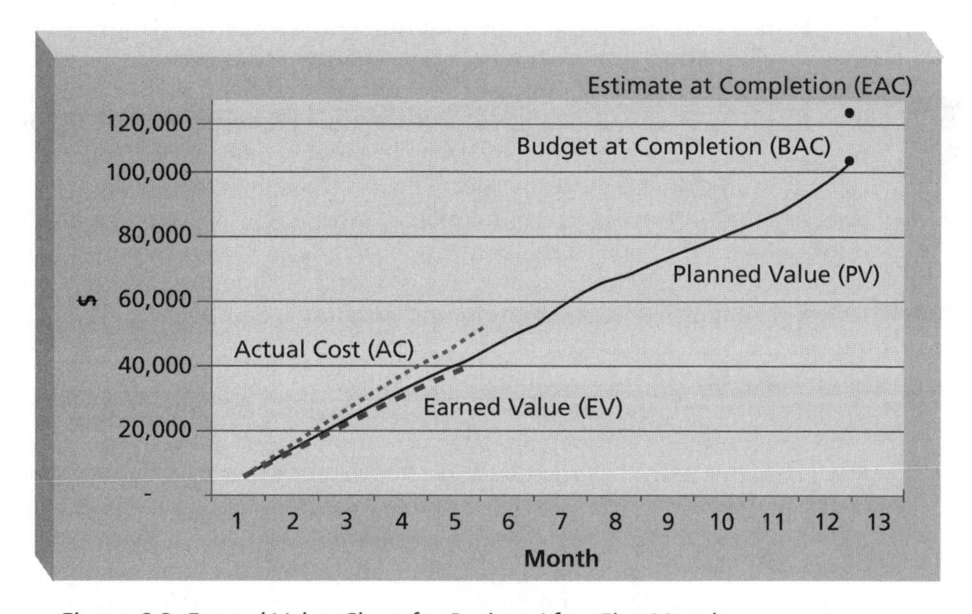

Figure 6-3. Earned Value Chart for Project After Five Months

Viewing earned value information in chart form helps you to visualize how the project is performing. For example, you can see the planned performance by looking at the planned value line. If the project goes as planned, it will finish in twelve months and cost $100,000, represented by budget at completion, or BAC. Notice in this example that the actual cost line is always right on or above the earned value line. When the actual cost line is right on or above the earned value line, costs are equal to or more than planned. The planned value line is pretty close to the earned value line, just slightly higher in the last month. This relationship means that the project has been right on schedule until the last month, when the project got a little behind schedule.

Senior managers overseeing multiple projects often like to see performance information in a graphical form such as this earned value chart. For example, in the opening case, the government officials were reviewing earned value charts and estimates at completion for several different projects. Earned value charts allow you to quickly see how projects are performing. If there are serious cost and schedule performance problems, senior management may decide to terminate projects or take other corrective action. The estimates at completion are important inputs to budget decisions, especially if total funds are limited. Earned value management is an important technique because, when used effectively, it helps senior management and project managers evaluate progress and make sound management decisions.

However, earned value management is not used on many projects outside of government agencies and their contractors. Two reasons earned value management is not used more widely are its focus on tracking actual performance versus planned performance and the importance of percentage completion data in making calculations. Many projects, particularly information technology projects, do not have good planning information, so tracking performance against a plan might produce misleading information. Several estimates are usually made on information technology projects, and keeping track of the most recent estimate and the actual costs associated with it could be cumbersome. In addition, estimating percentage completion of tasks might produce misleading information. What does it really mean to say that a task is 75 percent complete after three months? Such a statement is often not synonymous with saying the task will be finished in one more month or after spending an additional 25 percent of the planned budget.

To make earned value management simpler to use, organizations can modify the level of detail and still reap the benefits of the technique. For example, you can use percentage completion data such as 0 percent for items not yet started, 50 percent for items in progress, and 100 percent for completed tasks. As long as the project is defined in enough detail, this simplified percentage completion data should provide enough summary information to allow managers to see how well a project is doing overall. You can get very accurate total project performance information using these simple percentage complete amounts. For example, using simplified percentage complete amounts for a one-year project with weekly reporting and an average task or work packet size of one week, you can expect about a 1 percent error rate.[8]

Earned value management is the primary method available for integrating performance, cost, and schedule data. It can be a powerful tool for project managers and senior management to use in evaluating project performance.

USING SOFTWARE TO ASSIST IN COST MANAGEMENT

Most organizations use software to assist in various activities related to project cost management. Spreadsheets are a common tool for performing resource planning, cost estimating, cost budgeting, and cost control. Many companies also use more sophisticated and centralized financial applications software to provide important cost-related information to accounting and finance personnel. This section will focus specifically on how you can use project management software in cost management. Appendix A includes a brief guide to the cost management features in Microsoft Project 2000.

Project management software can be a very helpful tool during each project cost management process. It can help you study overall project information or focus on tasks that are over a specified cost limit. You can use the software to assign costs to resources and tasks, prepare cost estimates, develop cost budgets, and monitor cost performance. Microsoft Project 2000 has several standard cost reports: cash flow, budget, over budget tasks, over budget resources, and earned value reports. For several of these reports, you must enter percentage completion information and actual costs, just as you need this information when manually calculating earned value or other analyses.

Although Microsoft Project 2000 has a fair amount of cost management features, many information technology project managers use other tools to manage cost information; they do not know that they can use Microsoft Project 2000 for cost management. As with most software packages, users need training to use the software effectively and understand the features available. Instead of using dedicated project management software for cost management, some information technology project managers use company accounting systems; others use spreadsheet software to achieve more flexibility. Project managers who use other software often do so because these other systems are more generally accepted in their organizations and more people know how to use them. On the other hand, some companies have developed methods to link data between their project management software and their main accounting software systems.

[8] Brandon, Daniel M., Jr, "Implementing Earned Value Easily and Effectively," *Project Management Journal* (June 1998) 29, no. 2, 11-18.

CASE WRAP-UP

After talking to his colleagues about the meeting, Juan had a better idea of the importance of project cost management. He understood the value of doing detailed studies before making major expenditures on new projects, especially after learning about the high cost of correcting defects late in a project. He also learned how important it is to develop good cost estimates and keep costs on track. Government officials cancelled several projects when the project managers showed how poorly they were performing and admitted that they did not do much planning and analysis early in the projects. Juan knew that he could not focus on just the technical aspects of projects if he wanted to move ahead in his career. He began to wonder whether several projects the city was considering were really worth the taxpayers' money. Issues of cost management added a new dimension to Juan's job.

CHAPTER SUMMARY

Project cost management is a traditionally weak area of information technology projects. Information technology professionals must acknowledge the importance of cost management and take responsibility for improving resource planning, cost estimating, budgeting, and cost control.

Project managers must understand several basic principles of cost management in order to be effective in managing project costs. Important concepts include profits and profit margins, life cycle costing, cash flow analysis, sunk costs, and learning curve theory.

Resource planning involves determining what people, equipment, and materials should be used to perform project activities. It is important to evaluate alternatives and use expert judgement in determining resource requirements.

Cost estimating is a very important part of project cost management. There are several types of cost estimates: rough order of magnitude, budgetary, and definitive. Each type of estimate is done during different stages of the project life cycle, and each has a different level of accuracy. There are four basic tools and techniques for developing cost estimates: analogous, bottom-up, parametric modeling, and computerized tools. The main parts of a cost estimate include an objective statement, scope, assumptions, cost/benefit analysis, cash flow analysis, budget breakdown, and explanations or supporting detail.

Cost budgeting involves allocating costs to individuate work items over time. It is important to understand how particular organizations prepare budgets so estimates are made accordingly.

Project cost control includes monitoring cost performance, reviewing changes, and notifying project stakeholders of changes related to costs. Many basic accounting and finance principles are related to project cost management. However, earned value management is the main method used for measuring project performance. Earned value management integrates scope, cost, and schedule information.

You can use several software products to assist with project cost management. Microsoft Project 2000 has many cost management features, including earned value management.

Discussion Questions

1. Discuss why many information technology professionals have a poor attitude toward project cost management.
2. Do most colleges and companies provide adequate education and training in cost management for information technology project managers? Give examples to support your answer.
3. Give examples of rough order of magnitude, budgetary, and definitive cost estimates for an information technology project.
4. Give an example of using each of the following techniques for creating a cost estimate: analogous, parametric, bottom-up, and computerized tools.
5. What is earned value management? Why is it the preferred method for measuring project performance? Why isn't it used more often?
6. What are some general rules of thumb for deciding if cost variance, schedule variance, cost performance index, and schedule performance index numbers are good or bad?

Exercises

1. Given the following information for a one-year project, answer the following questions:

   PV= $23,000

   EV= $20,000

   AC= $25,000

   BAC= $120,000

Recall that PV is the planned value, EV is the earned value, AC is the actual cost, and BAC is the budget at completion.

a. What is the cost variance, schedule variance, cost performance index (CPI), and schedule performance index (SPI) for the project?
b. How is the project doing? Is it ahead of schedule or behind schedule? Is it under budget or over budget?
c. Use the CPI to calculate the estimate at completion (EAC) for this project. Is the project performing better or worse than planned?
d. Use the schedule performance index (SPI) to estimate how long it will take to finish this project.

2. Using the data in Figure 6-2 for cumulative PV, AC, and EV, create an earned value chart similar to the one shown in Figure 6-3.
3. Create a cost estimate for building a new state-of-the art multimedia classroom for your organization within the next six months. The classroom should include twenty high-end personal computers with appropriate software for your organization, a network server, Internet access for all machines, a teacher station, and a projection system. Be sure to include personnel costs associated with the project management for this project. Document the assumptions you made in preparing the estimate and provide explanations for key numbers.
4. Research online information, textbooks, and local classes for using the cost management features of Microsoft Project 2000. Ask three different people in different information technology organizations that use Microsoft Project 2000 if they use the cost management features and in what ways they use them. Write a brief report of what you learned from your research.
5. Do the project cost management part of one of the exercises provided in Appendix A. Print the reports mentioned and write a one-page paper interpreting the results.

MINICASE

You've decided that you are now an expert in project cost estimating, and you want to hold some public classes or workshops on this topic. You have done some training in the past and are confident in your abilities. Your first project is to develop materials, then market and hold your first seminar. You plan to do all of the work within the next six months.

You want to create a good cost estimate for this project. Below are some of your assumptions:

- You will charge $600 per person for a two-day class.
- Your most-likely estimate of how many people will attend is 30.
- Your fixed costs include $500 to rent a room for both days, setup fees of $400 for registration, and $300 for designing a postcard for advertising.
- You will not include any of your labor costs for this estimate, but you

estimate that you will spend at least 150 hours developing materials, managing the project, and giving the actual class.

- You will order 5,000 postcards, mail 4,000, and distribute the rest to friends and colleagues.

Your variable costs include the following:

- $5/registration plus 4% of the class fee per person to handle credit card processing
- $.40/postcard if you order 5,000 or more; $.60 if you order fewer
- $.25/postcard for mailing and postage
- $25/person for beverages and lunch
- $30/person for class handouts

Assume 30 people register for and attend the class.

Part 1: Using Excel, create a spreadsheet to calculate your projected total costs, total revenues, and total profits given the preceding information. Be sure to have input cells for any variables that might change such as the cost of postage, handouts, and so on. Calculate your profits based on the following number of people who attend: 10, 20, 30, 40, 50, and 60. Try to use the Excel scenario feature (under the Tools menu) to create a scenario pivot table showing total profits based on the number of people who attend.

Part 2: Research the local market (or the location of your choice) for short classes on topics like project cost estimating. Find at least three organizations that offer estimating courses. You can search the Registered Education Provider's database for organizations from PMI's Web site (www.pmi.org), using "estimating" in the title. Given your research into similar classes and the work you think is required for this project, what assumptions, variables, or costs would you change from the information provided above? Justify your changes to the estimate, and create a spreadsheet with your best estimate for this project. Then write a paragraph describing whether or not you think this particular project is worth pursuing.

Suggested Readings

1. Brandon, Daniel M., Jr. "Implementing Earned Value Easily and Effectively." *Project Management Journal* (June 1998): 29, no. 2, 11-18.

    Brandon describes the benefits of using earned value management and the reasons that more organizations do not use it. This paper provides suggestions for overcoming some of the problems associated with this technique.

2. Hamaker, Joseph W. "But What Will It Cost? The History of NASA Cost Estimating." http://www.jsc.nasa.gov/bu2/hamaker.html (2000).

   This article provides a wealth of information on how NASA develops cost estimates. Sections of the Web site include cost estimating models, handbooks to assist in cost estimating, salaries, and many links to great references and software.

3. Kemerer, Chris F. "Software Cost Estimation Models." *Software Project Management Readings and Cases*. Chicago: The McGraw-Hill Companies, Inc. (1997).

   This article summarizes the three basic types of software cost estimating models. The earliest models are sometimes called the economic stream and are based on work done at TRW and IBM. Boehm's COCOMO is the most popular of these economic stream models. The second stream, called the Rayleigh stream, is based on the work of Putnam. The third stream, called the function point stream, first appeared in 1979 and is based on work done by Albrecht.

4. Meyer, N. Dean and Mary E. Boone. *The Information Edge*. 2d ed. Toronto, CA: Educational Publishing Company (1989).

   This book provides over sixty case histories to demonstrate how to measure the value or benefits of information systems. It provides detailed techniques for justifying and evaluating information technology projects.

5. The COCOMO Suite, http://sunset.usc.edu/research/cocomosuite/index.html.

   The University of Southern California's Center for Software Engineering provides a detailed Web site with references to COCOMO, COCOMO II, and other models.

Key Terms

- **Actual cost (AC)** — the total of direct and indirect costs incurred in accomplishing work on an activity during a given period, formerly called the **actual cost of work performed (ACWP)**
- **Analogous estimates** — a cost estimating technique that uses the actual cost of a previous, similar project as the basis for estimating the cost of the current project, also called **top-down estimates**
- **Baseline** — the original project plan plus approved changes
- **Bottom-up estimates** — a cost estimating technique based on estimating individual work items and summing them to get a project total
- **Budget at Completion (BAC)** — the original total budget for a project
- **Budgetary estimate** — a cost estimate used to allocate money into an organization's budget

- **Cash flow analysis** — a method for determining the estimated *annual* costs and benefits for a project
- **COCOMO II** — a newer, computerized cost-estimating tool based on Boehm's original model that allows one to estimate the cost, effort, and schedule when planning a new software development activity
- **Computerized tools** — cost-estimating tools that use computer software, such as spreadsheets and project management software
- **Constructive Cost Model (COCOMO)** — a parametric model developed by Barry Boehm for estimating software development costs
- **Contingency reserves** — dollars included in a cost estimate to allow for future situations that may be partially planned for (sometimes called **known unknowns**) and are included in the project cost baseline
- **Cost budgeting** — allocating the overall cost estimate to individual work items to establish a baseline for measuring performance
- **Cost control** — controlling changes to the project budget
- **Cost estimating** — developing an approximation or estimate of the costs of the resources needed to complete the project
- **Cost management plan** — a document that describes how cost variances will be managed on the project
- **Cost performance index (CPI)** — the ratio of earned value to actual cost; can be used to estimate the projected cost to complete the project
- **Cost variance (CV)** — the earned value minus the actual cost
- **Definitive estimate** — a cost estimate that provides an accurate estimate of project costs
- **Direct costs** — costs that are related to a project and can be traced back in a cost-effective way
- **Earned value (EV)** — the percentage of work actually completed multiplied by the planned cost, formerly called the **budgeted cost of work performed (BCWP)**
- **Earned value management (EVM)** — a project performance measurement technique that integrates scope, time, and cost data
- **Estimate at completion (EAC)** — an estimate of what it will cost to complete the project based on performance to date
- **Function points** — technology-independent assessments of the functions involved in developing a system
- **Indirect costs** — costs that are related to the project but cannot be traced back in a cost-effective way
- **Intangible costs or benefits** — costs or benefits that are difficult to measure in monetary terms
- **Internal rate of return (IRR)** — the discount rate that makes the net present value equal to zero, also called **time-adjusted rate of return**

- **Learning curve theory** — a theory that states that when many items are produced repetitively, the unit cost of those items normally decreases in a regular pattern as more units are produced
- **Legacy systems** — older information systems that usually ran on an old mainframe computer
- **Life cycle costing** — considers the total cost of ownership, or development plus support costs, for a project
- **Management reserves** — dollars included in a cost estimate to allow for future situations that are unpredictable (sometimes called **unknown unknowns**)
- **Parametric modeling** — a cost-estimating technique that uses project characteristics (parameters) in a mathematical model to estimate project costs
- **Planned value (PV)** — that portion of the approved total cost estimate planned to be spent on an activity during a given period, formerly called the budgeted cost of work scheduled (BCWS)
- **Profit margin** — the ratio between revenues and profits
- **Profits** — revenues minus expenses
- **Project cost management** — the processes required to ensure that the project is completed within the approved budget
- **Reserves** — dollars included in a cost estimate to mitigate cost risk by allowing for future situations that are difficult to predict
- **Resource planning** — determining what resources (people, equipment, and materials) and what quantities of each resource should be used to perform project activities
- **Rough order of magnitude (ROM) estimate** — a cost estimate prepared very early in the life of a project to provide a rough idea of what a project will cost
- **Schedule performance index (SPI)** — the ratio of earned value to planned value; can be used to estimate the projected time to complete a project
- **Schedule variance (SV)** — the earned value minus the planned value
- **Sunk cost** — money that has been spent in the past
- **Tangible costs** or **benefits** — costs or benefits that can be easily measured in dollars

7

Project Quality Management

Objectives

After reading this chapter you will be able to:

1. *Understand the importance of project quality management and the role of the project manager in assuring quality*
2. *Define quality and how it relates to various aspects of information technology projects*
3. *Discuss quality experts' views of modern quality management*
4. *Describe what is involved in quality planning, quality assurance, and quality control on projects*
5. *Explain quality control tools and techniques such as Pareto charts, statistical sampling, quality control charts, six sigma, and the seven run rule*
6. *Compare the different types of testing for information technology projects and how they relate to quality*
7. *Describe key issues related to improving quality in information technology projects*

A large medical instruments company just hired Scott Daniels, a senior consultant from a large consulting firm, to lead a project to resolve the quality problems with the company's new Executive Information System (EIS). A team of internal programmers and analysts worked with several company executives to develop this new system themselves. Many executives who had never used computers before were hooked on the new EIS. They loved the way the system allowed them to quickly and easily track sales of various medical instruments by product, country, hospital, and sales representative. The system was also very user-friendly. After successfully testing the new EIS with several executives, the company decided to make the system available to all levels of management.

Unfortunately, several quality problems developed with the new EIS after a few months of operation. People were complaining that they could not get into the system. The system started going down a couple of times a month, and the response time was reportedly getting slower. Users complained when they could not access information within a few seconds. Several people kept forgetting how to log in to the system, thus increasing the number of calls to the Help desk. There were complaints that some of the reports in the system gave inconsistent information. How could a summary report show totals that were not consistent with a detailed report on the same information? The executive sponsor of the EIS wanted the problems fixed quickly and accurately, so he decided to hire an expert in quality from outside the company whom he knew from past projects. Scott Daniels' job was to lead a team of people from both the medical instruments company and his own firm in identifying and resolving quality-related issues with the EIS and in developing a plan to help prevent quality problems from happening on future information technology projects.

QUALITY OF INFORMATION TECHNOLOGY PROJECTS

Most people have heard jokes about what cars would be like if they followed a development history similar to that of computers. A joke that has been traveling around the Internet goes as follows:

At a recent computer exposition (COMDEX), Bill Gates, the founder and CEO of Microsoft Corporation, stated: "If General Motors had kept up with technology like the computer industry has, we would all be driving twenty-five dollar cars that got 1000 miles to the gallon."

In response to Gates' comments, General Motors issued a press release stating: "If GM had developed technology like Microsoft, we would all be driving cars with the following characteristics:

1. For no reason whatsoever your car would crash twice a day.
2. Every time they repainted the lines on the road you would have to buy a new car.
3. Occasionally your car would die on the freeway for no reason, and you would just accept this, restart, and drive on.

4. Occasionally, executing a maneuver such as a left turn, would cause your car to shut down and refuse to restart, in which case you would have to reinstall the engine.
5. Only one person at a time could use the car, unless you bought 'Car95' or 'CarNT.' But then you would have to buy more seats.
6. Macintosh would make a car that was powered by the sun, reliable, five times as fast, and twice as easy to drive, but would run on only five percent of the roads.
7. The oil, water temperature, and alternator warning lights would be replaced by a single 'general car default' warning light.
8. New seats would force everyone to have the same size hips.
9. The airbag system would say 'Are you sure?' before going off.
10. Occasionally, for no reason whatsoever, your car would lock you out and refuse to let you in until you simultaneously lifted the door handle, turned the key, and grabbed hold of the radio antenna.
11. GM would require all car buyers to also purchase a deluxe set of Rand McNally road maps (now a GM subsidiary), even though they neither need them nor want them. Attempting to delete this option would immediately cause the car's performance to diminish by 50 percent or more. Moreover, GM would become a target for investigation by the Justice Department.
12. Every time GM introduced a new model car, buyers would have to learn how to drive all over again because none of the controls would operate in the same manner as the old car.
13. You would press the Start button to shut off the engine."[1]

Most people simply accept poor quality from many information technology products. So what if your computer crashes a couple of times a month? Just make sure you back up your data. So what if you cannot log in to the corporate intranet or the Internet right now? Just try a little later when it is less busy. So what if the latest version of your word-processing software was shipped with several known bugs? You like the software's new features, and all new software has bugs. Is quality a real problem with information technology projects?

Yes, it is! Information technology is not just a luxury available in some homes, schools, or offices. Many companies provide all employees with access to computers, and in 2000 over 50 percent of homes in the United States had at least one personal computer. It took only five years for fifty million people to use the Internet compared to twenty-five years for fifty million people to use telephones. In 1999 some experts estimated that the number of people using the Internet was doubling every one hundred days. Many aspects of our daily lives depend on high quality information technology products. Our food is produced and distributed with the aid of computers;

[1] This joke was found on hundreds of Web sites and printed in the *Consultants in Minnesota Newsletter*, Independent Computer Consultants Association, December 1998.

our cars have computer chips to track performance; our children use computers to help them learn in school; our corporations depend on computers for many business functions; and millions of people rely on computers for entertainment and personal communications. Many information technology projects develop mission-critical systems that are used in life and death situations, such as navigation systems on board airplanes and computer components built into medical equipment. When one of these systems does not function correctly, it is much more than a slight inconvenience.

What Went Wrong?

- In 1981, a small timing difference caused by a computer program change created a 1/67 (or 1 in 67) chance that the space shuttle's five on-board computers would not synchronize. The error caused a launch abort.[2]
- In 1986, two hospital patients died after receiving fatal doses of radiation from a Therac 25 machine. A software problem caused the machine to ignore calibration data.[3]
- In one of the biggest software errors in banking history, Chemical Bank mistakenly deducted about $15 million from more than 100,000 customer accounts one evening. The problem resulted from a single line of code in an updated computer program that caused the bank to process every withdrawal and transfer at its automated teller machines (ATMs) twice. For example, a person who withdrew $100 from an ATM had $200 deducted from his or her account, though the receipt only indicated a withdrawal of $100. The mistake affected 150,000 transactions from Tuesday night through Wednesday afternoon.[4]
- In 1996, Apple Computer's PowerBook 5300 model had problems with lithium-ion battery packs catching fire, causing Apple to halt shipments and replace all the packs with nickel-metal-hydride batteries. Other quality problems also surfaced, such as cracks in the PowerBook's plastic casing and a faulty electric power adapter.[5]
- Hundreds of newspapers and Web sites ran stories about the "Melissa" virus in March of 1999. The rapidly spreading computer virus forced several large corporations to shut down their e-mail servers as the virus rode the Internet on a global rampage, according to several leading network security companies.[6]

Before tackling how you can improve quality on information technology projects, it is important to understand basic concepts about project quality management in general and the development of modern quality management.

[2] *Design News* (February 1988).

[3] *Datamation* (May 1987).

[4] *The New York Times* (February 18, 1994).

[5] "Apple to Try, Try Again," *Information Week* (July 29,1996) 15–16.

[6] Richtel, Matt, "New Fast-Spreading Virus Takes Internet by Storm," *The New York Times* on the Web (March 28, 1999) http://www.nytimes.com.

WHAT IS PROJECT QUALITY MANAGEMENT?

Project quality management is a difficult knowledge area to define. The International Organization for Standardization (ISO) defines **quality** as "the totality of characteristics of an entity that bear on its ability to satisfy stated or implied needs." Many people spent many hours coming up with this definition, yet it is still very vague. Other experts define quality based on conformance to requirements and fitness for use. **Conformance to requirements** means the project's processes and products meet written specifications. For example, if the scope statement requires delivery of one hundred Pentium 4 computers, you could easily check whether the correct computers had been delivered. **Fitness for use** means a product can be used as it was intended. If these Pentium 4 computers were delivered without monitors or keyboards and were left in boxes in the customer's shipping dock, the customer might not be satisfied because the computers would not be fit for use. The customer may have assumed that the delivery included monitors and keyboards, the unpacking of the computers, and installation so they would be ready to use.

The main purpose of project quality management is to ensure that the project will satisfy the needs for which it was undertaken. Recall that project management involves meeting or exceeding stakeholder needs and expectations. The project team must develop good relationships with key stakeholders, especially the main customer for the project, to understand what quality means to them. *After all, the customer ultimately decides if quality is acceptable.* Many technical projects fail because the project team focuses only on meeting the written requirements for the main products being produced and ignores other stakeholder needs and expectations for the project. For example, the project team should know what successfully delivering one hundred Pentium 4 computers means to the customer.

Quality, therefore, must be viewed on an equal level with project scope, time, and cost. If a project's stakeholders are not satisfied with the quality of how the project was managed or the resulting products of the project, the project team will need to make adjustments to scope, time, and cost to satisfy stakeholder needs and expectations. Meeting written requirements for scope, time, and cost is not sufficient. To achieve stakeholder satisfaction, the project team must develop a good working relationship with all stakeholders and understand their stated or implied needs.

Project quality management involves three main processes:

- **Quality planning** includes identifying which quality standards are relevant to the project and how to satisfy them. Incorporating quality standards into project design is a key part of quality planning. For an information technology project, quality standards might include allowing for system growth, planning a reasonable response time for a system, or ensuring that consistent and accurate information is produced. Quality standards can also apply to information technology services. For example, you can set standards for how long it should take to get a reply from a Help desk or how long it should take to ship a replacement part for a hardware item under warranty.

- **Quality assurance** involves periodically evaluating overall project performance to ensure the project will satisfy the relevant quality standards. The quality assurance process involves taking responsibility for quality during the project as well as at the end of the project. Senior management must take the lead in emphasizing the roles all employees play in quality assurance, especially senior managers' roles.
- **Quality control** involves monitoring specific project results to ensure that they comply with the relevant quality standards while identifying ways to improve overall quality. This process is often associated with the technical tools and techniques of quality management such as Pareto charts, quality control charts, and statistical sampling. These tools and techniques are covered later in this chapter.

MODERN QUALITY MANAGEMENT

Modern quality management requires customer satisfaction, prefers prevention to inspection, and recognizes management responsibility for quality. Several noteworthy people, including W. Edwards Deming, Joseph M. Juran, Philip B. Crosby, Koaru Ishikawa, Genichi Taguchi, and Armand V. Feigenbaum, helped develop modern quality management.[7] The suggestions from these quality experts led to many projects to improve quality.

Deming

Dr. W. Edwards Deming is primarily known for his work on quality control in Japan. Dr. Deming went to Japan after World War II, at the request of the Japanese government, to assist them in improving productivity and quality. Deming, a statistician and former professor at New York University, taught the Japanese that higher quality meant greater productivity and lower cost. By the 1980s, after seeing the excellent work coming out of Japan, U.S. corporations vied for Deming's expertise to help them establish quality improvement programs in their own factories. Many people are familiar with the Deming Prize, an award given to recognize high-quality organizations, and Deming's Cycle for Improvement: plan, do, check, and act. Many people are also familiar with Deming's 14 Points for Management, summarized below from Deming's text *Out of the Crisis*:

1. Create constancy of purpose for improvement of product and service.
2. Adopt the new philosophy.

[7] Kerzner, Harold, *Project Management*, 6th ed. New York: Van Nostrand Reinhold, 1998, 1048.

3. Cease dependence on inspection to achieve quality.
4. End the practice of awarding business on the basis of price tag alone. Instead, minimize total cost by working with a single supplier.
5. Improve constantly and forever every process for planning, production, and service.
6. Institute training on the job.
7. Adopt and institute leadership.
8. Drive out fear.
9. Break down barriers between staff areas.
10. Eliminate slogans, exhortations, and targets for the workforce.
11. Eliminate numerical quotas for the workforce and numerical goals for management.
12. Remove barriers that rob people of workmanship. Eliminate the annual rating or merit system.
13. Institute a vigorous program of education and self-improvement for everyone.
14. Put everybody in the company to work to accomplish the transformation.

Juran

Joseph M. Juran, like Deming, helped Japanese manufacturers learn how to improve their productivity. He, too, was later discovered by U.S. companies. He wrote the first edition of the *Quality Control Handbook* in 1974, stressing the importance of top management commitment to continuous product quality improvement. In 1999, at the age of 94, Juran published the fifth edition of this famous handbook. He also developed the Juran Trilogy: quality improvement, quality planning, and quality control. Juran stressed the difference between the manufacturer's view of quality and the customer's view. Manufacturers often focus on adherence to specifications, but customers focus on fitness-for-use. Most definitions of quality now use fitness-for-use to stress the importance of satisfying stated or implied needs and not just meeting stated requirements or specifications. Juran developed 10 steps to quality improvement:

1. Build awareness of the need and opportunity for improvement.
2. Set goals for improvement.
3. Organize to reach the goals (establish a quality council, identify problems, select projects, appoint teams, designate facilitators).
4. Provide training.
5. Carry out projects to solve problems.
6. Report progress.
7. Give recognition.
8. Communicate results.

9. Keep score.
10. Maintain momentum by making annual improvement part of the regular systems and processes of the company.

Crosby

Philip B. Crosby wrote *Quality Is Free* in 1979 and is best known for suggesting that organizations strive for zero defects. He stressed that the costs of poor quality should include all the costs of not doing the job right the first time, such as scrap, rework, lost labor hours and machine hours, customer ill will and lost sales, and warranty costs. Crosby suggests that the cost of poor quality is so understated that companies can profitably spend unlimited amounts of money on improving quality. Crosby developed the following 14 steps for quality improvement:

1. Make it clear that management is committed to quality.
2. Form quality improvement teams with representatives from each department.
3. Determine where current and potential quality problems lie.
4. Evaluate the cost of quality and explain its use as a management tool.
5. Raise the quality awareness and personal concern of all employees.
6. Take actions to correct problems identified through previous steps.
7. Establish a committee for the zero-defects program.
8. Train supervisors to actively carry out their part of the quality improvement program.
9. Hold a "zero-defects day" to let all employees realize that there has been a change.
10. Encourage individuals to establish improvement goals for themselves and their groups.
11. Encourage employees to communicate to management the obstacles they face in attaining their improvement goals.
12. Recognize and appreciate those who participate.
13. Establish quality councils to communicate on a regular basis.
14. Do it all over again to emphasize that the quality improvement program never ends.

Ishikawa

Kaoru Ishikawa is best known for his 1972 book *Guide to Quality Control*. He developed the concept of quality circles and pioneered the use of fishbone diagrams. **Quality circles** are groups of non-supervisors and work leaders in a single company department who volunteer to conduct group studies on how to improve the effectiveness of work in their department. Ishikawa suggested that Japanese managers and workers were totally committed to quality, but that most U.S. companies delegated the responsibility for quality to a few staff members. **Fishbone diagrams**, sometimes called Ishikawa diagrams, trace complaints about quality problems back to the responsible production operations. In other words, they help you find the root cause of quality problems.

Figure 7-1 provides an example of a fishbone or Ishikawa diagram that Scott Daniels, the consultant in the opening case, might create to trace the real cause of the problem of users not being able to get into the EIS. Notice that it resembles the skeleton of a fish, hence its name. This fishbone diagram lists the main areas that could be the cause of the problem: the EIS system's hardware, the individual user's hardware, software, or training. Two of these areas, the individual user's hardware and training, are described in more detail on this figure. The root cause of the problem would have a significant impact on the actions taken to solve the problem. If many users could not get into the system because their computers did not have enough memory, the solution might be to upgrade memory for those computers. If many users could not get into the system because they forgot their passwords, there might be a much quicker, less expensive solution.

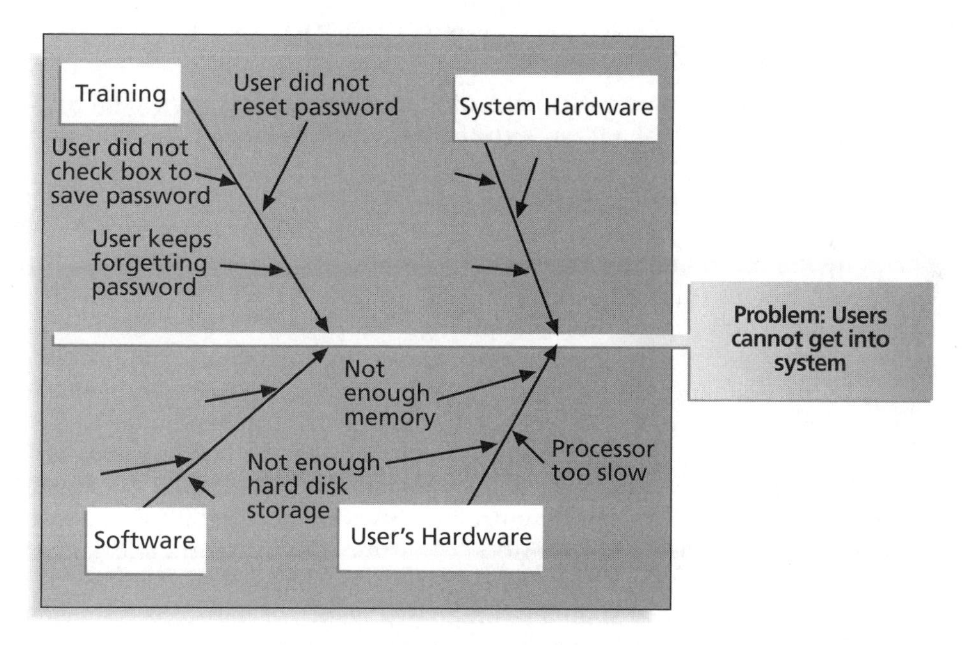

Figure 7-1. Sample Fishbone or Ishikawa Diagram

Taguchi

Genichi Taguchi is best known for developing the Taguchi methods for optimizing the process of engineering experimentation. Key concepts in the Taguchi methods are that quality should be designed into the product and not inspected into it and that quality is best achieved by minimizing deviation from the target value. For example, if the target response time for accessing the EIS described in the opening case is half a second, there should be little deviation from this time. A 1998 article in *Fortune* states that "Japan's Taguchi is America's new quality hero."[8] Many companies, including Xerox, Ford, Hewlett-Packard, and Goodyear, have recently used Taguchi's Robust Design methods to design high quality products. **Robust Design methods** focus on eliminating defects by substituting scientific inquiry for trial-and-error methods.

Feigenbaum

Armand V. Feigenbaum developed the concept of total quality control (TQC) in his 1983 book *Total Quality Control: Engineering and Management*. He proposed that the responsibility for quality should rest with the people who do the work. In TQC, product quality is more important than production rates, and workers are allowed to stop production whenever a quality problem occurs.

Malcolm Baldrige Award and ISO 9000

These quality experts helped define modern quality management. Today many people are familiar with the Malcolm Baldrige National Quality Award and ISO 9000. The **Malcolm Baldrige Award** was started in 1987 to recognize companies that have achieved a level of world-class competition through quality management. **ISO 9000**, a quality system standard developed by the International Organization for Standardization (ISO), is a three-part, continuous cycle of planning, controlling, and documenting quality in an organization. The International Organization for Standardization, based in Geneva, Switzerland, is a consortium of approximately one hundred industrial nations. ISO 9000 provides minimum requirements needed for an organization to meet their quality certification standards.

Modern quality management, quality awards, and quality standards are important parts of project quality management. The Project Management Institute was proud to announce in 1999 that their certification department had become the first certification department in the world to earn ISO 9000 certification, and that the *PMBOK Guide* 1996 had been recognized as an American National Standard. Emphasizing quality in project management helps ensure that projects produce products or services that meet customer needs and expectations.

[8] Bylinsky, Gene, "How to Bring Out Better Products Faster," *Fortune* (November 23, 1998) 238[B].

QUALITY PLANNING

Project managers today have a vast knowledge base of information related to quality, and the first step to ensuring project quality management is planning. Planning implies the ability to anticipate situations and prepare actions that bring about the desired outcome. The current thrust in modern quality management is the prevention of defects through a program of selecting the proper materials, training and indoctrinating people in quality, and planning a process that ensures the appropriate outcome. In project quality planning, it is important to identify relevant quality standards for each unique project and to design quality into the products of the project and the processes involved in managing the project.

Design of experiments is a quality technique that helps identify which variables have the most influence on the overall outcome of a process. Understanding which variables affect outcome is a very important part of quality planning. For example, computer chip designers might want to determine which combination of materials and equipment will produce the most reliable chips at a reasonable cost. You can also apply design of experiments to project management issues such as cost and schedule trade-offs. For example, junior programmers or consultants cost less than senior programmers or consultants, but you cannot expect them to complete the same level of work in the same amount of time. An appropriately designed experiment to compute project costs and durations for various combinations of junior and senior programmers or consultants can allow you to determine an optimal mix of personnel, given limited resources.

Planning also involves communicating the correct actions for ensuring quality in a form that is understandable and complete. In quality planning for projects, it is important to describe important factors that directly contribute to meeting the customer's requirements. Organizational policies related to quality, the particular project's scope statement and product descriptions, and related standards and regulations are all important inputs to the quality planning process. The main outputs of quality planning are a quality management plan and checklists for ensuring quality throughout the project life cycle.

As mentioned in the discussion of project scope management (*see* Chapter 4), it is often difficult to completely understand the performance dimension of information technology projects. Even if hardware, software, and networking technology would stand still for a while, it is often difficult for customers to explain exactly what they want in an information technology project. Important scope aspects of information technology projects that affect quality include functionality and features, system outputs, performance, and reliability and maintainability.

- **Functionality** is the degree to which a system performs its intended function. **Features** are the special characteristics that appeal to users. It is important to clarify what business functions and features the system *must* perform, and what functions and features are *optional*. In the EIS example, the mandatory functionality of the system might be that it allows users to track sales of specific medical instruments by predetermined categories such as the product group, country, hospital, and sales representative.

Mandatory features might be a graphical user interface with icons, drop-down menus, on-line help, and so on. One of the problems users experienced was forgetting passwords. Perhaps a required feature was missed that could alleviate this problem.

- **System outputs** are the screens and reports the system generates. It is important to clearly define what the screens and reports look like for a system. Can the users easily interpret these outputs? Can users get all of the reports they need in a suitable format?
- **Performance** addresses how well a product or service performs the customer's intended use. In order to design a system with high quality performance, project stakeholders must address many issues. What volumes of data and transactions should the system be capable of handling? How many simultaneous users should the system be designed to handle? What is the projected growth rate in the number of users? What type of equipment must the system run on? How fast must the response time be for different aspects of the system under different circumstances? For the EIS, several of the quality problems appear to be related to performance issues. The system is going down a couple of times a month, and users are unsatisfied with the response time. The project team may not have had specific performance requirements or tested the system under the right conditions to deliver the expected performance. Buying faster hardware might address these performance issues. Another performance problem that might be more difficult to fix is the fact that some of the reports are generating inconsistent results. This appears to be a quality problem with the software that may be difficult and costly to correct since the system is already in operation.
- **Reliability** is the ability of a product or service to perform as expected under normal conditions without unacceptable failures. **Maintainability** addresses the ease of performing maintenance on a product. Most information technology products cannot reach 100 percent reliability, but stakeholders must define what their expectations are. For the EIS, what are the normal conditions for operating the system? Should reliability tests be based on one hundred people accessing the system at once and running simple queries? Maintenance for the EIS might include uploading new data into the system or performing maintenance procedures on the system hardware and software. Are the users willing to have the system be unavailable several hours a week for system maintenance? Providing Help desk support could also be viewed as a maintenance function. How fast a response do users expect for help desk support? How often can system failure be tolerated? Are the stakeholders willing to pay more for higher reliability and less failures?

These aspects of project scope are just a few of the requirements issues related to quality planning. Project managers and their teams need to consider all of these project scope issues in determining quality goals for the project. The main customers for the project must also realize their role in defining the most critical quality needs for the project and constantly communicate these needs and expectations to the project team. Since most information technology projects involve requirements that are not set in stone, it is important for all project stakeholders to work together to balance the quality, scope, time, and cost dimensions of the project. *Project managers, however, are ultimately responsible for quality management on their projects.*

QUALITY ASSURANCE

It is one thing to develop a plan for ensuring quality on a project; it is another to ensure that quality products and services are actually delivered. Quality assurance includes all of the activities related to satisfying the relevant quality standards for a project. Another goal of quality assurance is continual quality improvement.

Senior management and project managers can have the greatest impact on the quality of projects by doing a good job of quality assurance. The importance of leadership in improving information technology project quality is discussed in more detail later in this chapter.

Several tools used in quality planning can also be used in quality assurance. Design of experiments, as described under quality planning, can also help ensure and improve product quality. **Benchmarking** can be used to generate ideas for quality improvements by comparing specific project practices or product characteristics to those of other projects or products within or outside the performing organization. For example, if a competitor has an EIS with an average downtime of only one hour a week, that might be a benchmark to strive for. Fishbone or Ishikawa diagrams, as described earlier, can assist in ensuring and improving quality by finding the root causes of quality problems.

One of the main tools and techniques for quality assurance is a quality audit. **Quality audits** are structured reviews of specific quality management activities that help identify lessons learned that can improve performance on current or future projects. Quality audits can be scheduled or random, and they can be performed by in-house auditors or by third parties with expertise in specific areas. Industrial engineers often perform quality audits by helping to design specific quality metrics for a project and then applying and analyzing the metrics throughout the project. For example, the Northwest Airlines ResNet project (described in Chapters 12-16) provides an excellent example of using quality audits to emphasize the main goals of a project and then track progress in reaching those goals. The main objective of the ResNet project was to develop a new reservation system to increase direct airline ticket sales and reduce the time it took for sales agents to handle customer calls. The measurement techniques for monitoring these goals helped ResNet's project manager and team supervise various aspects of the project by focusing on meeting those goals. Measuring progress toward increasing direct sales and reducing call times also helped the project manager justify continued investments in ResNet.

QUALITY CONTROL

Many people only think of quality control when they think of quality management. Perhaps it is because there are many popular tools and techniques in this area. Before describing some of these tools and techniques, it is important to distinguish quality control from quality planning and quality assurance.

Although one of the main goals of quality control is also to improve quality, the main outputs of this process are acceptance decisions, rework, and process adjustments.

- **Acceptance decisions** determine if the products or services produced as part of the project will be accepted or rejected. If project stakeholders reject some of the products or services produced as part of the project, there must be rework. For example, the executive who sponsored development of the EIS in the opening case was obviously not satisfied with the system and hired an outside consultant, Scott Daniels, to lead a team to address and correct the quality problems.
- **Rework** is action taken to bring rejected items into compliance with product requirements or specifications or other stakeholder expectations. Rework can be very expensive, so the project manager must strive to do a good job of quality planning and assurance to avoid this need. Since the EIS did not meet all of the stakeholders' expectations for quality, the medical instruments company was spending a lot of money for rework.
- **Process adjustments** correct or prevent further quality problems based on quality control measurements. For example, Scott Daniels, the consultant in the opening case, might recommend that the medical instruments company purchase a faster server for the EIS application to correct the response time problems. He was also hired to develop a plan to help prevent future information technology project quality problems.

TOOLS AND TECHNIQUES FOR QUALITY CONTROL

Many general tools and techniques are used in quality control. This section focuses on just a few of them—Pareto analysis, statistical sampling, and quality control charts—and discusses how they can be applied to information technology projects. Testing is also discussed since it is used extensively on information technology projects to ensure quality.

Pareto Analysis

Pareto analysis involves identifying the vital few contributors that account for most quality problems in a system. It is sometimes referred to as the 80-20 rule, meaning that 80 percent of problems are often due to 20 percent of the causes.

Pareto diagrams are histograms that help identify and prioritize problem areas. The variables described by the histogram are ordered by frequency of occurrence.

For example, suppose there was a history of user complaints about the EIS described in the opening case. Many organizations log user complaints, but many do not know how to analyze them. A Pareto diagram is an effective tool for analyzing user complaints and helping project managers decide which ones to address first. Figure 7-2 shows a Pareto diagram of user complaints by category that could apply to the EIS project. The bars represent the number of complaints for each category, and the line shows the cumulative percent of the complaints. For example, Figure 7-2 shows that the most frequent user complaint was log-in problems, followed by the system locking up, the system being too slow, the system being hard to use, and the reports being inaccurate. The first complaint accounts for 55 percent of the total complaints. The first and second complaints have a cumulative percentage of almost 80 percent, meaning these two areas account for 80 percent of the complaints. Therefore, the company should focus on making it easier to log in to the system, to improve quality, since the majority of complaints fall under that category. They should also address why the system locks up. Because the problem of inaccurate reports was rarely mentioned, the project manager should investigate who made this complaint before spending a lot of effort on addressing that potentially critical problem with the system. He/she should also find out if complaints about the system being too slow were actually due to the user not being able to log in or the system locking up.

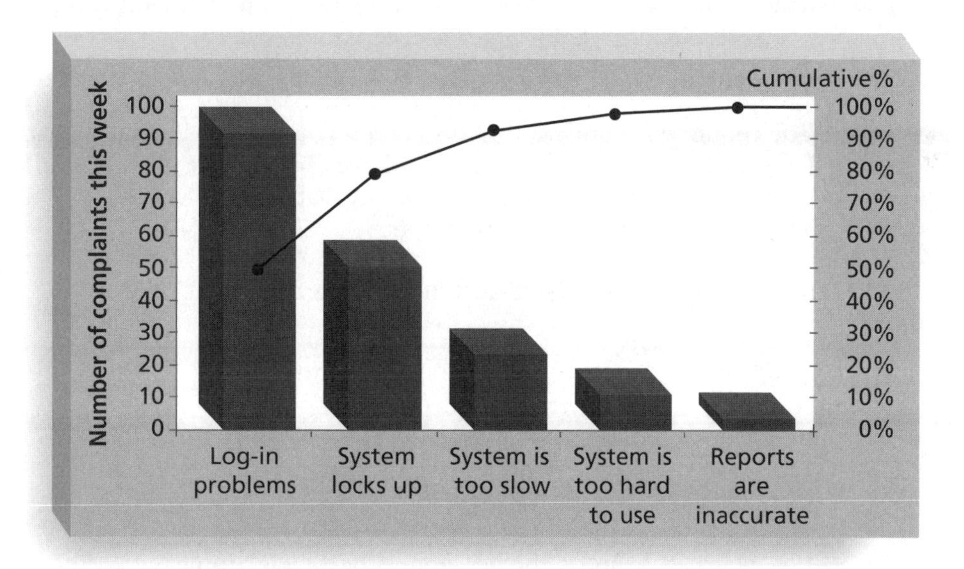

Figure 7-2. Sample Pareto Diagram

Statistical Sampling and Standard Deviation

Statistical sampling is a key concept in project quality management. Members of a project team who focus on quality control must have a strong understanding of statistics, but other project team members need to understand only the basic concepts. These concepts include statistical sampling, certainty factor, standard deviation, and variability. Standard deviation and variability are fundamental concepts for understanding quality control charts. This section briefly describes these concepts and how they might be applied on information technology projects.

Statistical sampling involves choosing part of a population of interest for inspection. For example, suppose a company wants to develop an electronic data interchange (EDI) system for handling data on invoices from all of its suppliers. Also assume that in the past year, the total number of invoices was fifty thousand from two hundred different suppliers. It would be very time-consuming and expensive to review every single invoice to determine data requirements for the new system. Even if the system developers did review all two hundred invoice forms, the data might be entered differently on every form. It is impractical to study every member of a population, such as all fifty thousand invoices, so statisticians have developed techniques to help determine an appropriate sample size. If the system developers used statistical techniques, they might find that by studying only one hundred invoices, they would have a good sample of the type of data they would need in designing the system.

The size of the sample depends on how representative you want the sample to be. A simple formula for determining sample size is

Sample size = .25 × (certainty factor/acceptable error)^2

The certainty factor denotes how certain you want to be that the data sampled will not include variations that do not naturally exist in the population. You calculate the certainty factor from tables available in statistics books. Table 7-1 shows commonly used certainty factors.

Table 7-1: Commonly Used Certainty Factors

Desired Certainty	Certainty Factor
95%	1.960
90%	1.645
80%	1.281

For example, suppose the developers of the electronic data interchange system described earlier would accept a 95 percent certainty that a sample of invoices would contain no variation unless it was present in the population of total invoices. They would then calculate the sample size as:

Sample size = 0.25 × (1.960/.05)^2 = 384

If the developers wanted 90 percent certainty, they would calculate the sample size as:

Sample size = $0.25 \times (1.645/.10)^2 = 68$

If the developers wanted 80 percent certainty, they would calculate the sample size as:

Sample size = $0.25 \times (1.281/.20)^2 = 10$

Assume the developers decide on 90 percent for the certainty factor. Then they would need to examine 68 invoices to determine the type of data the EDI system would need to capture.

Another key concept in statistics related to quality control is standard deviation. **Standard deviation** measures how much variation exists in a distribution of data. A small standard deviation means that data cluster closely around the middle of a distribution and there is little variability among the data. A large standard deviation means that data are spread out around the middle of the distribution and there is relatively greater variability. Statisticians use the Greek symbol σ (sigma) to represent the standard deviation.

Figure 7-3 provides an example of a **normal distribution**—a bell-shaped curve that is symmetrical about the **mean** or average value of the population. In any normal distribution, 68.3 percent of the population is within one standard deviation (1σ) of the mean. Ninety-five point five percent of the population is within two standard deviations (2σ), and 99.7 percent of the population is within three standard deviations (3σ) of the mean. Note that being within three sigma of the mean in this example means plus or minus three sigma to mean six sigma.

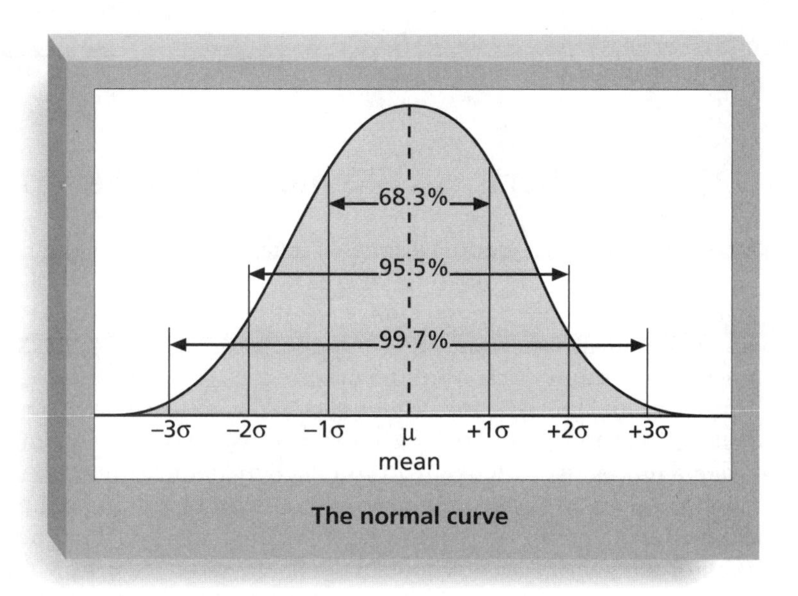

Figure 7-3. Normal Distribution and Standard Deviation

Standard deviation is important in quality control because it is a key factor in determining the acceptable number of defective units. Some companies, such as Motorola, GE, and Polaroid, are setting high quality standards by using six sigma (6σ) as a quality control standard instead of four sigma like most companies use. The lead article of the March 6, 2001 issue of 3M's *Stemwinder* newspaper announced that 3M leaders met to launch six sigma activities across the company. The objectives of the 3M Six Sigma Initiative are to accelerate 3M's growth, significantly improve productivity, and produce a substantial increase in cash flow. Six sigma is considered to be one of the best-known American contributions to quality improvement.

Table 7-2 further illustrates the relationship between sigma, the percentage of the population within that sigma range, and the number of defective units per billion. For example, plus or minus three sigma, or six sigma, means that there would be 2.7 million defective units per billion, or 2.7 defective units per million. The next section describes how sigma is used on quality control charts.

Table 7-2: Sigma and Defective Units

Specification Range (in +/- Sigmas)	Percent of Population Within Range	Defective Units Per Billion
1	68.27	317,300,000
2	95.45	45,400,000
3	99.73	2,700,000
4	99.9937	63,000
5	99.999943	57
6	99.9999998	2

Quality Control Charts, Six Sigma, and the Seven Run Rule

A **control chart** is a graphic display of data that illustrates the results of a process over time. The main use of control charts is to prevent defects, rather than to detect or reject them. Quality control charts allow you to determine whether a process is in control or out of control. When a process is in control, any variations in the results of the process are created by random events. Processes that are in control do not need to be adjusted. When a process is out of control, variations in the results of the process are caused by nonrandom events. When a process is out of control, you need to identify the causes of those nonrandom events and adjust the process to correct or eliminate them. Control charts are often used to monitor manufactured lots, but they can also be used to monitor the volume and frequency of change requests, errors in documents, cost and schedule variances, and other items related to project quality management.

Figure 7-4 illustrates an example of a control chart for a process that manufactures twelve-inch rulers. Assume that these are wooden rulers created by machines on an assembly line. Each point on the chart represents a length measurement for a ruler that comes off the assembly line. The scale on the vertical axis goes from 11.90 to 12.10. These numbers represent the lower and upper specification limits for the ruler. In this case, this would mean that the customer for the rulers has specified that all rulers they buy must be between 11.90 and 12.10 inches long, or twelve inches plus or minus 0.10 inches. The lower and upper control limits on the quality control chart are 11.91 and 12.09 inches, respectively. This means the manufacturing process is designed to produce rulers between 11.91 and 12.09 inches long.

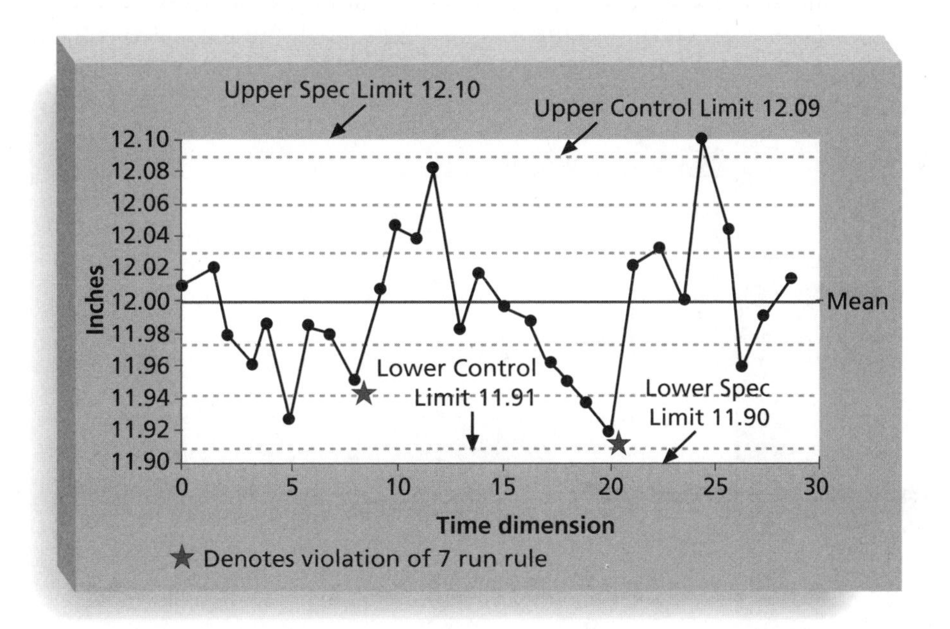

Figure 7-4. Sample Quality Control Chart

Note the dotted lines on the chart at 12.03 inches, 12.06 inches, and 12.09 inches. These dotted lines represent the points at one, two, and three standard deviations above the mean. The dotted lines at 11.97 inches, 11.94 inches, and 11.91 inches represent the points at one, two, and three standard deviations below the mean. Based on the definition of $\pm 3\sigma$ described earlier, 99.73 percent of the manufactured rulers should come off the assembly line with a measurement between 11.91 and 12.09 inches, if the manufacturing process is operating in control.

Figure 7-5 illustrates the concept of moving from a quality control process operating at $\pm 3\sigma$ to one operating at $\pm 6\sigma$. Important goals of quality are to reduce defects and process variability. By reducing process variability, the standard deviation of the process distribution becomes smaller. As you continue to reduce process variability, it is possible for the product tolerance or control limits that formerly included only plus or minus three sigmas to eventually include plus or minus six sigmas.[9]

You must be careful, however, to apply higher quality where it makes sense. A recent article in *Fortune* states that companies that have implemented six sigma initiatives have not necessarily boosted their stock values. Although some companies like GE boasted savings of more than $2 billion in 1999 due to its use of six sigma, other companies like Whirlpool cannot clearly demonstrate the value of their investments in six sigma. Why can't all companies benefit from six sigma? Because defects don't matter if you are making a product no one wants to buy. As one of six sigma's biggest supporters, Mikel Harry, puts it, "I could genetically engineer a Six Sigma goat, but if a rodeo is the marketplace, people are still going to buy a Four Sigma horse."[10]

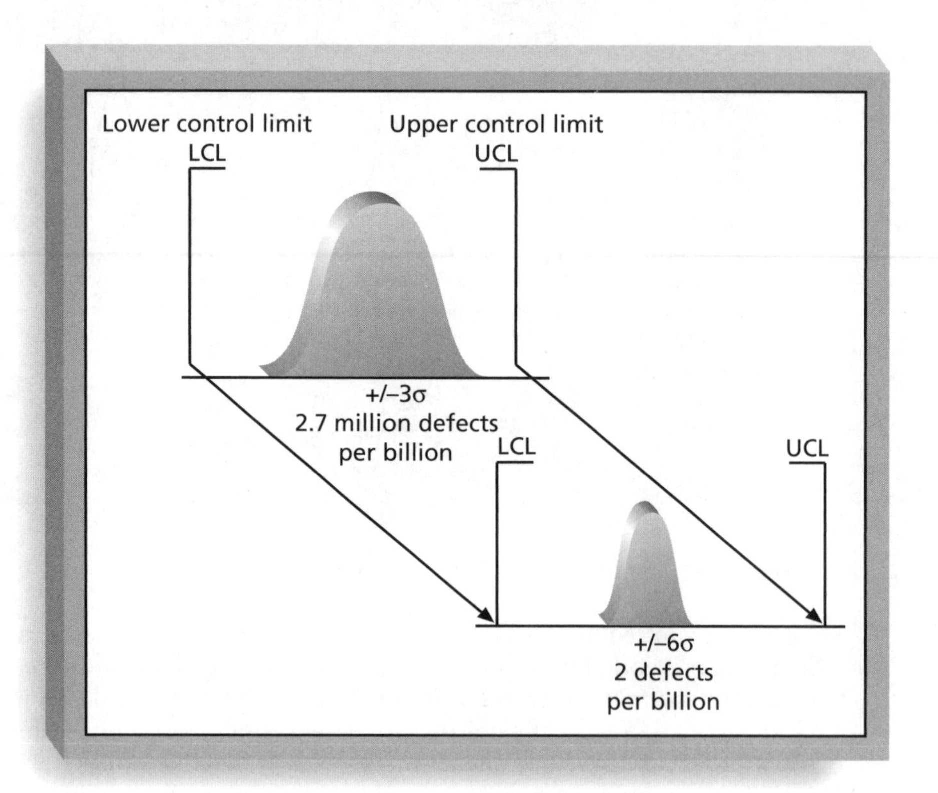

Figure 7-5. Reducing Defects with Six Sigma

[9] Ireland, Lewis R, *Quality Management for Projects and Programs*. PMI, 1991, cover page illustration and comments.

[10] Clifford, Lee, "Why You Can Safely Ignore Six Sigma," *Fortune* (January 22, 2001) 140.

Looking for and analyzing patterns in process data is an important part of quality control. You can use quality control charts and the seven run rule to look for patterns in data. The **seven run rule** states that if seven data points in a row are all below the mean, above the mean, or are all increasing or decreasing, then the process needs to be examined for non-random problems. In Figure 7-4, data points that violate the seven run rule are starred. In the ruler manufacturing process, these data points may indicate that a calibration device may need to be adjusted. For example, the machine that cuts the wood for the rulers might need to be adjusted or the blade on the machine might need to be replaced.

Testing

Many information technology professionals think of testing as a stage that comes near the end of information technology product development. Instead of putting serious effort into proper planning, analysis, and design of information technology projects, some companies rely on testing just before a product ships to ensure some degree of quality. In fact, testing needs to be done during almost every phase of the product development life cycle, not just before a product is shipped or handed over to the customer.

Figure 7-6 shows one way of portraying the software development project life cycle. This example includes seventeen main phases involved in a software development project and shows their relationships to each other. For example, every project should start by initiating the project, then doing a feasibility study, and then performing project planning. The figure then shows that the work involved in preparing detailed requirements and the detailed architecture for the system can be performed simultaneously. The oval-shaped phases represent areas where you should do testing to help ensure quality on software development projects.[11]

Several of the phases in Figure 7-6 include specific work related to testing. A **unit test** is done to test each individual component (often a program) to ensure it is as defect-free as possible. Unit tests are performed before moving on to the integration test. **Integration testing** occurs between unit and system testing to test functionally grouped components. It ensures a subset(s) of the entire system works together. **System testing** tests the entire system as one entity. It focuses on the big picture to ensure the entire system is working properly. **User acceptance testing** is an independent test performed by end users prior to accepting the delivered system. It focuses on the business fit of the system to the organization, rather than technical issues.

[11] Hollstadt & Associates, Inc., *Software Development Project Life Cycle Testing Methodology User's Manual*. Burnsville: MN, August 1998, 13.

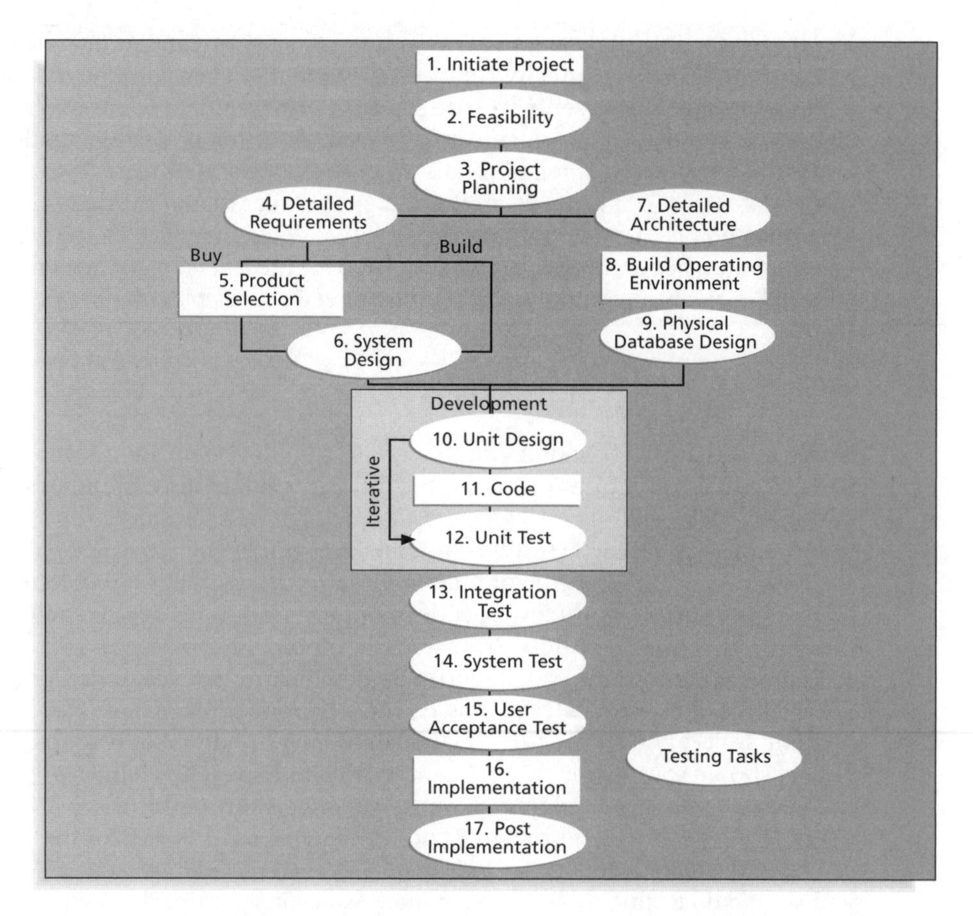

Figure 7-6. Testing Tasks in the Software Development Life Cycle

Figure 7-7 presents a Gantt chart that shows testing tasks that are appropriate for different phases of the software development project life cycle.[12] This is a simplified version of a detailed testing management plan used by a consulting firm. To help improve the quality of software development projects, it is important for organizations to follow a thorough and disciplined testing methodology. System developers and testers must also establish a partnership with all project stakeholders to make sure the system meets their needs and expectations and the tests are done properly.

[12] Ibid., 54–57.

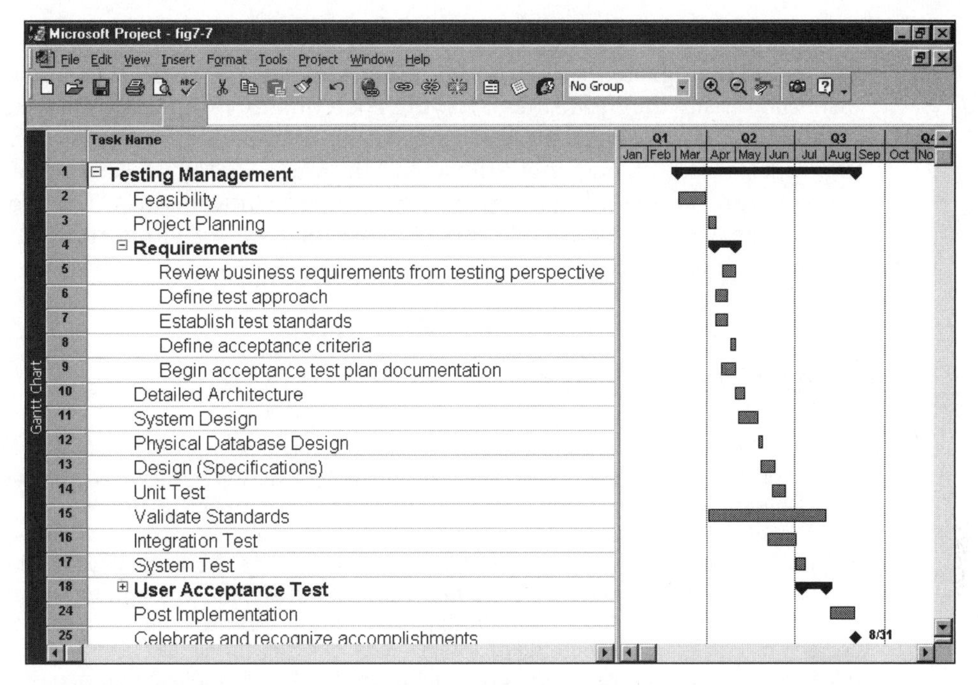

Figure 7-7. Gantt Chart for Building Testing into a Systems Development Project Plan

IMPROVING INFORMATION TECHNOLOGY PROJECT QUALITY

In addition to some of the suggestions provided for using good quality planning, quality assurance, and quality control, there are several other important issues involved in improving the quality of information technology projects. Strong leadership, understanding the cost of quality, providing a good workplace to enhance quality, and working toward improving the organization's overall maturity level in software development and project management can all assist in improving quality.

Leadership

As Joseph M. Juran said in 1945, "It is most important that top management be quality-minded. In the absence of sincere manifestation of interest at the top, little will happen below."[13] Juran and many other quality experts argue that the main cause of quality problems is a lack of leadership.

13 American Society for Quality (ASQ), Web site, http://www.asqc.org/about/history/juran.html.

As globalization continues to increase and customers become more and more demanding, creating quality products quickly at a reasonable price is essential for staying in business. Having good quality programs in place helps companies remain competitive. To establish and implement effective quality programs, senior management must lead the way. A large percentage of quality problems are associated with management, not technical issues. Therefore, senior management must take responsibility for creating, supporting, and promoting quality programs.

What Went Right?

In 1988, Motorola, Inc., an electronics equipment manufacturer with over 100,000 employees worldwide, won the Malcolm Baldrige National Quality Award. Just a few years earlier, however, Motorola was experiencing major quality problems in manufacturing electronic components.

In 1981, Motorola senior management set a goal to improve overall quality tenfold. The first step to improving quality involved defining a quality metric, a system for measuring quality. After gathering suggestions from people throughout the organization, Motorola decided on a single quality metric of Total Defects per Unit for all divisions. Motorola defined a unit as work applied to a product or service. For example, a unit of work could be an item of equipment, a circuit board assembly, a line of software code, a page in a technical manual, and any output a department produced. They defined a defect as anything that caused customer dissatisfaction, regardless of whether the problem was included in the original work specification.

By 1986, one sector of Motorola exceeded the ten-times-improvement goal. The following January, the company established its next goals. The first goal was reducing the number of defects per unit to one percent of the existing level of defects. The second goal was achieving six sigma capability by 1992. Motorola has surpassed all of these quality goals and continues to set new quality goals. Strong leadership improved quality at Motorola and helped the company succeed in a highly competitive business environment.[14]

Motorola provides an excellent example of a high technology company that truly emphasizes quality (*see* "What Went Right"). Leadership is one of the factors that helped Motorola achieve its great success in quality management. Senior management emphasized the need to improve quality and helped all employees take responsibility for customer satisfaction. Strategic objectives in Motorola's long-range plans included managing quality improvement in the same way that new products or technologies were managed. Senior management stressed the need to develop and use quality standards and provided resources such as staff, training, and customer inputs to help improve quality.

Leadership is needed to provide an environment conducive to producing quality. Management must publicly declare the company's philosophy and

[14] Smith, Bill, "The Motorola Story," A paper distributed by Motorola that describes the 1988 Malcolm Baldrige National Quality Award, 1989.

commitment to quality, implement company-wide training programs in quality concepts and principles, implement measurement programs to establish and track quality levels, and actively demonstrate the importance of quality. When every employee understands and insists on producing high quality products, then senior management has done a good job in promoting the importance of quality.

The Cost of Quality

The **cost of quality** is the cost of conformance plus the cost of nonconformance. **Conformance** means delivering products that meet requirements and fitness for use. Examples of these costs include the costs associated with developing a quality plan, costs for analyzing and managing product requirements, and costs for testing. The **cost of nonconformance** means taking responsibility for failures or not meeting quality expectations.

Table 7-3 summarizes the net bottom-line costs per hour of downtime caused by software defects for different businesses.[15] Both American Airlines and United Airlines have estimated that reservation system downtime costs them over $20,000 per minute in lost income. Table 7-3 illustrates more examples of the cost of nonconformance.

Table 7-3: Costs Per Hour of Downtime Caused by Software Defects

Business	Cost per Hour Downtime
Automated teller machines (medium-sized bank)	$14,500
Package shipping service	$28,250
Telephone ticket sales	$69,000
Catalog sales center	$90,000
Airline reservation center (small airline)	$89,500

The five major cost categories related to quality include:

1. **Prevention cost** — the cost of planning and executing a project so that it is error-free or within an acceptable error range. Preventive actions such as training, detailed studies related to quality, and quality surveys of suppliers and subcontractors fall under this category. Recall from the discussion of cost management (*see* Chapter 6) that detecting defects in information systems during the early phases of system development is much less expensive than during the later phases. One hundred dollars spent refining user requirements could save millions by finding a defect before

[15] Contingency Planning Research, White Plains, NY, 1995.

implementing a large system. The Year 2000 issue provides a good example of these costs. If companies had decided during the 1960s, 1970s, and 1980s that all dates would need four characters to represent the year instead of two characters, billions of dollars would have been saved.

2. **Appraisal cost** — the cost of evaluating processes and their outputs to ensure that a project is error-free or within an acceptable error range. Activities such as inspection and testing of products, maintenance of inspection and test equipment, and processing and reporting inspection data all contribute to appraisal costs of quality.
3. **Internal failure cost** — a cost incurred to correct an identified defect before the customer receives the product. Items like scrap and rework, charges related to late payment of bills, inventory costs that are a direct result of defects, costs of engineering changes related to correcting a design error, premature failure of products, and correcting documentation all contribute to internal failure cost.
4. **External failure cost** — a cost that relates to all errors not detected and corrected before delivery to the customer. Items such as warranty cost, field service personnel training cost, product liability suits, complaint handling, and future business losses are examples of external failure costs.
5. **Measurement and test equipment costs** — the capital cost of equipment used to perform prevention and appraisal activities.

Many industries tolerate a very low cost of non-conformance, but not the information technology industry. Tom DeMarco is famous for several studies done on the cost of non-conformance in the information technology industry. DeMarco found that the average large company devoted over 60 percent of its software development efforts to maintenance. Around 50 percent of development costs are typically spent on testing and debugging software.[16]

Senior management is primarily responsible for the high cost of nonconformance in information technology. Senior managers often rush their organizations to develop new systems and do not give project teams enough time or resources to do a project right the first time. To correct these quality problems, senior management must create a culture that embraces quality.

Organizational Influences, Workplace Factors, and Quality

A study done by Tom DeMarco and Timothy Lister produced interesting results related to organizations and relative productivity. Starting in 1984, DeMarco and Lister conducted "Coding War Games" over several years. Over the years, more than six hundred software developers from ninety-two organizations have participated in these games. The games are designed to examine programming quality and productivity over a wide range of organizations, technical

[16] DeMarco, Tom, *Controlling Software Projects: Management, Measurement and Estimation.* New York: Yourdon Press, 1982.

environments, and programming languages. The study demonstrated that organizational issues had a much greater influence on productivity than the technical environment or programming languages.

For example, DeMarco and Lister found that productivity varied by a factor of about one to ten across all participants. That is, one team may have finished a coding project in one day while another team took ten days to finish the same project. In contrast, productivity varied by an average of only twenty-one percent between pairs of software developers from the same organization. If one team from a specific organization finished the coding project in one day, the longest time it took for another team from the same organization to finish the project was 1.21 days.

DeMarco and Lister also found *no correlation* between productivity and programming language, years of experience, or salary. Furthermore, the study showed that providing a dedicated workspace and a quiet work environment were key factors in improving productivity. The results of the study suggest that senior managers must focus on workplace factors to improve productivity and quality.[17]

Maturity Models

Another approach to improving quality in software development projects and project management in general is the use of **maturity models**, which are frameworks for helping organizations improve their processes and systems. Three popular maturity models include the Software Quality Function Deployment (SQFD) model, the Capability Maturity Model (CMM), and the Project Management Maturity Model.

Software Quality Function Deployment Model

The Software Quality Function Deployment (SQFD) model is an adaptation of the quality function deployment model suggested in 1986 as an implementation vehicle for total quality management. SQFD focuses on defining user requirements and planning software projects. The end result of SQFD is a set of measurable technical product specifications and their priorities. Having clearer requirements can lead to fewer design changes, increased productivity, and, ultimately, software products that are more likely to satisfy stakeholder requirements. The idea of introducing quality early in the design stage was based on Taguchi's emphasis on robust quality.[18]

[17] DeMarco, Tom and Timothy Lister, *Peopleware: Productive Projects and Teams.* New York: Dorset House, 1987.

[18] Yilmaz, R. R. and Sangit Chatterjee, "Deming and the Quality of Software Development," *Business Horizons*, Foundation for the School of Business at Indiana University, 40, No. 6 (November-December 1997) 51(8).

Capability Maturity Model

Another popular maturity model is in continuous development at the Software Engineering Institute at Carnegie Mellon University. The Software Engineering Institute (SEI) is a federally funded research and development center established in 1984 by the U.S. Department of Defense with a broad mandate to address the transition of software engineering technology. The **Capability Maturity Model (CMM)** is a five-level model laying out a generic path to process improvement for software development in organizations. Watts Humphrey's book *Managing the Software Process* (1989) describes an earlier version of the CMM model.[19]

The five levels of the CMM model are:

1. *Initial*: The software development processes for organizations at this maturity level are ad hoc and occasionally even chaotic. Few processes are defined, and success often depends on individual effort.
2. *Repeatable*: Organizations at this maturity level have established basic project management processes to track cost, schedule, and functionality for software projects. Process discipline is in place to repeat earlier successes on similar projects.
3. *Defined*: At this level, the software processes for both management and software engineering activities are documented, standardized, and integrated into a standard software process for the organization. All projects use an approved, tailored version of the standard process in the organization.
4. *Managed*: At this maturity level, organizations collect detailed measures of the software process and product quality. Both the software processes and products are quantitatively understood and controlled.
5. *Optimizing*: Operating at the highest level of the maturity model, organizations can enable continuous process improvement by using quantitative feedback from the processes and from piloting innovative ideas and technologies.[20]

Project Management Maturity Model

In the late 1990s, several organizations began developing project management maturity models based on the Capability Maturity Model. Just as organizations realized the need to improve their software development processes and systems, they also realized the need to enhance their project management processes and systems.

The PMI Standards Development Program made substantial progress on an Organizational Project Management Maturity Model (OPM3) Standard in 1998. PMI is working with several companies to prepare guidelines for a maturity model for the entire project management profession. The OPM3 Web site (*see* Suggested Readings) provides information to assist organizations in implementing

[19] Humphrey, W. S, *Managing the Software Process*. Reading, MA: Addison-Wesley, 1989.

[20] Paulk, Mark C., Bill Curtis, Mary Beth Chrissis, and Charles V. Weber, *Capability Maturity Model for Software*, Version 1.1, Technical Report, CMU/SEI-93-TR-024, ESC-TR-93-177 (February 1993).

organization strategy through the successful, consistent, and predictable delivery of projects. The OPM3 will include a method for assessing organizations' project management maturity levels as well as a step-by-step method for increasing and maintaining an organization's ability to deliver projects as promised. One sample project management maturity model developed by Micro-Frame Technologies, Inc. and Project Management Technologies, Inc. in 1997 has the following basic levels:

1. *Ad-Hoc*: The project management process is described as disorganized and occasionally even chaotic. The organization has not defined systems and processes, and project success depends on individual effort. There are chronic cost and schedule problems.
2. *Abbreviated*: There are some project management processes and systems in place to track cost, schedule, and scope. Project success is largely unpredictable and cost and schedule problems are common.
3. *Organized*: There are standardized, documented project management processes and systems that are integrated into the rest of the organization. Project success is more predictable, and cost and schedule performance is improved.
4. *Managed*: Management collects and uses detailed measures of the effectiveness of project management. Project success is more uniform, and cost and schedule performance conforms to plan.
5. *Adaptive*: Feedback from the project management process and from piloting innovative ideas and technologies enables continuous improvement. Project success is the norm, and cost and schedule performance is continuously improving.[21]

William Ibbs and Young H. Kwak have been researching project management maturity in a series of studies, partially funded by the PMI Educational Foundation. They developed an assessment methodology to measure project management maturity, consisting of 148 multiple-choice questions based on the project management knowledge areas and process groups. Developing this systematic assessment methodology was a major contribution to the study of project management maturity. Companies can benchmark their project management maturity using factual, impartial techniques such as those contained in this questionnaire, giving them a good reference point from which to begin making process improvements. Preliminary findings of Ibbs' and Kwak's research show that certain industries, such as engineering and construction, have a higher maturity level than other industries, such as information systems/software development.[22]

Many organizations are assessing where they stand in terms of project management maturity, just as they did for software development maturity. Organizations are recognizing that they must make a commitment to the discipline of project management in order to improve project quality.

[21] Enterprise Planning Associates, Project Management Maturity Model, Interactive Quick Look (1998).

[22] Ibbs, C. William and Young Hoon Kwak, "Assessing Project Management Maturity," *Project Management Journal* (March 2000).

CASE WRAP-UP

Scott Daniels assembled a team to identify and resolve quality-related issues with the EIS and to develop a plan to help the medical instruments company prevent future quality problems. The first thing Scott's team did was to research the problems with the EIS. They created a fishbone diagram similar to the one in Figure 7-1 (see page 200). They also created a Pareto chart (see Figure 7-2, page 206) to help analyze the many complaints the Help desk received and documented about the EIS. After further investigation, Scott and his team found out that many of the managers using the system were very inexperienced in using computer systems. The system was designed to automatically generate a user password that included letters and numbers. Many managers did not know how to change their passwords to something they could remember or to have the system save their password. Scott also found out that several people who had just received passwords had misinterpreted the number one in their password for the lowercase letter L. There did not appear to be any major problems with the hardware for the EIS or the users' individual computers. The complaints about reports not giving consistent information all came from one manager, who had actually misread the reports, so there were no problems with the way the software was designed. Scott was very impressed with the quality of the entire project, except for the training. Scott reported his team's findings to the sponsor of the EIS, who was relieved to find out that the quality problems were not as serious as many people feared they were.

CHAPTER SUMMARY

Many news headlines regarding the poor quality of information technology projects demonstrate that quality is a serious issue in information technology projects. Several mission-critical information technology systems have caused deaths, and quality problems in many business systems have resulted in major financial losses.

Customers are ultimately responsible for defining quality. Important quality concepts include satisfying stated or implied stakeholder needs, conforming to requirements, and delivering items that are fit for use.

Project quality management includes quality planning, quality assurance, and quality control. Quality planning identifies which quality standards are relevant to the project and how to satisfy them. Quality assurance involves evaluating overall project performance to ensure the project will satisfy the relevant quality standards. Quality control includes monitoring specific project results to ensure that they comply with quality standards and identifying ways to improve overall quality.

Many people contributed to the development of modern quality management. Deming, Juran, Crosby, Ishikawa, Taguchi, and Feigenbaum all made significant contributions to the field.

There are many tools and techniques related to project quality management. Fishbone diagrams help find the root cause of problems. Pareto charts help identify the vital few contributors that account for the most quality problems. Statistical sampling helps define a realistic number of items to include in analyzing a population. Standard deviation measures the variation in data. Control charts display data to help keep processes in control. Six sigma has helped many companies reduce the number of defective items. Testing is very important in developing and delivering quality information technology products.

There is much room for improvement in information technology project quality. Strong leadership helps emphasize the importance of quality. Understanding the cost of quality provides an incentive for its improvement. Providing a good workplace can improve quality and productivity. Developing and following maturity models can help organizations systematically improve their project management processes to increase the quality and success rate of projects.

Discussion Questions

1. Discuss some of the examples of poor quality in information technology projects presented in the "What Went Wrong?" section. Could most of these problems have been avoided? Why do you think there are so many examples of poor quality in information technology projects?
2. Provide other examples of quality problems in information technology projects and discuss how these quality problems can be avoided.
3. Discuss how a project team can know if their project delivers good quality.
4. Discuss the history of modern quality management. How have experts such as Deming, Juran, Crosby, and Taguchi affected the quality movement?
5. Provide examples of improving information technology project quality through improved leadership, better understanding of customer requirements, the cost of quality, and improved testing.
6. What factors did DeMarco and Lister find to be correlated with improving productivity of programmers? Do these findings make sense to you?
7. Discuss three suggestions for improving information technology project quality that were not made in this chapter.

EXERCISES

1. Research progress that has been made in improving the quality of information technology projects. Write a one- to two-page paper discussing your findings.
2. To illustrate a normal distribution, shake and roll a pair of dice thirty times and graph the results. It is more likely for someone to roll a six, seven, or eight than a two or twelve, so these numbers should come up more often. To create the graph, use graph paper or draw a grid. Label the x-axis with the numbers two through twelve. Label the y-axis with the numbers one through eight. Fill in the appropriate grid for each roll of the dice. Do your results resemble a normal distribution? Why or why not?
3. Research the criteria for the Malcolm Baldrige National Quality Award. Investigate a company other than Motorola that has received this award. What steps did the company take to earn this prestigious quality award? What are the benefits of earning this award?
4. Research ISO 9000. What is involved in earning ISO 9000 certification? Is it important to have this certification? Why or why not?
5. Research an information technology project that used a quality tool or technique (robust design, quality audits, Pareto diagrams, statistical sampling, quality control charts, and so on). Write a one- to two-page paper describing the technique and how it was used on the information technology project.

MINICASE

You are part of a team analyzing quality problems that you have been having with your call center/customer service area. After studying customer complaints, you have categorized and logged them for a week as follows:

COMPLAINTS	FREQUENCY/WEEK
Customer is on hold too long	90
Customer gets transferred to wrong area or cut off	20
Service rep cannot answer customer's questions	120
Service rep does not follow through as promised	40

Part 1: Create a Pareto diagram based on this information. First create a spreadsheet in Excel, using the above data. Sort the problems so the most frequent ones are listed first. Add a column called "% of Total" and another one called "Cumulative %." Then enter formulas to calculate those items. Next use Excel's Chart Wizard to create a Pareto diagram based on this data. Use the Line – Column on 2 Axis custom type chart so your resulting chart looks similar to the one in Figure 7-2.

Part 2: Your team has decided to add some customer service features to your Web site. You believe that having "Frequently Asked Questions" and answers will help reduce the complaints about service reps not being able to answer customers' questions and will lower the volume of calls. Customers and service reps could access information through this new Web site. However, you know that just as people complain about call centers, they also complain about Web sites. Develop a list of at least four potential complaints that your customers and service reps might have about this new Web site, then describe your plans for preventing those problems. Also describe how this new Web site could help to continuously improve the quality of service provided by your call center/customer service area.

SUGGESTED READINGS

1. Bylinksky, Gene. "How to Bring Out Better Products Faster." *Fortune* (November 23, 1998) 238[B].

    This article describes how companies such as Xerox, Hewlett-Packard, Ford Motor Company, and Goodyear have developed outstanding new products by using Taguchi's Robust Design method. This method goes directly to the basic physics and thermodynamics of product design—torque, electrical charge, heat flux, and so on—that are causing trouble, and solves problems during the early design phase.

2. Deming, W. Edwards. *Out of the Crisis*. Cambridge, MA: Massachusetts Institute of Technology, 1988.

    Despite Deming's final book, The New Economics for Industry, Government, Education (1995), Out of the Crisis *is Deming's best-known classic and is the source of constant reference. Deming uses real-life case studies to share with the reader the "simplicity" of his profound knowledge, his 14 points, and the Deadly Diseases of quality. Deming has authored and coauthored several other books on quality.*

3. Ireland, Lewis R. *Quality Management for Projects and Programs*. Upper Darby, PA: Project Management Institute, 1991.

    This PMI publication includes seven chapters and several appendices with information on project quality management. Lew Ireland is well known for his expertise in quality and project management and his contributions to PMI. In 1999 he served as the president of PMI.

4. Juran, Joseph. *Juran's Quality Handbook*, 5th ed. New York: McGraw Hill, 1999.

    Juran published the fifth edition of this famous handbook at the age of 94. This major revision of the classic reference on quality provides a comprehensive body of knowledge needed to engineer and manage for quality into the 21st century. There is also an excellent Web site with information about quality and Juran at www.juran.com.

5. Project Management Institute, Organizational Project Management Maturity Model (OPM3) Web site, www.pmi.org/opm3/index.htm (2001).

This Web site provides information about PMI's Organizational Project Management Maturity Model (OPM3) program, created to assist organizations in implementing organization strategy through the successful, consistent, and predictable delivery of projects. The OPM3 will include a method for assessing organizations' project management maturity levels as well as a step-by-step method for increasing and maintaining an organization's ability to deliver projects as promised.

6. Yilmaz, R. R. and Sangit Chatterjee. "Deming and the Quality of Software Development," *Business Horizons*, Foundation for the School of Business at Indiana University, 40, No. 6 (November-December 1997).

This article discusses how Deming's philosophy can facilitate the continuous improvement of software quality. It also describes the Software Quality Function Deployment and Capability Maturity Models for improving software development.

KEY TERMS

- **Acceptance decisions** — decisions that determine if the products or services produced as part of the project will be accepted or rejected
- **Appraisal cost** — the cost of evaluating processes and their outputs to ensure that a project is error-free or within an acceptable error range
- **Benchmarking** — a technique used to generate ideas for quality improvements by comparing specific project practices or product characteristics to those of other projects or products within or outside the performing organization
- **Capability Maturity Model (CMM)** — a five-level model laying out a generic path to process improvement for software development in organizations
- **Conformance** — delivering products that meet requirements and fitness for use
- **Conformance to requirements** — the project processes and products meet written specifications
- **Control chart** — a graphic display of data that illustrates the results of a process over time
- **Cost of nonconformance** — taking responsibility for failures or not meeting quality expectations
- **Cost of quality** — the cost of conformance plus the cost of nonconformance
- **Design of experiments** — a quality technique that helps identify which variables have the most influence on the overall outcome of a process
- **External failure cost** — a cost related to all errors not detected and corrected before delivery to the customer

- **Features** — the special characteristics that appeal to users
- **Fishbone diagrams** — diagrams that trace complaints about quality problems back to the responsible production operations; sometimes called Ishikawa diagrams
- **Fitness for use** — a product can be used as it was intended
- **Functionality** — the degree to which a system performs its intended function
- **Integration testing** — testing that occurs between unit and system testing to test functionally grouped components to ensure a subset(s) of the entire system works together
- **Internal failure cost** — a cost incurred to correct an identified defect before the customer receives the product
- **ISO 9000** — a quality system standard developed by the International Organization for Standardization (ISO) that includes a three-part, continuous cycle of planning, controlling, and documenting quality in an organization
- **Maintainability** — the ease of performing maintenance on a product
- **Malcolm Baldrige Award** — an award started in 1987 to recognize companies that have achieved a level of world-class competition through quality management
- **Maturity model** — a framework for helping organizations improve their processes and systems
- **Mean** — the average value of a population
- **Measurement and test equipment cost** — the capital cost of equipment used to perform prevention and appraisal activities
- **Normal distribution** — a bell-shaped curve that is symmetrical about the mean of the population
- **Pareto analysis** — identifying the vital few contributors that account for most quality problems in a system
- **Pareto diagrams** — histograms that help identify and prioritize problem areas
- **Performance** — how well a product or service performs the customer's intended use
- **Prevention cost** — the cost of planning and executing a project so that it is error-free or within an acceptable error range
- **Process adjustments** — adjustments made to correct or prevent further quality problems based on quality control measurements
- **Quality** — the totality of characteristics of an entity that bear on its ability to satisfy stated or implied needs
- **Quality assurance** — periodically evaluating overall project performance to ensure the project will satisfy the relevant quality standards
- **Quality audits** — structured reviews of specific quality management activities that help identify lessons learned and can improve performance on current or future projects

- **Quality circles** — groups of nonsupervisors and work leaders in a single company department who volunteer to conduct group studies on how to improve the effectiveness of work in their department
- **Quality control** — monitoring specific project results to ensure that they comply with the relevant quality standards and identifying ways to improve overall quality
- **Quality planning** — identifying which quality standards are relevant to the project and how to satisfy them
- **Reliability** — the ability of a product or service to perform as expected under normal conditions without unacceptable failures
- **Rework** — action taken to bring rejected items into compliance with product requirements or specifications or other stakeholder expectations
- **Robust Design methods** — methods that focus on eliminating defects by substituting scientific inquiry for trial-and-error methods
- **Seven run rule** — if seven data points in a row on a quality control chart are all below the mean, above the mean, or are all increasing or decreasing, then the process needs to be examined for nonrandom problems
- **Standard deviation** — a measure of how much variation exists in a distribution of data
- **Statistical sampling** — choosing part of a population of interest for inspection
- **System outputs** — the screens and reports the system generates
- **System testing** — testing the entire system as one entity to ensure it is working properly
- **Unit test** — a test of each individual component (often a program) to ensure it is as defect-free as possible
- **User acceptance testing** — an independent test performed by end users prior to accepting the delivered system

8

Project Human Resource Management

Objectives

After reading this chapter you will be able to:

1. *Explain the importance of good human resource management on projects, especially on information technology projects for which experienced professionals are in high demand*
2. *Define the major processes involved in human resource management*
3. *Summarize crucial theories of human resource management, including the contributions of Abraham Maslow, Frederick Herzberg, and Douglas McGregor on motivation, H. J. Thamhain and D. L. Wilemon on influencing workers, and Stephen Covey on how people and teams can become more effective*
4. *Discuss organizational planning and be able to create a responsibility assignment matrix*
5. *Understand key issues involved in project staff acquisition and explain the concepts of resource loading and resource leveling*
6. *Explain some of the tools and techniques that assist in team building*
7. *Describe how software can assist in project human resource management*

This was the third time someone from the Information Technology Department tried to work with Ben, the head of the F-44 aircraft program. Ben, who had been with the company for almost thirty years, was known for being rough around the edges and very demanding. The company was losing money on the F-44 upgrade project because the upgrade kits were not being delivered on time. The Canadian government had written severe late penalty fees into their contract, and other customers were threatening to take their business elsewhere. The F-44 program manager blamed it all on the Information Technology Department for not letting staff access the upgrade project's information system directly so they could work with their customers and suppliers more effectively.

The information system was based on very old technology that only a couple of people in the company knew how to use. It often took days or even weeks for Ben's group to get the information they needed.

Ed Davidson, a senior programmer attended a meeting with Sarah Ellis, an internal information technology business consultant. Sarah, in her early thirties, had advanced quickly in her company, primarily due to her keen ability to work well with all types of people. Sarah's job was to uncover the real problems with the F-44 program's information technology support, and then develop a solution with Ben and his team. If she found it necessary to invest in more information technology hardware, software, or staff, Sarah would write a business case to justify these investments and then work with Ed, Ben, and his group to implement the suggested solution as quickly as possible. Ben and three of his staff entered the conference room. Ben threw his books on the table and started yelling at Ed and Sarah. Ed could not believe his eyes or ears when Sarah stood nose to nose with Ben and started yelling right back at him.

THE IMPORTANCE OF HUMAN RESOURCE MANAGEMENT

Many corporate executives have said, "People are our most important asset." People determine the success and failure of organizations and projects. Most project managers agree that managing human resources effectively is one of the toughest challenges they face. Project human resource management is a vital component of project management, especially in the information technology field—in which qualified people are often hard to find and keep. It is important to understand the current state of human resource management in the information technology industry and its implications for the future.

Current State of Human Resource Management

Beginning in the 1990s, there has been a growing shortage of personnel in information technology. In 1997, the Information Technology Association of America (ITAA), in cooperation with Virginia Polytechnic Institute and State University, conducted a survey of 1,500 information technology and non-information technology firms around the country. The results of the survey indicated that employers were having trouble finding, training, and retaining information technology workers. ITAA calculated that in 1997 there were over 345,000 openings for programmers, systems analysts, computer scientists, and computer engineers in information technology and non-information technology companies with one hundred or more employees. These estimated job openings represented about ten percent of current employment in these three occupations.[1] In April 2000, ITAA published its "Bridging the Gap" workforce study, which clearly documented the continued staffing crisis facing the information technology industry. This study revealed that one in fourteen American workers is involved in information technology, and that 844,000 information technology jobs were unfilled in 2000, with even more vacancies anticipated in future years. In 2000, the World Information Technology and Services Alliance (WITSA) released their study, "Digital Planet 2000: The Global Information Economy," which revealed that the global high-tech industry generated over $2.1 trillion in 1999, and is expected to surpass $3 trillion in 2003.[2] Many states or other geographic regions have conducted their own surveys. The Minnesota Department of Economic Security Research and Statistics Office published a report stating that the 1,000 students graduating each year from information technology-related post-secondary programs in Minnesota were not nearly enough to fill the 8,800 positions projected to be open each year between 1998 and 2006.[3]

These studies highlight what many people consider to be a serious national problem. High-tech firms, which add thousands of high-skilled, high-wage jobs to the U.S. economy and other nations' economies every year, cannot find the workers they need. Many U.S. firms have turned to other countries, such as India, to find information technology workers. Some companies are offering interest-free loans to people seeking education and training in information technology. Colleges, universities, and private firms are expanding course offerings and programs in information technology. As a consequence of these employment shortages, human resource management for information technology projects takes on an increasingly critical role.

In spite of the fact that people are a critical asset and good information technology employees are hard to find, employees are being forced to work over holidays. Jeannette Cabanis-Brewin, Editor-in-Chief of the Center for Business Practices publications and former editor for *PM Network* and the *Project Management Journal*, specializes in the organizational and human side of project management. Cabanis-Brewin, also a proud grandmother, wrote an article complaining that her grandchildren did not come for Christmas in 2000 because their parents could not take a few days off work.[4] Cabanis-Brewin and

[1] Information Technology Association of America (ITAA). "Help Wanted: The IT Workforce Gap at the Dawn of a New Century", 1997.

[2] Information Technology Association of America, "Global IT Spending to Rocket from Current $2 Trillion to $3 Trillion, New Study Finds," *Update for IT Executives* (2001) 6 (15) **www.itaa.org**.

[3] Minnesota Department of Economic Security Research and Statistics Office, *Beyond 2000: Information Technology Workers in Minnesota*, May 1998.

[4] Cabanis-Brewin, Jeannette, "An Open Letter to Scrooge, Inc.," *Best Practices e-Advisor* (1/7/01) 16.

many other people feel that too many bosses haven't learned their lessons from watching the movie *Scrooge*.

Working during the holidays isn't the only problem in this industry. In addition, the field of information technology has become unappealing to certain groups, such as women. The number of women entering the information technology field peaked in 1984 and continues to decline. More than half of all students attending colleges today are women, yet less than 20 percent of the graduates awarded bachelor's degrees in computer science and related fields are female. Anita Borg, a computer engineer at Xerox Palo Alto Research Center and founder of the Institute for Women and Technology, states that if women had been going into computer fields at the same rate as men since 1984, the current shortage of information technology workers would not exist.[5]

White House findings in 2000 showed that employment in the U.S. information technology industry grew 81 percent. Women, who represent 47 percent of the work force at large, made up only 29 percent of several information technology categories. Despite the demand for computing professionals outstripping the supply, more men than women are earning undergraduate, master's, and doctoral degrees in computer science. The Association for Computing Machinery Committee on Women in Computing (ACM-W), which surveyed the shrinking pipeline of women in computer science in 1998, released a study in 2000 that finds the pipeline is continuing to shrink. Females in high school, college, graduate schools, and beyond are not choosing computer science or related degrees. Why not? Denise Gurer, 3Com Corporation Technologist and ACM-W cochair says, "Girls and women are not turned off by technology, but by how it is used in our society.... We can attract more girls and women to technology areas if we make classrooms more inclusive of technology uses that interest them."[6] Girls and women are interested in using technology to save time, to be more productive and creative, and to develop relationships. Girls are often turned off by technology when they see computers used to play unproductive games, which are often violent or sexist, to track sports statistics, to download unappealing music, or to avoid having real relationships with other people.

The ACM-W studies found that factors contributing to the gender gap include the following:

- The stereotype that girls are not good at math
- Computer games that target boys
- Online discussion groups that are largely male-oriented and sexist
- The fact that few women hold positions of power in the computer industry[7]

[5] Olsen, Florence, "Institute for Women and Technology Works to Bridge Computing's Gender Gap," *The Chronicle of Higher Education* (February 25, 2000).

[6] DeBlasi, Patrick J, "The Shrinking Pipeline Unlikely to Reverse: Nearly 30 Percent Decline of Women Choosing Undergraduate Degrees in Computer Science," *Association of Computing Machinery (ACM) Pressroom* (6/5/2000).

[7] Ohlson, Kathleen, "ACM launches gender gap study," *Online News* (11/13/98).

Implications for the Future of Human Resource Management

It is crucial for organizations to practice what they preach about human resources. If people truly are their greatest asset, organizations must work to fulfill their human resource needs *and* the needs of individual people in their companies. If organizations want to be successful at implementing information technology projects, they need to understand the importance of project human resource management and take actions to make effective use of people. Proactive organizations are addressing current and future human resource needs, by, for example, improving benefits, redefining work hours and incentives, and finding future workers.

Many organizations have changed their benefits policies to meet worker needs. Most workers assume that their companies provide some perks, such as casual dress codes, flexible work hours, and tuition assistance. Other companies provide on-site day care, dry cleaning services, and staff to run errands such as walking your dog. Some companies go even further and use special perks as a competitive advantage. For example, they promote policies such as allowing workers to bring their pets to work, providing free vacation rentals, paying for children's college tuition, or providing workers with nice cars. Jeff Finney, a computer programmer in suburban Atlanta, accepted delivery of a BMW from his employer, Revenue Systems, Inc. All 45 employees at this firm get to lease BMWs at the bosses' expense. Finney, who normally drives a pickup, says, "It is probably one of the best perks I've seen any company give."[8]

Other implications for the future of human resource management relate to the hours many information technology professionals are expected to work and how performance is rewarded. Today people brag about the fact that they can work less than 40 hours a week or work from home several days a week—not that they haven't had a vacation in years. If companies plan their projects well, they can avoid the need for overtime, or they can make it clear that overtime is optional. Companies can also provide incentives that use performance, not hours worked, as the basis of rewards. If performance can be objectively measured, as it can in many aspects of information technology jobs, then it should not matter where employees do their work or how long it takes them to do it. For example, if a technical writer can produce a high-quality publication at home in one week, the company and the writer are better off than if the company insisted that the writer come to the office and take two weeks to produce the publication. Objective measures of work performance and incentives based on meeting those criteria are important considerations.

[8] Irvine, Martha, "Executive-style perks spread to rank and file," *Minneapolis Star Tribune* (9/29/98) D-2.

The need to develop future talent in information technology has important implications for everyone. Who will maintain the systems we have today? Who will continue to develop new products and services using new technologies? Some schools are requiring all students to take computer literacy courses. Some individuals and organizations are offering incentives to minorities and women to enter technical fields. For example, some wealthy alumni offer to pay all college expenses for minority students from the alumni's high school or grade school. Several colleges and government agencies have programs to help recruit more women into technical fields. All of these efforts will help to develop the human resources needed for future information technology projects.

WHAT IS PROJECT HUMAN RESOURCE MANAGEMENT?

Project human resource management includes the processes required to make the most effective use of the people involved with a project. Human resource management includes all project stakeholders: sponsors, customers, project team members, support staff, suppliers supporting the project, and so on. The major processes involved in human resource management include:

- **Organizational planning**, which involves identifying, assigning, and documenting project roles, responsibilities, and reporting relationships. Key outputs of this process include roles and responsibility assignments, often shown in a matrix form, and an organizational chart for the project.
- **Staff acquisition**, which involves getting the needed personnel assigned to and working on the project. Getting personnel is one of the crucial challenges of information technology projects.
- **Team development**, which involves building individual and group skills to enhance project performance. Building individual and group skills is also a challenge for many information technology projects.

Some topics related to human resource management (understanding organizations, stakeholders, and different organizational structures) were introduced in Chapter 2, *The Project Management Context and Processes.* This chapter will expand on some of those topics and introduce other important concepts in project human resource management, including theories about managing people, resource loading, and resource leveling. Using software to assist in project resource management will also be discussed.

KEYS TO MANAGING PEOPLE

Industrial-organizational psychologists and management theorists have devoted much research and thought to the field of managing people at work. Psychosocial issues that affect how people work and how well they work include motivation, influence and power, and effectiveness. This section will review Abraham Maslow's, Frederick Herzberg's, and Douglas McGregor's contributions to an understanding of motivation; H. J. Thamhain's and D. L. Wilemon's work on influencing workers and reducing conflict; the effect of power on project teams; and Stephen Covey's work on how people and teams can become more effective. Finally, you will look at some implications and recommendations for project managers.

Motivation Theories

Abraham Maslow, a highly respected psychologist who rejected the dehumanizing negativism of psychology in the 1950s, is best known for developing a hierarchy of needs. In the 1950s, proponents of Sigmund Freud's psychoanalytic theory were promoting the idea that human beings were not the masters of their destiny and that all their actions were governed by unconscious processes dominated by primitive sexual urges. During the same period, behavioral psychologists saw human beings as controlled by the environment. Maslow argued that both schools of thought failed to recognize unique qualities of human behavior: love, self-esteem, belonging, self-expression, and creativity. He argued that these unique qualities enable people to make independent choices, which gives them full control of their destiny.

Figure 8-1 shows the basic pyramid structure of Maslow's **hierarchy of needs**, which states that people's behaviors are guided or motivated by a sequence of needs. At the bottom of the hierarchy are physiological needs. Once physiological needs are satisfied, safety needs guide behavior. Once safety needs are satisfied, social needs come to the forefront, and so on up the hierarchy. The order of these needs and their relative sizes in the pyramid are significant. Maslow suggests that each level of the hierarchy is a prerequisite for the levels above. For example, it is not possible for a person to consider self-actualization if he or she has not addressed basic needs concerning security and safety. People in an emergency situation, such as a flood or hurricane, are not going to worry about personal growth. Personal survival will be their main motivation. Once a particular need is satisfied, however, it no longer serves as a potent motivator of behavior.

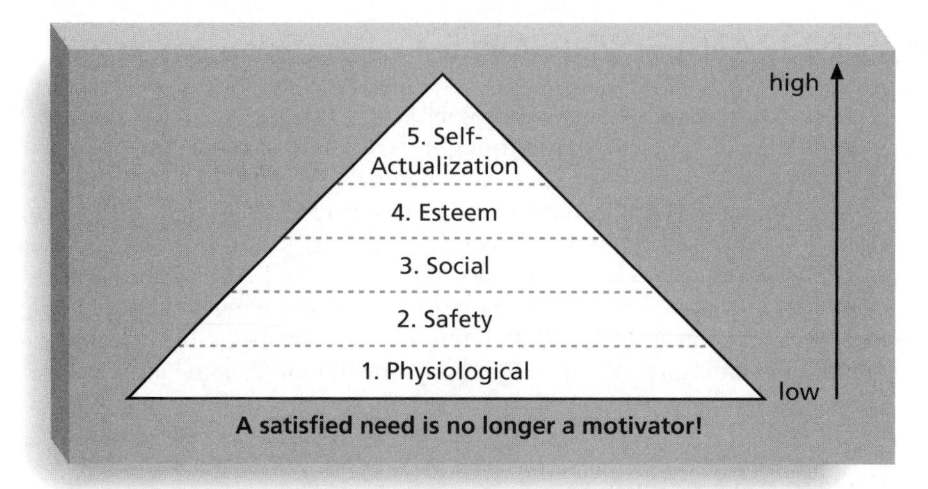

Figure 8-1. Maslow's Hierarchy of Needs

Notice in Figure 8-1 that each layer in the pyramid is smaller than the previous layer. The issues in each level are of greater value than issues in the preceding level, which presumably have been satisfied. Issues higher in the hierarchy that are associated with needs that are currently not satisfied are more important than earlier issues, although, by definition, they are less urgent than issues lower in the hierarchy.

The bottom four needs in Maslow's hierarchy—physiological, safety, social, and esteem needs—are referred to as deficiency needs, and the highest level, self-actualization, is considered to be a growth need. Only after meeting deficiency needs can individuals act upon growth needs. Self-actualized people are characterized as being problem-focused, having an appreciation for life, being concerned about personal growth, and having the ability to have peak experiences.

Most people working on an information technology project will probably have their basic physiological and safety needs met. To motivate project team members, the project manager needs to understand each person's motivation with regard to social, esteem, and self-actualization or growth needs. Team members new to a company and city might be motivated by social needs. To address social needs, some companies organize gatherings and social events for new workers in information technology. Other project members may find these events to be an invasion of personal time they would rather spend with their friends and family or working on an advanced degree.

Maslow's hierarchy conveys a message of hope and growth. People can work to control their own destinies and naturally strive to achieve higher and higher needs in Maslow's hierarchy. Project managers should know something about team members' professional and personal lives so they can provide motivational incentives that meet their team members' needs.

Frederick Herzberg is best known for distinguishing between motivational factors and hygiene factors when considering motivation in work settings. Head of Case Western University's Psychology Department, Herzberg wrote the book *Work and the Nature of Man* in 1966 and the famous *Harvard Business Review* article, "One More Time: How Do You Motivate Employees?" in 1968. Herzberg analyzed the factors that affected productivity among a sample of 1,685 employees. Popular beliefs at that time were that work output was most improved through larger salaries, more supervision, or a more attractive work environment. According to Herzberg, these hygiene factors would cause dissatisfaction if not present, but would not motivate workers to do more if present. Herzberg found that people were motivated to work mostly by feelings of personal achievement and recognition. Motivators, Herzberg concluded, included achievement, recognition, the work itself, responsibility, advancement, and growth.

In his books and articles, Herzberg explained why attempts to use positive factors such as reducing time spent at work, upward spiraling wages, offering fringe benefits, providing human relations and sensitivity training, and so on did not instill motivation. He argued that people want to actualize themselves. They need stimuli for their growth and advancement needs. Factors such as achievement, recognition, responsibility, advancement, and growth produce job satisfaction and are work motivators.[9]

Douglas McGregor was one of the great popularizers of a human relations approach to management, and is best known for developing Theory X and Theory Y. In his research, documented in his 1960 book *The Human Side of Enterprise*, McGregor found that although many managers spouted the right ideas, they actually followed a set of assumptions about worker motivation that he called Theory X (sometimes referred to as classical systems theory). Managers who believe in Theory X assume that workers dislike and avoid work if possible, so managers must use coercion, threats, and various control schemes to get workers to make adequate efforts to meet objectives. They assume that the average worker wants to be directed and prefers to avoid responsibility, has little ambition, and wants security above all else. Research seemed to clearly demonstrate that these assumptions were not valid. McGregor suggested a different series of assumptions about human behavior that he called Theory Y (sometimes referred to as human relations theory). Managers who believe in Theory Y assume that individuals do not inherently dislike work, but consider it as natural as play or rest. The most significant rewards are the satisfaction of esteem and self-actualization needs, as described by Maslow. McGregor urged managers to motivate people on the basis of these more valid Theory Y notions.

In 1981 William Ouchi introduced another approach to management in his book *Theory Z: How American Business Can Meet the Japanese Challenge*. Theory Z is based on the Japanese approach to motivating workers, which emphasizes

[9] Herzberg, Frederick, "One More Time: How Do You Motivate Employees?" *Harvard Business Review* (February 1968) 51–62.

trust, quality, collective decision making, and cultural values. Whereas Theory X and Theory Y emphasize how management views employees, Theory Z also describes how workers perceive management. Theory Z workers, it is assumed, can be trusted to do their jobs to their utmost ability, as long as management can be trusted to support them and look out for their well-being. Theory Z emphasizes things such as job rotation, broadening of skills, generalization versus specialization, and the need for continuous training of workers.

Influence and Power

Many people working on a project do not report directly to project managers, and project managers often do not have control over project staff who do report to them. For example, people are free to change jobs. If they are given work assignments they do not like, many workers will simply quit or transfer to other departments or projects. H. J. Thamhain and D. L. Wilemon investigated the approaches project managers use to deal with workers and how those approaches relate to project success. They identified nine influence bases available to project managers:

- Authority—the legitimate hierarchical right to issue orders
- Assignment—the project manager's perceived ability to influence a worker's later work assignments
- Budget—the project manager's perceived ability to authorize others' use of discretionary funds
- Promotion—the ability to improve a worker's position
- Money—the ability to increase a worker's pay and benefits
- Penalty—the project manager's perceived ability to dispense or cause punishment
- Work challenge—the ability to assign work that capitalizes on a worker's enjoyment of doing a particular task, which taps an intrinsic motivational factor
- Expertise—the project manager's perceived special knowledge that others deem important
- Friendship—the ability to establish friendly personal relationships between the project manager and others

Senior management grants authority to the project manager. Assignment, budget, promotion, money, and penalty influence bases may or may not be inherent in a project manager's position. Unlike authority, they are not automatically available to project managers as part of their position. Others' perceptions are important in establishing the usefulness of these influence bases. For example, any manager can influence workers by providing challenging work; the ability to provide challenging work (or take it away) is not a special ability of project managers. In addition, project managers must earn the ability to influence by using expertise and friendship.

Thamhain and Wilemon found that projects were more likely to fail when project managers relied too heavily on using authority, money, or penalty to influence people. When project managers used work challenge and expertise to influence people, projects were more likely to succeed. The effectiveness of work challenge in influencing people is consistent with Maslow's and Herzberg's research on motivation. The importance of expertise as a means of influencing people makes sense on projects that involve special knowledge, like most information technology projects.

Influence is related to the topic of power. Power is the potential ability to influence behavior to get people to do things they would not otherwise do. Power has a much stronger connotation than influence, especially since it is often used to force people to change their behavior.

The five main types of power include:

- **Coercive power**, which involves using punishment, threats, or other negative approaches to get people to do things they do not want to do. This type of power is similar to Thamhain's and Wilemon's influence category called penalty. For example, a project manager can threaten to fire workers or subcontractors to try to get them to change their behavior. If the project manager really has the power to fire people, he or she could follow through on the threat. Recall, however, that influencing through the use of penalties is correlated with unsuccessful projects.
- **Legitimate power**, which is getting people to do things based on a position of authority. This type of power is similar to the authority basis of influence. If senior management gives project managers organizational authority, project managers can use legitimate power in several situations. They can make key decisions without involving the project team, for example. Overemphasis of legitimate power or authority is also correlated with project failure.
- **Expert power**, which involves using one's personal knowledge and expertise to get people to change their behavior. If people perceive that project managers are experts in certain situations, they will follow their suggestions. For example, if a project manager has expertise in working with a particular information technology supplier and their products, the project team will be more likely to follow the project manager's suggestions on how to work with that vendor and its products.
- **Reward power**, which involves using incentives to induce people to do things. Rewards can include money, status, recognition, promotions, special work assignments, or other means of rewarding someone for desired behavior. Many motivation theorists suggest that only certain types of rewards, such as work challenge, achievement, and recognition, truly induce people to change their behavior or work hard.

- **Referent power**, which is based on an individual's personal charisma. People hold someone with referent power in very high regard and will do what they say based on their regard for the person. People such as Martin Luther King, Jr. and John F. Kennedy had referent power. Very few people possess the natural charisma that underlies referent power.

It is important for project managers to understand what types of influence and power they can use in different situations. New project managers often overemphasize their position—their legitimate power or authority influence—especially when dealing with project team members or support staff. They also neglect the importance of reward power or work challenge influence. People often respond much better to a project manager who motivates them with challenging work and provides positive reinforcement for doing a good job. It is important for project managers to understand the basic concepts of influence and power, and to practice using them to their own and their project team's advantage.

Improving Effectiveness

Stephen Covey, author of *The 7 Habits of Highly Effective People*, expanded on the work done by Maslow, Herzberg, and others to develop an approach for helping people and teams become more effective. Covey's first three habits of effective people—be proactive, begin with the end in mind, and put first things first—help people achieve a private victory by becoming independent. After achieving independence, people can then strive for interdependence by developing the next three habits—think win/win, seek first to understand then to be understood, and synergize. (**Synergy** is the concept that the whole is equal to more than the sum of its parts.) Finally, everyone can work on Covey's seventh habit—sharpen the saw—to develop and renew their physical, spiritual, mental, and social/emotional selves.

Project managers can apply Covey's seven habits to improve effectiveness on projects, as follows:

1. *Be proactive.* Covey, like Maslow, believes that people have the ability to be proactive and choose their responses to different situations. Project managers must be proactive and anticipate and plan for problems and inevitable changes on projects. They can also encourage their team members to be proactive in working on their project activities.
2. *Begin with the end in mind.* Covey suggests that people focus on their values, what they really want to accomplish, and how they really want to be remembered in their lives. He suggests writing a mission statement to help achieve this habit. Many organizations and projects have mission statements that help them focus on their main purpose.
3. *Put first things first.* Covey developed a time management system and matrix to help people prioritize their time. He suggested that most people need to spend more time doing things that are important, but not urgent.

Important but not urgent activities include planning, reading, and exercising. Project managers need to spend a lot of time working on important and not urgent activities, such as developing the project plan, building relationships with major project stakeholders, and mentoring project team members. They also need to avoid focusing only on important yet urgent activities—putting out fires.

4. *Think win/win.* Covey presents several paradigms of interdependence, with think win/win being the best choice in most situations. When you use a win/win paradigm, parties in potential conflict work together to develop new solutions that make them all winners. Project managers should strive to use a win/win approach in making decisions, but sometimes, especially in competitive situations, they must use a win-lose paradigm.
5. *Seek first to understand, then to be understood.* **Empathic listening** is listening with the intent to understand. It is even more powerful than active listening because you forget your personal interests and focus on truly understanding the other person. To really understand other people, you must learn to focus on others first. When you practice empathic listening, you can begin two-way communication. This habit is critical for project managers so they can really understand their stakeholders' needs and expectations.
6. *Synergize.* In projects, a project team can synergize by creating collaborative products that are much better than a collection of individual efforts. Covey also emphasizes the importance of valuing differences in others to achieve synergy. Synergy is essential to many highly technical projects; in fact, several major breakthroughs in information technology occurred because of synergy. For example, in his Pulitzer Prize-winning book, *The Soul of a New Machine*, Tracy Kidder documented the 1970s synergistic efforts of a team of Data General researchers to create a new 32-bit superminicomputer.[10]
7. *Sharpen the saw.* When you practice sharpening the saw, you take time to renew yourself physically, spiritually, mentally, and socially. The practice of self-renewal helps people avoid burnout. Project managers must make sure that they and their project team have time to retrain, reenergize, and occasionally even relax to avoid burnout.

Douglas Ross, author of *Applying Covey's Seven Habits to a Project Management Career*, related Covey's seven habits to project management. Ross suggests that Habit 5—Seek first to understand, then to be understood—differentiates good project managers from average or poor project managers. People have a tendency to focus on their own agendas instead of first trying to understand other people's points of view. Empathic listening can help project managers and team members find out what motivates different people. Understanding what motivates key stakeholders and customers can mean the difference between project success and project failure. Once project managers and team members begin to

[10] Kidder, Tracy, *The Soul of a New Machine*. New York: Modern Library, 1997.

practice empathic listening, they can communicate and work together to tackle problems more effectively.

Before you can practice empathic listening, you have to first get people to talk to you. In many cases, you must work on developing a rapport with the other person before he or she will really talk to you. **Rapport** is a relation of harmony, conformity, accord, or affinity. Without rapport, people cannot begin to communicate. For example, in the opening case, Ben was not ready to talk to anyone from the Information Technology Department, even if Ed and Sarah were ready to listen. Ben was angry about the lack of support he received from the Information Technology Department and bullied anyone who reminded him of that group. Before Sarah could begin to communicate with Ben, she had to establish rapport.

One technique for establishing rapport is using a process called mirroring. **Mirroring** is the matching of certain behaviors of the other person. You can mirror someone's voice tone and/or tempo, breathing, movements, or body postures. After Ben started yelling at her, Sarah quickly decided to mirror his voice tone, tempo, and body posture. She stood nose to nose with Ben and started yelling right back at him. This action made Ben realize what he was doing, and it also made him notice and respect this colleague from the Information Technology Department. Once Ben overcame his anger, he could start communicating his needs. In most cases, such extreme measures are not needed.

Recall the importance of getting users involved in information technology projects. For organizations to truly be effective in information technology project management, they must find ways to help users and developers of information systems work together. It is widely accepted that companies make better decisions about information technology projects when business professionals and information technology staff collaborate. It is also widely accepted that this task is easier said than done. Many companies have been very successful in integrating technology and business departments, but many other companies continue to struggle with this issue.

What Went Right?

Forrester Research, a well-known technology research company in Cambridge, Massachusetts, interviewed twenty-five Chief Information Officers (CIOs) at companies including L.L. Bean, Shell, GE, Taco Bell, Texas Instruments, and Coca-Cola to identify the best practices for ensuring partnerships between people in business and technology areas. Their research study identified three successful practices:

- Businesspeople, not information technology people, take the lead in determining and justifying investments in new computer systems. At several companies, business managers defined the expected benefits of new systems, with information technology support, and made the pitch for the investment to senior management. At some companies, senior business managers are running information technology departments. At GE, for example, the head of the corporate strategy unit assumed responsibility for the information technology division.

- CIOs continually push their staff to recognize that the needs of the business must drive all technology decisions. The CIOs at Textron and Texas Instruments, for example, go so far as to insist that everyone express all presentations, reports, and even discussions about information technology projects in business terms, not in the technical language used by information technology insiders.
- Several companies that are successful at creating partnerships between people in business and technology areas are reshaping their information technology units to look and perform like consulting firms. For instance, Texas Instruments borrowed an idea from McKinsey & Company by assigning central information technology employees to one of seventeen centers of excellence for specific technology disciplines. When a business challenge arises, managers pull together ad hoc teams of people drawn from those centers and the business itself to work on the problem. When the problem is solved, managers reassign team members to other problems. This approach allows managers to use resources more flexibly and keeps employees' skills fresh.

The Forrester Research study also found that successful CIOs downplayed the "T" (technology) of IT, and played up the "I" (information). The CIOs believed that choosing the right technology was important, but success rested more on having the right people involved in the right project, putting the right processes in place, and having a business-results focus at all times.[11]

It is important to understand and pay attention to concepts of motivation, influence, power, and improving effectiveness in all project processes. It is likewise important to remember that projects operate within an organizational environment. The challenge comes from applying these theories to the many unique individuals involved in particular projects in particular organizations.

ORGANIZATIONAL PLANNING

Organizational planning for a project involves identifying, documenting, and assigning project roles, responsibilities, and reporting relationships. This process generates an organizational chart for the project, roles, and responsibility assignments, often shown in a matrix form called a responsibility assignment matrix (RAM), and a staffing management plan.

[11] Kiely, Thomas, "Making Collaboration Work," *Harvard Business Review* (January-February 1997) 10–11.

Before creating an organizational chart for a project, senior management and the project manager must identify what types of people are really needed to ensure project success. If the key to success lies in having the best Java programmers you can find, the organizational planning should reflect that need. If the real key to success is having a top-notch project manager and team leaders whom people respect in the company, that need should drive organizational planning.

What Went Wrong?

The organizational culture of a company has a tremendous influence on how effectively human resource are utilized. For example, one author describes a well-run software development organization where programmers received top pay and status. As a result, the ratio of programmers to managers was about ten to one. In that organization, the company developed and installed a customized publication system for less than $1 million.
The author then described the next software development organization where he worked. Programming was considered a job of underlings. The typical software project had a manager, a project leader, a project manager, a marketing manager, an architect, a system engineer, and one or two programmers. In that environment, the company spent over $30 million in personnel costs to try to develop an almost identical publication system, without a single line of usable code ever being written.[12]

After identifying important skills and the types of people needed to staff a project, the project manager should work with senior management and project team members to create an organizational chart for the project. Figure 8-2 provides a sample organizational chart for a large information technology project. Note that the project personnel include a deputy project manager, subproject managers, and teams. **Deputy project managers** fill in for project managers in their absence and assist them as needed, which is similar to the role of a vice president. **Subproject managers** are responsible for managing the subprojects into which a large project might be divided. This structure is typical for large projects. With many people working on a project, clearly defining and allocating project work is essential. Smaller information technology projects usually do not have deputy project managers or subproject managers. On smaller projects, the project managers might have just team leaders reporting directly to them.

[12] James, Geoffrey, "IT Fiascoes . . . and How to Avoid Them," *Datamation* (November 1997).

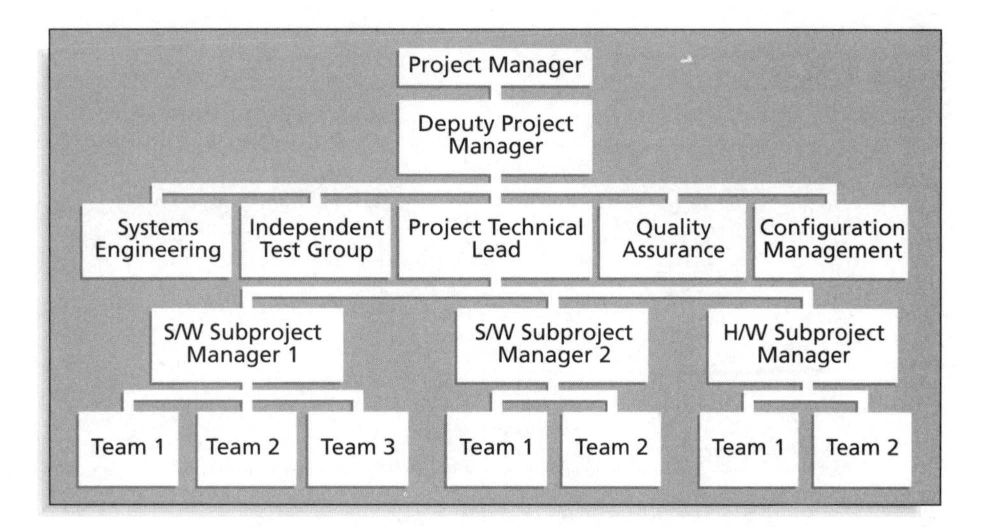

Figure 8-2. Sample Organizational Chart for a Large Information Technology Project

Figure 8-3 provides a framework for defining and assigning work. This process consists of four steps:

1. Finalizing the project requirements
2. Defining how the work will be accomplished
3. Breaking down the work into manageable elements
4. Assigning work responsibilities

The work definition and assignment process is carried out during the proposal and startup phases of a project. Note that the process is iterative, meaning it often takes more than one pass to refine it. A Request for Proposal (RFP) or draft contract often provides the basis for defining and finalizing work requirements, which are then documented in a final contract and technical baseline. If there is not an RFP, then the internal project charter and scope statement would provide the basis for defining and finalizing work requirements, as described in Chapter 4, *Project Scope Management*. The project team leaders then decide on a technical approach for how to do the work. Should work be broken down using a product-oriented approach or a phased approach? Will some of the work be outsourced or subcontracted to other companies? Once the project team has decided on a technical approach, they develop a work breakdown structure (WBS) to establish manageable elements of work (*see* Chapter 4, *Project Scope Management*). They then develop activity definitions to further define the work involved in each activity on the WBS (*see* Chapter 5, *Project Time Management*). The last step is assigning the work.

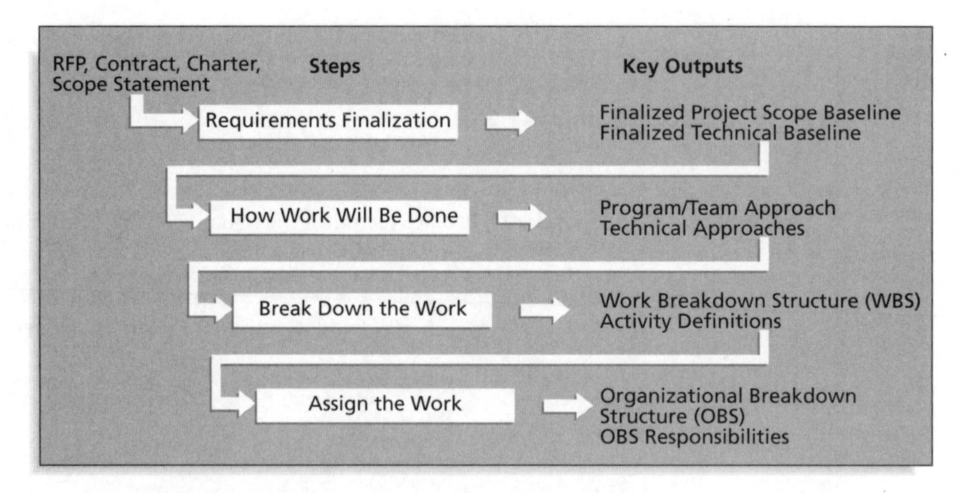

Figure 8-3. Work Definition and Assignment Process

Once the project manager and project team have broken down the work into manageable elements, the project manager assigns work to organizational units. The project manager often bases these work assignments on where the work fits in the organization and uses an organizational breakdown structure to conceptualize the process. An **organizational breakdown structure (OBS)** is a specific type of organizational chart that shows which organizational units are responsible for which work items. The OBS can be based on a general organizational chart and then broken down into more detail based on specific units within departments in the company or units in any subcontracted companies.

After developing an OBS, the project manager is in a position to develop a responsibility assignment matrix (RAM). A **responsibility assignment matrix (RAM)** is a matrix that maps the work of the project as described in the WBS to the people responsible for performing the work as described in the OBS. Figure 8-4 shows an example of a RAM. The RAM allocates work to responsible and performing organizations, teams, or individuals, depending on the desired level of detail. For smaller projects, it would be best to assign individual people to WBS activities. For very large projects, it is more effective to assign the work to organizational units or teams.

WBS activities →

OBS units ↓	1.1.1	1.1.2	1.1.3	1.1.4	1.1.5	1.1.6	1.1.7	1.1.8
Systems Engineering	R	R P					R	
Software Development			R P					
Hardware Development				R P				
Test Engineering	P							
Quality Assurance					R P			
Configuration Management						R P		
Integrated Logistics Support							P	
Training								R P

R = Responsible organizational unit
P = Performing organizational unit

Figure 8-4. Sample Responsibility Assignment Matrix (RAM)

In addition to using a RAM to assign detailed work activities, you can also use a RAM to define general roles and responsibilities on projects. This type of RAM can include the stakeholders in the project. Figure 8-5 provides a RAM that shows whether stakeholders are accountable or just participants in part of a project and whether they are required to provide input, review, or sign off on parts of a project. This simple tool can be a very effective way for the project manager to communicate roles and expectations of important stakeholders on projects.

	Stakeholders				
Items	A	B	C	D	E
Unit Test	S	A	I	I	R
Integration Test	S	P	A	I	R
System Test	S	P	A	I	R
User Acceptance Test	S	P	I	A	R

A = Accountable
P = Participant
R = Review Required
I = Input Required
S = Sign-off Required

Figure 8-5. RAM Showing Stakeholder Roles

Another output of organization planning is a **staffing management plan**. A staffing management plan describes when and how people will be added to and taken off the project team. It can be a formal or informal plan, and the level of detail may vary based on the type of project. For example, if an information technology project is projected to need one hundred people on average over a year, the staffing management plan would describe the types of people needed to work on the project, such as Java programmers, business analysts, technical writers, and so on, and the number of each type of person needed each month. The staffing management plan often includes a **resource histogram**, which is a column chart that shows the number of resources assigned to a project over time. Figure 8-6 provides an example of a histogram that might be used for a large, one-year information technology project. Notice that the columns represent the number of people needed in each area—Java programmers, business analysts, technical writers, managers, administrative staff, database analysts, and testing specialists. After determining the project staffing needs, the next steps in project human resource management are to acquire the necessary staff and then develop the project team.

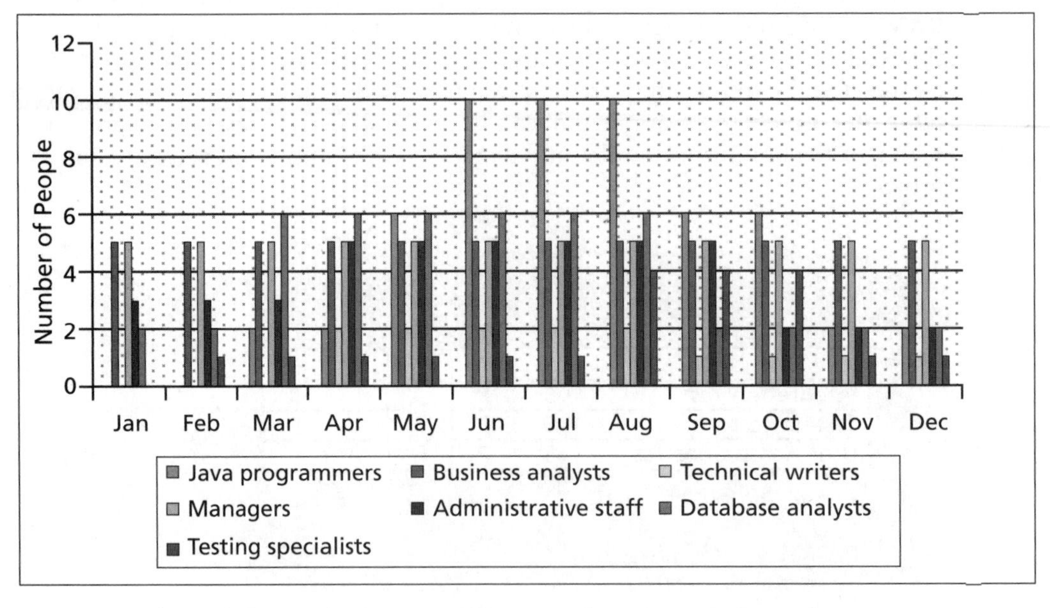

Figure 8-6. Sample Resource Histogram for a Large Information Technology Project

ISSUES IN PROJECT STAFF ACQUISITION AND TEAM DEVELOPMENT

During the late 1990s, the information technology job market became extremely competitive. It has since become a seller's market with corporations competing fiercely for a shrinking pool of qualified, experienced information technology professionals. Job market projections indicate that this highly competitive job market is likely to continue well into the 21st century. Thus, finding qualified information technology professionals—staff acquisition—is critical. It is also important to assign the appropriate type and number of resources to work on projects at the appropriate times. Resource loading and leveling is a project human resource management technique that addresses this issue. Even if you can find great workers and assign them well, if those professionals cannot work together as a team, the project will not be successful. Once professionals have been hired to staff a project, team development becomes an important issue.

Staff Acquisition

After developing a staffing management plan, project managers must work with other people in their organizations to assign particular personnel to their projects or to acquire additional human resources needed to staff the project. Project managers with strong influencing and negotiating skills are often good at getting internal people to work on their projects. However, the organization must ensure that people are assigned to the projects that best fit their skills and the needs of the organization. The main outputs of the staff acquisition process are project staff assignments and a project team directory.

Organizations that do a good job of staff acquisition have good staffing plans. These plans describe the number and type of people who are currently in the organization and the number and type of people anticipated to be needed for the project based on current and upcoming activities. An important component of staffing plans is maintaining a complete and accurate inventory of employees' skills. If there is a mismatch between the current mix of people's skills and needs of the organization, it is the project manager's job to work with senior management, human resource managers, and other people in the organization to address staffing and training needs.

It is also important to have good procedures in place for hiring subcontractors and recruiting new employees. Since the Human Resource Department is normally responsible for hiring people, project managers must work with their human resource managers to address any problems in recruiting appropriate people. It is also a priority to address retention issues, especially for information technology professionals.

One innovative approach to hiring and retaining information technology staff is to offer existing employees incentives for helping recruit and retain personnel. For example, several consulting companies give their employees one dollar for every hour a new person they helped hire works. This provides an incentive for current employees to help attract new people and to keep both them and the person they recruited working at that company. Another approach that several companies are taking to attract and retain information technology professionals is to provide benefits based on their personal needs. For example, some people might want to work only four days a week or have the option of working a couple of days a week from home. As it gets more and more difficult to find good information technology professionals, organizations must become more innovative and proactive in addressing this issue.

Several organizations, publications, and Web sites address the need for good staff acquisition and retention. For example, ICEX, Inc. conducts several research studies to identify best practices. They have found that world-class information technology organizations have groups dedicated to managing information technology staff recruiting. They have also found that the most effective recruiting message emphasizes that "this is a great place to work and you'll have a chance to work on innovative information technology projects."[13] William C. Taylor, cofounder of *Fast Company* magazine and a public speaker, also believes that people today are more demanding and have higher expectations of their jobs than just earning a paycheck. His company's research has found that people leave their jobs because:

- They feel they don't make a difference.
- They do not get proper recognition.
- They are not learning anything new or growing as a person.
- They don't like their coworkers.
- They want to earn more money.[14]

It is very important to consider the needs of individuals and the organization when making recruiting and retention decisions and to study the best practices of leading companies in these areas.

[13] Quillard, Judith, "Leading Practices for Recruiting IT Staff," ICEX Web site (1998) www.icex.com.

[14] Taylor, William C, Keynote address at the Project Management Institute's 2000 Annual Seminars & Symposium (September 2000).

Resource Loading and Leveling

Chapter 5, *Project Time Management*, described using network diagrams to help manage a project's schedule. One of the problems or dangers inherent in scheduling processes is that they often do not address the issues of resource utilization and availability. Schedules tend to focus exclusively on time rather than on both time and resources, which includes people. An important measure of a project manager's success is how well he or she balances the trade-offs among performance, time, and cost. During a period of crises, it is occasionally possible to add additional resources—such as additional staff—to a project at little or no cost. Most of the time, however, resolving performance, time, and cost trade-offs entails additional costs to the organization. The project manager's goal must be to achieve project success without increasing the costs or time required to complete the project. The key to accomplishing this goal is effectively managing human resources on the project.

Once people are assigned to projects, there are two techniques available to project managers that help them use project staff most effectively: resource loading and resource leveling. **Resource loading** refers to the amount of individual resources an existing schedule requires during specific time periods. Resource loading helps project managers develop a general understanding of the demands a project will make on the organization's resources, as well as on individual people's schedules. Project managers often use histograms, as described in Figure 8-6, to depict period-by-period variations in resource loading. A histogram can be very helpful in determining staffing needs or in identifying staffing problems.

A resource loading histogram can show when work is being over-allocated to a certain person or group. **Over-allocation** means more resources than are available are assigned to perform work at a given time. For example, Figure 8-7 provides a sample resource histogram for a fictitious software launch project. This histogram illustrates how much one individual, Joe Franklin, is assigned to work on the software launch project each week. The percentage numbers on the vertical axis represent the percentage of Joe's available time that is allocated for him to work on the project. Time is represented in weeks along the top horizontal axis. Note that Joe Franklin is over allocated most of the time. For example, for most of March and April and part of May, Joe's work allocation is 300 percent of his available time. If Joe is normally available eight hours per day, this means he would have to work twenty-four hours a day to meet this staffing projection!

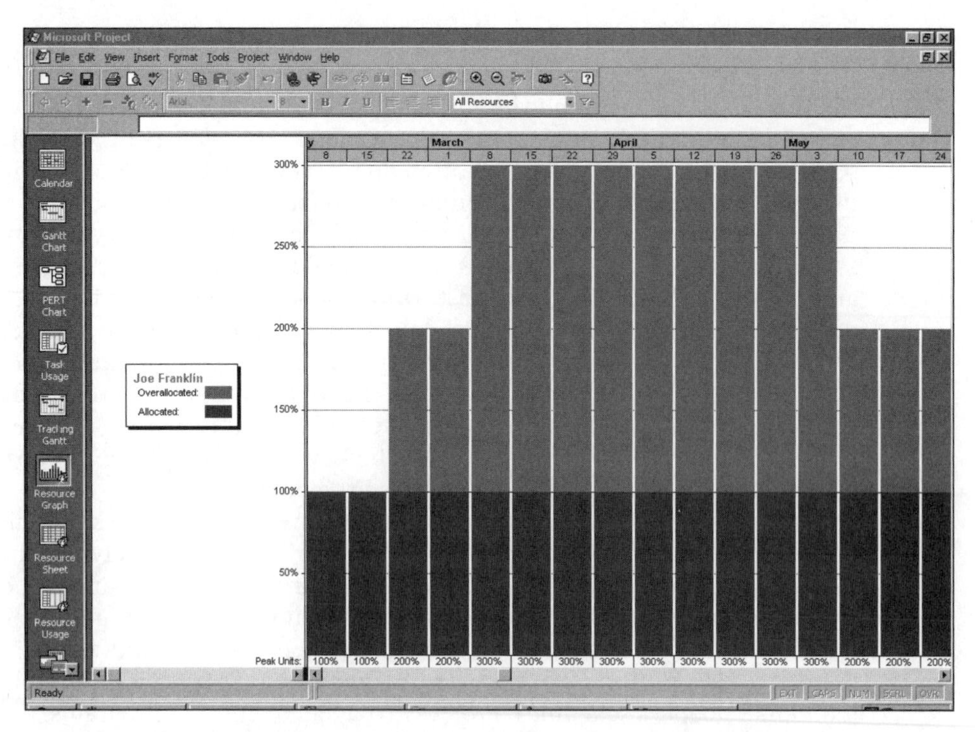

Figure 8-7. Sample Histogram Showing an Overallocated Individual

Resource leveling is a technique for resolving resource conflicts by delaying tasks. It is a form of network analysis in which resource management concerns drive scheduling decisions (start and finish dates). The main purpose of resource leveling is to create a smoother distribution of resource usage. Project managers examine the network diagram for areas of slack or float, and to identify resource conflicts. You can also view resource leveling as addressing the resource constraints described in critical chain scheduling (*see* Chapter 5, *Project Time Management*).

Over allocation is one type of resource conflict. If a certain resource is over allocated, the project manager can change the schedule to remove resource over allocation. If a certain resource is under allocated, the project manager can change the schedule to try to improve the use of the resource. Resource leveling, therefore, aims to minimize period-by-period variations in resource loading by shifting tasks within their slack allowances.

Figure 8-8 illustrates a simple example of leveling resources. The network diagram at the top of this figure shows that Activities A, B, and C can all start at the same time. Activity A has a duration of two days and will take two people to complete; Activity B has a duration of five days and will take four people to complete; and Activity C has a duration of three days and will take two people to complete. The histogram on the lower left of this figure shows the resource usage if all activities start on day one. The histogram on the lower right of Figure 8-8 shows the resource usage if Activity C is delayed two days, its total slack allowance. Notice that the lower-right histogram is flat or leveled; that is, its pieces (activities) are arranged to take up the least space (saving days and numbers of workers). You may recognize this strategy from the computer game Tetris, in which you earn points for keeping the falling shapes as level as possible. The player with the most points (most level shape allocation) wins. Resources are also used best when they are leveled.

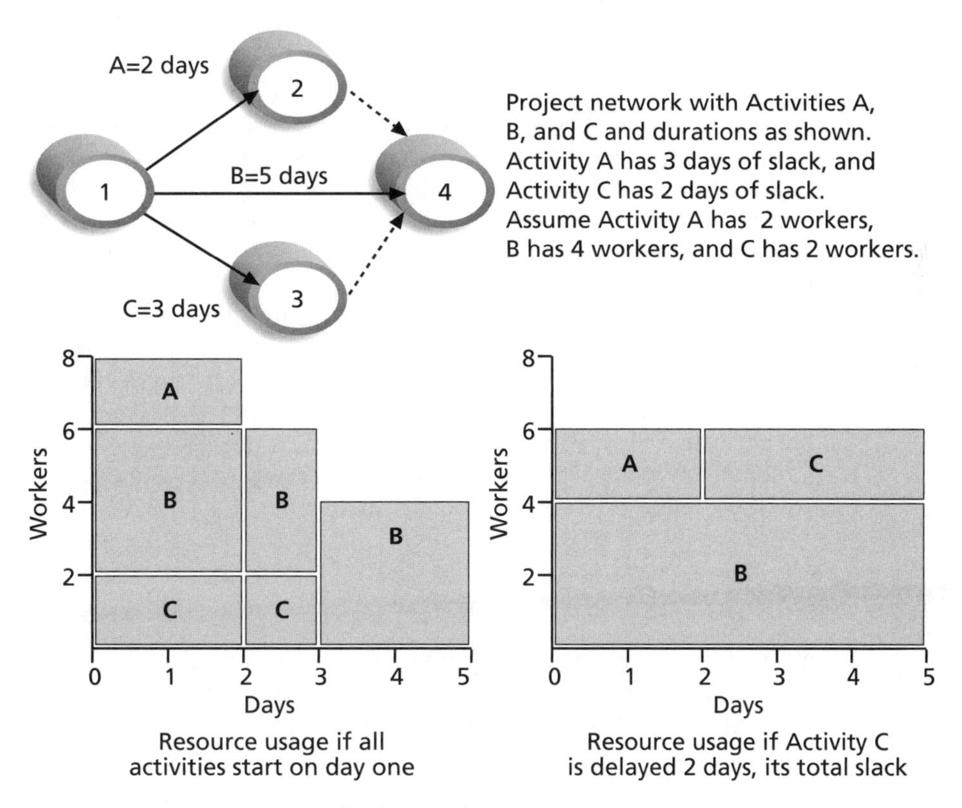

Figure 8-8. Resource Leveling Example

Resource leveling has several benefits. First, when resources are used on a more constant basis, they require less management. For example, it is much easier to manage a part-time project member who is scheduled to work twenty hours per week on a project for the next three months than it is to manage the same person who is scheduled to work ten hours one week, forty the next, five the next, and so on.

Second, resource leveling may enable project managers to use a just-in-time inventory type of policy for using subcontractors or other expensive resources. For example, a project manager might want to level resources related to work that must be done by particular subcontractors such as testing consultants. This leveling might allow the project to use four outside consultants full-time to do testing for four months instead of spreading the work out over more time or needing to use more than four people. The latter approach is usually more expensive.

Third, resource leveling results in fewer problems for project personnel and Accounting Departments. Increasing and decreasing labor levels and particular human resources often produce additional work and confusion. For example, if a person with expertise in a particular area is only assigned to a project two days a week and another person they need to work with is not assigned to the project those same days, they cannot work well together. The Accounting Department might complain when subcontractors charge a higher rate for billing less than twenty hours a week on a project. The accountants will remind project managers to strive for getting the lowest rates possible.

Finally, resource leveling often improves morale. People like to have some stability in their jobs. It is very stressful for people to not know from week to week or even day to day what projects they will be working on and with whom they will be working.

Project management software can automatically level resources. However, the project manager must be careful in using the results without making adjustments. Automatic leveling often pushes out the project's completion date. Resources may also be reallocated to work at times that are inappropriate with other constraints. A wise project manager would have one of his or her team members who is proficient in using the project management software ensure that the leveling is done appropriately.

Team Development

Even if you have successfully recruited enough skilled people to work on a project, you must ensure that people can work together as a team to achieve project goals. Many information technology projects have had very talented individuals working on them. However, it takes teamwork to successfully complete most projects. The main goal of team development is to help people work together more effectively to improve project performance. There is an extensive body of literature on team development. This section will highlight a few important tools and techniques for team development, including training, team-building activities, and reward and recognition systems. It will also provide general advice on the effective use of teams.

Training

Project managers often recommend that people take specific training courses to improve individual and team development. For example, Sarah from the opening case had gone through training in dealing with difficult people. She was familiar with the mirroring technique and felt comfortable using that approach with Ben. Many other people would not have reacted so quickly and effectively in the same situation. If Ben and Sarah did reach agreement on what actions they could take to resolve the F-44 program's information technology problems, it might result in a new project to develop and deliver a new system for Ben's group. If Sarah became the project manager for this new project, she would understand the need for special training in interpersonal skills for specific people in her and Ben's departments. Individuals could take special training classes to improve their personal skills. If Sarah thought the whole project team could benefit from taking training together to learn to work as a team, she could arrange for a special team-building session for the entire project team and key stakeholders.

Team-Building Activities

Many companies provide in-house team-building training activities, and many also use specialized services provided by external companies that specialize in this area. Two common approaches to team-building activities include using physical challenges and psychological preference indicator tools.

Several organizations have teams of people go through certain physically challenging activities to help them develop as a team. Military basic training or boot camps provide one example. Men and women who wish to join the military must first make it through basic training, which often involves several strenuous physical activities such as rappelling off towers, running and marching in full military gear, going through obstacle courses, passing marksmanship training, and mastering survival training. Many companies use a similar approach by sending teams of people to special locations where they work as a team to navigate white water rapids, climb mountains or rocks, participate in paintball exercises, and so on.

Even more companies have teams participate in mental team-building activities in which they learn about themselves, each other, and how to work as a group most effectively. It is important for people to understand and value each other's differences in order to work effectively as a team. Two common exercises used in mental team building include the Myers-Briggs Type Indicator and the Wilson Learning Social Styles Profile.

The **Myers-Briggs Type Indicator (MBTI)** is a popular tool for determining personality preferences. During World War II, Isabel B. Myers and Katherine C. Briggs developed the first version of the MBTI based on psychologist Carl Jung's theory of psychological type. The four dimensions of psychological type in the MBTI include:

- Extrovert/Introvert (E/I): The first dimension determines if you are generally extroverted or introverted. The dimension also signifies whether people draw their energy from other people (extroverts) or from inside themselves (introverts).

- Sensation/Intuition (S/N): The second dimension relates to the manner in which you gather information. Sensation type people take in facts, details, and reality and describe themselves as practical. Intuitive type people are imaginative, ingenious, and attentive to hunches or intuition. They describe themselves as innovative and conceptual.
- Thinking/Feeling (T/F): The third dimension represents thinking judgment and feeling judgment. Thinking judgment is objective and logical, and feeling judgment is subjective and personal.
- Judgment/Perception (J/P): The fourth dimension concerns people's attitude toward structure. Judgment type people like closure and task completion. They tend to establish deadlines and take them seriously, expecting others to do the same. Perceiving types prefer to keep things open and flexible. They regard deadlines more as a signal to start rather than complete a project and do not feel that work must be done before play or rest begins.[15]

There is much more involved in personality types, and many books are available on this topic. In 1998, David Keirsey and Ray Choiniere published *Please Understand Me II: Temperament Character Intelligence*. This book includes an easy-to-take and interpret test called The Keirsey Temperament Sorter, which is a personality type preference test based on the work done by Jung, Myers, and Briggs.

In 1985, an interesting study of the MBTI types of the general population in the United States and information systems (IS) developers revealed some significant contrasts.[16] Both groups of people were most similar in the judgment/perception dimension, with slightly more than half of each group preferring the judgment type (J). There were significant differences, however, in the other three dimensions. Most people would not be surprised to hear that most information systems developers are introverts. This study found that 75 percent of IS developers were introverts (I), and only 25 percent of the general population were introverts. This personality type difference might help explain some of the problems users have communicating with developers. Another sharp contrast found in the study was that almost 80 percent of IS developers were thinking types (T) compared to 50 percent of the general population. IS developers were also much more likely to be intuitive (N) (about 55 percent) than the general population (about 25 percent). These results fit with Keirsey's classification of NT (Intuitive/Thinking types) people as rationals. Educationally,

[15] Briggs, Isabel Myers, with Peter Myers, *Gifts Differing: Understanding Personality Type*. Palo Alto, CA: Consulting Psychologists Press, 1995.

[16] Lyons, Michael L, "The DP Psyche," *Datamation* (August 15, 1985).

they tend to study the sciences, enjoy technology as a hobby, and pursue systems work. Keirsey also suggests that no more than 7 percent of the general population are NTs. Would you be surprised to know that Bill Gates is classified as a rational?[17]

Many organizations also use the Social Styles Profile in team-building activities. Psychologist David Merril, who helped develop the Wilson Learning Social Styles Profile, describes people as falling into four approximate behavioral profiles, or zones. People are perceived as behaving primarily in one of four zones, based on their assertiveness and responsiveness:

- "Drivers" are proactive and task-oriented. They are firmly rooted in the present, and they strive for action. Adjectives to describe drivers include pushy, severe, tough, dominating, harsh, strong-willed, independent, practical, decisive, and efficient.
- "Expressives" are proactive and people-oriented. They are future-oriented and use their intuition to look for fresh perspectives on the world around them. Adjectives to describe expressives include manipulating, excitable, undisciplined, reacting, egotistical, ambitious, stimulating, wacky, enthusiastic, dramatic, and friendly.
- "Analyticals" are reactive and task-oriented. They are past-oriented and strong thinkers. Adjectives to describe analyticals include critical, indecisive, stuffy, picky, moralistic, industrious, persistent, serious, expecting, and orderly.
- "Amiables" are reactive and people-oriented. Their time orientation varies depending on whom they are with at the time, and they strongly value relationships. Adjectives to describe amiables include conforming, unsure, ingratiating, dependent, awkward, supportive, respectful, willing, dependable, and agreeable.[18]

Figure 8-9 shows these four social styles and how they relate to assertiveness and responsiveness. Note that the main determinants of the social style are your levels of assertiveness—if you are more likely to tell people what to do or ask what should be done—and how you respond to tasks—by focusing on the task itself or on the people involved in performing the task.

[17] The Web Site for the Keirsey Temperament Sorter and Keirsey Temperament Theory, http://keirsey.com/personality/nt.html.

[18] Robbins, Harvey A. and Michael Finley, *The New Why Teams Don't Work: What Goes Wrong and How to Make It Right.* San Francisco, CA: Berrett-Koehler Publishers, 1999.

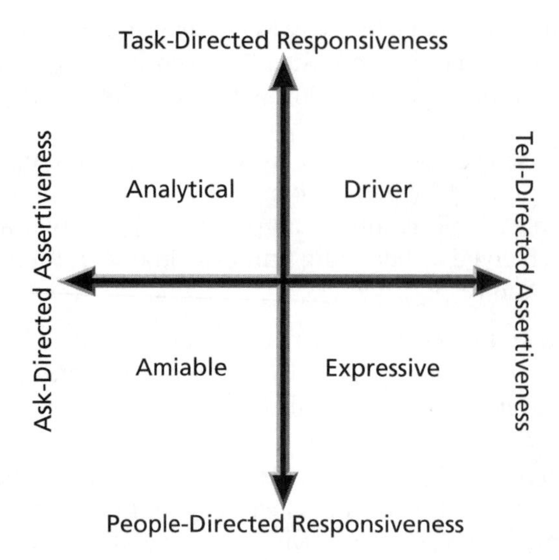

Figure 8-9. Social Styles

Reward and Recognition Systems

Another important tool for promoting team development is the use of team-based reward and recognition systems. If management rewards teamwork, they will promote or reinforce people to work more effectively in teams. Some companies offer bonuses, trips, or other rewards to workgroups that meet or exceed company or project goals. In a project setting, project managers can recognize and reward people who willingly work overtime to meet an aggressive schedule objective or go out of their way to help a teammate. Project managers should not reward people who work overtime just to get extra pay or as a result of their own poor work or planning.

General Advice on Teams

Effective project managers must be good team builders. Suggestions for ensuring that teams are productive include the following:

- Be patient and kind with your team, and assume the best about people. Do not assume that your team members are lazy and careless.
- Fix the problem instead of blaming people. Help people work out problems by focusing on behaviors.
- Establish regular, effective meetings. Focus on meeting project objectives and producing positive results.
- Limit the size of work teams to three to seven members.
- Plan some social activities to help project team members and other stakeholders get to know each other better. Make the social events fun and not mandatory.

- Stress team identity. Create traditions that team members enjoy.
- Nurture team members and encourage them to help each other. Identify and provide training that will help individuals and the team as a whole become more effective.
- Acknowledge individual and group accomplishments.

As you can imagine, staffing and team development are critical concerns on many information technology projects. Information technology project managers must break out of their rational/NT preference and focus on empathically listening to other people to address their concerns and create an environment in which individuals and teams can grow and prosper.

USING SOFTWARE TO ASSIST IN HUMAN RESOURCE MANAGEMENT

Earlier in this chapter you saw that a simple responsibility assignment matrix (Figures 8-4 and 8-5) or histograms (Figures 8-6 and 8-7) are useful tools that can help you effectively manage human resources on projects. You can use several different software packages, including spreadsheets or project management software such as Microsoft Project 2000, to create matrixes and histograms. Many people do not realize that Project 2000 provides a variety of human resource management tools, some of which include assigning and tracking resources, resource leveling, resource usage reports, overallocated resource reports, and to-do lists. You can find an overview of how to use many of these functions and features in Appendix A. (Also note that collaborative software to help people communicate is described in Chapter 9, *Project Communications Management*.)

You can use Project 2000 to assign resources—including equipment, materials, facilities, or people—to tasks. Project 2000 enables you to allocate individual resources to individual projects or to pool resources and share them across multiple projects. By defining and assigning resources in Project 2000, you can:

- Keep track of the whereabouts of resources through stored information and reports on resource assignments.
- Identify potential resource shortages that could force a project to miss scheduled deadlines and possibly extend the duration of a project.
- Identify underutilized resources and reassign them, which may enable you to shorten a project's schedule and possibly reduce costs.
- Use automated leveling to make level resources easier to manage.

Two particularly helpful features of Project 2000 are the resource usage view and the resource usage report. Both of these features help you see who is scheduled to work how many hours on which project activities. Figure 8-10 shows an example of a resource usage view, and Figure 8-11 shows an example of a resource usage report. Both of these figures use information from the software launch sample file used to generate the histogram on overallocation in Figure 8-7. The resource usage view shows the names of the people or resources working on the project, the total number of hours they are scheduled to work, and a calendar indicating how many hours they are scheduled to work each month for each task. If someone is scheduled to work more than his or her allowable number of hours, Project 2000 automatically puts a warning symbol (an exclamation point) in the column to the left of that person's name. Note the exclamation point next to Joe Franklin's name in the first row of Figure 8-10. Upon reviewing the number of hours more closely, you can see that Joe Franklin is overallocated February through May. These numbers are displayed in red in Project 2000 to help them stand out. You can adjust the calendar by clicking an icon to see hours scheduled by months, days, and so on.

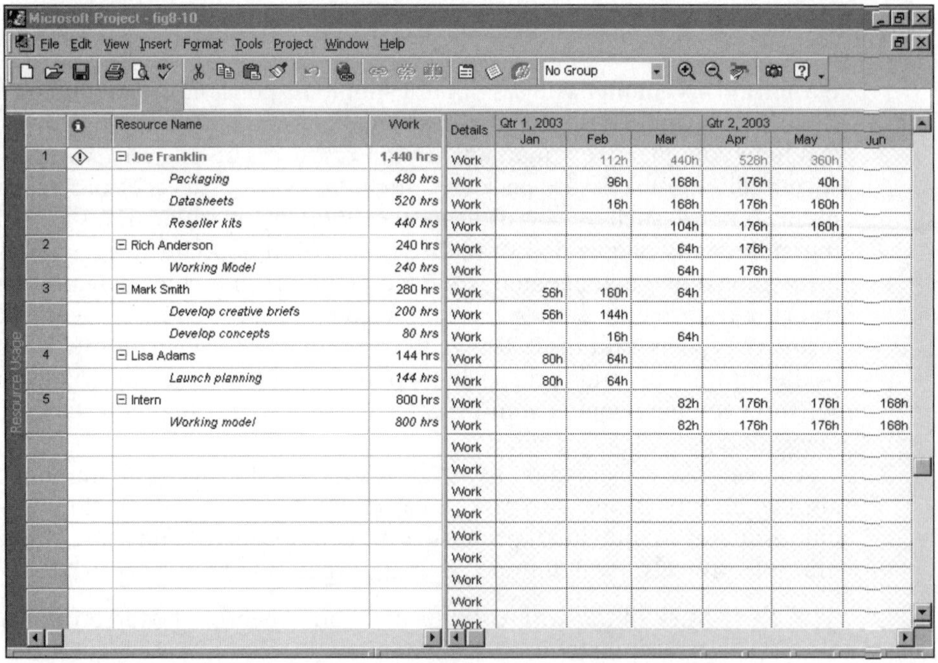

		Resource Name	Work	Details	Qtr 1, 2003			Qtr 2, 2003		
					Jan	Feb	Mar	Apr	May	Jun
1	!	Joe Franklin	1,440 hrs	Work		112h	440h	528h	360h	
		Packaging	*480 hrs*	Work		96h	168h	176h	40h	
		Datasheets	*520 hrs*	Work		16h	168h	176h	160h	
		Reseller kits	*440 hrs*	Work			104h	176h	160h	
2		Rich Anderson	240 hrs	Work			64h	176h		
		Working Model	*240 hrs*	Work			64h	176h		
3		Mark Smith	280 hrs	Work	56h	160h	64h			
		Develop creative briefs	*200 hrs*	Work	56h	144h				
		Develop concepts	*80 hrs*	Work		16h	64h			
4		Lisa Adams	144 hrs	Work	80h	64h				
		Launch planning	*144 hrs*	Work	80h	64h				
5		Intern	800 hrs	Work			82h	176h	176h	168h
		Working model	*800 hrs*	Work			82h	176h	176h	168h
				Work						
				Work						
				Work						
				Work						
				Work						
				Work						
				Work						
				Work						
				Work						

Figure 8-10. Resource Usage View from Microsoft Project

Figure 8-11 is a resource usage report based on the same software launch file. This standard report shows similar information, but the data is automatically formatted to show the number of hours each person is scheduled to work each week during the project. In this figure, you can see that Joe Franklin is scheduled to work 120 hours the weeks of March 16, March 23, and every week in April. He is also scheduled to work 104 hours the week of May 4. If a standard work week is 40 or even 60 hours per week, the project manager can easily see from this report that Joe Franklin cannot do all of the work assigned to him.

	Mar 16,'03	Mar 23,'03	Mar 30,'03	Apr 6,'03	Apr 13,'03	Apr 20,'03	Apr 27,'03	May 4,'03	May '
Joe Franklin	120 hrs	120 hrs	120 hrs	120 hrs	120 hrs	120 hrs	120 hrs	104 hrs	
Packaging	40 hrs	40 hrs	40 hrs	40 hrs	40 hrs	40 hrs	40 hrs	24 hrs	
Datasheets	40 hrs	40 hrs	40 hrs	40 hrs	40 hrs	40 hrs	40 hrs	40 hrs	
Reseller kits	40 hrs	40 hrs	40 hrs	40 hrs	40 hrs	40 hrs	40 hrs	40 hrs	
Rich Anderson	16 hrs	40 hrs	40 hrs	40 hrs	40 hrs	40 hrs	24 hrs		
Working Model	16 hrs	40 hrs	40 hrs	40 hrs	40 hrs	40 hrs	24 hrs		
Mark Smith									
Develop creative briefs									
Develop concepts									
Lisa Adams									
Launch planning									
Intern	34 hrs	40 hrs	40 hrs	40 hrs	40 hrs	40 hrs	40 hrs	40 hrs	
Working model	34 hrs	40 hrs	40 hrs	40 hrs	40 hrs	40 hrs	40 hrs	40 hrs	
Total	170 hrs	200 hrs	200 hrs	200 hrs	200 hrs	200 hrs	184 hrs	144 hrs	

Figure 8-11. Resource Usage Report from Microsoft Project

Just as many project management professionals are not aware of the powerful cost-management features of Microsoft Project 2000, many are also unaware of its powerful human resource management features. With the aid of this type of software, project managers can have more information available in useful formats to help them decide how to most effectively manage human resources.

Project resource management involves much more than using software to assess and track resource loading, level resources, and so on. People are the most important asset on most projects, and human resources are very different from other resources. You cannot simply replace people the same way that you would replace a piece of equipment. People need far more than a tune-up now and then to keep them performing well. It is essential to treat people with consideration and respect, to understand what motivates them, and to communicate carefully with them. What makes good project managers great is not their use of tools, but rather their ability to enable project team members to deliver the best work they possibly can on a project.

CASE WRAP-UP

After Sarah yelled right back at Ben, he said "You're the first person who's had the guts to stand up to me." After that brief introduction, Sarah and Ben and the other meeting participants had a good discussion about what was really happening on the F-44 project. Sarah was able to write a justification to get Ben's group special software and support to download key information from the old system so they could manage their project better. When Sarah stood nose to nose with

Ben and yelled at him, she used a technique for establishing rapport called "mirroring." Although Sarah was by no means a loud and obnoxious person, she saw that Ben was and decided to mirror his behavior and attitude. She put herself in his shoes for a while, and doing so helped break the ice so Sarah, Ben, and the other people at the meeting could really start communicating and working together as a team to solve their problems.

CHAPTER SUMMARY

People are the most important assets in organizations and on projects. Therefore, it is essential for project managers to be good human resource managers.

Psychosocial issues that affect how people work and how well they work include motivation, influence and power, and effectiveness.

Maslow developed a hierarchy of needs that suggests physiological, safety, social, esteem, and self-actualization needs motivate behavior. Once a need is satisfied, it no longer serves as a motivator.

Herzberg distinguished between motivators and hygiene factors. Hygiene factors such as larger salaries or a more attractive work environment will cause dissatisfaction if not present, but do not motivate workers to do more if present. Achievement, recognition, the work itself, responsibility, and growth are factors that contribute to work satisfaction and motivate workers.

McGregor developed Theory X and Theory Y to describe different approaches to managing workers, based on assumptions of worker motivation. Research supports the use of Theory Y, which assumes that people see work as natural and indicates that the most significant rewards are the satisfaction of esteem and self-actualization needs that work can provide.

According to Ouchi's Theory Z, workers can be trusted to do their jobs to their utmost ability, as long as management can be trusted to support them and look out for their well-being. Theory Z emphasizes things such as job rotation, broadening of skills, generalization versus specialization, and the need for continuous training of workers.

Thamhain and Wilemon identified nine influence bases available to project managers: authority, assignment, budget, promotion, money, penalty, work challenge, expertise, and friendship. Their research found that project success is associated with project managers who use work challenge and expertise to influence workers. Project failure is associated with using too much influence by authority, money, or penalty.

Power is the potential ability to influence behavior to get people to do things they would not otherwise do. The five main types of power are coercive power, legitimate power, expert power, reward power, and referent power.

Project managers can use Steven Covey's seven habits of highly effective people to help themselves and project teams become more effective. The seven habits include being proactive; beginning with the end in mind; putting first things first; thinking win/win; seeking first to understand, then to be understood; achieving synergy; and sharpening the saw. Using empathic listening is a key skill of good project managers.

The major processes involved in project human resource management include organizational planning, staff acquisition, and team development. Organizational planning involves identifying, assigning, and documenting project roles, responsibilities, and reporting relationships. A responsibility assignment matrix (RAM) is a key tool for defining roles and responsibilities on projects.

Staff acquisition is getting the appropriate staff assigned to and working on the project. This is an important issue in today's competitive environment. Companies must use innovative approaches to find and retain good information technology staff.

Resource loading shows the amount of individual resources an existing schedule requires during specific time frames. Histograms are often used to show resource loading and to identify overallocation of resources.

Resource leveling is a technique for resolving resource conflicts, such as overallocated resources, by delaying tasks. Leveled resources require less management, lower costs, produce fewer personnel and accounting problems, and often improve morale.

One of the most important skills of a good project manager is team development. Teamwork helps people work more effectively to achieve project goals. Project managers can recommend individual training to improve skills related to teamwork, organize team-building activities for the entire project team and key stakeholders, and provide reward and recognition systems that encourage teamwork.

Spreadsheets and project management software such as Microsoft Project 2000 can help project managers in project human resource management. Software makes it easy to produce responsibility assignment matrixes, create resource histograms, identify overallocated resources, level resources, and provide various views and reports related to project human resource management.

Project human resource management involves much more than using software to facilitate organizational planning and assign resources. What makes good project managers great is their ability to enable project team members to deliver the best work they possibly can on a project.

Discussion Questions

1. Discuss the evidence pointing to a shortage of information technology workers. Why are there not more students majoring in computer science or MIS, if there is such a labor shortage? Speculate about whether the information technology labor shortage will continue in the near future.

2. Briefly summarize the work done by Maslow, Herzberg, McGregor, Ouchi, Thamhain and Wilemon, and Covey. How do their theories relate to project management?
3. Give examples of different ways to have influence on projects. Which do you think would be most effective with you?
4. Describe the work definition and assignment process.
5. Discuss key issues related to staff acquisition and team building.
6. How could you use resource loading and resource leveling to ensure that a project runs smoothly?

EXERCISES

1. Develop a responsibility assignment matrix for building a new state-of-the-art multimedia classroom for your university. Assume the faculty in the Computer Science/MIS Department received a grant to lead this project. The grant includes funds for all of the hardware and software plus some money for faculty and student labor. Also assume the university must match the funds provided by the granting agency. Key stakeholders are faculty who might use the facility, students, the Dean, the head of facilities, the head of finance, and the head of information technology. Major activities might include deciding where to put the classroom, defining the classroom configuration (number and type of computers, projections system software, layout, and so on), writing a request for proposal, evaluating proposals, overseeing the winning contractor, obtaining matching funds from the college, surveying faculty and students on the use of the classroom, providing training on using the classroom, and scheduling use of the classroom. Use the responsibility assignment matrix in Figures 8-4 and 8-5 as models.
2. Research recent work on what motivates workers, especially in information technology. Summarize at least three articles on this subject.
3. Read Douglas Ross's article "Applying Covey's Seven Habits to a Project Management Career" (*see* Suggested Readings). Summarize three of Covey's habits in your own words and give examples of how these habits would apply to a project you are familiar with.
4. Research recruiting and retention strategies at three different companies. What distinguishes one company from another in this area? Are strategies such as signing bonuses, tuition reimbursement, and business casual dress codes standard for new information technology workers? What strategies appeal most to you?
5. Write a paper summarizing the main features of Microsoft Project 2000 that can assist project managers in human resource management. Include your opinion on the usefulness of specific features.

MINICASE

Your company is planning to launch an important new project starting January 1. You estimate that you will need one full-time project manager for this one-year project, two business analysts full-time for the first six months, two senior programmers full-time for the whole year, four junior programmers full-time for the months of July, August, and September, and one technical writer full-time for the last three months. For simplicity's sake, assume that the main tasks these people will perform are as follows, respectively:

- Project management
- Systems analysis
- Systems design
- Coding and testing
- Documentation

Part 1: Use Microsoft Excel to create a column chart showing a resource histogram for this project, similar to the one shown in Figure 8-6. Be sure to include a legend to label the types of resources needed, and include a column showing the total number of people for each month of the project. Use appropriate titles and axis labels.

Part 2: Enter the five project tasks—project management, systems analysis, systems design, coding and testing, and documentation—and their given durations in Microsoft Project 2000. Assume that one year is fifty-two weeks, six months is twenty-six weeks, and so on. Assume the first three tasks will all start at the same time, that coding and testing can start when systems design is 50 percent completed, and that documentation starts when the coding and testing is completed. Then enter the resource information, as provided in the first paragraph of the minicase. Instead of entering names for each resource, put the job title. For two people with the same job title, enter business analyst 1 and business analyst 2, for example. Then assign the resources to the appropriate tasks. Print the Gantt chart view of your work on one page when you are finished.

SUGGESTED READINGS

1. Brooks, Frederick P., Jr. *The Mythical Man-Month: Essays on Software Engineering*. Reading, MA: Addison-Wesley Publishing Co., 1995.

    The Mythical Man-Month *is one of the classics in the field of software program management. Brooks wrote the first version of this book in 1975. He draws on his experience as the head of operating systems development for IBM's famous 360 mainframe and some of his most recent experiences.*

2. Covey, Stephen. *The 7 Habits of Highly Effective People*. New York: Simon & Schuster, 1990.

   Stephen Covey is famous for his books, audiotapes, videotapes, and seminars on improving effectiveness. This book continues to be a business bestseller, with more than ten million copies sold.

3. Herzberg, Frederick. "One More Time: How Do You Motivate Employees?" *Harvard Business Review* (February 1968): 51–62.

   This classic article describes the results of Herzberg's studies on what motivates people. He distinguishes between hygiene factors and motivating factors in a work setting and offers suggestions on how to effectively motivate workers.

4. Keirsey, David, and Ray Choiniere. *Please Understand Me II: Temperament, Character, Intelligence*. Del Mar, CA: Prometheus Nemesis Book Co., May 1998.

   Keirsey and Bates's Please Understand Me, *first published in 1978, sold nearly two million copies in its first twenty years, becoming a bestseller all over the world. Keirsey has continued to investigate personality differences, and recently published this sequel to refine his theory of the four temperaments and to define the facets of character that distinguish one person from another. You can get more information on this book and take an online version of the Keirsey Temperament Sorter test at* www.keirsey.com.

5. Ross, Douglas. "Applying Covey's Seven Habits to a Project Management Career." *PM Network*, Project Management Institute (April 1996): 26–30.

   Ross summarizes key points from Covey's seven habits and relates them to improving effectiveness in project management. This article helps project managers translate Covey's principles to their own teams, projects, and personal lives.

6. Verma, Vijay K. *Organizing Projects for Success: The Human Aspects of Project Management*. Upper Darby, PA: Project Management Institute, 1995.

   This book focuses specifically on managing human resource in a project environment. There are separate chapters on interfacing with major stakeholders, designing organizational structures, addressing important issues in project organization design, designing a project organizational structure, and making matrix structures work. Verma has also written several other books related to project human resource management.

KEY TERMS

- **Coercive power** — using punishment, threats, or other negative approaches to get people to do things they do not want to do
- **Deputy project managers** — people who fill in for project managers in their absence and assist them as needed, similar to the role of a vice president

- **Empathic listening** — listening with the intent to understand
- **Expert power** — using one's personal knowledge and expertise to get people to change their behavior
- **Hierarchy of needs** — a pyramid structure illustrating Maslow's theory that people's behaviors are guided or motivated by a sequence of needs
- **Legitimate power** — getting people to do things based on a position of authority
- **Mirroring** — the matching of certain behaviors of the other person
- **Myers-Briggs Type Indicator (MBTI)** — a popular tool for determining personality preferences
- **Organization planning** — identifying, assigning, and documenting project roles, responsibilities, and reporting relationships
- **Organizational breakdown structure (OBS)** — a specific type of organizational chart that shows which organizational units are responsible for which work items
- **Overallocation** — when more resources than are available are assigned to perform work at a given time
- **Power** — the potential ability to influence behavior to get people to do things they would not otherwise do
- **Rapport** — a relation of harmony, conformity, accord, or affinity
- **Referent power** — getting people to do things based on an individual's personal charisma
- **Resource histogram** — a column chart that shows the number of resources assigned to a project over time
- **Resource leveling** — a technique for resolving resource conflicts by delaying tasks
- **Resource loading** — the amount of individual resources an existing schedule requires during specific time periods
- **Responsibility assignment matrix (RAM)** — a matrix that maps the work of the project as described in the WBS to the people responsible for performing the work as described in the OBS
- **Reward power** — using incentives to induce people to do things
- **Staff acquisition** — getting the needed personnel assigned to and working on the project
- **Staffing management plan** — a document that describes when and how people will be added to and taken off the project team
- **Subproject managers** — people responsible for managing the subprojects that a large project might be broken into
- **Synergy** — an approach where the whole is greater than the sum of the parts
- **Team development** — building individual and group skills to enhance project performance

9

Project Communications Management

Objectives

After reading this chapter, you will be able to:

1. *Understand the importance of good communication on projects and describe the major components of a communications management plan*
2. *Discuss the elements of communications planning, including information distribution, performance reporting, and administrative closure*
3. *Discuss various methods for distributing project information and the advantages and disadvantages of each*
4. *Understand the relationship between the number of people involved in a project and the complexity of communications*
5. *Discuss sources of conflict on projects and strategies for managing them*
6. *Understand important aspects of verbal, nonverbal, and written communication and how to develop these skills*
7. *Describe various types of project documentation and understand the value of using templates for aiding in project communications*
8. *Describe how software can enhance project communications*

Peter Gumpert worked his way up the corporate ladder in a large telecommunications company. He was intelligent, competent, and a strong leader, but a new undersea fiber-optic telecommunications systems program was much larger and more complicated than anything he had worked on, let alone managed. The undersea systems were separated into several distinct projects, and Peter was the program manager in charge of overseeing them all. The changing marketplace for undersea telecommunications systems and the large number of projects involved made communications and flexibility critical concerns for Peter. For missing milestone and completion dates, his company would suffer huge financial penalties ranging from thousands of dollars per day for smaller projects to over

$250,000 per day for larger projects. Many projects depended on the success of other projects, so Peter had to understand and actively manage those critical interfaces.

Peter held several informal and formal discussions with the project managers reporting to him on this program. He worked with them and his project executive assistant, Christine Braun, to develop a communications plan for the program. He was still unsure, however, of the best way to distribute information and manage all of the inevitable changes that would occur. He also wanted to develop consistent ways for all of the project managers to develop their plans and track performance without stifling their creativity and autonomy. Christine suggested that they consider using some new communications technologies to keep important project information up to date and synchronized. Although Peter knew a lot about telecommunications and laying fiber-optic lines, he was not an expert in using information technology to improve the communication process. In fact, that was part of the reason why he asked Christine to be his assistant. Could they really develop a process for communicating that would be flexible and easy to use? Time was of the essence as more projects were being added to the undersea telecommunications program every week.

THE IMPORTANCE OF PROJECT COMMUNICATIONS MANAGEMENT

Many experts agree that the greatest threat to the success of any project, especially information technology projects, is a failure to communicate. Recall that the 1995 Standish Group study found the three major factors related to information technology project success were user involvement, executive management support, and a clear statement of requirements.[1] All of these factors depend on having good communications skills, especially with non-information technology personnel.

Our culture often portrays computer professionals as nerds who like to sit in dark corners hacking away on computers. When computer professionals have to communicate with non-computer people, it can be as if they were talking to someone from another planet. The information technology field is constantly changing, and these changes bring with them a great deal of technical jargon. Even though more and more people use computers, the gap between users and developers increases as technology advances.

[1] The Standish Group, "CHAOS" (1995) www.standishgroup.com/chaos.html.

Compounding the problem, our educational system for information technology graduates promotes strong technical skills over strong communication and social skills. Most information technology-related degree programs have many technical requirements, but few require courses in communication (speaking, writing, listening), psychology, sociology, and the humanities. People often assume it is easy to pick up these "soft skills," but they *are* skills and, as such, they must be learned and developed.

Many studies have shown that it is these soft skills that are most needed by information technology professionals. You cannot totally separate technical skills and soft skills when working on information technology projects. For projects to succeed, every project member needs both types of skills, and both types of skills need to be continuously developed through formal education and on-the-job training.

An article in the *Journal of Information Systems Education* on the importance of communications skills for information technology professionals presented the following conclusion:

> Based on the results of this research we can draw some general conclusions. First, it is evident that IS professionals engage in numerous verbal communication activities that are informal in nature, brief in duration, and with a small number of people at a time. Second, we can infer that most of the communication is indeed verbal in nature but sometimes it is supported by notes or graphs on a board or a handout and also by computer output. Third, it is clear that people expect their peers to listen carefully during a conversation and respond correctly to the issues at hand. Fourth, all IS professionals must be aware of the fact that they will have to engage in some form of informal public speaking. Fifth, it is evident that IS professionals must be able to communicate effectively in order to be successful in their current position but they must also be able to do so in order to move to higher positions. Since our respondents, on average, seem to have moved throughout their IS career, from lower to higher positions, and they ranked verbal skills more important for their advancement than for their current job, the ability to communicate verbally seems to be a key factor in career advancement.[2]

This chapter will highlight key aspects of project communications management, provide some suggestions for improving communications management, and describe how software can assist in project communications management.

The goal of project communications management is to ensure timely and appropriate generation, collection, dissemination, storage, and disposition of project information. Project communications management processes include:

- **Communications planning**, which involves determining the information and communications needs of the stakeholders: who needs what information, when will they need it, and how the information will be given to them.
- **Information distribution**, which involves making needed information available to project stakeholders in a timely manner.

[2] Sivitanides, Marcos P., James R. Cook, Roy B. Martin, and Beverly A. Chiodo, "Verbal Communication Skills Requirements for Information Systems Professionals," *Journal of Information Systems Education* (Spring 1995) 7(1).

- **Performance reporting**, which involves collecting and disseminating performance information, including status reports, progress measurement, and forecasting.
- **Administrative closure**, which involves generating, gathering, and disseminating information to formalize phase or project completion.

COMMUNICATIONS PLANNING

Because communication is so important on projects, every project should include a **communications management plan**—a document that guides project communications. This plan should be part of the overall project plan (described in Chapter 3, *Project Integration Management*). The type of communications management plan will vary with the needs of the project, but such a plan should always be prepared. The main parts of a communications management plan include:

- *A description of a collection and filing structure for gathering and storing various types of information.* For example, if a project team member attends a conference and brings back valuable information, where should it be filed and stored? If a supplier sends a new product brochure, where should it be filed and stored? If the same supplier sends a new product brochure six months later, how should the new brochure and the old brochure be handled? If the new product is used on the project and might affect other areas such as accounting or engineering, how will that new product information be communicated to those areas? Everyone knows it can be difficult to organize his or her individual work, so it is imperative to develop and follow a system for filing important project-related work for many people. In addition, several government agencies require a detailed filing system and hold inspections to ensure their instructions for filing are followed.
- *A distribution structure describing what information goes to whom, when, and how.* For example, are all status reports written or are some oral? Does every stakeholder receive every master schedule update? Do executives receive different formats of status reports?
- *A format for communicating key project information.* Is there a template for project team members to follow in preparing written and oral status reports? Is there a master list of all acronyms and definitions, or do they need to be repeated on different project documentation? Many hours of confusion can be avoided by providing templates and examples of key project reports. You will find several examples later in this chapter.
- *A production schedule for producing the information.* Have resources been assigned to create, assemble, and disseminate key project information? Do stakeholders know when to expect different information, when they need to attend key meetings, and so on? Has time been allowed for review and approval of key project documentation? Many people procrastinate when it comes to documenting their work. It is important to allow time for creating key project information and ensuring its quality.

- *Access methods for obtaining the information.* Who can see a draft document? Can everyone access all project documentation? Which information is kept online and which is kept only in hard copy or other formats? Can anyone check out hard copies of documents? Who can attend what meetings?
- *A method for updating the communications management plans as the project progresses and develops.* Who will update the communications management plan as changes are made? How will the new plan be distributed?
- *A stakeholder communications analysis.* It is important to know what kinds of information will be distributed to which stakeholders. By analyzing stakeholder communications, you can avoid wasting time or money on creating or disseminating unnecessary information. The project's organizational chart is a starting point for identifying internal stakeholders. You must also include key stakeholders outside of the project organization such as the customer, the customer's senior management, and subcontractors.

Table 9-1 provides a sample stakeholder communications analysis that shows which stakeholders should get which written communications. Note that the analysis includes information such as the contact person for the information, when the information is due, and the preferred format for the information. You can create a similar table to show which stakeholders should attend which project meetings. It is always a good idea to include comment sections with these types of tables to record special considerations or details related to each stakeholder, document, meeting, and so on. Having stakeholders review and approve all stakeholder analysis materials will ensure that the information is correct and useful.

Table 9-1: Sample Stakeholder Analysis for Project Communications

Stakeholders	Document Name	Document Format	Contact Person	Due
Customer Management	Monthly Status Report	Hard copy	Gail Feldman, Tony Silva	First of month
Customer Business Staff	Monthly Status Report	Hard copy	Julie Grant, Jeff Martin	First of month
Customer Technical Staff	Monthly Status Report	E-mail	Evan Dodge, Nancy Michaels	First of month
Internal Management	Monthly Status Report	Hard copy	Bob Thomson	First of month
Internal Business and Technical Staff	Monthly Status Report	Intranet	Angie Liu	First of month
Training Subcontractor	Training Plan	Hard Copy	Jonathan Kraus	11/1/2002
Software Subcontractor	Software Implementation Plan	E-mail	Barbara Gates	6/1/2002

Many projects do not include enough initial information on communications. Project managers, senior management, and team members assume using existing communications channels to relay project information is sufficient. The problem with using existing communication channels is that each of these groups, as well as other stakeholders, has different communication needs. Creating some sort of communications management plan and reviewing it with project stakeholders early in a project helps prevent or reduce later communication problems. If organizations work on many projects, developing some consistency in handling project communications helps the organization run smoothly.

Consistent communication helps organizations improve project communications, especially for programs composed of multiple projects. For example, Peter Gumpert, the undersea telecommunications program manager in the opening case, would benefit greatly from having a communications management plan that all of the project managers under him helped develop and follow. Since several of the projects probably have some of the same stakeholders, it is even more important to develop a coordinated communications plan. For example, if customers receive status reports from Peter's company that have totally different formats and do not coordinate information from related projects within the same company, they will question the ability of Peter's company to manage large programs.

Information regarding the content of essential project communications comes from the work breakdown structure (WBS). In fact, many WBSs include a section for project communications to ensure that reporting key information is a deliverable of the project. If reporting essential information is an activity defined in the WBS, it becomes even more important to develop a clear understanding of what project information is to be reported, when it is to be reported, how it is to be reported, who is responsible for generating the report, and so on.

INFORMATION DISTRIBUTION

Getting project information to the right people at the right time and in a useful format is just as important as developing the information in the first place. The stakeholder analysis for project communications serves as a good starting point for information distribution. Project managers and their teams must decide who receives what information, but they must also decide the best way to distribute the information. Is it sufficient to send written reports for project information? Are meetings alone effective in distributing project information? Are meetings and written communications both required for project information?

After answering these questions, project managers and their teams must decide the best way to distribute the information. Important considerations for information distribution include the use of technology, formal and information communications, and the complexity of communications.

Using Technology to Enhance Information Distribution

Technology can facilitate the process of distributing information. Using an internal project management information system, you can organize project documents, meeting minutes, customer requests, requests to change status, and so on, and make them available in an electronic format. You can store this information in local software or make it available on an intranet, an extranet, or the Internet, if the information is nonsensitive. Storing templates and samples of project documents electronically can make accessing standard forms easier, thus making the information distribution process easier. A section later in this chapter, on using software to assist in project communications management, shows examples of using Microsoft Project 2000 and the Internet to improve information distribution.

Formal and Informal Methods for Distributing Information

It is not enough for project team members to submit status reports to their project managers and other stakeholders and assume that everyone who needs to know that information will read the reports. Many technical professionals assume that submitting the appropriate status reports is sufficient. Informal verbal communications are equally effective ways to distribute information, but technical professionals tend to neglect more informal techniques. In contrast to many technical professionals, many non-technical professionals—from colleagues to managers—prefer to hear important project information informally and have a two-way conversation about it. They prefer not to read stacks of paperwork, e-mails, or Web pages to try to find pertinent information.

Instead of focusing on getting information by reading technical documents, many colleagues and managers want to know the people working on their projects and develop a trusting relationship with them. They use informal discussions about the project to develop these relationships. As a consequence, project managers must be good at nurturing relationships through good communication. Many experts believe that the difference between good project managers and excellent project managers is their ability to nurture relationships, as described in Chapter 8, *Project Human Resource Management*.

Effective distribution of information depends on project managers and project team members having good communication skills. Communicating includes many different dimensions such as writing, speaking, and listening, and project personnel need to use all of these dimensions in their daily routines. In addition, different people respond positively to different levels of or types of communication. For example, a project sponsor may prefer to stay informed through informal discussions held once a week over coffee. The project manager needs to be aware of and take advantage of this special communication need. The project sponsor will give better feedback about the project during these informal talks than he or she could give through some other form of communication. Informal conversations can allow the project sponsor to exercise his or her role of leadership and provide insights and information that are critical to the success of the project and the organization as a whole. Short face-to-face meetings are often more effective than electronic communications, particularly for sensitive information.

What Went Wrong?

A well-publicized example of misuse of e-mail comes from the 1998 Justice Department's high-profile, antitrust suit against Microsoft. E-mail emerged as a star witness in the case. Many executives sent messages that should never have been put in writing. The court used e-mail as evidence, even though the senders of the notes said the information was being interpreted out of context.[3]

Some companies, such as Amazon.com, have established policies to encourage employees to watch their use of e-mail and delete it often. Their "Sweep and Clean" program instructed employees to purge e-mails that were no longer required for business or legal purposes. They even offered free café lattes to employees who complied immediately.[4]

Another, more amusing, example of miscommunication comes from a director of communications at a large firm:

I was asked to prepare a memo reviewing our company's training programs and materials. In the body of the memo in one of the sentences I mentioned the "pedagogical approach" used by one of the training manuals. The day after I routed the memo to the executive committee, I was called into the HR director's office, and told that the executive vice president wanted me out of the building by lunch. When I asked why, I was told that she wouldn't stand for perverts (pedophiles?) working in her company. Finally, he showed me her copy of the memo, with her demand that I be fired—and the word "pedagogical" circled in red. The HR manager was fairly reasonable, and once he looked the word up in his dictionary and made a copy of the definition to send back to her, he told me not to worry. He would take care of it. Two days later, a memo to the entire staff came out directing us that no words which could not be found in the local Sunday newspaper could be used in company memos. A month later, I resigned. In accordance with company policy, I created my resignation memo by pasting words together from the Sunday paper.[5]

Many written reports neglect to provide the important information that good managers and technical people have a knack for asking about. For example, it is important to include detailed technical information that will affect critical performance features of products the company is producing as part of a project. It is even more important to document any changes in technical specifications that might impact product performance. For example, if the undersea telecommunications program included a project to purchase and provide special diving gear, and the supplier who provided the oxygen tanks enhanced

[3] Harmon, Amy, "E-mail Comes Back to Haunt Companies," *Minneapolis Star Tribune* (from the *New York Times*) (November 29, 1998).

[4] *Ibid.*

[5] Projectzone Humor www.projectzone.com/memos.html (2001).

the tanks so divers could stay underwater longer, it would be very important to let other people know about this new capability. The information should not be buried in an attachment with the supplier's new product brochure. People also have a tendency to not want to report bad information. If the oxygen tank vendor was behind on production, the person in charge of the project to purchase the tanks might wait until the last minute to report this critical information. Oral communication via meetings and informal talks helps bring important information—positive or negative—out into the open.

Oral communication also helps build stronger relationships among project personnel and project stakeholders. People make or break projects, and people like to interact with each other to get a true feeling for how a project is going. Studies show that less than 10 percent of communications consist of the actual content or words communicated. A person's tone of voice and body language say a lot about how they really feel.

Since information technology projects often require a lot of coordination, it is a good idea to have short, frequent meetings. For example, some information technology project managers require all project personnel to attend a "stand-up" meeting every week or even every morning, depending on the project needs. Stand-up meetings have no chairs, and the lack of chairs forces people to focus on what really needs to be communicated.

To encourage more face-to-face, informal communications, some companies have instituted policies that workers cannot use e-mail between certain hours of the business day. Sending frequent e-mails to a large group of people has a much different effect than sending frequent e-mails within a small team of people working on a particular task.

Table 9-2 provides guidelines from Practical Communications, Inc., a communications consulting firm, about how well different types of media, such as hard copy, telephone calls, voice mail, e-mail, meetings, and Web sites, are suited to different communication needs. For example, if you were trying to assess commitment of project stakeholders, a meeting would be the most appropriate medium to use. A telephone call would be adequate, but the other media would not be appropriate. Project managers must assess the needs of the organization, the project, and individuals in determining which communication medium to use, and when.

Determining the Complexity of Communications

Another important aspect of information distribution is the number of people involved in a project. As the number of people involved increases, the complexity of communications also increases. In addition to needing to understand more individuals' personal preferences for communications, there are more communications channels as more people are involved in projects.

Table 9-2: Media Choice Table

KEY: 1 = EXCELLENT	2 = ADEQUATE			3 = INAPPROPRIATE		
How well medium is suited to:	**Hard Copy**	**Telephone Call**	**Voice Mail**	**E-mail**	**Meeting**	**Web Site**
Assessing commitment	3	2	3	3	1	3
Building consensus	3	2	3	3	1	3
Mediating a conflict	3	2	3	3	1	3
Resolving a misunderstanding	3	1	3	3	2	3
Addressing negative behavior	3	2	3	2	1	3
Expressing support/appreciation	1	2	2	1	2	3
Encouraging creative thinking	2	3	3	1	3	3
Making an ironic statement	3	2	2	3	1	3
Conveying a reference document	1	3	3	3	3	1
Reinforcing one's authority	1	2	3	3	1	2
Providing a permanent record	1	3	3	1	3	1
Maintaining confidentiality	2	1	2	3	1	3
Conveying simple information	3	2	1	1	2	3
Asking an informational question	3	2	1	1	3	3
Making a simple request	3	3	1	1	3	3
Giving complex instructions	3	3	3	2	1	2
Addressing many people	2	3	3 or 1*	2	3	1

Galati, Tess. Email Composition and Communication (EmC2) Practical Communications, Inc. (www.praccom.com) (2001).
*Depends on system functionality

There is a simple formula for determining the number of communications channels as the number of people involved in a project increases. You can calculate the number of communications channels as follows:

$$\text{number of communications channels} = \frac{n(n-1)}{2}$$

where *n* is the number of people involved. For example, two people have one communications channel: (2(2-1))/2 = 1. Three people have three channels: ((3(3-1))/2 = 3. Four people have six channels, five people have ten, and so on. Figure 9-1 illustrates this concept. You can see that as the number of people communicating increases above three, the number of communications channels increases rapidly. The lesson is a simple one: If you want to enhance communications, you must consider the interactions among different project team members and stakeholders. If you raise an issue by sending e-mail to 100 people, the result will be quite different than if you bring up the item at a large review meeting or at smaller team meetings. Sending e-mail to 100 people is likely to result in more problems than would arise if you discussed the same issue in a meeting.

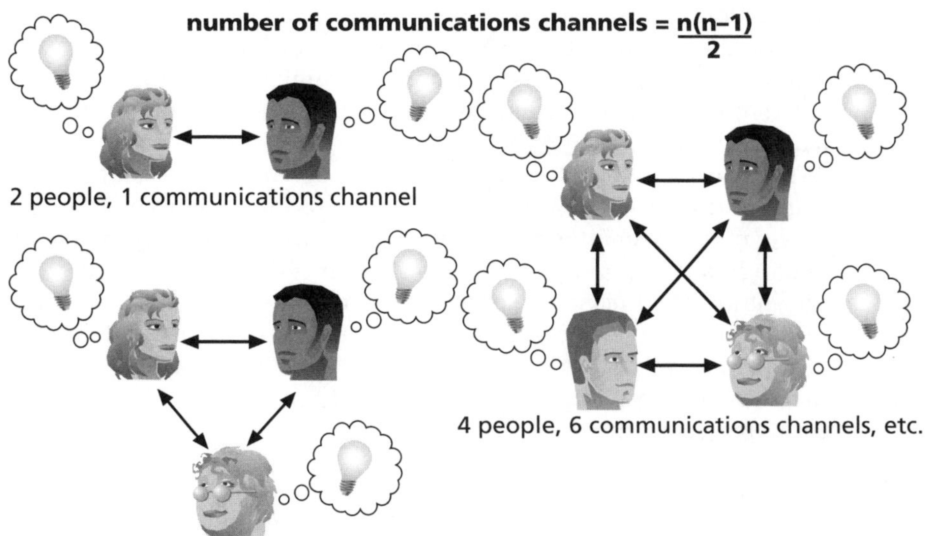

Figure 9-1. The Impact of the Number of People on Communications Channels

When asked why you cannot send an e-mail to a team of 100 people, just as you would to a team of five, one CIO answered, "As a group increases in size, you have a whole slew of management challenges. Communicating badly exponentially increases the possibility of making fatal mistakes. A large-scale project has a lot of moving parts, which makes it that much easier to break down. Communication is the oil that keeps everything working properly. It's much easier to address an atmosphere of distrust among a group of five team members than it is with a team of 500 members."[6]

Rarely does the receiver interpret a message exactly as the sender intended. Therefore, it is important to provide many methods of communication and an environment that promotes open dialogue. It is important for project managers and their teams to be aware of their own communication styles and preferences and those of other project stakeholders. As described in the previous chapter, many information technology professionals have different personality traits than the general population, such as being more introverted, intuitive, and oriented to thinking (as opposed to feeling). These personality differences can lead to miscommunication with people who are extroverted, sensation-oriented, and feeling types. For example, information technology people are notorious for writing

[6] Hildenbrand, Carol, "Loud and Clear," *CIO Magazine* (April 15, 1996) http://www.cio.com/archive/041596_qa.html.

poor documentation. Their version of a user guide might not provide the detailed steps most users need. Many users also prefer face-to-face meetings to learn how to use a new system instead of trying to follow a cryptic user guide.

Geographic location and cultural background also affect the complexity of project communications. For example, if project stakeholders are in different countries, it is often difficult or impossible to schedule times for two-way communications during normal working hours. Language barriers can also cause communications problems. The same word may have very different meanings in different languages. Times, dates, and other units of measure are also interpreted differently. People from some cultures also prefer to communicate in ways that may be uncomfortable to others. For example, managers in some countries still do not allow workers of lower ranks or women to give formal presentations. Some cultures also reserve written documents for binding commitments.

As you can see, information distribution involves more than creating and sending status reports or holding periodic meetings. Many good project managers know their personal strengths and weaknesses in this area and surround themselves with people who complement their skills, just as Peter Gumpert in the opening case did in asking Christine to be his assistant. It is good practice to share the responsibility for project communications management with the entire project team.

PERFORMANCE REPORTING

Performance reporting keeps stakeholders informed about how resources are being used to achieve project objectives. The project plan and work results are important inputs to performance reporting. The main outputs of performance reporting include status reports, progress reports, forecasts, and change requests.

Status reports describe where the project stands at a specific point in time. Recall the importance of the triple constraint. Status reports address where the project stands in terms of meeting scope, time, and cost goals. How much money has been spent to date? How long did it take to do certain tasks? Is work being accomplished as planned? Status reports can take various formats depending on the stakeholders' needs. Chapter 6, *Project Cost Management*, provides detailed information on earned value management, a technique used for reporting project performance that integrates scope, time, and cost data.

Progress reports describe what the project team has accomplished during a certain period of time. Many projects have each individual team member prepare a monthly or sometimes weekly progress report. Team leaders often create consolidated progress reports based on the information received from team members. A sample template for a monthly progress report is provided later in this chapter.

Project forecasting predicts future project status and progress based on past information and trends. How long will it take to finish the project based on how things are going? How much more money will be needed to complete the project? Earned value management can also be used to answer these questions by estimating the budget at completion, based on how the project is progressing.

Another important technique for performance reporting is the status review meeting. Status review meetings, as described in Chapter 3, *Project Integration Management*, are a good way to highlight information provided in important project documents, empower people to be accountable for their work, and have face-to-face discussions about important project issues. Many program and project managers hold monthly status review meetings to exchange important project information and motivate people to make progress on their parts of the project. Likewise, many senior managers hold monthly or quarterly status review meetings where program and project managers must report overall status information.

Status review meetings sometimes become battlegrounds where conflicts between different parties come to a head. Project managers or higher level senior managers should set ground rules for review meetings to control the amount of conflict and should work to resolve any potential problems. It is important to remember that project stakeholders should work together to address performance problems.

ADMINISTRATIVE CLOSURE

A project or phase of a project (concept, development, implementation, or close-out) requires closure. Administrative closure consists of verifying and documenting project results. This process formalizes sponsor or customer acceptance of the project's products. Administrative closure also allows time to collect project records, ensure those records reflect final specifications, analyze project effectiveness, and archive information for future use.

The main outputs of administrative closure are project archives, formal acceptance, and lessons learned. **Project archives** include a complete set of organized project records that provide an accurate history of the project. **Formal acceptance** is documentation that the project's sponsor or customer signs to show they have accepted the products of the project. **Lessons learned** are reflective statements written by project managers and their team members. Many projects are done under contract, and contracts usually specify what each of these outputs should consist of. If projects are not done under contract, it is still very important to prepare these items.

Project archives often come in handy many years after the project is completed. For example, a new project manager might want to know more details on some of the tools or techniques used in a certain aspect of an older project.

Documentation contained in the project archives could save time and money on the current project. There might be an audit of the organization, and good project archives could provide valuable information very quickly.

Formal acceptance should be provided on internal as well as external projects. This process helps to formally end the project and avoids dragging out project termination. In a contract situation, the buyer must legally accept the products produced as part of the contract so the seller can receive payment. There are usually added costs if the contract does not end as planned. In a noncontract situation, all parties must agree on completion of the work so that people and other resources can be reassigned.

The project manager and project team members should all prepare a lessons-learned report after completing a project. Some items discussed in lessons learned reports include the causes of variances on the project, the reasoning behind corrective actions chosen, the use of different project management tools and techniques, and personal words of wisdom based on team members' experiences. Some projects require all project members to write a brief lessons learned report. These reports provide valuable reflections by people who know what really worked or did not work on the project. Everyone learns in different ways and has different insights into a project. These reports can be an excellent resource and help future projects run more smoothly.

To reinforce the benefits of lessons learned reports, some companies require new project managers to read several past project managers' lessons learned reports and discuss how they will incorporate some of their ideas into their own projects.

SUGGESTIONS FOR IMPROVING PROJECT COMMUNICATIONS

You've seen that good communication is vital to the management and success of information technology projects; you've also learned that project communications management can ensure that essential information reaches the right people at the right time, that feedback and reports are appropriate and useful, and that there is a formalized process of administrative closure. This section highlights a few areas that all project managers and team members should consider in their quests to improve project communications. Tips are provided for managing conflict, developing better communication skills, running effective meetings, using templates for project communications, and developing a communications infrastructure.

Using Communication Skills to Manage Conflict

Most large information technology projects are high-stake endeavors that are highly visible within organizations. They require tremendous effort from team members, are expensive, commandeer significant resources, and can have an extensive impact on the way work is done in an organization. When the stakes are high, conflict is never far away; when the potential for conflict is high, good communication is a necessity.

Chapter 5, *Project Time Management*, explained that schedule issues cause the most conflicts over the project life cycle and provided suggestions for improving project scheduling. Other common conflicts occur over project priorities, staffing, technical issues, administrative procedures, personalities, and cost. It is crucial for project managers to develop and use their human resources and communication skills to help identify and manage conflict on projects. Project managers should lead their teams in developing norms for dealing with various types of conflicts that might arise on their projects. For example, team members should know that disrespectful behavior toward any project stakeholder is inappropriate, and that team members are expected to try to work out small conflicts themselves before elevating them to higher levels.

Blake and Mouton (1964) delineated five basic modes for handling conflicts: confrontation, compromise, smoothing, forcing, and withdrawal.

1. *Confrontation*. When using the **confrontation mode**, project managers directly face a conflict using a problem-solving approach that allows affected parties to work through their disagreements. This approach is also called the problem-solving mode.
2. *Compromise*. With the **compromise mode**, project managers use a give and-take approach to resolving conflicts. They bargain and search for solutions that bring some degree of satisfaction to all the parties in a dispute.
3. *Smoothing*. When using the **smoothing mode**, the project manager de-emphasizes or avoids areas of differences and emphasizes areas of agreement.
4. *Forcing*. The **forcing mode** can be viewed as the win-lose approach to conflict resolution. Project managers exert their viewpoint at the potential expense of another viewpoint. Managers who are very competitive or autocratic in their management style might favor this approach.
5. *Withdrawal*. When using the **withdrawal mode**, project managers retreat or withdraw from an actual or potential disagreement. This approach is the least desirable conflict-handling mode.

Research indicates that project managers favor using confrontation for conflict resolution over the other four modes. The term confrontation may be misleading. This mode really focuses on addressing conflicts using a problem-solving approach. Using Stephen Covey's paradigms of interdependence, this mode focuses on a win/win approach. All parties work together to find the best way to solve the conflict. The next most favored approach to conflict resolution is compromise. Successful project managers are less likely to use smoothing, forcing, or withdrawal than they are to use confrontation or compromise.

Project managers must also realize that not all conflict is bad. In fact, conflict can often be good. Conflict often produces important results, such as new ideas, better alternatives, and motivation to work harder and more collaboratively. Project team members may become stagnant or develop **groupthink**—conformance to the values or ethical standards of a group—if there are no conflicting viewpoints on various aspects of a project. Research by Karen Jehn, Professor of Management at Wharton, suggests that task-related conflict, which derives from differences over team objectives and how to achieve them, often improves team performance. Emotional conflict, however, which stems from personality clashes and misunderstandings, often depresses team performance.[7] Project managers should create an environment that encourages and maintains the positive and productive aspects of conflict.

Several organizations are emphasizing the importance of conflict management. For example, an innovative program at California State University, Monterey Bay, has incorporated conflict resolution as one of the program's eleven major learning objectives that students must know and understand in order to graduate. The academic program, based at the Institute for Community Collaborative Studies, focuses on preparing students for careers in the field of health and human services. The program's belief is that collaboration is an essential aspect of success for workers in the modern health and human services delivery field. Core competencies for developing collaboration skills include conflict resolution, negotiation, and mediation. These skills are also important for project managers in any field.

Developing Better Communication Skills

Some people seem to be born with great communication skills. Others seem to have a knack for picking up technical skills. It is rare to find someone with a natural ability for both. Both communication and technical skills, however, can be developed. Most information technology professionals enter the field because of their technical skills. Most find, however, that communication skills are the key to advancing in their careers, especially if they want to become good project managers.

Most companies spend a lot of money on technical training for their employees, even when employees might benefit more from communication training. Individual employees are also more likely to enroll voluntarily in classes on the latest technology than those on developing their soft skills. Communication skills training usually includes role-playing activities in which participants learn concepts such as building rapport, as described in Chapter 8,

[7] *Wharton Leadership Digest*, "Constructive Team Conflict" (March, 1997) 1(6).

Project Human Resource Management. Training sessions also give participants a chance to develop specific skills in small groups. Training sessions that focus on presentation skills usually videotape the participants. Most people are surprised to see some of their mannerisms on tape and enjoy the challenge of improving their skills. A minimal investment in communication and presentation training can have a tremendous payback to individuals, their projects, and their organizations. These skills also have a much longer shelf life than many of the skills learned in technical training courses.

As organizations become more global, they realize that they must also invest in ways to improve communication with people from different countries and cultures. For example, many Americans are raised to speak their minds, while in some other cultures people are offended by outspokenness. Not understanding other cultures and how to effectively communicate with people of diverse backgrounds hurts projects and businesses. Many training courses are available to educate people in cultural awareness, international business, and international team building.

It takes leadership to help improve communication. If senior management lets employees give poor presentations, write sloppy reports, offend people from different cultures, or behave poorly at meetings, the employees will not want to improve their communication skills. Senior management must set high expectations and lead by example. Some organizations send all information technology professionals to training that includes development of technical *and* communication skills. Some organizations allocate time in project schedules for preparing drafts of important reports and presentations and incorporating feedback on the drafts. Some include time for informal meetings with customers to help develop relationships. Some organizations even provide staff to assist in relationship management. As with any other goal, improving communication can be achieved with proper planning, support, and leadership from senior management.

Running Effective Meetings

A well-run meeting can be a vehicle for fostering team building and reinforcing expectations, roles, relationships, and commitment to the project. However, a poorly run meeting can have a detrimental effect on a project. For example, a terrible **kickoff meeting**—a meeting held at the beginning of a project or project phase where all major project stakeholders discuss project objectives, plans, and so on—may cause some important stakeholders to decide not to support the project any further. Many people complain about the time they waste in unnecessary or poorly planned and poorly executed meetings. Following are some guidelines to help improve time spent at meetings:

- *Determine if a meeting can be avoided.* Do not have meetings if there is a better way of achieving the objective at hand. For example, a project manager might know that he or she needs approval from a senior manager to hire another person for the project team. It could take a week or longer to schedule even a ten-minute meeting on the senior manager's calendar. Instead, an e-mail or phone call describing the situation and justifying the request is a faster, more effective approach than having a meeting.

- *Define the purpose and intended outcome of the meeting.* Be specific about what should happen as a result of the meeting. Is the purpose to brainstorm ideas, provide status information, or solve a problem? Make the purpose of meetings very clear to all meeting planners and participants. For example, if a project manager calls a meeting of all project team members without knowing the true purpose of the meeting, everyone will start focusing on their own agendas and very little will be accomplished. All meetings should have a purpose and intended outcome.
- *Determine who should attend the meeting.* Do certain stakeholders have to be at a meeting to make it effective? Should only the project team leaders attend a meeting, or should the entire project team be involved? Many meetings are most effective with the minimum number of participants possible, especially if decisions must be made. Other meetings require many attendees. It is important to determine who should attend a meeting based on the purpose and intended outcome of the meeting.
- *Provide an agenda to participants before the meeting.* Meetings are most effective when the participants come prepared. Did they read reports before the meeting? Did they collect necessary information? Some professionals refuse to attend meetings if they do not have an agenda ahead of time. Insisting on an agenda forces meeting organizers to plan the meeting and gives potential attendees the chance to decide whether they really need to attend the meeting.
- *Prepare handouts, visual aids, and make logistical arrangements ahead of time.* By creating handouts and visual aids, the meeting organizers must organize their thoughts and ideas. This usually helps the entire meeting run more effectively. It is also important to make logistical arrangements by booking an appropriate room, having necessary equipment available, and providing refreshments or entire meals, if appropriate. It takes time to plan for effective meetings. Project managers and their team members must take time to prepare for meetings, especially important ones with key stakeholders.
- *Run the meeting professionally.* Introduce people, restate the purpose of the meeting, and state any ground rules that should be followed. Have someone facilitate the meeting to make sure important items are discussed, watch the time, encourage participation, summarize key issues, and clarify decisions and action items. Designate someone to take minutes and send the minutes out soon after the meeting. Minutes should be short and focus on the crucial decisions and action items from the meeting.
- *Build relationships.* Depending on the culture of the organization and project, it may help to build relationships by making meetings fun experiences. For

example, it may be appropriate to use humor, refreshments, or prizes for good ideas to keep meeting participants actively involved. If used effectively, meetings are a good way to build relationships.

Using Templates for Project Communications

Many intelligent people have a hard time writing a performance report or preparing a ten-minute technical presentation for a customer review. Some people in these situations are too embarrassed to ask for help. To make preparing project communications easier, project managers need to provide examples and templates for common project communications items such as project descriptions, project charters, monthly performance reports, oral status reports, and so on. Good documentation from past projects can be an ample source of examples. Samples and templates of both written and oral reports are particularly helpful for people who have never before had to write project documents or give project presentations. Finding, developing, and sharing relevant templates is an important task for many project managers. Several examples of project documentation such as a project charter, stakeholder analysis, WBS, Gantt chart, and cost estimate are provided throughout this book. Other templates and guidelines for preparing project documents are provided in this section.

Figure 9-2 shows a sample template for a one-page project description. This form could be used to show a "snapshot" of an entire project on one page. For example, senior managers might require all project managers to provide a brief project description as part of a quarterly management review meeting. Peter Gumpert, the program manager in the opening case, might request this type of document from all of the project managers working for him to get a big picture of what each project involves. According to Figure 9-2, a project description should include the project objective, scope, assumptions, cost information, and schedule information. This template suggests including information from the project's Gantt chart to highlight key deliverables and other milestones.

Table 9-3 shows a template for a monthly progress report. Sections of the progress report include accomplishments from the current period, plans for the next period, issues, and project changes.

Table 9-4 provides a template for writing a letter of agreement for a class project. Many projects, such as class projects, do not have a contract or official project charter. Instead, they have a letter of agreement, which provides similar information in a friendlier and less formal manner than a contract or project charter. A letter of agreement can also be used to document and clarify goals and expectations for projects. Table 9-4 suggests that a letter of agreement for a class project include a project description (see Figure 9-2), the sponsoring organization's goals and expectations, the students' goals and expectations, meeting information, contact information and communications plans, and signatures of the main sponsor, student project manager, and all students on the project team.

Project X Descripton

Objective: Describe the objective of the project in one or two sentences. Focus on the business benefits of doing the project.

Scope: Briefly describe the scope of the project. What business functions are involved, and what are the main products the project will produce?

Assumptions: Summarize the most critical assumptions for the project.

Cost: Provide the total estimated cost of the project. If desired, list the total cost each year.

Schedule: Provide summary information from the project's Gantt chart, as shown. Focus on summary tasks and milestones.

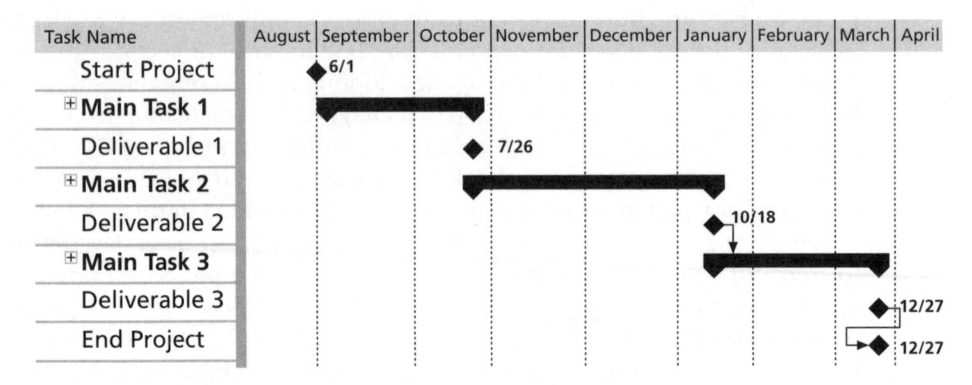

Figure 9-2. Sample Template for a Project Description

Table 9-3: Sample Template for a Monthly Progress Report

I. Accomplishments for Month of January (or appropriate month):

- Describe most important accomplishments. Relate to project's Gantt chart.
- Describe other important accomplishments, one bullet for each. If any issues were resolved from the previous month, list them as accomplishments.

II. Plans for February (or following month):

- Describe most important items to be accomplished in the next month. Again, relate tothe project's Gantt chart.
- Describe other important items to accomplish, one bullet for each.

III. Issues: Briefly list important issues that surfaced or are still important. Managers hate surprises and want to help the project succeed, so be sure to list issues.

IV. Project Changes (Date and Description): List any approved or requested changes to the project. Include the date of the change and a brief description.

Table 9-4: Sample Template for a Letter of Agreement for a Class Project

I.	**Project Description:** Describe the project's objective, scope, assumptions, cost information, and schedule information, as shown in Figure 9-2. Be sure to include important dates that the project's sponsor needs to be aware of.
II.	**Organizational Goals and Expectations:** Have the main sponsor from the organization briefly state his or her goals and expectations for the project.
III.	**Student Goals and Expectations:** Students on the project team should briefly state their goals and expectations for the project.
IV.	**Meeting Information:** The project's sponsor and all students on the project team should agree to meet for at least one hour per week when all parties can work on this project. The meeting place should be a convenient location without distractions. Virtual meetings may be an option for some people and projects.
V.	**Contact Information and Communications Plan:** List the sponsor's and students' names, phone numbers, e-mail addresses, and important procedures for communications. It is a good idea to set up a Web site for all project information.
VI.	**Signatures:** Have the main sponsor and students on the project team sign the letter of agreement. Designate which student is the project manager. This student should be the main contact for all project information.

Table 9-5 provides an outline for a final project report that could be used for a class project or other projects. It is important to include a cover page and table of contents. For a long report, it is also a good idea to include a one-page executive summary that highlights the most important information in the report. Other major sections of a final project report include a description of the need for the project, the original project description and letter of agreement (see Figure 9-2 and Table 9-4), the overall outcome of the project and reasons for success or failure, project management tools and techniques used and an assessment of them, project team recommendations and future considerations, the final project Gantt chart, and attachments with all project deliverables.

For very large projects, the contents of a final project report might be similar to those listed in Table 9-5, but the report would probably include many more items, depending on the nature of the project. It is important to write a final report and archive all important project documents. As you can imagine, a large project can generate a lot of documentation. Table 9-6 provides an exhaustive list of all of the documentation that should be organized and filed at the end of a major project. From this list, you can see that a large project can generate a lot of documentation. In fact, some project professionals have observed that documentation for designing an airplane usually weighs more than the airplane itself. (Smaller projects usually generate much less documentation!)

Table 9-5: Outline for a Final Project Report

I.	Cover page
II.	Table of contents and executive summary (for a long report)
III.	Need for the project
IV.	Project description and letter of agreement
V.	Overall outcome of the project and reasons for success or failure
VI.	Project management tools and techniques used and assessment of them
VII.	Project team recommendations and future considerations
VIII.	Final project Gantt chart
IX.	Attachments with all deliverables

Table 9-6: Final Project Documentation Items

I.	Project description
II.	Project proposal and backup data (request for proposal, statement of work, proposal correspondence, and so on)
III.	Original and revised contract information and client acceptance documents
IV.	Original and revised project plans and schedules (WBS, Gantt charts and network diagrams, cost estimates, communications management plan, etc.)
V.	Design documents
VI.	Final project report
VII.	Deliverables, as appropriate
VIII.	Audit reports
IX.	Lessons learned reports
X.	Copies of all status reports, meeting minutes, change notices, and other written and electronic communications

Figure 9-3 provides a template file created in Project 2000 that students could use for class projects. People working in companies could develop and follow a similar template for their projects. Notice the main tasks in this template follow the project management process of project initiating, planning, executing, controlling, and closing. Milestones are included to highlight significant events, such as having the letter of agreement signed, or the completion of major deliverables. Hyperlinks are included to template files for creating a letter of agreement, progress report, and final report. Hyperlinks could also be included to other project documentation such as meeting minutes, product information, and presentations.

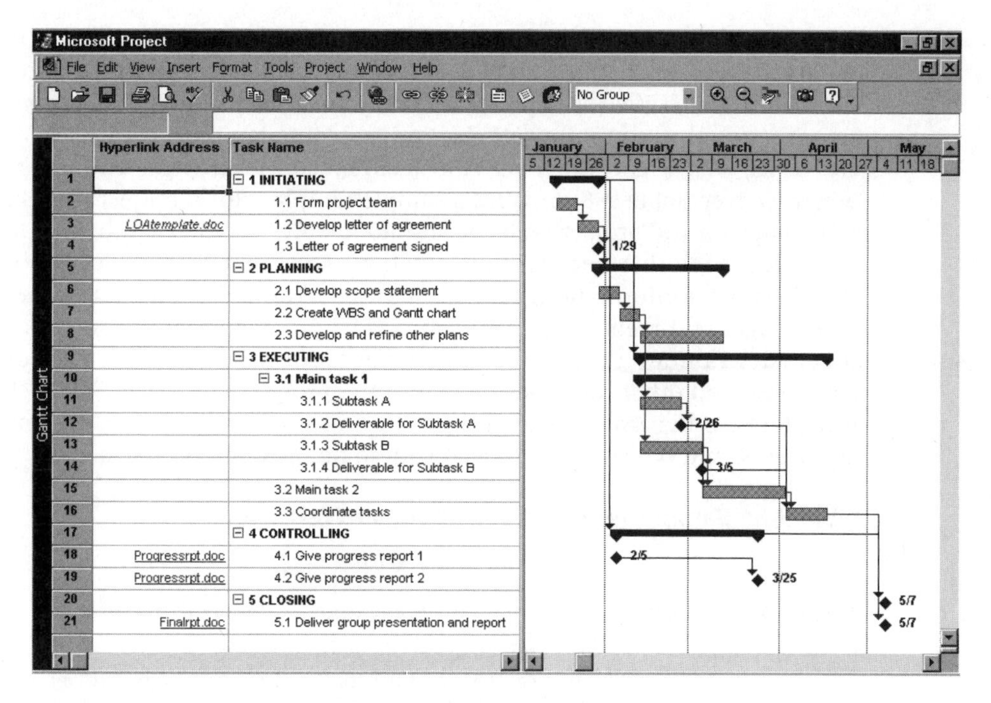

Figure 9-3. Gantt Chart Template for a Class Project

Table 9-7 provides guidance for writing a lessons learned report. These particular guidelines apply to a project for students taking a class in project management. As you can see, templates, outlines, and guidelines for project documentation often need to be tailored to the particular organization and project.

Table 9-7: Guidance for Students' Lessons Learned Report

Every two weeks or after a major event in your group project, write a brief journal entry describing what happened and how you felt about it. At the end of the term, write a 2- to 3-page paper describing your lessons learned based on your group project. Answer the following questions:

- What were your roles and responsibilities on the team? How were they decided?
- What did you like/dislike about the project?
- What did you learn about project management and yourself by doing the group project?
- What did you learn about teamwork and yourself by doing the group project?
- What would you have done differently? What will you remember to do on the next project you work on after this experience?

In the past few years, more and more project teams have started putting all or part of their project information, including various templates, on project Web sites. Project Web sites provide a centralized way of delivering project documents and other communications. Project teams can develop project Web sites using Web-authoring tools, such as Microsoft FrontPage or Macromedia Dreamweaver. Table 9-8 provides a sample template for a project Web site. The home page for the project Web site should include summary information about the project and other pertinent information. The home page should also include contact information, such as names and e-mail addresses for the project manager and Webmaster. Links should be provided to items such as the project description, team member list, project documents, templates, a discussions area, and other materials related to the project. The project team should also address other issues in creating and using a project Web site, such as security, access, and type of content that should be included on the site.

Table 9-8: Sample Template for a Project Web Site

	PROJECT X WEB SITE
Project Description	Welcome to the home page for Project X.
Team Members	Provide contact information, such as the names, e-mail addresses, and telephone numbers of the project manager and Webmaster for the project Web site.
Project Documents **Charter** **Gantt Chart** **Progress Reports** **Final Report**	Provide summary information about the project and other pertinent information.
Templates	
Discussions	
Related Links	

For more sophisticated Web sites, project teams can also use one of the many software products created specifically to assist in project communications through the Web. These products vary considerably in price and functionality. Sample products include Welcom's WelcomHome and eProject.com, Inc.'s eProject Enterprise 3.0.[8]

When the project team develops their communications management plan, they should also determine what templates could be used for key documentation. To make it even more convenient to use templates, the organization should make project templates readily available online for all projects. The project team should also be certain to understand senior management's and customers' documentation expectations for each particular project. For example, if a project sponsor or customer wants a one-page monthly progress report

[8] Vandersluis, Chris, "Web-based Project Management Systems for Everyman," *PMNetwork*, (May 2000).

for a specific project, but the project team delivers a twenty-page report, there are communication problems. In addition, if particular customers or senior managers want specific items in all final project reports, they should make sure the project team is aware of those expectations and modify any templates for those reports to take these requirements into account.

Developing a Communications Infrastructure

A **communications infrastructure** is a set of tools, techniques, and principles that provides a foundation for the effective transfer of information among people. Tools include e-mail, project management software, groupware, fax machines, telephones, teleconferencing systems, document management systems, and word processors. Techniques include reporting guidelines and templates, meeting ground rules and procedures, decision-making processes, problem-solving approaches, conflict resolution and negotiation techniques, and the like. Principles include providing an environment for open dialogue using "straight talk" and following an agreed upon work ethic.

Since the defense industry has been involved in project management for a long time, many government and defense contractors have formal communications infrastructures already in place. For example, in the early 1980s—before most people ever used personal computers—the U.S. Air Force had standard forms for reporting project progress information, outlines for developing project final reports, regulations describing the progression of major projects, inspections of project archives, and forms and procedures for creating project Gantt charts. There were even special stickers you could place on Gantt charts to show changes, such as a white diamond for a slipped milestone. (No project manager wanted to ask for that particular sticker.)

Today, some organizations have designed customized systems for collecting and reporting project information, and most use several forms of information technology as part of their project communications infrastructure. Some companies have intranets to keep track of all project information. Others use Lotus Notes or other groupware to maintain consistent and complete project information. Still other industries are developing communications infrastructures to help in managing projects.

In his 1999 book, *Business @ the Speed of Thought: Using a Digital Nervous System,* Microsoft CEO Bill Gates suggests that organizations must develop a communications infrastructure or "digital nervous system" that allows for rapid movement of information inside a company as well as with customers, suppliers, and other business partners. Gates suggests that how organizations gather, manage, and use information to empower people will determine whether they win or lose in a competitive business environment.[9] In one of his interviews, Gates mentioned systems to aid project management as an example of a great use of technology that really makes a difference in the companies that implement such systems.[10]

[9] Gates, Bill, *Business @ the Speed of Thought:: Using a Digital Nervous System.* New York: Warner Books, 1999.

[10] Schlender, Brent, "E-Business According to Gates," *Fortune* (April 12, 1999) 73.

USING SOFTWARE TO ASSIST IN PROJECT COMMUNICATIONS

Even though information technology companies routinely use many types of hardware and software to enhance communications, they need to adjust already existing systems to serve the special communications needs of a project environment. Many organizations develop their own systems, and products such as Involv, MobileManager, and CSI Project (described in the following "What Went Right?") are available to assist companies with enterprise-wide project communications. Many other products have been developed or enhanced in the late 1990s and early 2000s to address the problem of providing fast, convenient, consistent, and up-to-date project information. Microsoft Project 2000 has many features to enhance project communications management.

What Went Right?

A start-up company specializing in European mergers and acquisitions needed a quick and effective way to communicate among participants from several different firms. They investigated several possible approaches and decided to use a subscription-based Internet product that had many of the features they needed and did not require any significant technology implementation. The product (*Involv* by Changepoint of Ontario, Canada) cost $12 per month per user. It allowed the designated project manager and staff to build a project task list, and everyone else was given read access to the plan and other documents and write access to their individual tasks. The product also provided a means for group discussions among all project members, independent of time and place. The system created automatic e-mail messages to notify team members when tasks had been assigned, and team members updated tasks electronically. The software also interfaced with Microsoft Project to allow full project planning and reporting. This system made the project history and upcoming task information available to all team members at any time from anywhere. Because the entire project team was involved in selecting the product, it was necessary to explicitly discuss project communications. This explicit discussion of project communications allowed the project team to avoid many of the problems that can arise when there are different views about what, how, when, and with whom to communicate.[11]

One of the biggest problems on large projects has often been providing the most recent project plans, Gantt charts, specifications, meeting information, change requests, and so on to all or selective stakeholders in a timely fashion. The World Wide Web has provided a common format for distributing information quickly, and e-mail is a good vehicle for notifying specific project stakeholders of information in a timely fashion. Microsoft Project 2000 includes several features that take advantage of these technologies.

[11] Pitagorsky, George, "Building a Communications Infrastructure," *PM Network* (August 1998).

To take advantage of the Web, Microsoft Project 98 introduced a "Save-as-HTML" feature that allowed users to select specific information for conversion to HTML—Hypertext Markup Language, the formatting language that allows documents to be displayed on the Web. Figure 9-4 provides an example of a Project 98 sample file saved as HTML that shows information about tasks on the critical path for launching a software product called Cyclone. Project 2000 includes a similar "Save as Web page" feature that includes several built-in formats for extracting and converting data to HTML, and users can design their own HTML formats as well.

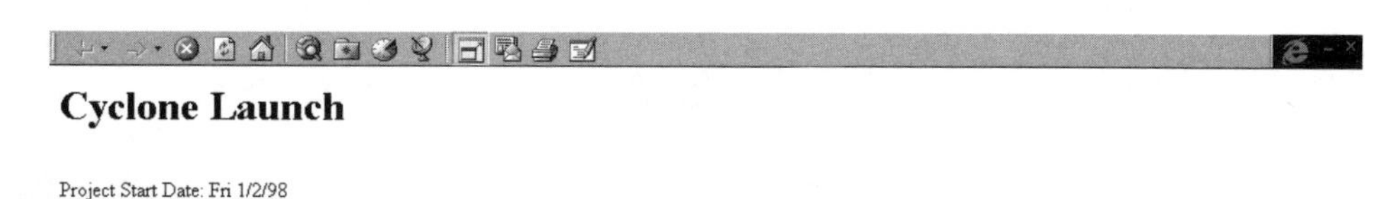

Cyclone Launch

Project Start Date: Fri 1/2/98
Project Finish Date: Thu 6/4/98

Software Launch Critical Tasks

ID	Task_Name	Start_Date	Finish_Date	Late_Start	Late_Finish	Free_Slack	Total_Slack
1	Marketing Plan distributed	Tue 3/17/98	Tue 3/17/98	Tue 3/17/98	Tue 3/17/98	0 days	0 days
2	**Corp Comm**	Mon 1/5/98	Thu 5/28/98	Mon 1/5/98	Thu 6/4/98	0 days	0 days
3	Corp Comm Kickoff	Mon 1/5/98	Mon 1/5/98	Mon 1/5/98	Mon 1/5/98	0 days	0 days
4	Comm. Plan delivered	Fri 1/23/98	Fri 1/23/98	Fri 1/23/98	Fri 1/23/98	0 days	0 days
8	Competitive comparison	Fri 4/3/98	Fri 4/3/98	Fri 4/3/98	Fri 4/3/98	0 days	0 days
14	Creative Concepts	Wed 3/11/98	Wed 3/11/98	Wed 3/11/98	Wed 3/11/98	0 days	0 days
16	**Public Relations**	Mon 1/19/98	Fri 5/29/98	Mon 1/19/98	Wed 6/3/98	0 days	0 days
17	PR Kickoff Meeting	Mon 1/19/98	Mon 1/19/98	Mon 1/19/98	Mon 1/19/98	0 days	0 days
19	PR Plan delivered	Thu 3/5/98	Thu 3/5/98	Thu 3/5/98	Thu 3/5/98	0 days	0 days
25	Mail Beta Copies	Thu 3/26/98	Thu 3/26/98	Thu 3/26/98	Thu 3/26/98	0 days	0 days
32	**Hardware Relationships**	Wed 4/22/98	Tue 5/5/98	Wed 4/22/98	Tue 5/5/98	0 days	0 days
33	Meet with Evangelist	Wed 4/22/98	Wed 4/22/98	Wed 4/22/98	Wed 4/22/98	0 days	0 days
34	Meet with Engineering/Tech	Mon 5/4/98	Mon 5/4/98	Mon 5/4/98	Mon 5/4/98	0 days	0 days
35	Meet with Government Evang.	Tue 5/5/98	Tue 5/5/98	Tue 5/5/98	Tue 5/5/98	0 days	0 days
36	Meet with Aerospace Evang.	Tue 5/5/98	Tue 5/5/98	Tue 5/5/98	Tue 5/5/98	0 days	0 days
37	**Programs**	Thu 3/5/98	Thu 6/4/98	Thu 4/9/98	Thu 6/4/98	0 days	0 days
39	Mail Eval. to key consultants	Thu 6/4/98	Thu 6/4/98	Thu 6/4/98	Thu 6/4/98	0 days	0 days
43	Release to manufacturing	Mon 5/18/98	Mon 5/18/98	Mon 5/18/98	Mon 5/18/98	0 days	0 days
45	Project announced	Wed 6/3/98	Wed 6/3/98	Wed 6/3/98	Wed 6/3/98	0 days	0 days

Figure 9-4. Microsoft Project Information Saved as HTML File

Project 2000 users can use the "Insert Hyperlink" function to add links to other project-related files. For example, project managers or one of their team members can insert a hyperlink to a Word file that contains the project charter or an Excel file that contains a staffing management plan or cost estimate. The Project 2000 file and all associated hyperlinked files could then be placed on a local area network server or Web server so all project stakeholders could easily access important project information.

Microsoft Project 2000 also includes many workgroup functions that allow a team of people at different locations to work together on projects and share project information. Workgroup functions allow the exchange of messages through e-mail, an intranet, or the World Wide Web. You can also use Project 2000 to alert members about new or changed task assignments, and members can return status information and notify other workgroup members about changes in the schedule or other project parameters. Appendix A provides details on using these features. Many other products, such as Microsoft Project Central, interface with Microsoft Project 2000 to assist in project communications.

Microsoft Project Central is a companion product of Project 2000. Project Central facilitates collaboration and communication of project information by enabling project managers, team members, and other stakeholders to view schedule and task data and interact with the project manager over an intranet. Figure 9-5 provides a view of the Project Central home page, which can be customized with corporate branding or custom controls.[12]

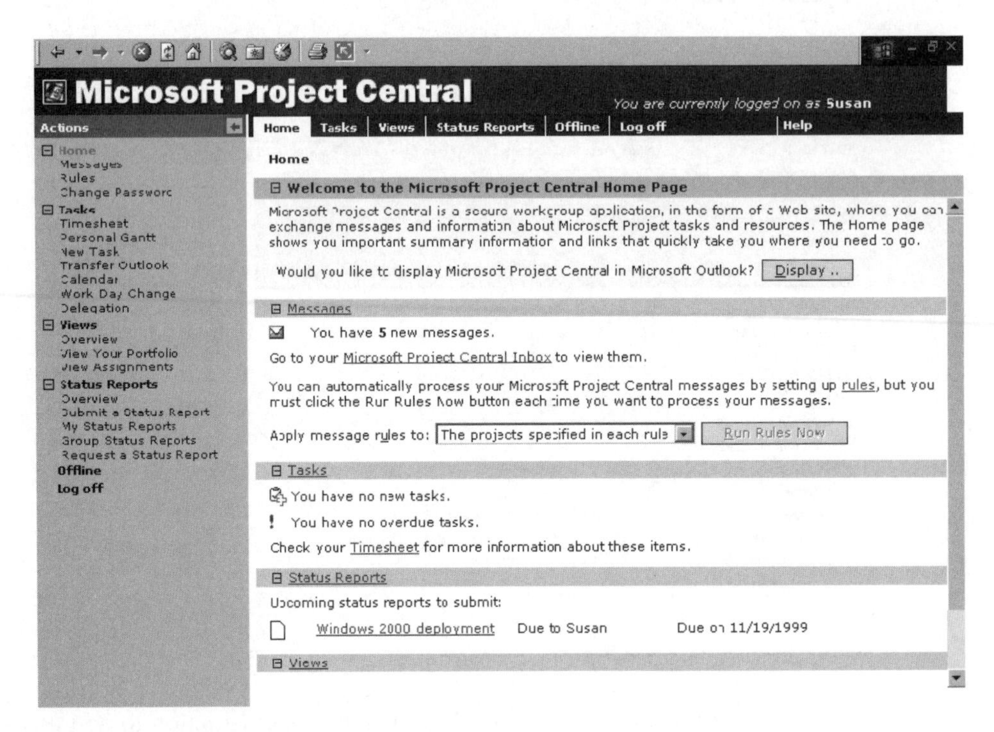

Figure 9-5. Microsoft Project Central

Communication is among the more important factors for success in project management. Whereas technology can aid in the communications process and can be the easiest aspect of the process to address, it is not the most important aspect of the communications process. Far more important is improving an organization's ability to communicate. Improving the ability to communicate often requires a cultural change in an organization that takes a lot of time, hard work, and patience. Information technology personnel, in particular, often need special

[12] Microsoft, Microsoft Project Central (November 1999) www.microsoft.com/office/project/ProCenWP.htm.

coaching to improve their communications skills. The project manager's chief role in the communications process is that of facilitator. Project managers must educate all stakeholders—management, team members, and customers—on the importance of good project communications and ensure that the project has a communications plan to help make good communication happen.

CASE WRAP-UP

Christine Braun worked closely with Peter Gumpert and his project managers to develop a communications plan for all of the undersea fiber-optic telecommunications systems projects. Peter was very skilled at running effective meetings, so everyone focused on meeting specific objectives. Peter stressed the importance of keeping himself, the project managers, and other major stakeholders informed about the status of all projects. He emphasized that the project managers were in charge of their projects, and that he had no intentions of telling them how to do their jobs. He just wanted to have accurate and consistent information to help coordinate all of the projects and make everyone's jobs easier. When some of the project managers balked at the additional work of providing more project information in different formats, Peter openly discussed the issues with them in more detail. He then authorized each project manager to use additional staff to help develop and follow standards for all project communications.

Christine used her strong technical and communications skills to create a Web site that included samples of important project documents and presentations and templates that other people could download and use on their own projects. After determining the need for more remote communications, Christine and other staff members researched the latest hardware and software products for updating project information. Peter authorized funds for a new project led by Christine to evaluate and then purchase several hand-held devices and software that would link to their company's project information on the Web. Any project stakeholder could check out one of these hand-held devices and get one-on-one training on how to use it. Even Peter learned how to use one and took it along on all of his business trips.

CHAPTER SUMMARY

A failure to communicate is often the greatest threat to the success of any project, especially information technology projects. Communication is the oil that keeps a project running smoothly.

Communications planning involves information distribution, performance reporting, and administrative closure. It requires determining the information and communications needs of the project stakeholders. A communications management plan should be created for all projects. A stakeholder analysis for project communications helps determine communications needs for different people involved in a project.

The various methods for distributing project information include formal, informal, written, and verbal. It is important to determine the most appropriate means for distributing different types of project information. Project managers and their teams should focus on the importance of building relationships as they communicate project information. As the number of people that need to communicate increases, the complexity of communications also increases.

Performance reporting involves collecting and disseminating information about how well a project is moving toward meeting its goals. Project teams can use earned value charts and other forms of progress information to communicate and assess project performance. Status review meetings are an important part of communicating, monitoring, and controlling projects.

Administrative closure involves generating, gathering, and disseminating information to formalize phase or project completion. The main customer for the project should formally accept the product or products produced by the project. Project archives and lessons learned reports should be generated as part of administrative closure.

To improve project communications, project managers and their teams must develop good conflict management skills, as well as other communication skills. Conflict resolution is an important part of project communications management. The main causes of conflict during a project are schedules, priorities, staffing, technical opinions, procedures, cost, and personalities. A problem-solving approach to managing conflict is often the best approach. Other suggestions for improving project communications include learning how to run more effective meetings, using templates for project communications, and developing a communications infrastructure.

Microsoft Project 2000 has several project communications features. Users can convert information to HTML for posting on the World Wide Web or insert hyperlinks to related project files. Workgroup features can be used to make it easier for project teams to disseminate information via e-mail. Microsoft Project Central is a companion product of Project 2000 that facilitates collaboration and communication of project information.

Discussion Questions

1. Discuss examples of media that poke fun at the communications skills of technical professionals. How does poking fun at technical professionals' communications skills impact the industry and educational programs?
2. What are the key components of a project communications plan? How can a stakeholder analysis assist in preparing and implementing parts of this plan?
3. Discuss the advantages and disadvantages of different ways to distribute project information.
4. What are some of the ways to create and distribute project performance information?
5. Do you agree with some of the suggestions provided for improving project communications? What other suggestions do you have?
6. How can software assist in project communications? How can it hurt project communications?

Exercises

1. Create a stakeholder analysis for project communications. Assume your organization has a project to determine employees' training needs and then provide in-house and external sources for courses in developing communications skills for its employees. Stakeholders could be various levels and types of employees, vendors, the HR Department in charge of the project, and so on. Determine specific stakeholders for the project. List at least one type of project communication for each stakeholder and the format for disseminating information to each. Use Table 9-1 as an example.
2. How many different communications channels does a project team with six people have? How many more channels would there be if the team grew to ten people?
3. Review the templates for various project documents provided in this chapter. Pick one of them and apply it to a project of your choice. Make suggestions for improving the template.
4. Write a lessons learned report for a project of your choice using Table 9-7 as a guide. Do you think it is important for all project managers and team members to write lessons learned reports? Would you take the time to read them if they were available in your organization?
5. Research software products that assist in managing large projects. Write a one- to two-page paper summarizing your findings. Include Web sites for software vendors and your opinion of some of the products.
6. Write a one- to two-page paper summarizing one of the suggested readings.

MINICASE

You are a member of a project team with twelve members. Several team members work at remote locations, so good virtual communications are crucial. Your team has decided to use a project Web site to store all project documentation and facilitate other communications. You are assigned the task of determining the requirements for the project Web site, researching alternate solutions, making a recommendation, getting team consensus on what to implement, and then implementing the project Web site and acting as the Webmaster.

Part 1: Create a survey that you could use to solicit inputs on requirements for the project Web site from your team members. First research information on how to design and administer surveys. One potential source of information is www.surveysystem.com. Then create the survey, either using word-processing software or Web-based survey software. Be sure to address a wide range of requirements, such as functionality, cost, security, and maintenance.

Part 2: Your project team has decided to use Web-authoring software to create its project Web site. Using the template provided in Table 9-8, create a project Web site, using your choice of Web-authoring software (for example, Microsoft FrontPage, Macromedia Dreamweaver, Allaire Homesite, or even Microsoft Word). Create the home page and pages for all of the links mentioned in the template. Test the Web pages to make sure they work, then post them on the Web site.

SUGGESTED READINGS

1. Bohner, Shawn. "PMOs: Projects in Harmony." META Group, www.gantthead.com (June 2000).

   As companies are faced with increasingly complex initiatives involving multiple information technology projects, executives are turning to project management offices (PMOs) to improve communication, manage multiple risks, and target their vital resources. This article describes some of the patterns of evolution and maturity that are emerging.

2. Gates, Bill. *Business @ The Speed of Thought: Using a Digital Nervous System.* New York: Warner Books, 1999.

   The main premise of the Microsoft CEO's 1999 book is that the speed of business is accelerating at an ever-increasing rate, and to survive, organizations must develop a "digital nervous system" that allows for rapid movement of information. Gates suggests that how organizations gather, manage, and use information to empower people will determine whether they win or lose in the competitive business environment.

3. Kirchof, Nicki S. and John R. Adams. *Conflict Management for Project Managers*. Upper Darby, PA: Project Management Institute, 1989.

This 53-page booklet contains information on the theory of conflict management, conflict in project organizations, conflict management and the project manager, and two-party conflict management. It also provides more information on Blake and Mouton's five basic modes for handling conflict and many references on the subject of conflict management.

4. Sivitanides, Marcos P., James R. Cook, Roy B. Martin, and Beverly A. Chiodo. "Verbal Communication Skills Requirements for Information Systems Professionals." *Journal of Information Systems Education*, http://www.gise.org/JISE/Vol7/v71_7.htm (Spring 1995): 7, no.1.

This research investigates the verbal communication skills requirements for practicing professionals and new university graduates in information technology. Organizations can use the findings to identify areas of weakness in their information technology personnel's communication skills and improve them with continuing education. Universities can use the findings to identify and make adjustments for skills that they may not be currently emphasizing in their curricula.

5. Van Slyke, Erik J. "Resolving Team Conflict." *PMNetwork* (June 2000).

This article states that conflict is a regular part of managing projects. Project managers should encourage the positive and productive aspects of conflict and approach all conflict by seeking constructive resolution.

6. Wetherbe, Bond and James C. Wetherbe. *So, What's Your Point?: A Practical Guide to Learning and Applying Effective Techniques for Interpersonal Communication*. Houston, TX: Mead Publishing, 1996.

James Wetherbe is well known in the information technology profession for several of his books and entertaining speeches at various conferences. This short book uses a compelling and humorous approach to provide practical techniques for improving communication skills.

Key Terms

- **Administrative closure** — generating, gathering, and disseminating information to formalize phase or project completion
- **Communications infrastructure** — a set of tools, techniques, and principles that provide a foundation for the effective transfer of information among people
- **Communications management plan** — a document that guides project communications
- **Communications planning** — determining the information and communications needs of the stakeholders: who needs what information, when will they need it, and how will the information be given to them

- **Compromise mode** — using a give-and-take approach to resolving conflicts. Bargaining and searching for solutions that bring some degree of satisfaction to all the parties in a dispute
- **Confrontation mode** — directly facing a conflict using a problem-solving approach that allows affected parties to work through their disagreements
- **Forcing mode** — using a win-lose approach to conflict resolution to get one's way
- **Formal acceptance** — documentation that the project's sponsor or customer signs to show they have accepted the products of the project
- **Groupthink** —conformance to the values or ethical standards of a group
- **Information distribution** — making needed information available to project stakeholders in a timely manner
- **Kickoff meeting** — a meeting held at the beginning of a project or project phase where all major project stakeholders discuss project objectives, plans, and so on
- **Lessons learned** — reflective statements written by project managers and their team members
- **Performance reporting** — collecting and disseminating performance information, which includes status reports, progress measurement, and forecasting
- **Progress reports** — reports that describe what the project team has accomplished during a certain period of time
- **Project archives** — a complete set of organized project records that provide an accurate history of the project
- **Project forecasting** — predicting future project status and progress based on past information and trends
- **Smoothing mode** — deemphasizing or avoiding areas of differences and emphasizing areas of agreements
- **Status reports** — reports that describe where the project stands at a specific point in time
- **Withdrawal mode** — retreating or withdrawing from an actual or potential disagreement

10

Project Risk Management

Objectives

After reading this chapter, you will be able to:

1. *Understand the importance of good project risk management*
2. *Understand what risk is and describe different tolerances for risk*
3. *Describe each of the processes involved in project risk management, including risk management planning, risk identification, qualitative risk analysis, quantitative risk analysis, risk response planning, and risk response control*
4. *Identify common sources of risk on information technology projects and develop strategies for reducing them*
5. *Describe common risk conditions that occur in each project management knowledge area and techniques for identifying potential risks on specific projects*
6. *Use tools and techniques for qualitative risk analysis, such as probability/impact matrices and the Top 10 Risk Item Tracking approach, and quantitative risk analysis, such as expected monetary value and simulation*
7. *Describe how software, such as Monte Carlo simulation software, can assist in project risk management*
8. *Explain the results of good project risk management*

Cliff Branch was the president of a small information technology consulting firm that specialized in developing Internet applications and providing full-service support. The staff consisted of programmers, business analysts, database specialists, Web designers, project managers, and so on. The firm had a total of fifty people and planned to hire at least ten more in the next year. The company had done very well the past few years, but was recently having difficulty winning contract awards. Spending time and resources to respond to various requests for proposals from prospective clients was becoming expensive. Many clients were starting to require presentations and even some prototype development before awarding a contract.

Cliff knew he had a fairly aggressive approach to risk and liked to bid on the projects with the highest payoff. He did not use a systematic approach to evaluating the risks involved in various projects before bidding on them. He focused on the profit potentials and how challenging the projects were. His strategy was now causing problems for his company because they were investing heavily in the preparation of proposals, yet winning few contract awards. Several consultants were not currently assigned to projects but were still on the payroll. What could Cliff and his company do to better understand project risks? Should Cliff adjust his strategy for deciding what projects to pursue? How?

THE IMPORTANCE OF PROJECT RISK MANAGEMENT

Project risk management is the art and science of identifying, analyzing, and responding to risk throughout the life of a project and in the best interests of meeting project objectives. A frequently overlooked aspect of project management, risk management can often result in significant improvements in the ultimate success of projects. Risk management can have a positive impact on selecting projects, determining the scope of projects, and developing realistic schedules and cost estimates. It helps project stakeholders understand the nature of the project, involves team members in defining strengths and weaknesses, and helps to integrate the other project management knowledge areas.

All industries, especially the software development industry, tend to neglect the importance of project risk management. As mentioned in Chapter 7, *Project Quality Management*, William Ibbs and Young H. Kwak performed a study to assess project management maturity. The thirty-eight organizations participating in the study were divided into four industry groups: engineering and construction, telecommunications, information systems/software development, and hi-tech manufacturing. Survey participants answered 148 multiple-choice questions to assess how mature their organization was in the project management knowledge areas of scope, time, cost, quality, human resources, communications, risk, and procurement. The rating scale ranged from 1 to 5, with 5 being the highest maturity rating. Table 10-1 shows the results of the survey. Notice that risk management was the only knowledge area for which all ratings were less than 3. This study shows that all companies should put more effort into project risk management, especially the information systems industry, which had the lowest rating of 2.75.[1]

[1] Ibbs, C. William and Young Hoon Kwak, "Assessing Project Management Maturity," *Project Management Journal* (March 2000).

Table 10-1. Project Management Maturity by Industry Group and Knowledge Area

KEY: 1 = LOWEST MATURITY RATING, 5 = HIGHEST MATURITY RATING

KNOWLEDGE AREA	ENGINEERING/ CONSTRUCTION	TELECOMMUNICATIONS	INFORMATION SYSTEMS	HI-TECH MANUFACTURING
Scope	3.52	3.45	3.25	3.37
Time	3.55	3.41	3.03	3.50
Cost	3.74	3.22	3.20	3.97
Quality	2.91	3.22	2.88	3.26
Human Resources	3.18	3.20	2.93	3.18
Communications	3.53	3.53	3.21	3.48
Risk	**2.93**	**2.87**	**2.75**	**2.76**
Procurement	3.33	3.01	2.91	3.33

Ibbs, C. William and Young Hoon Kwak. "Assessing Project Management Maturity," *Project Management Journal* (March 2000).

Other studies also note the lack of attention paid to project risk management. A KPMG study found that 55 percent of **runaway projects**—projects that have significant cost or schedule overruns—did no risk management at all, 38 percent did some (but half didn't use their risk findings after the project was underway), and 7 percent didn't know whether they did risk management or not.[2] This study suggests that performing risk management is important to improving the likelihood of project success and preventing runaway projects.

Before you can improve project risk management, you must understand what risk is. A basic dictionary definition says that **risk** is "the possibility of loss or injury." This definition highlights the negativity often associated with risk and suggests that uncertainty is involved. Project risk involves understanding potential problems that might occur on the project and how they might impede project success.

In many respects, risk management is like a form of insurance. It is an activity undertaken to lessen the impact of potentially adverse events on a project. It is important to note that risk management *is* an investment—there are costs associated with it. The investment a project is willing to make in risk management activities depends on the nature of the project, the experience of the project team, and the constraints imposed on both. In any case, the cost for risk management should not exceed the potential benefits.

Why do organizations continue to introduce new systems, then, if they often entail the possibility of loss or injury? If the information technology project failure rate is so high, why do companies continue to start new information technology projects? Many companies are in business today because they took

[2] McConnell, Steve, "10 Keys to Successful Software Projects Presentation," *Construx* (2000).

risks that created great opportunities. Organizations survive over the long term when they pursue opportunities. Information technology is often a key part of a business's strategy; without it, many businesses might not survive. Given that all projects involve risks and opportunities, the question is how to decide which projects to pursue and how to manage project risk throughout a project's life cycle.

Several risk experts suggest that organizations and individuals strive to find a balance between risks and opportunities in all aspects of projects and their personal lives. The idea of striving to balance risks and opportunities suggests that different organizations and people have different tolerances for risk. Some organizations or people have a neutral tolerance for risk, some have an aversion to risk, and others are risk-seeking. These three preferences for risk are part of the utility theory of risk.

Risk utility or **risk tolerance** is the amount of satisfaction or pleasure received from a potential payoff. Figure 10-1 shows the basic difference between risk-averse, risk-neutral, and risk-seeking preferences. The y-axis represents utility, or the amount of pleasure received from taking a risk. The x-axis shows the amount of potential payoff, opportunity, or dollar value of the opportunity at stake. Utility rises at a decreasing rate for the person who is **risk-averse**. In other words, when more payoff or money is at stake, a person or organization that is risk-averse gains less satisfaction from the risk, or has lower tolerance for the risk. Those who are **risk-seeking** have a higher tolerance for risk, and their satisfaction increases when more payoff is at stake. A risk seeker prefers more uncertain outcomes and is often willing to pay a penalty to take risks. The **risk-neutral** approach achieves a balance between risk and payoff.

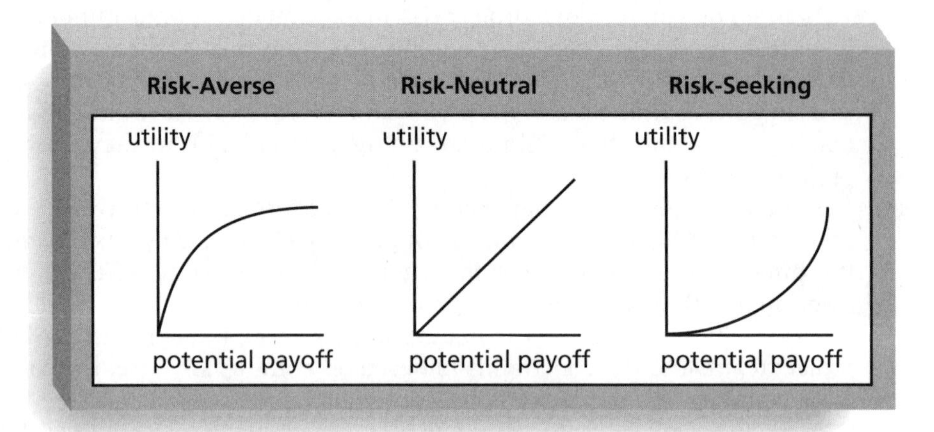

Figure 10-1. Risk Utility Function and Risk Preference

The goal of project risk management can be viewed as minimizing potential risks while maximizing potential opportunities or payoffs. The major processes involved in risk management include:

- **Risk management planning**, which involves deciding how to approach and plan the risk management activities for the project. By reviewing the project charter, WBS, roles and responsibilities, stakeholder risk tolerances, and the organization's risk management policies, the project team can formulate a risk management plan.
- **Risk identification**, which involves determining which risks are likely to affect a project and documenting the characteristics of each. The risk management plan, outputs from project planning, risk categories, and historical information are key inputs to the process of identifying risks.
- **Qualitative risk analysis**, which involves characterizing and analyzing risks and prioritizing their effects on project objectives. After identifying risks, project teams can use various tools and techniques to develop an overall risk ranking for the project.
- **Quantitative risk analysis**, which involves measuring the probability and consequences of risks and estimating their effects on project objectives. After identifying risks, project teams can use additional tools and techniques to prioritize quantified risks and estimate probabilities of achieving project objectives.
- **Risk response planning**, which involves taking steps to enhance opportunities and reduce threats to meeting project objectives. Using outputs from the preceding risk management processes, project teams can develop a risk response plan.
- **Risk monitoring and control**, which involves monitoring known risks, identifying new risks, reducing risks, and evaluating the effectiveness of risk reduction throughout the life of the project. The main outputs of this process include corrective actions in response to risks and updates to the risk response plan.

The first step in project risk management is deciding how to address this knowledge area for your particular project by performing risk management planning.

RISK MANAGEMENT PLANNING

Risk management planning is the process of deciding how to approach and plan for risk management activities for a project, and the main output of this process is a risk management plan. Project teams should hold several planning meetings to help develop the risk management plan. The project team should review project documents such as the project charter, WBS, and definitions of roles and responsibilities, as well as organizational documents such as corporate risk management policies and templates for creating a risk management plan. It is also important to review the risk tolerances of various stakeholders. For example, if the project sponsor is risk-averse, the project might require a different approach to risk management than if the project sponsor were a risk seeker.

A **risk management plan** documents the procedures for managing risk throughout the project. A risk management plan summarizes the results of the risk identification, qualitative analysis, quantitative analysis, response planning, and monitoring and control processes. Table 10-2 lists the questions that a risk management plan should address. It is important to define specific deliverables for the project related to risk, assign people to work on those deliverables, and evaluate milestones associated with the risk mitigation approach. **Risk mitigation** is reducing the impact of a risk event by reducing the probability of its occurrence. The level of detail included in the risk management plan will vary with the needs of the project.

Table 10-2: Questions Addressed in a Risk Management Plan

• *Why* is it important to take/not take this risk in relation to the project objectives?
• *What* is the specific risk, and what are the risk mitigation deliverables?
• *How* is the risk going to be mitigated? (What risk mitigation approach is to be used?)
• *Who* are the individuals responsible for implementing the risk management plan?
• *When* will the milestones associated with the mitigation approach occur?
• *How much* is required in terms of resources to mitigate risk?

The risk management plan can include a methodology for risk management, roles and responsibilities for activities involved in risk management, budgets and schedules for the risk management activities, descriptions of scoring and interpretation methods used for the qualitative and quantitative analyses of risk, threshold criteria for risks, reporting formats for risk management activities, and a description of how the team will track and document risk activities.

In addition to a risk management plan, many projects also include contingency plans, fallback plans, and contingency reserves. **Contingency plans** are predefined actions that the project team will take if an identified risk event occurs. For example, if the project team knows that a new release of a software package may not be available in time for them to use it for their project, they might have a contingency plan to use the existing, older version of the software. **Fallback plans** are developed for risks that have a high impact on meeting project objectives, and are put into effect if attempts to reduce the risk are not effective. **Contingency reserves** or **contingency allowances** are provisions held by the project sponsor that can be used to mitigate cost or schedule risk, if changes in project scope or quality occur. For example, if a project appears to be off course because the staff is inexperienced with some new technology, the project sponsor may provide additional funds from contingency reserves to hire an outside consultant to train and advise the project staff in using the new technology.

Before you can begin to understand and use the other project risk management processes on information technology projects, you must be able to recognize and understand the common sources of risk.

COMMON SOURCES OF RISK ON INFORMATION TECHNOLOGY PROJECTS

Several studies have shown that information technology projects share some common sources of risk. For example, the Standish Group did a follow-up study to the 1995 CHAOS research, which they called Unfinished Voyages. This study brought together sixty information technology professionals to elaborate on how to evaluate a project's overall likelihood of being successful. Table 10-3 shows the Standish Group's success potential scoring sheet and shows the relative importance of the project success criteria factors.[3]

Table 10-3: Information Technology Success Potential Scoring Sheet

SUCCESS CRITERION	RELATIVE IMPORTANCE
User Involvement	19
Executive Management Support	16
Clear Statement of Requirements	15
Proper Planning	11
Realistic Expectations	10
Smaller Project Milestones	9
Competent Staff	8
Ownership	6
Clear Visions and Objectives	3
Hardworking, Focused Staff	3
Total	100

[3] The Standish Group, "Unfinished Voyages" (1996) www.standishgroup.com/voyages.html.

The Standish Group provides specific questions for each success criterion to help decide the number of points to assign to a project. For example, the five questions related to user involvement include the following:

- Do I have the right user(s)?
- Did I involve the user(s) early and often?
- Do I have a quality user(s) relationship?
- Do I make involvement easy?
- Did I find out what the user(s) need(s)?

The number of questions corresponding to each success criterion determines the number of points each positive response is assigned. For example, in the case of user involvement there are five questions. For each positive reply, you would get 3.8 (19/5) points; 19 comes from the weight of the criterion, and 5 comes from the number of questions. Therefore, you would assign a value to the user involvement criterion by adding 3.8 points to the score for each question you can answer positively.

Another risk questionnaire developed by F. W. McFarlan and the Dayton Tire Co. can be used to identify the major sources of risk in the categories of people, structure, and technology.[4] Risks in the people category include inadequate skills (both technical and managerial), inexperience in general, and inexperience in a specific application area or technology. Structural risk includes the degree of change a new project will introduce into user areas and business procedures, the number of distinct user groups the project must satisfy, and the number of other systems the new project must interact with. The experience of the organization with the project's technology and with other similar projects also affects structural risk. Technological risk involves using new or untried technology.

Table 10-4 provides samples of questions from McFarlan's risk questionnaire. You can select more than one answer for a question. After you answer, multiply the total points by the question's weight (not shown here). McFarlan suggests comparing your scores in each risk area—people, structure, and technology—to the norms for the organization. High scores should warn you that high risk is involved. Review your answers to the questions in the high-risk areas to determine the causes for increased risk. For example, perhaps the project could be restructured to use less new technology or to reduce the size of the project or number of user areas involved.

[4] McFarlan, F. W., "Portfolio Approach to Information Systems," *Harvard Business Review* (September-October 1981): 142–150.

Table 10-4: McFarlan's Risk Questionnaire

WHAT IS THE PROJECT ESTIMATE IN CALENDAR (ELAPSED) TIME?	
( ) 12 months or less	Low = 1 point
( ) 13 months to 24 months	Medium = 2 points
( ) Over 24 months	High = 3 points
WHAT IS THE ESTIMATED NUMBER OF PERSON DAYS FOR THE SYSTEM?	
( ) 12 to 375	Low = 1 point
( ) 375 to 1875	Medium = 2 points
( ) 1875 to 3750	Medium = 3 points
( ) Over 3750	High = 4 points
NUMBER OF DEPARTMENTS INVOLVED (EXCLUDING INFORMATION TECHNOLOGY)	
( ) One	Low = 1 point
( ) Two	Medium = 2 points
( ) Three or more	High = 3 points
IS ADDITIONAL HARDWARE REQUIRED FOR THE PROJECT?	
( ) None	Low = 0 points
( ) Central processor type change	Low = 1 point
( ) Peripheral/storage device changes	Low = 1
( ) Terminals	Med = 2
( ) Change of platform, for example PCs replacing mainframes	High = 3

Other broad categories of risk include:

- *Market risk*: If the information technology project is to produce a new product or service, will it be useful to the organization or marketable to others? Will users accept and use the product or service? Will someone else create a better product or service faster, making the project a waste of time and money?
- *Financial risk*: Can the organization afford to undertake the project? How confident are stakeholders in the financial projections? Will the project meet NPV, ROI, and payback estimates? If not, can the company afford to continue the project? Is this project the best way to use the company's financial resources?
- *Technology risk*: Is the project technically feasible? Will hardware, software, and networks function properly? Will the technology be available in time to meet project objectives? Could the technology be obsolete before a useful product can be produced?

What Went Wrong?

Many information technology projects fail because of technology risk. One project manager learned an important lesson on a large information technology project: focus on business needs first, not technology. David Anderson, a project manager for Kaman Sciences Corp., shared his experience with a project failure in an article for *CIO Enterprise Magazine*. After spending two years and hundreds of thousands of dollars on a project to provide new client/server-based financial and human resources information systems for their company, Anderson and his team finally admitted they had a failure on their hands. Anderson revealed that he had been too enamored of the use of cutting-edge technology and had taken a high-risk approach on the project. He "ramrodded through" what the project team was going to do and then admitted that he was wrong. The company finally decided to switch to a more stable technology to meet their business needs.[5]

Reviewing a proposed project in terms of the Standish Group's success criteria, McFarlan's risk questionnaire, or any other similar tool is a good method for understanding common sources of risk on information technology projects. Understanding common sources of risk also helps in risk identification, the next step in project risk management.

RISK IDENTIFICATION

Identifying risks is the process of gaining an understanding of what potential unsatisfactory outcomes are associated with a particular project. By reviewing a project's risk management plan, other planning documents, and the broad categories of risk, project managers and their teams can often identify several potential risks. A review of historical information related to risks on similar projects is also an important input to the risk identification process. Understanding historical risks can help project managers learn from the past and thereby identify and reduce risks on current and future projects.

In addition to identifying risk based on the nature of the project or products produced, it is also important to identify potential risk according to project management knowledge areas, such as scope, time, cost, and quality. Table 10-5 lists potential risk conditions that can exist within each knowledge area.[6]

[5] Hildebrand, Carol, "If At First You Don't Succeed," *CIO Enterprise Magazine* (April 15, 1998).

[6] Wideman, R. Max, *Project and Program Risk Management: A Guide to Managing Project Risks and Opportunities*, Upper Darby, PA: Project Management Institute, 1992, II–4.

Table 10-5: Potential Risk Conditions Associated with Each Knowledge Area

Knowledge Area	Risk Conditions
Integration	Inadequate planning; poor resource allocation; poor integration management; lack of post-project review
Scope	Poor definition of scope or work packages; incomplete definition of quality requirements; inadequate scope control
Time	Errors in estimating time or resource availability; errors in determining the critical path; poor allocation and management of float; early release of competitive products
Cost	Estimating errors; inadequate productivity, cost, change, or contingency control; poor maintenance, security, or purchasing
Quality	Poor attitude toward quality; substandard design/materials/workmanship; inadequate quality assurance program
Human Resources	Poor conflict management; poor project organization and definition of responsibilities; absence of leadership
Communications	Carelessness in planning or communicating; lack of consultation with key stakeholders
Risk	Ignoring risk; unclear analysis of risk; poor insurance management
Procurement	Unenforceable conditions or contract clauses; adversarial relations

There are several tools and techniques for identifying risks. Project teams often begin the risk identification process by reviewing project documentation, analyzing project assumptions, and gathering risk-related information. For example, after reviewing project documents and assumptions, project team members and outside experts often hold meetings to discuss these documents and assumptions and ask important questions about them as they relate to risk. After identifying potential risks at this initial meeting, the project team might then use different information-gathering techniques to further identify risks. Four common information-gathering techniques include brainstorming, the Delphi Technique, interviewing, and SWOT analysis.

Brainstorming is a technique by which a group attempts to generate ideas or find a solution for a specific problem by amassing ideas spontaneously and without judgment. This approach could be used to create a comprehensive list of risks that could be addressed later in the qualitative and quantitative risk analysis processes. Care must be taken, however, not to overuse or misuse brainstorming. Although businesses use brainstorming widely to generate new ideas, the psychology literature shows that individuals, working alone, produce a greater number of ideas than the same individuals produce through brainstorming in small face-to-face groups. Group effects, such as fear of social disapproval, the effects of authority hierarchy, and domination of the session by one or two very vocal people often inhibit idea generation for many participants.[7]

[7] Couger, J. Daniel, *Creative Problem Solving and Opportunity Finding*, Boyd & Fraser Publishing Company, 1995.

One common approach to gathering information from experts is the Delphi Technique. The basic concept of the **Delphi Technique** is to derive a consensus among a panel of experts who make predictions about future developments. Developed by the Rand Corporation for the U.S. Air Force in the late 1960s, the Delphi Technique is a systematic, interactive forecasting procedure based on independent and anonymous inputs regarding future events. The Delphi Technique uses repeated rounds of questioning and written responses, including feedback to earlier-round responses, to take advantage of group input, while avoiding the biasing effects possible in oral panel deliberations. To use the Delphi Technique, you must select a panel of experts for the particular area in question. For example, Cliff Branch from the opening case could use the Delphi Technique to help him understand why his company is no longer winning many contracts. Cliff could assemble a panel of people with knowledge in his business area. Each expert would answer questions related to Cliff's scenario, then Cliff or a facilitator would evaluate their responses, together with opinions and justifications, and provide that feedback to each expert in the next iteration. You continue this process until the group responses converge to a specific solution. If the responses diverge, the facilitator of the Delphi Technique needs to determine if there is a problem with the process.

Interviewing is a fact-finding technique for collecting information in face-to-face or telephone discussions. Some interviews are conducted through e-mail and instant messaging, too. Interviewing people with similar project experience is an important tool for identifying potential risks. For example, if a new project involves using a particular type of hardware or software, someone with recent experience with that hardware or software could describe problems he or she had on a past project. If someone has worked with a particular customer, he or she might provide insight into potential risks involved in working for that group again.

As described in Chapter 4, *Project Scope Management*, strengths, weaknesses, opportunities, and threats (SWOT) analysis is often used in strategic planning. It can also assist in risk identification by having project teams focus on the broad perspectives of potential risks for particular projects. For example, before writing a particular proposal, Cliff Branch could have a group of his employees discuss in detail what their company's strengths are, what their weaknesses are related to that project, and what opportunities and threats exist. Do they know that several competing firms are much more likely to win a certain contract? Do they know that winning a particular contract award will likely lead to future awards and help expand their business? Applying SWOT to specific potential projects can help identify the broad risks and opportunities that apply in that scenario.

Two other techniques for risk identification include the use of checklists and diagramming. Checklists based on risks that have been encountered in previous projects provide a meaningful template for understanding risks in a current project. You can use checklists like those developed by the Standish Group, McFarlan, or other groups to help identify risks on information technology projects.

Diagramming techniques include using cause-and-effect diagrams or fishbone diagrams, flow charts, and influence diagrams. Recall from Chapter 7, *Project Quality Management*, that fishbone diagrams help you trace problems back to their root cause. **System or process flow charts** are diagrams that show how different parts of a system interrelate. For example, many programmers create flow charts to show programming logic. Another type of diagram, an **influence diagram**, represents decision problems by displaying essential elements, including decisions, uncertainties, and objectives, and how they influence each other.[8]

The main outputs of the risk identification process are identified risk events for your project, triggers or risk symptoms, and inputs to other processes, such as updates to the WBS or schedule based on identified risks. **Risk events** are specific things that may occur to the detriment of the project. Examples of risk events are significant changes in scope, the performance failure of products produced as part of a project, specific delays in the project due to rejection of work or labor unavailability, supply shortages, litigation against your company, strikes, and so on. **Triggers,** or **risk symptoms**, are indicators of actual risk events. For example, cost overruns on early activities may be symptoms of poor cost estimates. Defective products may be symptoms of a low-quality supplier. Documenting potential risk symptoms for projects also helps the project team identify potential risk events and determine what response to take. In the opening case, the fact that Cliff Branch's firm was not winning more contracts may have been a symptom that there was more competition in their market or that they were not writing competitive proposals.

QUALITATIVE RISK ANALYSIS

Qualitative risk analysis involves assessing the likelihood and impact of identified risks, to determine their magnitude and priority. This section describes examples of using a probability/impact matrix to produce a prioritized list of risks and the use of the Top 10 Risk Item Tracking technique to produce an overall ranking for project risks and to track trends in qualitative risk analysis. It also discusses the importance of expert judgment in performing risk analysis.

Calculating Risk Factors Using Probability/Impact Matrixes

People often describe a risk probability or consequence as being high, moderate, or low. For example, a meteorologist might predict that there is a high probability of severe rain showers on a certain day. If that day happens to be your wedding day and you are planning a large outdoor wedding, the consequences of severe showers might also be high. To quantify risk probability and consequence, the Defense Systems Management College (DSMC) developed a

[8] Lumina Analytical, Influence Diagrams, (2001) www.lumina.com/software/influencediagrams.html.

technique for calculating **risk factors**—numbers that represent the overall risk of specific events, based on their probability of occurring and the consequences to the project if they do occur. The technique makes use of a probability/impact matrix that shows the probability, or likelihood, of risks occurring and the impact or consequences of the risks.

Probabilities of a risk occurring can be estimated on the basis of several factors, as determined by the unique nature of each project. For example, factors evaluated for potential hardware or software technology risks could include the technology not being mature, the technology being too complex, and an inadequate support base for developing the technology. The impact of a risk occurring could include factors such as the availability of fallback solutions or the consequences of not meeting performance, cost, and schedule estimates.

Table 10-6 provides an example of an actual probability/impact matrix that was used in assessing the risk of various technologies that could make aircraft more reliable. The Air Force sponsored a multimillion-dollar research project in the mid-1980s, the High Reliability Fighter study, to evaluate potential technologies for improving the reliability of fighter aircraft. Many aircraft were too often unfit for flying because of reliability problems with various components of the planes and problems with their maintenance. The purpose of the High Reliability Fighter study was to help the Air Force decide what technologies to invest in to make planes more reliable and maintainable, thereby decreasing their downtime. The matrix in Table 10-6 was used to assess the risk of proposed technologies, such as providing radial tires for planes, using a more efficient fuel system, or developing a sophisticated onboard computer system to monitor and adjust various systems in the aircraft. The top matrix in Table 10-6 was used to assign a Probability of Failure (Pf) value to each proposed technology, and the bottom matrix was used to assign a Consequence of Failure (Cf) value. Experts used their judgment to assign a value for both the probability and impact of each proposed technology. For example, an expert assigned a Pf value of .1 for the radial tires technology since the hardware existed, had a simple design, and had multiple programs and services that used it. Likewise, the expert assigned a Cf value of .9 to a risk that had no acceptable alternatives, would increase life cycle costs, was unlikely to meet schedule dates, and might not reduce the aircraft downtime factor. These values were then used in a formula to calculate an overall risk factor. A risk factor is defined as the probability of failure (Pf) plus the consequence of failure (Cf) minus the product of the two.[9] For example, a technology with a Pf of .1 and a Cf of .9 would have a risk factor of .01, or (.1 + .9) - (.1 * .9) = .01.

[9] Defense Systems Management College, *Systems Engineering Management Guide*, Washington, DC, 1989.

Table 10-6: Sample Probability/Impact Matrixes for Qualitative Risk Assessment

PROBABILITY OF FAILURE (P_F) ATTRIBUTES OF SUGGESTED TECHNOLOGY			
VALUE	**MATURITY HARDWARE/SOFTWARE**	**COMPLEXITY HARDWARE/SOFTWARE**	**SUPPORT BASE**
0.1	Existing	Simple Design	Multiple Programs And Services
0.3	Minor Redesign	Somewhat Complex	Multiple Programs
0.5	Major Change Feasible	Fairly Complex	Several Parallel Programs
0.7	Complex HW Design/ New SW Similar to Existing	Very Complex	At Least One Other Program
0.9	Some Research Completed/ Never Done Before	Extremely Complex	No Additional Programs

CONSEQUENCE OF FAILURE (C_F) ATTRIBUTES OF SUGGESTED TECHNOLOGY				
VALUE	**FALLBACK SOLUTIONS**	**LIFE CYCLE COST (LCC) FACTOR**	**SCHEDULE FACTOR (INITIAL OPERATIONAL CAPABILITY = IOC)**	**DOWNTIME (DT) FACTOR**
0.1	Several Acceptable Alternatives	Highly Confident Will Reduce LCC	90–100% Confident Will Meet IOC Significantly	Highly Confident Will Reduce DT
0.3	A Few Known Alternatives	Fairly Confident Will Reduce LCC	75–90% Confident Will Meet IOC	Fairly Confident Will Reduce DT Significantly
0.5	Single Acceptable Alternative	LCC Will Not Change Much	50–75% Confident Will Meet IOC	Highly Confident Will Reduce DT Somewhat
0.7	Some Possible Alternatives	Fairly Confident Will Increase LCC	25–50% Confident Will Meet IOC	Fairly Confident Will Reduce DT Somewhat
0.9	No Acceptable Alternatives	Highly Confident Will Increase LCC	0–25% Confident Will Meet IOC	DT May Not Be Reduced Much

Figure 10-2 provides an example of how the risk factors were used to graph the probability and consequence of failure for proposed technologies in this study. The figure classifies potential technologies (dots on the chart) as high-, medium-, or low-risk, based on the probability and consequences of failure. The researchers for this study highly recommended that the Air Force invest in the low- to medium-risk technologies and suggested that they not pursue the high-risk technologies.[10] You can see that the rigor behind using the probability/impact matrix and risk factors provides a much stronger argument than simply stating that risk probabilities or consequences are high, medium, or low.

[10] McDonnell Douglas Corporation, "Hi-Rel Fighter Concept," Report MDC B0642, 1988.

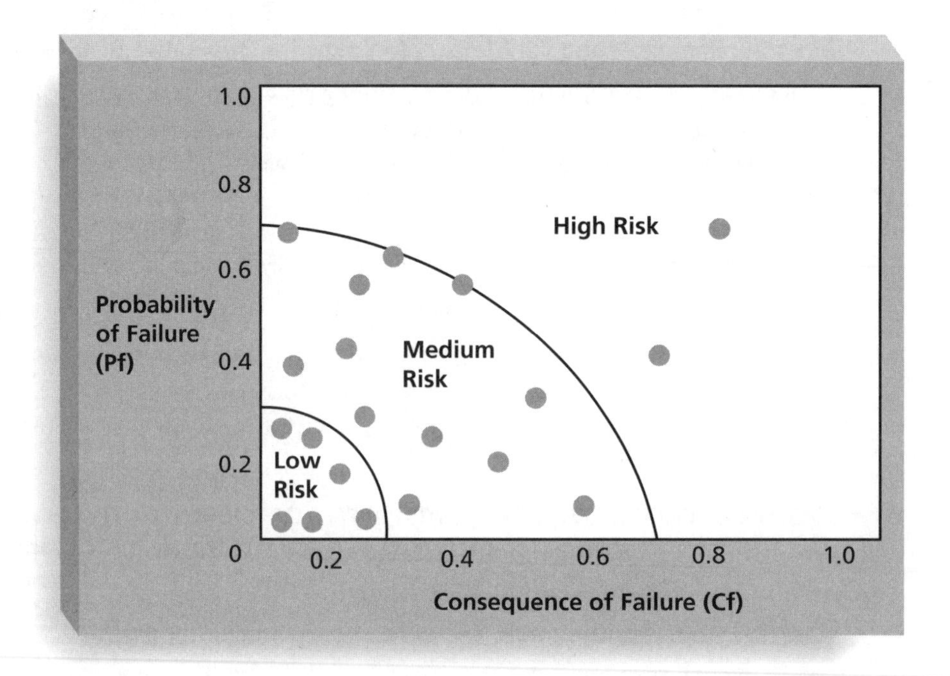

Figure 10-2. Chart Showing High-, Medium-, and Low-Risk Technologies

Top 10 Risk Item Tracking

Top 10 Risk Item Tracking is a qualitative risk analysis tool, and in addition to identifying risks, it also maintains an awareness of risks throughout the life of a project. It involves establishing a periodic review of the project's most significant risk items with management and, optionally, with the customer. The review begins with a summary of the status of the top ten sources of risk on the project. The summary includes each item's current ranking, previous ranking, number of times it appears on the list over a period of time, and a summary of progress made in resolving the risk item since the previous review. The Microsoft Solution Framework (MSF) includes a risk management model that includes developing and monitoring a Top 10 master list of risks.

Table 10-7 provides an example of a Top 10 Risk Item Tracking chart that could be used at a management review meeting for a project. This particular example includes only the top five risk items. Notice that each risk item is ranked on the basis of the current month, previous month, and how many months it has been in the top 10. The last column briefly describes the progress for resolving each particular risk item.

Table 10-7: Example of Top 10 Risk Item Tracking

	Monthly Ranking			
Risk Item	This Month	Last Month	Number of Months	Risk Resolution Progress
Inadequate planning	1	2	4	Working on revising the entire project plan
Poor definition	2	3	3	Holding meetings with project customer and sponsor to clarify scope
Absence of leadership	3	1	2	After previous project manager quit, assigned a new one to lead the project
Poor cost estimates	4	4	3	Revising cost estimates
Poor time estimates	5	5	3	Revising schedule estimates

A risk management review accomplishes several objectives. First, it keeps management and the customer (if included) aware of the major influences that could prevent the project from being a success. Second, by involving the customer, the project team may be able to consider alternatives that could mitigate the risk, such as reducing the scope of a project by postponing some work for a later project in order to meet cost and schedule goals. Third, it is a means of promoting confidence in the project team by demonstrating to management and the customer that the team is aware of the significant risks, has a mitigation strategy in place, and is effectively carrying out that strategy.

Expert Judgment

Many organizations rely on the intuitive feelings and past experience of experts in performing qualitative risk analyses. They may use expert judgment in lieu of or in addition to other techniques for analyzing risks. For example, experts can categorize risks as being high, medium, or low with or without more sophisticated techniques such as calculating risk factors, as described earlier.

Using sophisticated risk analysis tools has a number of disadvantages. For example, the outputs are only as good as the inputs, and the people using the tools may be using poor assumptions. Many people get confused when someone tries to explain the math and statistics behind the various techniques. Because of these disadvantages, it is important to include expert opinions when using both qualitative and quantitative risk assessment techniques.

QUANTITATIVE RISK ANALYSIS

Quantitative risk analysis often follows qualitative risk analysis, yet both processes can be done together or separately. On some projects, the team may only perform qualitative risk analysis. The nature of the project and availability of time and money affect the type of risk analysis techniques to use. Large, complex projects involving leading-edge technologies often require extensive quantitative risk analysis. The main techniques for quantitative risk analysis include decision tree analysis and simulation, as described in the following sections. Teams can also use simpler approaches, such as interviewing and sensitivity analysis, to assist in quantitative risk analysis.

Decisions Trees and Expected Monetary Value

A **decision tree** is a diagramming method used to help you select the best course of action in situations in which future outcomes are uncertain. A common application of decision tree analysis involves calculating expected monetary value. **Expected monetary value (EMV)** is the product of a risk event probability and the risk event's monetary value. Figure 10-3 uses the decision of which project(s) an organization might pursue to illustrate this concept. Suppose Cliff Branch's company was trying to decide if it should submit a proposal for Project 1, Project 2, both projects, or neither project. They could draw a decision tree with two branches, one for Project 1 and one for Project 2. The company could then calculate the expected monetary value to help make this decision.

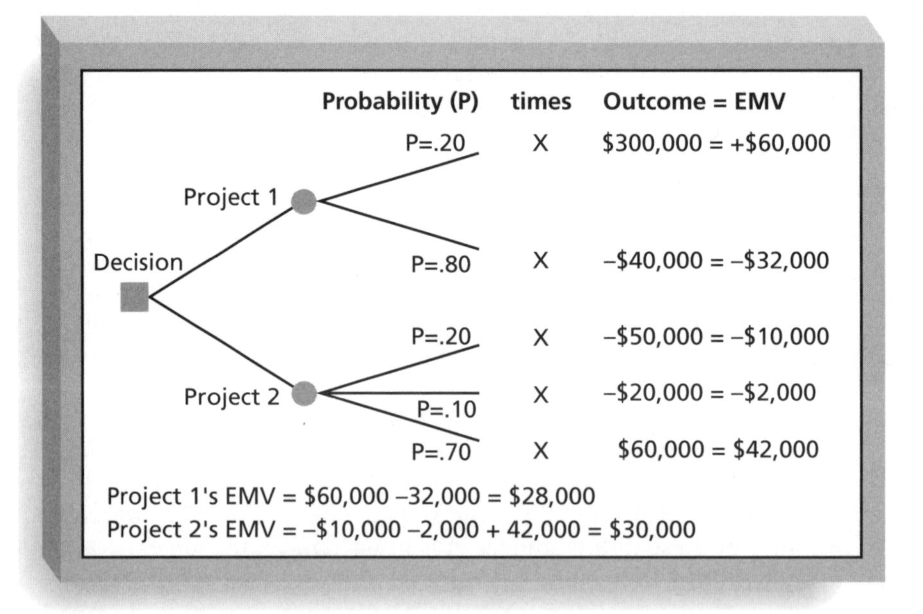

Figure 10-3. Expected Monetary Value (EMV) Example

In order to create a decision tree, and to calculate expected monetary value specifically, you must estimate the probabilities, or chances, of certain events occurring. For example, in Figure 10-3 there is a 20 percent probability or chance (P=.20) that Cliff's firm will win Project 1, which is estimated to be worth $300,000 in profits, the outcome of the top branch in the figure. There is an 80 percent probability (P=.80) that they will not win the competition for Project 1, and the outcome is estimated to be –$40,000, meaning that the firm will have to invest $40,000 into Project 1 with no reimbursement if they are not awarded the contract. The sum of the probabilities for outcomes for each project must equal one (for Project 1, 20 percent plus 80 percent). Probabilities are normally determined on the basis of expert judgment. Cliff or other people in his firm should have some sense of their likelihood of winning certain projects.

Figure 10-3 also shows probabilities and outcomes for Project 2. Suppose there is a 20 percent probability that Cliff's firm will lose $50,000 on Project 2, a 10 percent probability that they will lose $20,000, and a 70 percent probability that they will earn $60,000. Again, experts would need to estimate these dollar amounts and probabilities.

To calculate the expected monetary value for each project, you multiply the probability by the outcome value for each potential outcome for each project and sum the results. To calculate expected monetary value (EMV) for Project 1, going from left to right, multiply the probability times the outcome for each branch and sum the results. In this example, the EMV for Project 1 is $28,000.

.2($300,000) + .8(–$40,000) = $60,000 – $32,000 = $28,000

The EMV for Project 2 is $30,000.

.2(–$50,000) + .1(–$20,000) + .7($60,000)= –$10,000 –$2,000 + $42,000 = $30,000

Because the EMV provides an estimate for the total dollar value of a decision, you want to have a positive number; the higher the EMV, the better. Since the EMV is positive for both Projects 1 and 2, Cliff's firm would expect a positive outcome from each and could bid on both projects. If they had to choose between the two projects, perhaps because of limited resources, Cliff's firm should bid on Project 2 because it has a higher EMV.

Also notice in Figure 10-3 that if you just looked at the potential outcome of the two projects, Project 1 looks more appealing. You could earn $300,000 in profits from Project 1, but you can only earn $60,000 for Project 2. If Cliff were a risk seeker, he would naturally want to bid on Project 1. However, there is only a 20 percent chance of getting that $300,000 on Project 1, and there is a 70 percent chance of earning $60,000 on Project 2. Using EMV helps account for all possible outcomes and their probabilities of occurrence, thereby reducing the tendency to pursue overly aggressive or conservative risk strategies.

Simulation

A more sophisticated quantitative risk analysis technique is simulation. Simulation uses a representation or model of a system to analyze the expected behavior or performance of the system. Most simulations are based on some form of Monte Carlo analysis. **Monte Carlo analysis** simulates a model's outcome many times to provide a statistical distribution of the calculated results. For example, a Monte Carlo simulation can determine if a project will meet its schedule or cost goals given a 10 percent, 50 percent, or 90 percent probability.

The basic steps of a Monte Carlo simulation are:

1. Assess the range for the variables being considered and determine the probability distribution for each. In other words, collect the most likely, optimistic, and pessimistic estimates for the variables in the model and determine the probability of each variable falling between the optimistic and most likely estimates.
2. For each variable, select a random value based on the probability distribution for the occurrence of the variable. For example, suppose an optimistic estimate is 10 (units can be days, dollars, labor hours, or whatever unit is in the model). Also, suppose the most likely estimate is 20, and the pessimistic estimate is 50. If there is a 30 percent probability of being between 10 and 20 (the optimistic and most likely estimates), then 30 percent of the time, select a random number between 10 and 20, and 70 percent of the time, select a number between 20 and 50 (the pessimistic estimate).
3. Run a deterministic analysis or one pass through the model using the combination of values selected for each one of the variables.
4. Repeat steps 2 and 3 a large number of times to obtain the probability distribution of the results. The number of iterations depends on the number of variables and the degree of confidence required in the results, but it typically lies between 100 and 1,000.

Another technique for quantifying risk that was described in Chapter 5, *Project Time Management*, is Program Evaluation and Review Technique (PERT) analysis. Recall that PERT analysis involves making three estimates of each activity's duration. The three estimates represent a pessimistic or worst case estimate, an optimistic or best case estimate, and a most likely estimate, similar to those described in a Monte Carlo simulation. However, the PERT formula simply weights the most likely estimate four times more than the pessimistic or optimistic estimates, instead of assigning a probability for the estimate falling between the optimistic and most likely estimates and then running a simulation of the model. Although this approach can be better than using one discrete estimate, it does not provide the flexibility or accuracy of a Monte Carlo simulation.

As you can imagine, people now use software to perform the steps required for a Monte Carlo simulation. Several PC-based software packages are available that perform Monte Carlo simulations. Examples of using Monte Carlo simulation software are discussed in the section of this chapter on using software to assist in project risk management.

What Went Right?

A large aerospace firm used Monte Carlo simulation to help quantify risks on several advanced-design engineering projects. The National Aerospace Plan (NASP) project involved many risks. The purpose of this multibillion dollar project was to design and develop a vehicle that could fly into space using a single-stage-to-orbit approach. A single-stage-to-orbit approach meant the vehicle would have to achieve a speed of Mach 25 (25 times the speed of sound) without a rocket booster. A team of engineers and business professionals worked together in the mid-1980s to develop a software model for estimating the time and cost of developing the NASP. This model was then linked with Monte Carlo simulation software to determine the sources of cost and schedule risk for the project. The results of the simulation were then used to determine how the company would invest its internal research and development funds. Although the NASP project was terminated, the resulting research has helped develop more advanced materials and propulsion systems used on many modern aircraft.

RISK RESPONSE PLANNING

After risks are identified and quantified, an organization must develop a response to them. Developing a response to risks involves defining steps for enhancing opportunities and developing plans for handling risks or threats to project success. The four basic response strategies are avoidance, acceptance, transference, and mitigation. Important outputs of the risk response development process include a risk management plan, contingency plans, and reserves.

Risk avoidance involves eliminating a specific threat or risk, usually by eliminating its causes. Of course, all risks cannot be eliminated, but specific risk events can be. For example, a project team may decide to continue using a specific piece of hardware or software on a project because they know it works. Other products that could be used on the project may be available, but if the team is unfamiliar with them, they could cause significant risk. Using familiar hardware or software eliminates this risk.

Risk acceptance means accepting the consequences should a risk occur. For example, a project team planning a big project review meeting could take an active approach to risk by having a contingency or backup plan and contingency reserves if they cannot get approval for a specific site for the meeting. On the other hand, they could take a passive approach and accept whatever facility their organization provides them.

Risk transference is shifting the consequence of a risk and responsibility for its management to a third party. For example, risk transference is often used in dealing with financial risk exposure. A project team may purchase special insurance or warranty protection for specific hardware needed for a project. If the hardware fails, the insurer must replace it within an agreed-upon period of time.

Risk mitigation involves reducing the impact of a risk event by reducing the probability of its occurrence. Suggestions for reducing common sources of risk on information technology projects were provided at the beginning of this chapter. Other examples of risk mitigation include using proven technology, having competent project personnel, using various analysis and validation techniques, and buying maintenance or service agreements from subcontractors.

Table 10-8 provides general mitigation strategies for technical, cost, and schedule risks on projects.[11] Note that increasing the frequency of project monitoring and using a work breakdown structure (WBS) and Critical Path Method (CPM) are strategies for all three areas. Increasing the project manager's authority is a strategy for mitigating technical and cost risks, and selecting the most experienced project manager is recommended for reducing schedule risks. Improving communication is also an effective strategy for mitigating risks.

Table 10-8: General Risk Mitigation Strategies for Technical, Cost, and Schedule Risks

TECHNICAL RISKS	COST RISKS	SCHEDULE RISKS
Emphasize team support and avoid stand-alone project structure	Increase the frequency of project monitoring	Increase the frequency of project monitoring
Increase project manager authority	Use WBS and CPM	Use WBS and CPM
Improve problem handling and communication	Improve communication, project goals understanding, and team support	Select the most experienced project manager
Increase the frequency of project monitoring	Increase project manager authority	
Use WBS and CPM		

Important outputs from risk response planning include development of a risk response plan, analysis of residual risks, and analysis of secondary risks. The risk response plan describes identified risks, people assigned responsibilities for managing those risks, results from risk analyses, response strategies, budget and schedule estimates for responses, and contingency and fallback plans. **Residual risks** are risks that remain after all of the response strategies have been implemented. For example, even though a more stable hardware product may have been used on a project, there may still be some risk of it failing to function properly. **Secondary risks** are a direct result of implementing a risk response. For example, using the more stable hardware may have caused a risk

[11] Couillard, Jean, "The Role of Project Risk in Determining Project Management Approach," *Project Management Journal*, Project Management Institute (December 1995).

of peripheral devices failing to function properly. Other outputs of risk response planning include contractual agreements, estimates of needed contingency reserve, and inputs to other processes and the project plan.

RISK MONITORING AND CONTROL

Risk monitoring and control involves executing the risk management processes and the risk management plan to respond to risk events. Executing the risk management processes means ensuring that risk awareness is an ongoing activity performed by the entire project team throughout the entire project. Project risk management does not stop with the initial risk analysis. Identified risks may not materialize, or their probabilities of occurrence or loss may diminish. Previously identified risks may be determined to have a greater probability of occurrence or a higher estimated loss value. Similarly, new risks will be identified as the project progresses. Newly identified risks need to go through the same process as those identified during the initial risk assessment. A redistribution of resources devoted to risk management may be necessary because of relative changes in risk exposure.

Carrying out individual risk management plans involves monitoring risks on the basis of defined milestones and making decisions regarding risks and mitigation strategies. It may be necessary to alter a mitigation strategy if it becomes ineffective, implement a planned contingency activity, or eliminate a risk from the list of potential risks when it no longer exists. Project teams sometimes use **workarounds**—unplanned responses to risk events—when they do not have contingency plans in place.

Project risk audits, periodic risk reviews such as the Top 10 Risk Item Tracking method, earned value management, technical performance measurement, and additional risk response planning are all tools and techniques for performing risk monitoring and control. Outputs of this process are corrective action, project change requests, and updates to other plans.

USING SOFTWARE TO ASSIST IN PROJECT RISK MANAGEMENT

As mentioned earlier, you can use software tools to enhance various risk management processes. Databases can keep track of risks, spreadsheets can aid in tracking and quantifying risks, and more sophisticated risk management software can help you develop models and use simulations to analyze and respond to various risks.

You can use Microsoft Project 2000 to perform PERT analysis, and you can also use add-on software to perform Monte Carlo simulations. For example, Risk+ (by C/S Solutions, Inc.) is a comprehensive risk analysis tool that integrates with Project 2000 to quantify the cost and schedule uncertainty associated with projects. It uses Monte-Carlo-based simulation techniques to answer questions such as: What are the chances of completing the project by December 1, 2002? How confident are we that costs will remain below $10 million? What are the chances that this task will end up on the critical path?

To use a Monte Carlo simulation to estimate the probability of meeting specific schedule goals, you would collect optimistic, pessimistic, and most likely duration estimates for project tasks on a project network diagram, which is similar to the PERT technique. You must also collect estimates for the probability of completing each task between the optimistic and most likely times. The same approach can be used for cost estimates. You would collect optimistic, pessimistic, and most likely estimates for factors that determine project costs and the probability of the cost factors being between the optimistic and most likely values.

For example, an expert might estimate that a certain task will most likely take three months to complete, but it could take as little as one month or up to nine months. When asked the probability of completing the task between one and three months, the expert might honestly reply that the probability is only 20 percent. Another expert might estimate that another project task will take five months to complete, but it could take as little as two months, or it could take up to seven months. This expert might estimate that the probability of completing that task between two and five months is 80 percent. Estimating the probability of completing tasks between the optimistic and most likely times helps to account for estimating bias. Compared with a PERT calculation, the Monte Carlo approach simulates various probability distributions for each estimate instead of applying the same simple PERT variation for all estimated durations. Unlike PERT, which focuses on schedule estimates, Monte Carlo simulation can also be used to estimate cost risks.

Figure 10-4 illustrates the results from a Monte Carlo simulation of a project schedule. The simulation was done using Microsoft Project and the Risk+ add-on software. On the left side of Figure 10-4 is a chart displaying columns and an S-shaped curve. The height of each column, read by the scale on the left of the chart, indicates how many times the project was completed in a given time interval during the simulation run, which is the sample count. In this example, the time interval is two working days, and the simulation was run 250 times. The first column shows that the project was completed by 1/29/02 only two times during the simulation. The S-shaped curve, read from the scale on the right of the chart, shows the cumulative probability of completing the project on or before a given date. The information is also shown in tabular form on the right side of Figure 10-4. For example, there is a 10 percent probability that the project will be completed by 2/8/02, a 50 percent chance of completion by 2/17/02, and a 90 percent chance of completion by 2/25/02.

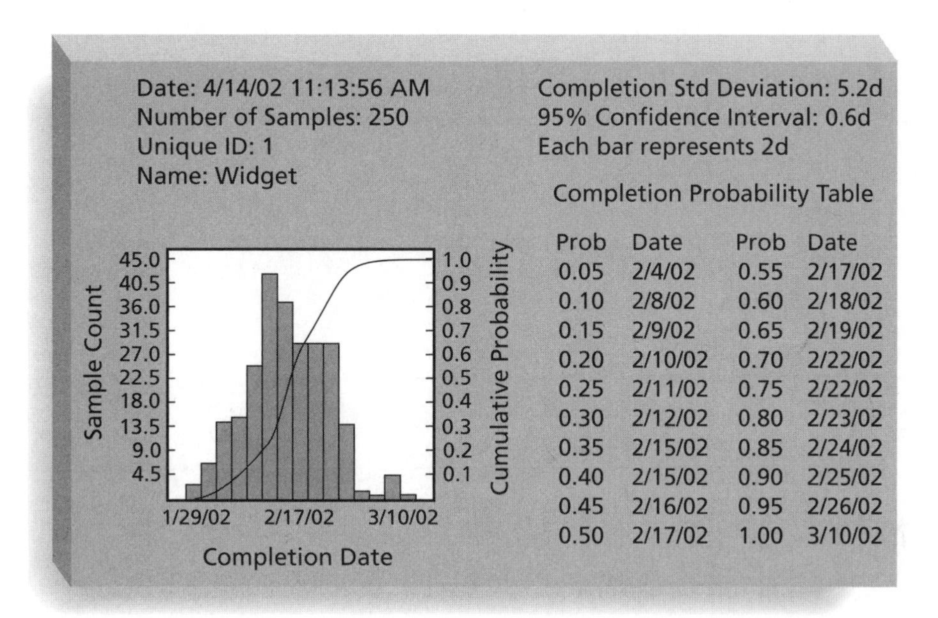

Prob	Date	Prob	Date
0.05	2/4/02	0.55	2/17/02
0.10	2/8/02	0.60	2/18/02
0.15	2/9/02	0.65	2/19/02
0.20	2/10/02	0.70	2/22/02
0.25	2/11/02	0.75	2/22/02
0.30	2/12/02	0.80	2/23/02
0.35	2/15/02	0.85	2/24/02
0.40	2/15/02	0.90	2/25/02
0.45	2/16/02	0.95	2/26/02
0.50	2/17/02	1.00	3/10/02

Figure 10-4. Sample Monte Carlo Simulation Results for Project Schedule

You can also use Monte Carlo simulations to help estimate project costs. First you would develop a model for estimating the total project cost. Suppose project costs could be estimated based on the number of pounds of a certain material, the cost per pound of the material, the number of hours of specific workers, and the cost per hour for each category of worker (such as managers, programmers, and electrical engineers). You could run a Monte Carlo simulation of the total project cost based on estimates of the optimistic, pessimistic, and most likely number of pounds of material, costs per pound of the material, number of hours of specific workers, and the cost per hour for the various workers.

Figure 10-5 shows the results of a Monte Carlo simulation to estimate total project cost. These simulation results show that there is a 20 percent chance of the project costing less than $175,693, a 65 percent chance of it costing less than $180,015, and a 95 percent chance of the total project cost being under $184,528. For example, you can use this information to decide how much to bid on a project, if you are the seller, or how much to budget for the project, if you are the buyer, given your risk tolerance. For example, if you are risk-averse, you might want to bid $185,000 to be extremely confident that you will not go over budget.

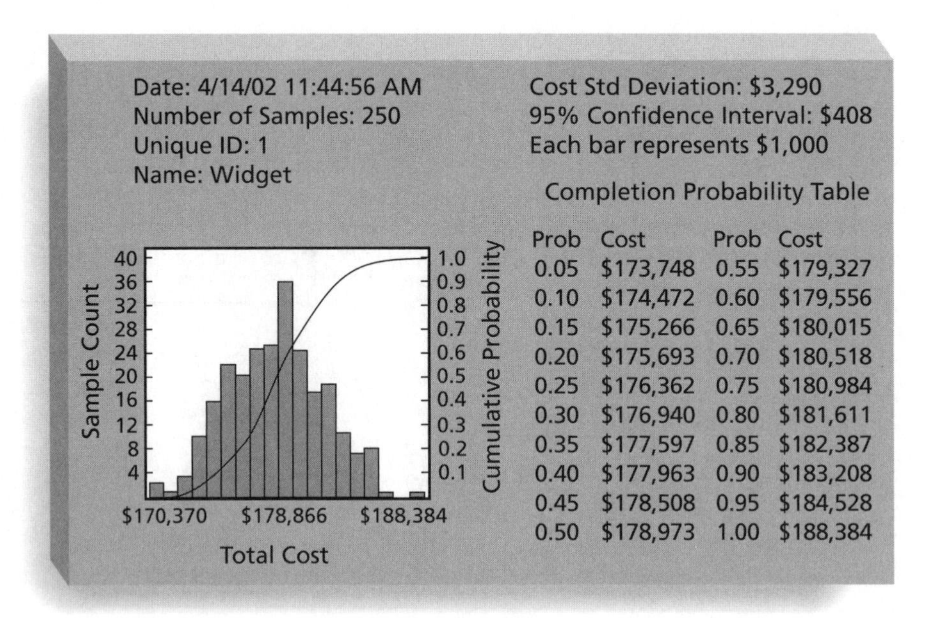

Figure 10-5. Sample Monte Carlo Simulation Results for Project Cost

In addition to estimating overall probabilities of project goals such as completion dates or cost estimates, top sources of risk (risk drivers) can also be found. For example, a cost simulation might show that the number of labor hours for the electrical engineers was the main source of cost risk for a project.

Simulations are very powerful tools, and it is important that people who use them understand all of the variables, inputs, and outputs involved. As with any software product, the information that comes out is only as good as the information that goes in. It is important to collect data from people who understand the project or specific tasks involved. It is also important to test the model used in simulations to ensure that it provides realistic results.

RESULTS OF GOOD PROJECT RISK MANAGEMENT

Good project risk management often goes unnoticed, unlike crisis management. With crisis management, there is an obvious danger to the success of a project. The crisis, in turn, receives the intense interest of the entire project team. Resolving a crisis has much greater visibility, often accompanied by rewards from management, than successful risk management. In contrast, when risk management is effective, it results in fewer problems, and for the few problems that exist, it results in more expeditious resolutions. It may be difficult for outside observers to tell whether risk management or luck was responsible for the smooth development of a new system, but teams will always know that their projects worked out better because of good risk management.

Well-run projects, like a master violinist's performance, an Olympic athlete's gold medal win, or a Pulitzer-Prize-winning book, appear to be almost effortless. Those on the outside—whether audiences, customers, or managers—cannot observe the effort that goes into a superb performance. They cannot see the hours of practice, the edited drafts, or the planning, management, and foresight that create the appearance of ease. To improve information technology project management, project managers should strive to make their jobs look easy—it reflects the results of a well-run project.

CASE WRAP-UP

Cliff Branch and two of his senior people attended a seminar on risk management where the speaker discussed several techniques such as estimating the expected monetary value of projects, Monte Carlo simulations, and so on. Cliff asked the speaker how he might use these techniques to help his company decide which projects to bid on, since bidding on projects often required up-front investments with the possibility of no payback. The speaker walked through an example of EMV and then ran a quick Monte Carlo simulation. Cliff did not have a strong math background and had a hard time understanding the EMV calculations. He thought the simulation was much too confusing to have any practical use for him. He believed in his gut instincts much more than any math calculation or computer output.

The speaker finally sensed that Cliff was not impressed, so she explained the importance of looking at the odds of winning project awards and not just at the potential profits. She suggested using a risk-neutral strategy by bidding on projects that the firm had a good chance of winning (50 percent or so) and that had a good profit potential, instead of focusing on projects that they had a small chance of winning and that had a larger profit potential. Cliff disagreed with this advice, and he continued to bid on high-risk projects. The two other managers who attended the seminar now understood why the firm was having problems—their leader loved taking risks, even if it hurt the firm. They soon found jobs with competing firms, as did several others.

CHAPTER SUMMARY

Risk is the possibility of loss or injury. Projects, by virtue of their unique nature, involve risk.

Risk management is an investment; that is, there are costs associated with identifying risks, analyzing those risks, and establishing plans to mitigate those risks. Those costs must be included in cost, schedule, and resource planning.

Organizations take risks to benefit from potential opportunities.

Risk utility or risk tolerance is the amount of satisfaction or pleasure received from a potential payoff. Risk seekers enjoy high risks, risk-averse people do not like to take risks, and risk-neutral people seek to balance risks and potential payoff.

Risk management is an ethic in which the project team continually assesses what may negatively impact the project, determines the probability of such events occurring, and determines the impact should such events occur. It also involves analyzing and determining alternate strategies to deal with risks. The six main processes involved in risk management are risk management planning, risk identification, qualitative risk analysis, quantitative risk analysis, risk response planning, and risk monitoring and control.

Risk management planning is the process of deciding how to approach and plan for risk management activities for a particular project. A risk management plan is a key output of risk management. Contingency plans are predefined actions that a project team will take if an identified risk event occurs. Fallback plans are developed for risks that have a high impact on meeting project objectives, and are implemented if attempts to reduce the risk are not effective. Contingency reserves or allowances are provisions held by the project sponsor that can be used to mitigate cost or schedule risk, if changes in project scope or quality occur.

Information technology projects often involve several risks: lack of user involvement, lack of executive management support, unclear requirements, poor planning, and so on. Lists developed by the Standish Group, McFarlan, and other organizations can help you identify potential risks on information technology projects. Lists of common risk conditions in project management knowledge areas can also be helpful in identifying risks, as can information-gathering techniques such as brainstorming, the Delphi Technique, interviewing, and SWOT analysis.

Risks can be assessed qualitatively and quantitatively. Tools for qualitative risk analysis include a probability/impact matrix, the Top 10 Risk Item Tracking technique, and expert judgment. Tools for quantitative risk analysis include decision trees and simulation. Expected monetary value uses decision trees to evaluate potential projects on the basis of their expected value. Simulations are a more sophisticated method for creating estimates to help you determine the likelihood of meeting specific project schedule or cost goals.

The four basic responses to risk are avoidance, acceptance, transference, and mitigation. Risk avoidance involves eliminating a specific threat or risk. Risk acceptance means accepting the consequences of a risk, should it occur. Risk transference is shifting the consequence of a risk and responsibility for its management to a third party. Risk mitigation is reducing the impact of a risk event by reducing the probability of its occurrence.

Risk monitoring and control involves executing the risk management processes and the risk management plan to respond to risks. Outputs of this process include corrective action, project change requests, and updates to other plans.

Several types of software can assist in project risk management. Monte Carlo simulation software is a particularly useful tool for helping you get a better idea of project risks and top sources of risk or risk drivers.

Discussion Questions

1. Discuss the risk utility function and risk preference chart in Figure 10-1. Would you rate yourself as being risk-averse, risk-neutral, or risk-seeking? Give examples of each approach from different aspects of your life, such as your current job, your personal finances, romances, and eating habits.
2. What is your organization's (your employer's or your college's) risk preference when it comes to information technology projects? Give evidence to support your position.
3. Discuss the common sources of risk on information technology projects and suggestions for managing them. Which suggestions do you find most useful? Which do you feel would not work in your organization? Why?
4. Describe the Top 10 Risk Item Tracking approach. How could you use this technique in your organization?
5. Discuss the various techniques for quantifying risk. Give an example of how you could use each on an information technology project.
6. Provide examples of risk avoidance, risk acceptance, risk transference, and risk mitigation as responses to risks.
7. Describe the tools and techniques for performing risk monitoring and control.

Exercises

1. Suppose your college or organization is considering a new project that would involve developing an information system that would allow all employees and students/customers to access and maintain their own human resources-related information, such as address, marital status, tax information, and so on. The main benefits of the system would be a reduction in human resources personnel and more accurate information. For example, if an employee, student, or customer had a new telephone number or e-mail address, he or she would be responsible for entering the new data in the new system. The new system would also allow employees to change their tax withholdings or pension plan contributions. Identify five potential risks for this new project. Provide a detailed description of each risk and propose strategies for mitigating each risk.

2. Review a project risk management plan. Did the organization do a good job of answering the questions that should be addressed in a risk management plan? If you cannot find a risk management plan, write one of your own for an information technology project your organization is considering.
3. Research risk management software. Are many products available? What are the main advantages of using them? What are the main disadvantages? Write a paper discussing your findings, and include at least three references.
4. Suppose your organization is deciding which of four projects to bid on. Information on each is in Table 10-9. Assume that all up-front investments are not recovered, so they are shown as negative profits. Draw a diagram and calculate the EMV for each project. Write a few paragraphs explaining which projects you would bid on. Be sure to use the EMV information and your personal risk tolerance to justify your answer.

Table 10-9: Information on Four Potential Projects

PROJECT	CHANCE OF WINNING	ESTIMATED PROFITS
Project 1	50 percent	$120,000
	50 percent	–$50,000
Project 2	30 percent	$100,000
	40 percent	$50,000
	30 percent	–$60,000
Project 3	70 percent	$20,000
	30 percent	–$5,000
Project 4:	30 percent	$40,000
	30 percent	$30,000
	20 percent	$20,000
	20 percent	–$50,000

5. Find an example of a company that took a big risk on an information technology project and succeeded. Also find an example of a company that took a big risk and failed. Summarize each project and situation. Did anything besides luck make a difference between success and failure?

MINICASE

Your college or organization implemented a project, described in Exercise 1, to provide an information system for all employees, students, and customers, allowing them to access and maintain their own human resources-related information, such as address, marital status, and tax information. Many problems arose after implementation. You knew there was a potential risk that people would not use the system, and you've discovered that less than half of all potential users are using it. This low usage percentage has caused bottlenecks in the human resources department, because the staff was reduced to pay for the new system. You also anticipated some security problems if users shared their passwords with friends or family members, but now your organization is facing potential litigation.

Part 1: List four alternatives to address the problem of eligible users not using the system. Describe each option in detail. Include estimates of how long it would take to implement, how much it would cost, and how effective it would be. Also list at least one potential secondary risk that might arise after implementing each option.

Part 2: You have confirmed the fact that the new system has several security problems, including users sharing their passwords with friends or family members. This behavior has resulted in several pranks, such as entries of bogus names and addresses. More serious compromises of security have caused financial and legal problems. One woman is threatening to sue the organization because someone gathered a list of e-mail addresses from the system and sent a mass mailing of a multilevel marketing spam. Write a one- to two-page policy statement that your organization could implement to transfer the risk of these inappropriate uses of the system to the users of the system or to a third party.

Suggested Readings

1. Boehm, Barry W. "Software Risk Management: Principles and Practices." *IEEE Software* (January 1991).

   Boehm describes detailed approaches for several risk management techniques, including risk identification checklists, risk prioritization, risk management planning, and risk monitoring. This article includes an example of Top 10 Risk Item Tracking.

2. Microsoft, White Papers on Microsoft Solutions Framework (MSF), www.microsoft.com/enterpriseservices (2001).

   Microsoft has written several white papers describing their approach to managing risks. The Microsoft Solutions Framework includes three core disciplines, one being risk management. Their "Process of Risk Management" white paper describes techniques that project managers can use to create an environment in which teams successfully identify and manage risks.

4. Project Management Institute. *PM Network* (February 1998).

   This entire issue of the magazine focuses on managing risk in projects. Topics of articles include analyzing risk decisions, using risk-based scheduling and analysis, and encouraging team-based risk assessment.

5. Riskdriver.com.

   This Web site provides information regarding best practices in project risk management, risk assessment tools, and mailing lists to provide the latest news and events related to risk management. The site also links to other Web sites, such as PMI's Risk Management Specific Interest Group and Carnegie Mellon's SEI Web site.

6. Van Scoy, Roger L. "Software Development Risk: Opportunity, Not Problem." *Technical Report CMU/SEI-92-TR-30,* Software Engineering Institute, Carnegie Mellon University (September 1992).

   This report summarizes risk management for software projects. The Software Engineering Institute (SEI) provides this and several other publications on the topic of risk management and other subjects related to software development. More recent articles are available on their Web site at www.sei.cmu.edu.

7. Wideman, R. Max (editor) and Rodney J. Dawson. *Project & Program Risk Management: A Guide to Managing Project Risks and Opportunities.* Upper Darby, PA: Project Management Institute, 1998.

   Wideman has written several books and articles on project risk management. This textbook includes practical advice for managing project risks and opportunities.

8. Williams, Ray C., Julie A. Walker, and Audrey J. Dorofee. "Putting Risk Management into Practice." *IEEE Software* (May/June 1997): 75–82.

   The authors of this article use a road map provided by the Software Engineering Institute (SEI) to discuss effective and ineffective risk management methods based on six years' experience with software-intensive Department of Defense programs. These programs followed the SEI approach of continuous risk management, selecting processes and methods that would work best for their work cultures.

Key Terms

- **Brainstorming** — a technique by which a group attempts to generate ideas or find a solution for a specific problem by amassing ideas spontaneously and without judgment
- **Contingency plans** — predefined actions that the project team will take if an identified risk event occurs
- **Contingency reserves** or **allowances** — provisions held by the project sponsor that can be used to mitigate cost and/or schedule risk, should possible changes in project scope or quality occur
- **Delphi Technique** — an approach used to derive a consensus among a panel of experts, to make predictions about future developments
- **Expected monetary value (EMV)** — the product of the risk event probability and the risk event's monetary value
- **Fallback plans** —plans developed for risks that have a high impact on meeting project objectives, to be implemented if attempts to reduce the risk are not effective
- **Flowcharts** — diagrams that show how various elements of a system relate to each other

- **Influence diagrams** — diagrams that represent decision problems by displaying essential elements, including decisions, uncertainties, and objectives, and how they influence each other
- **Interviewing** —a fact-finding technique that is normally done face-to-face or via telephone
- **Monte Carlo analysis** — a risk quantification technique that simulates a model's outcome many times, to provide a statistical distribution of the calculated results
- **Qualitative risk analysis** — qualitatively analyzing risks and prioritizing their effects on project objectives
- **Quantitative risk analysis** — measuring the probability and consequences of risks and estimating their effects on project objectives
- **Residual risks** — risks that remain after all of the response strategies have been implemented
- **Risk** — the possibility of loss or injury
- **Risk acceptance** — accepting the consequences should a risk occur
- **Risk-averse** — having a low tolerance for risk
- **Risk avoidance** — eliminating a specific threat or risk, usually by eliminating its causes
- **Risk events** — specific circumstances that may occur to the detriment of the project
- **Risk factors** — numbers that represent overall risk of specific events, given their probability of occurring and the consequence to the project if they do
- **Risk identification** — determining which risks are likely to affect a project and documenting the characteristics of each
- **Risk management plan** — a plan that documents the procedures for managing risk throughout the project
- **Risk management planning** —deciding how to approach and plan the risk management activities for a project, by reviewing the project charter, WBS, roles and responsibilities, stakeholder risk tolerances, and the organization's risk management policies and plan templates
- **Risk mitigation** — reducing the impact of a risk event by reducing the probability of its occurrence
- **Risk monitoring and control** —monitoring known risks, identifying new risks, reducing risks, and evaluating the effectiveness of risk reduction throughout the life of the project
- **Risk-neutral** — a balance between risk and payoff
- **Risk response planning** —taking steps to enhance opportunities and reduce threats to meeting project objectives

- **Risk-seeking** — having a high tolerance for risk
- **Risk symptoms** or **triggers** — indications for actual risk events
- **Risk transference** — shifting the consequence of a risk and responsibility for its management to a third party
- **Risk utility** or **risk tolerance** — the amount of satisfaction or pleasure received from a potential payoff
- **Runaway projects** —projects that have significant cost or schedule overruns
- **Secondary risks** —risks that are a direct result of implementing a risk response
- **Top 10 Risk Item Tracking** — a qualitative risk analysis tool for identifying risks and maintaining an awareness of risks throughout the life of a project

11

Project Procurement Management

Objectives

After reading this chapter, you will be able to:

1. *Understand the importance of good procurement management and the increasing use of outsourcing for information technology projects*
2. *Describe the main processes and deliverables of procurement management*
3. *Perform a simple make-or-buy analysis*
4. *Explain the various types of contracts, the risks involved in using each, and provide examples of when each might be used for an information technology project*
5. *Describe the basic contents of a Request for Proposal*
6. *Create and use a proposal evaluation worksheet*
7. *Understand the importance of having good contracts and managing them well*
8. *Discuss types of software available to assist in procurement management*

Marie McBride could not believe how much money her company was paying for outside consultants to help them finish an important operating system conversion project. The consulting company's proposal said they would provide experienced professionals who had completed similar conversions and that the job would be finished in six months or less with four consultants working full-time. Nine months later her company was still paying high consulting fees, and half of the original consultants on the project had been replaced with new people. One new consultant had graduated from college only two months before. Marie's internal staff complained that they were wasting time training some of these "experienced professionals." Marie talked to her company's purchasing manager about the contract, fees, and special clauses that might be relevant to the problems they were experiencing.

Marie was dismayed at how difficult it was to interpret the contract. It was very long and obviously written by someone with a legal background. When she asked what her company could do since the consulting firm was not following their proposal, the purchasing manager stated that the proposal was not part of the official contract. Marie's company was paying for time and materials, not specific deliverables. There was no clause stating the minimum experience level required for the consultants, nor were there penalty clauses for not completing the work on time. There was a termination clause, however, meaning the company could terminate the contract. Marie wondered why her company had signed such a poor contract. Was there a better way to deal with procuring services from outside the company?

IMPORTANCE OF PROJECT PROCUREMENT MANAGEMENT

Procurement means acquiring goods and/or services from an outside source. The term procurement is widely used in the government; many private companies use the term *purchasing*; and information technology professionals use the term *outsourcing*. Organizations or individuals who provide procurement services are referred to as suppliers, vendors, subcontractors, or sellers, with suppliers being the most widely used term. Many information technology projects involve the use of goods and services from outside the organization. In fact, the U.S. market for information technology outsourcing is projected to reach $110 billion by 2003.[1] Because outsourcing is a growing area, it is important for project managers to understand project procurement management.

Many organizations are turning to outsourcing to:

- Reduce both fixed and recurrent costs. Outsourcing suppliers are often able to use economies of scale that may not be available to the client alone. For example, clients can save on information technology hardware and software costs through consolidation of operations.
- Allow the client organization to focus on its core business. Most companies are not in business to provide information technology services, yet many have spent valuable time and resources on information technology functions when they should have focused on core competencies such as marketing, customer service, and new product design. By outsourcing many information technology functions, employees can focus on jobs that are critical to the success of the organization.

[1] Input.com, "Internet-related Spending to Propel the U.S. Outsourcing Market to $110 Billion by 2003, **www.input.com/public/article38.cfm** (2001).

- Access skills and technologies. Organizations can gain access to specific skills and technologies when they are required by using outside resources. For example, a project may require an expert in a particular field or use of expensive hardware or software for one particular month on a project. Planning for this procurement will ensure that the needed skills or technology will be available for the project.
- Provide flexibility. Using outsourcing to provide extra staff during periods of peak workloads can be much more economical than trying to staff entire projects with internal resources.
- Increase accountability. A well-written **contract**—a mutually binding agreement that obligates the supplier to provide the specified products or services and obligates the buyer to pay for them—can clarify responsibilities and sharpen focus on key deliverables of a project. Because contracts are legally binding, there is more accountability for delivering the work as stated in the contract.

Outsourcing of important information technology functions is growing tremendously. For example, many companies outsource their data operations or help desk support to other companies that specialize in those areas. In addition, many companies are outsourcing some of their information technology projects. For example, a company might hire a large consulting firm to facilitate development of a strategic information technology plan. Another company might hire an outside firm to develop and manage their e-commerce applications. These unique needs in an organization can be filled by an outside organization in a finite period of time.

Organizations must also consider reasons they might not want to outsource. When you outsource, you often do not have as much control over those aspects of projects that suppliers carry out. In addition, you could become too dependent on particular suppliers. If those suppliers went out of business or lost key personnel, it could cause great damage to your project. Organizations must also be careful to protect strategic information that could become vulnerable in the hands of suppliers. According to Scott McNeally, CEO of Sun Microsystems, Inc., "What you want to handle in-house is the stuff that gives you an edge over your competition—your core competencies. I call it your 'secret sauce.' If you're on Wall Street and you have your own program for tracking and analyzing the market, you'll hang onto that. At Sun, we have a complex program for testing microprocessor designs, and we'll keep it."[2] Project teams must think carefully about procurement issues and make wise decisions based on the unique needs of their projects and organizations.

The success of many information technology projects that use outside resources is often due to good project procurement management. **Project procurement management** includes the processes required to acquire goods

[2] McNeally, Scott, "The Future of the Net: Why We Don't Want You to Buy Our Software," Executive Perspectives, Sun Microsystems, Inc., **www.sun.com/dot-com/perspectives/stop.html** (1999): 1.

and services for a project from outside the performing organization. The main processes of project procurement management are:

- **Procurement planning**, which involves determining what to procure and when. This process involves deciding what to outsource, determining the type of contract, and creating a statement of work. The project team also creates a procurement management plan as part of the procurement planning process.
- **Solicitation planning**, which involves documenting product requirements and identifying potential sources. This process involves writing procurement documents, such as a Request for Proposal (RFP), and developing evaluation criteria. At the end of the solicitation planning process, the organization often issues an RFP.
- **Solicitation**, which involves obtaining quotations, bids, offers, or proposals as appropriate. This process usually involves finalizing procurement documents, advertising, holding a bidders' conference, and receiving proposals or bids for the work. Occasionally, work is outsourced without a formal solicitation.
- **Source selection**, which involves choosing from among potential suppliers. This process involves evaluating prospective suppliers, negotiating the contract, and awarding the contract.
- **Contract administration**, which involves managing the relationship with the supplier. This process involves monitoring contract performance, making payments, and awarding contract modifications. By the end of the contract administration process, project teams expect that a substantial amount of the contracted work has been completed.
- **Contract close-out**, which involves completion and settlement of the contract, including resolution of any open items. This process usually includes product verification, formal acceptance and closure, and a contract audit.

Figure 11-1 summarizes the major processes involved in procurement management and important milestones that often occur after completing each stage of the process.[3] For example, an important milestone after procurement planning is the make-or-buy decision. A **make-or-buy decision** is one in which an organization decides if it is in their best interests to make certain products or perform certain services inside the organization, or if it is better to procure them from an outside organization. If there is no need to buy any products or services from outside the organization, then there is no need to perform any of the other procurement management processes. At the end of solicitation planning, a key milestone is often the issuance of an RFP. After the solicitation process, the organization would receive proposals. At the end of source selection, a contract would be awarded. After the contract administration process, the supplier would complete a substantial amount of the work described in the contract. At the end of contract close-out, there would be formal acceptance and closure of the contract.

[3] Modified figure based on Rita Mulcahy's *Mastering Contracts for Project Managers* (training materials), 1998.

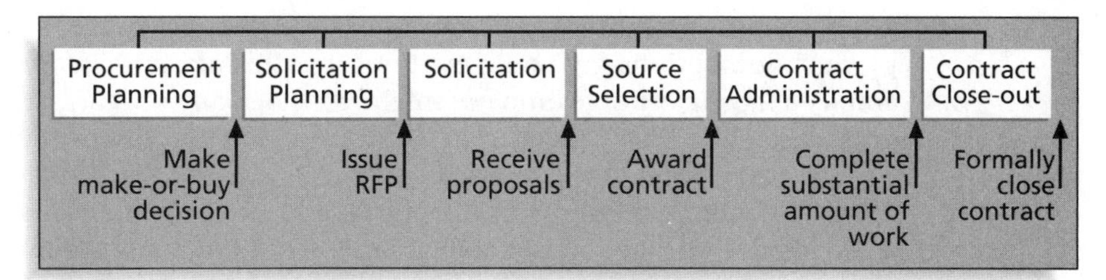

Figure 11-1. Project Procurement Management Processes and Key Outputs

PROCUREMENT PLANNING

Procurement planning is the process of identifying which project needs can be best met by using products or services outside the organization. It involves deciding whether to procure, how to procure, what to procure, how much to procure, and when to procure. For many projects, it is important to be thorough and creative in the procurement planning process. Even though many companies may be viewed as competitors, it often makes sense to collaborate on some projects, as Kodak did in the following example of *What Went Right.*

What Went Right?

Kodak's Advantix Advanced Photo System (APS) project was named the 1997 International Project of the Year by the Project Management Institute. This project was among Kodak's most ambitious ever. For the good of the industry, Kodak worked with five competitors to develop standard specifications for the APS. All parties involved worked together to build contracts and relationships based on shared understandings. The trust, respect, and shared commitment of all of the companies involved in this project led to unique approaches focused on quality and cycle time reduction. For example, the assembly equipment supplier delivered equipment for a given production line on a Friday, installed it over the weekend, and made it operational by Tuesday of the following week. The packaging equipment supplier, who subcontracted components to the other suppliers, set up a trailer on-site for all suppliers to use to coordinate the installation of each packaging flowline. Suppliers worked side by side, relying on their shared vision.[4]

Properly planning procurement can also save organizations millions of dollars. Many companies centralize purchasing for products such as personal computers, software, and printers to earn special pricing discounts. For example, when the Air Force decided to allow for a minimum contract award and flexible pricing strategy for its local online network system for automating fifteen Air Force Systems Command bases in the mid-1980s, the winning supplier lowered its bid by over $40 million.[5]

[4] Adams, Chris, "A Kodak Moment, Advantix Project Named 1997 International Project of the Year," *PM Network* (January 1998): 21–27.

[5] Schwalbe, Kathy, Air Force Commendation Medal Citation, 1986.

Inputs needed for procurement planning include the project scope statement, product descriptions, market conditions, constraints, and assumptions. For example, a large clothing company might consider outsourcing the delivery of, maintenance of, and basic user training and support for laptops supplied to its international sales and marketing force. This decision would be made according to the procurement planning inputs. If there are suppliers who can provide this service well at a reasonable price, it would make sense to outsource, because this could reduce fixed and recurring costs for the clothing company and let them focus on their core business of selling clothes.

It is important to understand why you would want to procure goods or services and what inputs you would need to do procurement planning. In the opening case, Marie's company hired outside consultants to help complete an operating system conversion project because they needed people with specialized skills for a short period of time. This is a common occurrence in many information technology projects. It can be more effective to hire skilled consultants to perform specific tasks for a short period of time than to keep employees on staff full-time.

However, it is also important to clearly define the scope of the project, the products, market conditions, and constraints and assumptions. In Marie's case, the scope of the project and products was fairly clear, but her organization may not have adequately discussed or documented the market conditions or constraints and assumptions involved in using the outside consultants. Were there many companies that provided consultants to do operating conversion projects similar to theirs? Did the project team investigate the background of the company that provided their consultants? Did they list important constraints and assumptions for using the consultants, such as limiting the time that the consultants had to complete the conversion project or the minimum years of experience for any consultant assigned to their project?

Procurement Planning Tools and Techniques

Tools and techniques of procurement management include performing make-or-buy analysis and consulting with experts.

Make-or-buy analysis is a general management technique used to determine whether a particular product or service should be made or performed inside the organization or purchased from someone else. It involves estimating the internal costs of providing a product or service and comparing that estimate to the cost of outsourcing. Consider a company that has 1,000 international salespeople who require laptops with maintenance, training, and user support. Using make-or-buy analysis, the company would estimate what it would cost to provide this product and service using only internal resources. The estimate would need to include costs for hardware and software, travel, shipping, and technical support. Suppose the company estimated an initial investment of $3 million and annual support costs of $2 million. Note that a make-or-buy analysis must include the life cycle cost, as described in Chapter 6, *Project Cost Management*. If the company received supplier quotes that were less than their internal estimates, they should definitely consider outsourcing the project.

Internal experts should be consulted as part of procurement planning. They might suggest that the company could not provide quality maintenance, training, and service for the laptops since the service involves so many people with different skill levels in so many different locations. Experts in the company might also know that most of their competitors outsource this type of work and know who the qualified outside suppliers are. Experts outside the company, including potential suppliers themselves, can also provide expert judgment. For example, suppliers might suggest an option for salespeople to purchase the laptops themselves at a reduced cost. This option would solve problems of employee turnover—exiting employees would own their laptops and new employees would purchase a laptop through the program. An internal expert might then suggest that employees receive a technology bonus to help offset what they might view as an added expense. Expert judgment, both internal and external, is an asset in making procurement decisions.

Types of Contracts

Contract type is an important consideration. Different types of contracts can be used in different situations. Four broad categories of contracts are fixed price or lump sum, cost reimbursable, time and material, and unit price. Descriptions of each contract type and examples for information technology projects follow.

Fixed price or **lump sum contracts** involve a fixed total price for a well-defined product or service. The buyer incurs little risk in this situation. For example, a company could award a fixed price contract to purchase 100 laser printers with a certain print resolution and print speed to be delivered to one location within two months. In this example, the product and delivery date are well defined. Fixed price contracts may also include incentives for meeting or exceeding selected project objectives. For example, the contract could include an incentive fee paid if the laser printers are delivered within one month. A firm-fixed price (FFP) contract has the least amount of risk for the buyer, followed by a fixed price incentive (FPI) contract.

Cost reimbursable contracts involve payment to the supplier for direct and indirect actual costs. Recall from Chapter 6 that direct costs are costs that are related to a project and can be traced back in a cost-effective way. Indirect costs are costs related to the project that cannot be traced back in a cost-effective way. For example, the salaries for people working directly on a project and hardware or software purchased for a specific project are direct costs, while the cost of providing a work space with electricity, a cafeteria, and so on are indirect costs. Indirect costs are often calculated as a percentage of direct costs. Cost reimbursable contracts often include fees such as a profit percentage or incentives for meeting or exceeding selected project objectives. These contracts are often used for projects that include providing goods and services that involve new technologies. The buyer absorbs more of the risk with cost reimbursable contracts than they do with fixed price contracts. Three types of cost reimbursable contracts, in order of lowest to highest risk to the buyer, include cost plus incentive fee, cost plus fixed fee, and cost plus percentage of costs.

- With a **cost plus incentive fee (CPIF) contract**, the buyer pays the supplier for allowable performance costs along with a predetermined fee and an incentive bonus. If the final cost is less than the expected cost, both the buyer and the supplier benefit from the cost savings, according to a prenegotiated share formula. For example, suppose the expected cost of a project is $100,000, the fee to the supplier is $10,000, and the share formula is 85/15, meaning that the buyer absorbs 85 percent of the uncertainty and the supplier absorbs 15 percent. If the final price is $80,000, the cost savings are $20,000. The supplier would be paid the final cost and the fee plus an incentive of $3,000 (15 percent of $20,000), for a total reimbursement of $93,000.
- With a **cost plus fixed fee (CPFF) contract**, the buyer pays the supplier for allowable performance costs plus a fixed fee payment usually based on a percentage of estimated costs. This fee does not vary, however, unless the scope of the contract changes. For example, suppose the expected cost of a project is $100,000 and the fixed fee is $10,000. If the actual cost of the contract rises to $120,000 and the scope of the contract remains the same, the contractor will still receive the fee of $10,000.
- With a **cost plus percentage of costs (CPPC) contract**, the buyer pays the supplier for allowable performance costs along with a predetermined percentage based on total costs. From the buyer's perspective, this is the least desirable type of contract because the supplier has no incentive to decrease costs. In fact, the supplier may be motivated to increase costs, since doing so will automatically increase profits based on the percentage of costs. This type of contract is prohibited for federal government use, but is sometimes used in private industry, particularly in the construction industry. All of the risk is borne by the buyer.

Time and material contracts are a hybrid of both fixed price and cost reimbursable contracts. For example, an independent computer consultant might have a contract with a company based on a fee of $80 per hour for his or her services plus a fixed price of $10,000 for providing specific materials for the project. The materials fee might also be based on approved receipts for purchasing items, with a ceiling of $10,000. The consultant would send an invoice to the company each week or month, listing the materials fee, the number of hours worked, and a description of the work produced. This type of contract is often used for services that are needed when the work cannot be clearly specified and total costs cannot be estimated in a contract. Many contract programmers and consultants, such as those Marie's company hired in the opening case, prefer time and material contracts.

Unit price contracts require the buyer to pay the supplier a predetermined amount per unit of service, and the total value of the contract is a function of the quantities needed to complete the work. Consider an information technology department that might have a unit price contract for purchasing computer hardware. If the company purchases only one unit, the cost might be $1,000. If it purchases ten units, the cost would be $10,000. This type of contract often involves volume discounts. For example, if the company purchases between ten and fifty units, the contracted cost might be $900 per unit. If it purchases over fifty units, the cost might go down to $800 per unit.

Any type of contract should include specific clauses that take into account issues unique to the project. For example, if a company uses a time and material contract for consulting services, the contract should stipulate different hourly rates based on the level of experience of each individual contractor. The services of a junior programmer with no bachelor's degree and less than three years' experience might be billed at $40 per hour, whereas the services of a senior programmer with at least a bachelor's degree and over ten years of experience might be billed at $80 per hour.

Figure 11-2 summarizes the spectrum of risk to the buyer and supplier for different types of contracts. Buyers have the lowest risk with firm-fixed price contracts because they know exactly what they will need to pay the supplier. Buyers have the most risk with cost plus percentage of costs contracts because they do not know what the supplier's costs will be in advance, and the suppliers may be motivated to keep increasing costs. From the supplier's perspective, there is the least risk with a cost plus percentage of costs contract and the most risk with the firm-fixed price contract.

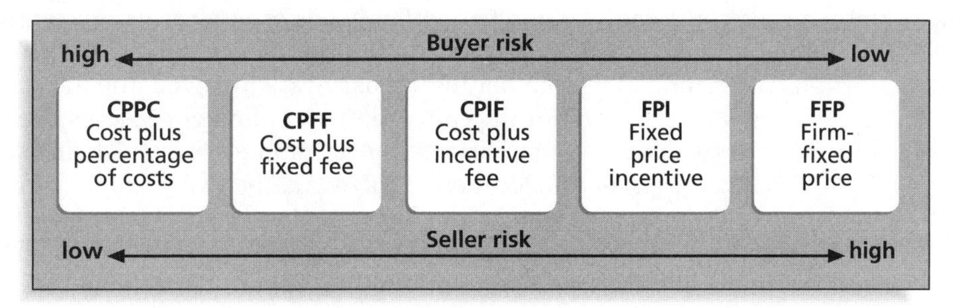

Figure 11-2. Contract Types Versus Risk

Time and material and unit price contracts can be viewed as high- or low-risk, depending on the nature of the project and other contract clauses. For example, if an organization is unclear on what work needs to be done, it cannot expect a supplier to sign a firm-fixed price contract. However, the buyer could find a consultant or group of consultants to work on specific tasks based on a predetermined hourly rate. The buying organization could evaluate the work produced every day or week to decide if it wants to continue using the consultants. In this case the contract would include a **termination clause**—a contract clause that allows the buyer or supplier to end the contract. Some termination clauses state that the buyer can terminate a contract for any reason and give the supplier only 24 hours' notice. Suppliers must often give a one-week notice to terminate a contract and must have sufficient reasons for the termination. The buyer could also include a contract clause specifying hourly rates based on education and experience of consultants. These contract clauses reduce the risk incurred by the buyer while providing flexibility for accomplishing the work.

Statement of Work

Many contracts often include a statement of work. The **statement of work (SOW)** is a description of the work required for the procurement. The SOW is a type of scope statement that describes the work in sufficient detail to allow prospective suppliers to determine if they are capable of providing the goods and services required and to determine an appropriate price. An SOW should be clear, concise, and as complete as possible. It should describe all services required and include performance reporting. It is important to use appropriate words in an SOW such as *must* instead of *may*. For example, *must* implies that something has to be done; *may* implies that there is a choice involved in doing something or not. The SOW should specify the product of the project, use industry terms, and refer to industry standards.

Many organizations use samples and templates to generate SOWs. Figure 11-3 provides a template for a SOW that Marie's organization could use when they hire outside consultants or purchase other goods or services. For example, for the operating system conversion project, Marie's company should specify the specific manufacturer and model number for the hardware involved, the former operating systems and new ones for the conversion, the number of pieces of each type of hardware involved (mainframes, midrange computers, or PCs), and so on. The SOW should also specify the location of the work, the expected period of performance, specific deliverables and when they are due, applicable standards, acceptance criteria, and special requirements. A good SOW gives bidders a better understanding of the buyer's expectations. A SOW can and should become part of the official contract to ensure that the buyer gets what the supplier bid on.

Statement of Work (SOW)

I. **Scope of Work:** Describe the work to be done in detail. Specify the hardware and software involved and the exact nature of the work.

II. **Location of Work:** Describe where the work must be performed. Specify the location of hardware and software and where the people must perform the work.

III. **Period of Performance:** Specify when the work is expected to start and end, working hours, number of hours that can be billed per week, where the work must be performed, and related schedule information.

IV. **Deliverables Schedule:** List specific deliverables, describe them in detail, and specify when they are due.

V. **Applicable Standards:** Specify any company or industry-specific standards that are relevant to performing the work.

VI. **Acceptance Criteria:** Describe how the buyer organization will determine if the work is acceptable.

VII. **Special Requirements:** Specify any special requirements such as hardware or software certifications, minimum degree or experience level of personnel, travel requirements, and so on.

Figure 11-3. Statement of Work (SOW) Template

SOLICITATION PLANNING

Solicitation planning involves preparing the documents needed for solicitation and determining the evaluation criteria for the contract award. Two common examples of solicitation documents include Requests for Proposal (RFPs) and Requests for Quotes (RFQs). A **Request for Proposal (RFP)** is a document used to solicit proposals from prospective suppliers. Many organizations issue RFPs to potential suppliers. For example, if a government department wants to automate its work practices, it writes and issues an RFP so suppliers can respond with proposals. Suppliers might propose various hardware, software, and networking solutions to meet the government's need. A **Request for Quote (RFQ)** is a document used to solicit quotes or bids from prospective suppliers. Organizations often use an RFQ for solicitations that involve specific items. For example, if the government wanted to purchase 100 personal computers with specific features, it might issue an RFQ to potential suppliers. RFQs usually do not take nearly as long to prepare as RFPs, nor do responses to them.

Writing a good RFP is a critical part of project procurement management. Many people have never had to write or respond to an RFP. To generate a good RFP, expertise is invaluable. Many examples of RFPs are available within different companies, from potential contractors, and from government agencies. There are often legal requirements involved in issuing RFPs and reviewing proposals, especially for government projects. It is important to consult with experts familiar with the solicitation planning process for particular organizations. To make sure the RFP has enough information to provide the basis for a good proposal, the buying organization should try to put themselves in the suppliers' shoes. Could you develop a good proposal based on the information in the RFP? Could you determine detailed pricing and schedule information based on the RFP? Developing a good RFP is difficult, as is writing a good proposal.

Figure 11-4 provides a basic outline for an RFP. The main sections of an RFP usually include a statement of the purpose of the RFP, background information on the organization issuing the RFP, the basic requirements for the products and/or services being proposed, the hardware and software environment (usually important information for information technology-related proposals), a description of the RFP process, the statement of work and schedule information, and possible appendices. A fairly simple RFP might be three to five pages long, while an RFP for a larger, more complicated procurement might be hundreds of pages long.

Request for Proposal Outline

I. Purpose of RFP

II. Organization's Background

III. Basic Requirements

IV. Hardware and Software Environment

V. Description of RFP Process

VI. Statement of Work and Schedule Information

VII. Possible Appendices

A. Current System Overview

B. System Requirements

C. Volume and Size Data

D. Required Contents of Vendor's Response to RFP

E. Sample Contract

Figure 11-4. Outline for a Request for Proposal (RFP)

Other documents often used in solicitation planning include invitations for bid, invitations for negotiation, and initial contractor responses. All solicitation documents should be written to facilitate accurate and complete responses from prospective suppliers. They should include background information on the organization and project, the relevant statement of work, a schedule, a description of the desired form of response, evaluation criteria, pricing forms, and any required contractual provisions. They should also be rigorous enough to ensure consistent, comparable responses, but flexible enough to allow consideration of supplier suggestions for better ways to satisfy the requirements.

It is very important for organizations to prepare some form of evaluation criteria, preferably before they issue a formal RFP or RFQ. Organizations use criteria to rate or score proposals, and they often assign a weight to each criterion to indicate how important it is. Some examples of criteria include the technical approach (30 percent weight), management approach (30 percent weight), past performance (20 percent weight), and price (20 percent weight). The criteria should be specific and objective. For example, if the buyer wants the supplier's project manager to be a certified Project Management Professional (PMP), that requirement should be stated clearly in the procurement documents and followed during the award process. Losing bidders may pursue legal recourse if the buyer does not follow a fair and consistent evaluation process.

INPUT, a Web-based IT market research and marketing services firm, provides market reports and buyer guides to assist organizations in information technology outsourcing. One buyer guide offers suggestions on selection criteria, which should include reputation and past performance, industry knowledge, strategic partnership, and ability to meet needs. INPUT's research shows that most contracts include provisions to safeguard against unsatisfactory partnering, but most companies hesitate to exercise the provisions. Therefore, buyers should look for a supplier with a solid record of excellence and a proven reputation for quality.[6]

Organizations should heed the saying, "Let the buyer beware." It is critical to evaluate proposals based on more than the appearance of the paperwork submitted. A key factor in evaluating bids, particularly for projects involving information technology, is the past performance record of the bidder. The RFP should require bidders to list other similar projects they have worked on and provide customer references for those projects. Reviewing performance records and references helps to reduce the risk of selecting a supplier with a poor track record. Suppliers should also demonstrate their understanding of the buyer's need, their technical and financial capabilities, their management approach to the project, and their price for delivering the desired goods and services.

Some information technology projects also require potential suppliers to deliver a technical presentation as part of their proposal. The proposed project manager should lead the potential supplier's presentation team. When the outside project manager leads the proposal presentation, the organization can begin

[6] INPUT, "Executive Guide to IT Outsourcing in the U.S.", **www.input.com** (March 2000).

building a relationship with the potential provider from the beginning. Visits to contractor sites can also help the buyer get a better feeling for the supplier's capabilities and management style.

SOLICITATION

Solicitation involves obtaining proposals or bids from prospective suppliers. Prospective suppliers do most of the work in this process, normally at no cost to the buyer or project. The buying organization is responsible for advertising the solicitation, and often holds some sort of bidders' conference to answer questions about the solicitation. The main output of this process is receipt of proposals or bids.

Organizations can advertise to procure outside goods and services in many different ways. Sometimes a specific supplier might be the number one choice for the buyer. In this case, the buyer gives solicitation information to just that company. If the preferred supplier responds favorably, both organizations proceed to work together. Many companies have formed good working relationships with certain suppliers, so they want to continue working with them.

In many cases, however, there may be more than one supplier qualified to provide the goods and services. Providing information and receiving bids from multiple sources often takes advantage of the competitive business environment. As a result of pursuing a competitive bidding strategy, the buyer receives better goods and services than expected at a lower price.

A bidders' conference, also called a supplier conference or pre-bid conference, is a meeting with prospective suppliers prior to preparation of a proposal. These conferences help ensure that everyone has a clear, common understanding of the buyer's desired products or services. Before, during, or after the conference, the buyer may incorporate responses to questions into the procurement documents as amendments.

SOURCE SELECTION

Once buyers receive proposals, they must select a supplier or decide to cancel the procurement. Source selection involves evaluating bidders' proposals, choosing the best one, negotiating the contract, and awarding the contract. It is often a long, tedious process. Stakeholders in the procurement process should be involved in selecting the best supplier for the project. Often, teams of people are responsible for evaluating various sections of the proposals. There might be a technical team, a management team, and a cost team to focus on each of those major areas. Often buyers develop a short list of the top three to five suppliers to reduce the work involved in selecting a source.

Experts in source selection highly recommend that buyers use formal proposal evaluation sheets during source selection. Figure 11-5 provides a sample proposal evaluation sheet that the project team might use to help create a short list of the best three to five proposals. Experts also recommend that technical criteria should not be given more weight than management or cost criteria. Many organizations have suffered the consequences of paying too much attention to the technical aspects of proposals. For example, the project might cost much more than expected or take longer to complete because the source selection team focused only on technical aspects of proposals. Paying too much attention to technical aspects of proposals is especially likely to occur on information technology projects. However, it is often the supplier's management team—not the technical team—that makes procurement successful.

		Proposal 1		Proposal 2		Proposal 3	
Criteria	**Weight**	**Rating**	**Score**	**Rating**	**Score**	**Rating**	**Score**
Technical Approach	30%						
Management Approach	30%						
Past Performance	20%						
Price	20%						
Total Score	100%						

Figure 11-5. Sample Proposal Evaluation Sheet

After developing a short list of possible suppliers, organizations often follow a more detailed proposal evaluation process. Figure 11-6 lists items that might be part of an evaluation of the top three suppliers for a large information technology project. This particular list focuses on the project management capabilities of each supplier. Notice that the criteria include the project manager's educational background and PMP certification, his/her presentation (meaning the suppliers had to give a formal presentation as part of the evaluation process), and the organization's project management methodologies. All of these criteria are assigned a certain number of possible points, and the project team members and other stakeholders performing the evaluation would then assign points to each of the supplier finalists for each criterion. In the example in Figure 11-6, Supplier 3 has the highest score (28 out of 30 points) based on the management approach criteria for the award. Similiar scores could be assigned for the selection criteria. The supplier with the most points based on all of the criteria for each category, therefore, should be offered the award.

Criteria	Possible Points	Supplier 1 Points	Supplier 2 Points	Supplier 3 Points
Project manager's educational background and experience	10	8	6	9
Project manager is PMP certified	5	5	0	5
Presentation on management approach	5	4	3	5
Organization's project management methodology	10	7	4	9
Total Score	30	24	13	28

Figure 11-6. Detailed Criteria for Selecting Suppliers

It is customary to have contract negotiations during the source selection process. Suppliers on the short list are often asked to prepare a best and final offer (BAFO). People who negotiate contracts for a living often conduct these negotiations for contracts involving large amounts of money. In addition, senior managers from both the buying and selling organizations often meet before making final decisions. The final output from the source selection process is a contract that obligates the supplier to provide the specified products or services and obligates the buyer to pay for them.

What Went Wrong?

The growth in information technology outsourcing may be one of the reasons for the creation and subsequent failure of several dot-com companies. Hundreds of dot-coms started in the late 1990s and early 2000s to provide products and services such as Internet hosting, e-commerce applications, and online publications. However, over 130 Internet companies and over 26 business-to-business companies shut down in 2000.[7] Several Web sites, such as "whytheyfailed.com,"attempt to track dot-com failures. Kevin Scully, President and CEO of a successful dot-com, Webcritical Technologies, suggests that most dot-com failures were due to poor business planning, lack of senior management operations experience, lack of leadership, and lack of vision.[8] Buyers should definitely consider the business stability of suppliers when making source selection decisions, and many dot-coms failed because buyers were not willing to take the risk of doing business with unstable dot-coms. Selecting a stable company does not guarantee procurement success, however. Department 56, Inc., a giftware company In Minnesota, filed a $6 billion lawsuit against Arthur

[7] Davis, Jessica, "Dot-com Failures Abound," **www.infoworld.com** (November 16, 2000).

[8] Scully, Kevin, "There Are Solid Reasons for Dot.com Failures, "Washington Techway, **www.washtech.com** (October 23, 2000).

Andersen Worldwide in March 2001. Department 56, Inc. claimed that the well-known consulting firm badly managed the overhaul of their computer systems in 1999. Arthur Andersen assured Department 56 officials that the new system would not cost more than $3.3 million, but costs ultimately exceeded $14 million. Computer system problems allegedly required Department 56 to spend an additional $4 million to hire other consultants to fix the problems. In addition, Department 56 allegedly lost over $8 million in revenues, took extraordinary charges to earnings, had bad debt of $18 million, and spent almost two years dealing with computer problems instead of focusing on company growth. Andersen claims the lawsuit was retaliation for a lawsuit they had filed five months earlier against Department 56 for nonpayment.[9]

CONTRACT ADMINISTRATION

Contract administration ensures that the supplier's performance meets contractual requirements. The contractual relationship is a legal relationship, and as such it is subject to state and federal contract laws. It is very important that appropriate legal and contracting professionals be involved in writing and administering contracts. Good contracts and contract administration can help organizations avoid problems such as those experienced by Department 56, Inc. and Arthur Andersen Worldwide, described in *What Went Wrong*.

Many project managers know very little about contract administration. Many technical professionals would prefer not to look at contracts at all. However, many project managers and technical people do not understand or trust their own purchasing departments, if their organization even has one, because of problems they may have encountered in the past. Ideally, the project manager and his or her team should be actively involved in writing and administering the contract so that everyone understands the importance of good procurement management. The project team should also seek expert advice in working with contractual issues.

Project team members must be aware of potential legal problems they might cause by not understanding a contract. For example, most projects involve changes, and these changes must be handled properly for items under contract. Without understanding the provisions of the contract, a project manager may not realize he or she is authorizing the contractor to do additional work at additional costs. Therefore, change control is an important part of the contract administration process.

Following are suggestions to ensure adequate change control on projects that involve outside contracts:

- Changes to any part of the project need to be reviewed, approved, and documented by the same people in the same way that the original part of the plan was approved.

[9] Moore, Janet, "Department 56 Sues Arthur Andersen for $6 Billion Over Computer System," *Minneapolis Star Tribune* (March 2, 2001).

- Evaluation of any change should include an impact analysis. How will the change affect the scope, time, cost, and quality of the goods or services being provided? There must also be a baseline to understand and analyze changes.
- Changes must be documented in writing. Project team members should document all important meetings and telephone calls.

When a project involves outside contracts, it is critical that project managers and team members watch for constructive change orders. **Constructive change orders** are oral or written acts or omissions by someone with actual or apparent authority that can be construed to have the same effect as a written change order. For example, if a member of the buyer's project team has met with the contractor on a weekly basis for three months to provide guidelines for performing work, he or she can be viewed as an apparent authority. If he or she tells the contractor to redo part of a report that has already been delivered and accepted by the project manager, that action can be viewed as a constructive change order and the contractor can legally bill the buyer for the additional work. Likewise, if this apparent authority tells the contractor to skip parts of a critical review meeting in the interests of time, the omission of that information is not the contractor's fault.

CONTRACT CLOSE-OUT

The final process in project procurement management is contract close-out. Contract close-out includes product verification to determine if all work was completed correctly and satisfactorily. It also includes administrative activities to update records to reflect final results and archiving information for future use. Procurement audits are often done during contract close-out to identify lessons learned in the procurement process. Outputs from contract close-out include a contract file and formal acceptance and closure. The organization responsible for contract administration should provide the supplier with formal written notice that the contract has been completed. The contract itself should include requirements for formal acceptance and closure.

USING SOFTWARE TO ASSIST IN PROJECT PROCUREMENT MANAGMENT

Over the years, organizations have used various types of productivity software to assist in project procurement management. For example, most organizations use word-processing software to write proposals or contracts, spreadsheet software to create proposal evaluation worksheets, databases to track suppliers, and presentation software to present procurement-related information.

It wasn't until the late 1990s and early 2000s, however, that many companies started using more advanced software to assist in procurement management. In fact, the term "e-procurement" is often used to describe various procurement functions that are now done electronically. "The purchasing function was neglected for many, many years and has only recently begun to be viewed as strategic," says Hank Adamany, of PricewaterhouseCoopers. "Purchasing was a place where you had people who spent 20 to 25 years in positions below the senior team, and as a result the department never got much attention or investment. That's odd when you consider that those departments can account for up to 70 percent of a company's total spending."[10]

Many different Web sites and software tools are now available to assist in various procurement functions. For example, most college students have discovered Web sites to help them purchase books, airline tickets, or even automobiles. Likewise, many businesses can now access free Web-based portals for purchasing numerous items, or they can buy more specialized software to help streamline their procurement activities. Companies such as Commerce One, Ariba, Concur Technologies, and others have started providing corporate procurement services over the Internet in the past few years. Other established companies, such as SAS and Baan, have developed new software products to assist in procurement management. Traditional procurement methods were very inefficient and costly, and new e-procurement services have proved to be very effective in reducing the costs and burdens of indirect procurement. As Hank Adamany of PricewaterhouseCoopers said, "In fact, most user organizations that we surveyed were able to realize more than a 300 percent return on investment in Internet procurement automation within the first year of deployment."[11]

Organizations can also take advantage of information available on the Web, in industry publications, or in various discussion groups offering advice on selecting suppliers. For example, many organizations invest millions of dollars in enterprise resource planning (ERP) systems. ERP systems attempt to integrate several business functions, such as manufacturing, finance, human resources, sales, and distribution across an enterprise into a single computer system that can serve all those different departments' particular needs. Before deciding which ERP suppliers to use, you can find information describing specific products provided by various suppliers, prices, and current customer information to assist in making procurement decisions.

As with any information or software tool, organizations must focus on using the information and tools to meet project and organizational needs. Many nontechnical issues are often involved in getting the most value out of new technologies, especially new e-procurement software. For example, companies must often develop partnerships and strategic alliances with other firms to take

[10] *Fortune*, "What Tools These Portals Be," www.fortune.com/fortune/sections/eprocurement (January 16, 2001).

[11] Ibid.

advantage of potential cost savings. Companies should practice good procurement management in selecting new software tools and managing relationships with the chosen suppliers.

The processes involved in project procurement management follow a clear, logical sequence. However, many project managers are not familiar with the many issues involved in purchasing goods and services from other organizations. If projects will benefit by procuring goods or services, then project managers and their teams must follow good project procurement management. As outsourcing for information technology projects increases, it is important for all project managers to have a fundamental understanding of this knowledge area.

CASE WRAP-UP

After reading the contract for their consultants carefully, Marie McBride found a clause giving her company the right to terminate the contract with a one-week notice. She met with her project team to get their suggestions. They still needed help completing the operating system conversion project. One team member had a friend who worked for a competing consulting firm. The competing consulting firm had experienced people available, and their fees were lower than the fees in the current contract. Marie asked this team member to help her research other consulting firms in the area who could work on the operating system conversion project. Marie then requested bids from these companies. She personally interviewed people from the top three suppliers' management teams and checked their references for similar projects.

Marie worked with the purchasing department to terminate the original contract and issue a new one with a new consulting firm that had a much better reputation and lower hourly rates. This time, she made certain the contract included a statement of work, specific deliverables, and requirements stating the minimum experience level of consultants provided. The contract also included incentive fees for completing the conversion work within a certain time period. Marie learned the importance of good project procurement management.

CHAPTER SUMMARY

Procurement, purchasing, or outsourcing, is acquiring goods and/or services from an outside source. Experts predict that information technology outsourcing will grow to over $110 billion by the year 2003. Organizations outsource to reduce costs, focus on their core business, access skills and technologies, provide flexibility, and increase accountability. It is becoming increasingly important for information technology professionals to understand project procurement management.

Project procurement management processes include procurement planning, solicitation planning, solicitation, source selection, contract administration, and contract close-out.

Procurement planning involves deciding what to procure or outsource, what type of contract to use, and how to describe the effort in a statement of work. A make-or-buy analysis helps an organization determine whether it can cost-effectively procure a product or service. Project managers should consult internal and external experts to assist them with procurement planning because many legal, organizational, and financial issues are often involved.

The basic types of contracts are fixed price, cost reimbursable, time and material, and unit price contracts. Fixed price contracts involve a fixed total price for a well-defined product and entail the least risk to buyers. Cost reimbursable contracts involve payments to suppliers for direct and indirect actual costs and require buyers to absorb some of the risk. Time and material contracts are a hybrid of fixed price and cost reimbursable contracts and are commonly used by consultants. Unit price contracts involve paying suppliers a predetermined amount per unit of service and impose different levels of risk on buyers, depending on how the contract is written. It is important to decide which contract type is most appropriate for a particular procurement. All contracts should include specific clauses that address unique aspects of a project and that describe termination requirements.

A statement of work describes the work required for the procurement in enough detail to allow prospective suppliers to determine if they are capable of providing the goods and services and to determine an appropriate price.

Solicitation planning involves writing procurement documents such as a Request for Proposal (RFP) and developing source selection evaluation criteria.

Solicitation involves finalizing procurement documents, advertising the work, holding a bidders' conference, and receiving proposals or bids.

Source selection is the process used to evaluate prospective suppliers and negotiate a contract. Organizations should use a formal proposal evaluation form when evaluating suppliers. Technical criteria should not be given more weight than management or cost criteria during evaluation.

Contract administration involves finalizing and awarding a contract, monitoring performance, and awarding contract modifications. The project manager and key team members should be involved in writing and administering the contract.

Project managers must be aware of potential legal problems they might cause when they do not understand a contract. Project managers and teams should use careful change control procedures when working with outside contracts and should be especially careful about constructive change orders.

Contract close-out includes product verification and administrative close-out, and outputs include a contract file, formal acceptance, and closure. Procurement audits identify lessons learned during the procurement process.

Several types of software can assist in project procurement management. A new category of e-procurement software has been developed in the past few years to help organizations save money in procuring various goods and services. Organizations can also use the Web, industry publications, and discussion groups to research and compare various suppliers.

Discussion Questions

1. Discuss the scenario in the opening case. Have you experienced similar situations? How did the parties involved handle them?
2. Provide examples of information technology goods and services that were outsourced. Which were for information technology projects and which were parts of ongoing operations? Was it advantageous for the organization to use outsourcing?
3. Discuss the make-or-buy process. What nonfinancial factors should be considered in make-or-buy decisions?
4. Suppose you decided to become an independent information technology consultant. What type of contract would you prefer for your services and why? Suppose you formed a business and had ten other consultants working for you. What type of contract would you use for your employees? What type of contract would you use for your customers? What are the advantages and disadvantages of each?
5. Do you think many information technology professionals have experience writing RFPs and evaluating proposals for information technology projects? What skills would be useful for these tasks?
6. Discuss examples of source selection for information technology procurements and the contract administration process. Do you think that most organizations have good processes in place for these functions?
7. How can software assist in procuring goods and services? What is e-procurement software?

Exercises

1. Research information on information technology outsourcing. Find at least three articles and summarize them. Answer the following questions:

- What are the main types of goods and services being outsourced?
- Who are some of the largest companies that provide information technology outsourcing services?
- Why is outsourcing growing so rapidly?
- Have most organizations benefited from outsourcing? Why or why not?

2. Interview someone who was involved in an information technology procurement process and have him or her explain the process that was followed. Write a paper describing the procurement and any lessons learned by the organization.
3. Suppose your company is trying to decide whether it should buy special equipment to prepare some of its high-quality publications itself or lease the equipment from another company. Suppose leasing the equipment costs $240 per day. If you decide to purchase the equipment, the initial investment is $6,800, and operations will cost $70 per day. After how many days will the lease cost be the same as the purchase cost for the equipment? Assume your company would only use this equipment for thirty days. Should your company buy the equipment or lease it?
4. Find an example of a contract for information technology services. Analyze the key features of the contract. What type of contract was used and why? Review the language and clauses in the contract. What are some of the key clauses? List questions you have about the contract and try to get answers from someone familiar with the contract.
5. Draft an RFP for an information technology project to provide laptops for all students, faculty, and staff at your college or university. Use the outline provided in Figure 11-4. List the assumptions you made in preparing the RFP.
6. Draft the source selection criteria that might be used for evaluating proposals for providing laptops for all students, faculty, and staff at your college or university. Use Figure 11-6 as a guide. Include at least ten criteria, and make the total scores add up to 100. Write a paper justifying the criteria and scores.

MINICASE

Your company is reevaluating its policies concerning cellular telephones, and its approach to providing cellular telephone services to its employees. All 1,500 consultants and salespeople are authorized to use their choice of cellular telephone services with unlimited calling privileges, and these expenses have increased dramatically. Several employees are using new features such as Internet access via their cell phones. To provide for new capabilities and to benefit from a corporate discount, your company is issuing a request for proposals to provide cellular telephone services for the entire company.

Part 1. Research at least three companies that provide cellular telephone services. Write a paragraph describing the purpose of the Request for Proposal that your company plans to issue and another paragraph describing the basic requirements. Write these paragraphs so that at least three potential suppliers would meet the basic requirements. Then develop a list of criteria that your company could use for evaluating a cellular telephone services supplier. Include at least ten criteria, and assign weights to each criterion to total 100%. Create a spreadsheet to accompany your proposal evaluation sheet, using Figure 11-5 as a guide.

Part 2: You want to include special clauses in your contract for cellular telephone services to provide for the following:

- To ensure that your company is getting the best rates possible. Assume that you will commit to a one-year renewable contract and a minimum of 1,000 cell phones.
- To be able to make special features, such as Internet access, available to only certain employees
- To receive special reports identifying employees who have excessive charges

Write a paragraph for each of these clauses, using appropriate contractual language.

Suggested Readings

1. *Fortune*, E-procurement (May 2, 2000), www.fortune.com/fortune/sections/, scroll down and click E-procurement (accessed on March 15, 2001).

   This Web site provides several articles related to e-procurement and how companies are using various technologies to transform corporate purchasing. Topics include portals to Internet procurement, supply chain management, and trading communities. Links are provided to several companies that supply e-procurement services.

2. Klepper, Robert, and Wendell Jones. *Outsourcing Information Technology, Systems, and Services*, Upper Saddle River, NJ: Prentice Hall, 1997.

   Klepper and Jones based this book on real-world experience and lessons learned from information technology executives, service providers, outsourcing consultants, attorneys, and others who have successfully and unsuccessfully outsourced. The book includes many lists, charts, facts, figures, and sample documents such as RFPs and contracts to guide you through outsourcing decisions and other processes.

3. Martin, Martin D., C. Claude Teagarden, and Charles F. Lambreth. *Contract Administration for the Project Manager*. Upper Darby, PA: Project Management Institute, 1990.

   This booklet provides basic information on contract types as well as practical advice for project managers on administering contracts.

4. Project Management Institute. *PM Network,* Special Topic Issue: Service and Outsourcing Projects (October 1998).

The special topic of this magazine is outsourcing. Articles include the growth in outsourcing, using tiger teams (an integrated team of subject matter experts) on outsourcing projects, developing a statement of work, enhancing supplier relationships, and financial constraints on outsourcing projects.

5. Ripin, Kathy M. and Leonard R. Sayles. *Insider Strategies for Outsourcing Information Systems: How to Build Productive Partnerships and Avoid Seductive Traps*. Oxford University Press, New York: 1999.

This book provides advice for negotiating with potential suppliers and facilitating information technology development projects. The authors provide several case studies to illustrate how client managers obtain new skills through project participation that enable them to make more effective use of new technologies.

Key Terms

- **Constructive change orders** — oral or written acts or omissions by someone with actual or apparent authority that can be construed to have the same effect as a written change order
- **Contract** — a mutually binding agreement that obligates the supplier to provide the specified products or services, and obligates the buyer to pay for them
- **Contract administration** — managing the relationship with the supplier
- **Contract close-out** — completion and settlement of the contract, including resolution of any open items
- **Cost plus fixed fee (CPFF) contract** — a contract in which the buyer pays the supplier for allowable performance costs plus a fixed fee payment usually based on a percentage of estimated costs
- **Cost plus incentive fee (CPIF) contract** — a contract in which the buyer pays the supplier for allowable performance costs along with a predetermined fee and an incentive bonus
- **Cost plus percentage of costs (CPPC) contract** — a contract in which the buyer pays the supplier for allowable performance costs along with a predetermined percentage based on total costs
- **Cost reimbursable contracts** — contracts involving payment to the supplier for direct and indirect actual costs
- **Fixed price** or **lump sum contracts** — contracts with a fixed total price for a well-defined product or service
- **Make-or-buy decision** — when an organization decides if it is in their best interests to make certain products or perform certain services inside the organization, or if it is better to buy them from an outside organization

- **Procurement** — acquiring goods and/or services from an outside source
- **Procurement planning** — determining what to procure and when
- **Project procurement management** — the processes required to acquire goods and services for a project from outside the performing organization
- **Request for Proposal (RFP)** — a document used to solicit proposals from prospective suppliers
- **Request for Quote (RFQ)** — a document used to solicit quotes or bids from prospective suppliers
- **Solicitation** — obtaining quotations, bids, offers, or proposals as appropriate
- **Solicitation planning** — documenting product requirements and identifying potential sources
- **Source selection** — choosing from among potential suppliers
- **Statement of work (SOW)** — a description of the work required for the procurement
- **Termination clause** — a contract clause that allows the buyer or supplier to end the contract
- **Time and material contracts** — a hybrid of both fixed price and cost reimbursable contracts
- **Unit price contract** — a contract where the buyer pays the supplier a predetermined amount per unit of service, and the total value of the contract is a function of the quantities needed to complete the work

12 Initiating

Objectives

After reading this chapter, you will be able to:

1. *Understand the importance of initiating projects that add value to an organization*
2. *Discuss the background of ResNet at Northwest Airlines*
3. *Distinguish among the three major projects involved in ResNet*
4. *Appreciate the importance of top management support on ResNet*
5. *Discuss key decisions made early in the project by the project manager*
6. *Relate some of the early events in ResNet to concepts described in previous chapters*
7. *Discuss some of the major events early in the project that helped set the stage for project success*

Fay Beauchine became Vice President of Reservations at Northwest Airlines (NWA) in 1992. One area that had continually lost money for the company was the reservations call center. Fay developed a new vision and philosophy for the reservations call center that was instrumental in turning this area around. She persuaded people to understand that they needed to focus on sales and not just service. Instead of monitoring the number of calls and length of calls, it was much more important to focus on the number of sales made through the call centers. If potential customers were calling NWA directly, booking the sale at that time was in the best interest of both the customer and the airline. Additionally, a direct sale with the customer saved NWA 13 percent on the commission fees paid to travel agents and another 18 percent for related overhead costs.

Fay knew that developing a new information system was critical to implementing a vision that focused on sales rather than service, and she wanted to sponsor this new information system. Although the Information Services (IS) Department had worked to improve the technology for call centers, past projects never went anywhere. The new reservation system project, ResNet, would be managed by business area leaders and not Information Services managers—a first in NWA's history and a major culture change for the company. Fay made Peeter Kivestu, a marketing director, the project manager for the ResNet Beta project in 1993. NWA was going through tremendous business changes at that time, and the airline almost went bankrupt in 1993. How could Fay and Peeter pull off the project?

WHAT IS INVOLVED IN PROJECT INITIATION?

In project management, initiating is the process of recognizing and starting a new project or project phase. This process seems simple enough, but a lot of thought should go into it to ensure that the right kinds of projects are being initiated for the right reasons. It is better to have moderate or even a small amount of success on an important project than huge success on an unimportant one. The selection of projects for initiation, therefore, is crucial, as is the selection of project managers.

Recall from Chapter 4, *Project Scope Management,* that strategic planning serves as the foundation for deciding which of several projects to pursue. The organization's strategic plan expresses the vision, mission, goals, objectives, and strategies of the organization. It also provides the basis for information technology project planning. Information technology is usually a support function in an organization, so it is critical that the people initiating information technology projects understand how those projects relate to current and future needs of the organization. For example, Northwest Airlines' main business is providing air transportation, not developing information systems. Information systems, therefore, must support the airline's major business goals, such as providing air transportation more effectively and efficiently.

Information technology projects are initiated for several reasons, but the most important one is to support explicit business objectives. As mentioned in the opening case, Northwest Airlines was having financial difficulties in the early 1990s, so reducing costs was a key business objective. Providing an information system to stop the financial drain caused by the reservation call centers was the primary objective of the ResNet project.

Table 12-1 lists the knowledge areas, processes, and outputs that are typically part of project initiation. Tasks often involved in the project initiation process include the completion of a stakeholder analysis and preparation of a feasibility study and an initial requirements document. The outputs or outcomes of project initiation generally include a project charter of some sort, selection of a project manager, and documentation of key project constraints and assumptions. This chapter provides background information on Northwest Airlines and ResNet and then describes the initiation tasks involved in this large information technology project.

You will find in this chapter, and the following process group chapters, that real projects often do not follow all of the guidelines found in this or other texts. For example, the initiating project management process group generally only includes the process of initiation, part of project scope management, and the outputs listed in Table 12-1. The first ResNet project, the ResNet Beta or Prototype project, included some but not all of these outputs plus several others as part of initiating and preproject planning. Many projects include groundwork that is done before they are considered to be official projects. Every project is unique, as is every organization, every project manager, and every project team. These variations are part of what makes project management such a diverse and challenging field.

Table 12-1: Initiating Processes and Outputs

Knowledge Area	Process	Outputs
Scope	**Initiation**	**Project Charter**
		Project Manager Identified/Assigned
		Constraints
		Assumptions

BACKGROUND ON NORTHWEST AIRLINES

Northwest Airlines is the world's fourth largest airline and America's oldest carrier. Northwest began on October 1, 1926, flying mail between Minneapolis/St. Paul and Chicago. Passenger service began the following year. On July 15, 1947, Northwest pioneered the "Great Circle" route to Asia, with service to Tokyo, Seoul, Shanghai, and Manila.

Today, Northwest Airlines, with its global travel partners, serves more than 750 destinations in 120 countries on six continents. In 2001, it had more than 53,000 employees worldwide. The U.S. system spans 49 states and the District of Columbia. Northwest has more than 2,600 daily flights and operates more than 200 nonstop flights between the United States and Asia each week. Hub cities include Detroit, Memphis, Minneapolis/St. Paul, and Tokyo.

In the early 1990s, Northwest Airlines' sales agents accessed a reservation system by using approximately 3,000 dumb terminals—display monitors, with no processing capabilities, connected to a mainframe computer. As the airline business became more complicated and competitive, so did the reservation process. Calls were taking longer to complete and few direct sales were being made. Therefore, the airline was losing money by providing this necessary function of the business. It was Fay Beauchine's intent to turn this situation around by initiating the ResNet project.

BACKGROUND ON RESNET

Arvid Lee had worked in the IS Department at Northwest Airlines since 1971. One of the project ideas he and his colleagues had kicked around for several years was improving the system interface for the sales agents in the call centers. Changes in the business were making the call center jobs more complicated, and sales agents were complaining about the old Passenger Airline Reservation System (PARS). The government had just deregulated the airline industry, and new marketing initiatives such as frequent flier programs complicated matters. The average length of calls in the call centers was increasing due to the complexity of the job and the inflexibility of the information system being used. The IS Department did some research on improving the interface of the reservation system, but no improvements were ever implemented.

Figure 12-1 shows a sample screen from the PARS reservation system used at Northwest Airlines in the early 1990s. Notice the unfriendly, character-based interface. There was only one window with no help or menus to assist sales agents in the reservation process. Sales agents attended special training classes to learn all of the codes and procedures for using the PARS reservation system. At times, the call center job would get very demanding as more and more people called to obtain flight information, and the PARS information system provided little flexibility in helping sales agents meet potential customers' needs.

In 1992, Fay Beauchine became the Vice President of Reservations at Northwest Airlines. She knew the call centers were losing money, and she knew their focus on improving service was not working. Fay realized that a major change was needed in the information systems used by the sales agents. They needed a system that would help them quickly give potential customers complete and accurate information and allow them to book flights directly with NWA. Fay also knew that several competing airlines had successfully implemented new reservation systems.

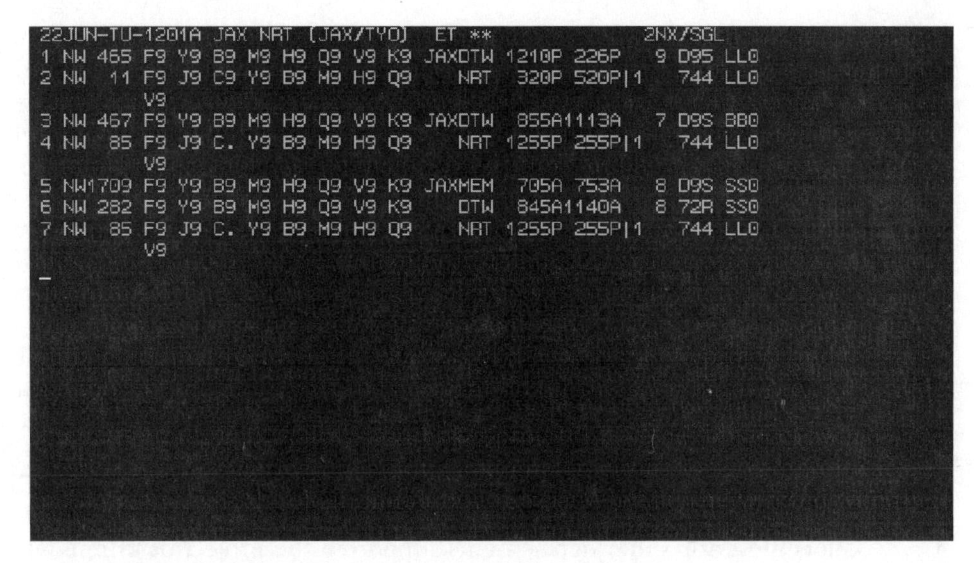

Figure 12-1. Sample Reservation Screen Before ResNet

Peeter Kivestu was a marketing director at NWA in 1993. He knew the company was having financial problems, and he had heard about Fay's vision of turning around the call centers. Peeter met with Fay to exchange ideas, and they both decided that Peeter could meet the challenges of being the ResNet project director. (NWA did not have a job title of project manager in 1993, so Peeter was named the project director, their title for a project manager. He will be referred to as the project manager in this book.)

To succeed, Peeter knew that he needed strong support from the IS Department. Peeter had discussed the project with the director of Information Services, and she agreed to have her department support the project. In May 1993, Peeter met with Arvid Lee, a technical specialist and senior member of the IS Department. Arvid had a great reputation in the company, and Peeter wanted to solicit his ideas and support for developing the new reservation system.

Arvid's first meeting with Peeter was quite an experience. Peeter exuded energy as he explained his overall strategy. He wanted to have a beta version of the new reservation system (ResNet) done in less than fifteen months—by early August, 1994. A rough estimate for creating a beta version of the system was about $500,000. Peeter asked Arvid to take the lead on developing a project plan for the beta system, and he wanted the plan done in one week.

Arvid would be in charge of all of the Information Services people supporting the ResNet Beta project, focusing primarily on the hardware, networking, and software integration efforts. His title would be the ResNet IS project manager. Arvid would also work with Kathy (Krammer) Christenson, a former marketing analyst and the new ResNet application development manager, to customize the ResNet software to meet the sales agents' needs. ResNet would use off-the-shelf software as much as possible, but NWA staff would have to do some customization, system software development, and new application development to have the new system work within their company's business environment.

After completing a prototype system that would prove the potential benefits of ResNet, Peeter would have to convince upper management to invest over $30 million in a new system involving over 3,000 personal computers. To get initial and continued funding, Peeter knew they needed to prepare convincing documentation for the project, especially since the airline was not in good financial health. Creating a way to measure the benefits of ResNet was a key part of his strategy from the start.

Although the budget plans for all 1993 projects were due in September, Peeter knew he could not sit around and wait for formal approval before he got people working on ResNet. By the time any official funds were approved in December of 1993, Peeter had about twenty people working on the project in various capacities. Several Information Services staff were redirected from other, lower-priority projects, to help support ResNet. Other NWA staff in the call centers and other departments supported the project part-time while maintaining their normal duties. These people worked on developing the plan for the beta test, researching various software and hardware options, documenting the work flow of the current reservation process, recruiting people to work on the project, and so on. After funds were approved and the ResNet Beta project was formally recognized in December of 1993, several people were officially assigned to the project.

Recall that every project is unique and has a definite beginning and a definite end. Many large information technology projects are also broken down into smaller projects. ResNet was really a series of three distinct projects. Table 12-2 provides an overview of the three distinct projects undertaken in creating ResNet—the ResNet Beta or Prototype project, ResNet 1995, and ResNet 1996. Each project had specific scope, time, and cost goals. Peeter Kivestu was the project manager for all three ResNet projects, with Arvid Lee and Kathy Christenson as team leaders.

The ResNet Beta project or prototype started in May of 1993 and ended in August of 1994. The Beta project involved customizing and testing new reservation software and hardware using sixteen personal computers. The project also involved developing a method for measuring the true benefits of ResNet before making major financial investments in new reservations systems. This project cost about $500,000.

The ResNet Phase I project, also called ResNet 1995, was approved by NWA's finance committee in November of 1994. This project involved installing personal computers in the call centers in Baltimore and Tampa and the international portion of the Minneapolis/St. Paul reservations offices. It also involved developing ResNet software for Reservations Sales and Support, the new Iron Range Reservations Center, and the Sales Action Center. The price of this 1995 project was about $8.3 million in capital costs: computer equipment, facilities, and purchased software. The total cost estimate for the ResNet 1995 project was about $13 million.

Table 12-2: Three Main ResNet Projects

	ResNet Beta or Prototype	ResNet 1995	ResNet 1996
Scope	Writing and testing new reservations system software, installing system on 16 PCs, developing measurement approach	Installing PCs, software, and networks in Baltimore, Tampa, and international portion of Minneapolis/ St. Paul reservations offices; developing more software	Completing the installation of PCs, software, and networks at other six reservations offices, developing more software
Time	May 1993 – August 1994	September 1994 – December 1995	August 1995 – May 1997
Cost	About $500,000	About $13 million	About $20 million

The ResNet Phase II project, also called ResNet 1996, completed the installation of the new reservations system at six other call centers and provided additional software development. This project cost another $10.7 million in capital and $20 million total. The total cost of all three ResNet projects was approximately $33.5 million. More detailed information on project costs is provided in Chapter 13, *Planning*.

What Went Wrong?

After finishing the ResNet Beta project, Peeter and his team saw the rest of ResNet as one large project to implement the new reservation system in all of the call centers. Senior management, however, broke the rest of ResNet into two separate projects, ResNet 1995 and ResNet 1996. Their goals were to avoid a huge investment commitment and to provide further incentives for the ResNet team to produce successful results. If the ResNet 1995 project was not successful, senior management would decide not to fund the 1996 project. Although this strategy reduced financial risk, Peeter and his team did not like the decision. If ResNet 1996 were not approved for some reason, they would be stuck with two totally different reservations systems in different sales offices. This situation would cause huge management, technical, and support problems. The ResNet 1995 team was under a lot of pressure to do a good job or the ResNet 1996 project would not be funded.

SELECTING THE PROJECT MANAGER

An important part of project initiation is selecting a project manager. Fay Beauchine asked Peeter to be the project manager for several reasons.

- Peeter knew the airline business and had over thirteen years experience in the industry. He joined Northwest Airlines in 1991 after holding several positions at Canadian Airlines International and American Airlines. Peeter had a bachelor's degree in engineering and a master's in aeronautics and finance.
- He understood the technology. Peeter was Vice President of Advanced Productivity Programs at Canadian Airlines International in the late 1980s. He was very successful at leading business and technical professionals in developing and applying new technologies.
- He knew that technologies could improve business productivity. Peeter was working in NWA's marketing and scheduling area and had some discussions with Fay about new reservation technologies. He made a case to Fay that big technology projects can be successful if managed properly. He convinced her that he was the right person for the job. His passion for the project was obvious to Fay, and Peeter used this passion to convince others how important ResNet was for Northwest Airlines.

PREPARING BUSINESS JUSTIFICATION FOR THE PROJECTS

Most projects require some form of justification to secure resources and funding. Whereas the ResNet projects addressed a broad organizational need to cut costs, they also required significant investments before any cost reductions would be realized. Peeter used different approaches for justifying each of the ResNet projects: the Beta project, ResNet 1995, and ResNet 1996.

ResNet Beta Project

Fay, Peeter, and Arvid took specific actions to convince senior management at Northwest Airlines to fund the ResNet Beta project. All three people understood the company's strategic plan and knew that it was important to cut costs, minimize risks, and remain competitive in handling reservations.

Fay convinced senior managers at numerous meetings that her vision of focusing on sales would turn around the poor financial performance of the reservation centers. She did an excellent job of analyzing the stakeholders and addressing their unique interests and concerns. She emphasized the fact that the company had to continue making some investments to improve its financial performance. The $500,000 they were asking for was a reasonable investment given the huge potential benefits of the project.

Peeter had a successful track record in implementing information systems to meet business goals, and he focused on using proven technology to reduce potential risk. Peeter collected facts showing that their competitors were increasing productivity by investing in new reservations systems. Peeter's experience and charismatic personality convinced people that ResNet was the project to work on and that failure was not an option. Having a strong project manager helps to justify investing in new projects and reduces project risk.

Arvid was very familiar with what competing airlines were doing in their call centers, so he knew what NWA had to do to remain competitive. Arvid's expertise was instrumental in planning the details of the beta project and making recommendations on hardware, software, networks, and staffing. The plans he helped create convinced senior management that the ResNet team knew what they were doing.

ResNet 1995 and ResNet 1996

After successfully completing the ResNet Beta version of the new reservation system on time and on budget, Peeter and his team had to convince upper management to make a major investment in the operational ResNet system. They developed a project plan in October 1994, for the 1995 and 1996 ResNet projects.

What Went Right?

Peeter and the ResNet team prepared outstanding business justification for all of the ResNet projects. Using his strong technical and business skills, Peeter wrote persuasive documents justifying ResNet, and he made a good case in his presentations. Peeter's justification strategy for the 1995 and 1996 ResNet projects included very strong financial analysis, which greatly impressed the finance committee. Peeter knew that the decision makers wanted to see the bottom-line numbers, and he provided those numbers with detailed rationale for how he got them. He prepared a five-year cash flow analysis of all costs and benefits, clearly defined major assumptions, and described the basis for all of his projections. It was obvious to the finance committee that Peeter knew what he was doing. They also appreciated the fact that someone leading a major information technology project was driven by business needs instead of technology needs.

Because of the company's poor financial condition in 1994, Peeter knew he had to have an extremely compelling argument to convince senior management to make any large investments in information technology. When communicating with upper management, Peeter focused on the key business objectives of the project, highlighted the impressive results from the beta project, and focused on the opportunity to make money with the new system.

Peeter provided detailed financial estimates in the ResNet 1995-1996 project plan. (Peeter viewed ResNet 1995 and 1996 as one large project in his initial project plans and business justifications.) Table 12-3 shows the financial summary from the October 1994 ResNet project plan. Peeter and his team estimated a net present value of $37.7 million for the project, based on a five-year system life and an 11.5 percent discount rate. Peeter also estimated the discounted payback period for ResNet to be 30 months or 2.48 years, and the internal rate of return as 45.2 percent. All of these financial projections provided strong support for investing in ResNet for 1995 and 1996.

Table 12-3: ResNet 1995-96 Financial Summary

FINANCIAL CATEGORY	EXPENSE (MILLIONS OF DOLLARS)
Net Present Value over 5 years @ 11.5%	37.7
Commitment (over 5 years)	
- One-time capital	21.5
- Chisholm equipment credit	(2.4)
- One-time operating expenses	2.6
- Recurring operating expenses	11.9

The majority of estimated benefits came from increasing the sales conversion, the percentage of calls that resulted in direct sales. This benefit is also called improving the call-to-booking ratio. Peeter prepared a detailed financial analysis of the estimated benefits of improving the call-to-booking ratio. Inputs for this calculation included the annual number of calls, the booking percentage with and without ResNet, the assumed percent of bookings flown, the average number of passengers per booking, the airfare, the savings due to direct sales from not paying the 13 percent commission fees to travel agents, and an additional 18 percent for overhead cost savings. Another major category of projected benefits was a reduction in headcount (sales agents) due to reduced call-handling times.

Peeter also explained the capabilities of the new reservation system in business terms. In the project plan to management he wrote:

> "The objective of the ResNet PC user presentation software is to convert agents from reservationists to salespeople through the use of intuitive software, which anticipates the information an agent will need, incorporates context-appropriate sales messages, highlights marketing programs and promotions, and ensures accuracy and consistency in call handling."[1]

Figure 12-2 shows a sample of the ResNet screen. Notice that much more information is available on the new reservation system screen than was available on the old screen (see Figure 12-1). A critical part of the software develop-

[1] Kivestu, Peeter, PR2 submitted to Robert E. Weil and Anne C. Carter, October 25, 1994.

ment for ResNet involved integrating information from different areas and putting that information all on one screen. The added context-appropriate sales messages also helped sales agents provide customers with information that would help close a sale. As Bill Hawkins, one of the ResNet Beta test sales agents, put it, "What ResNet is to PARS is like what AT&T is to two sticks and a hollow log. It's a world of difference."

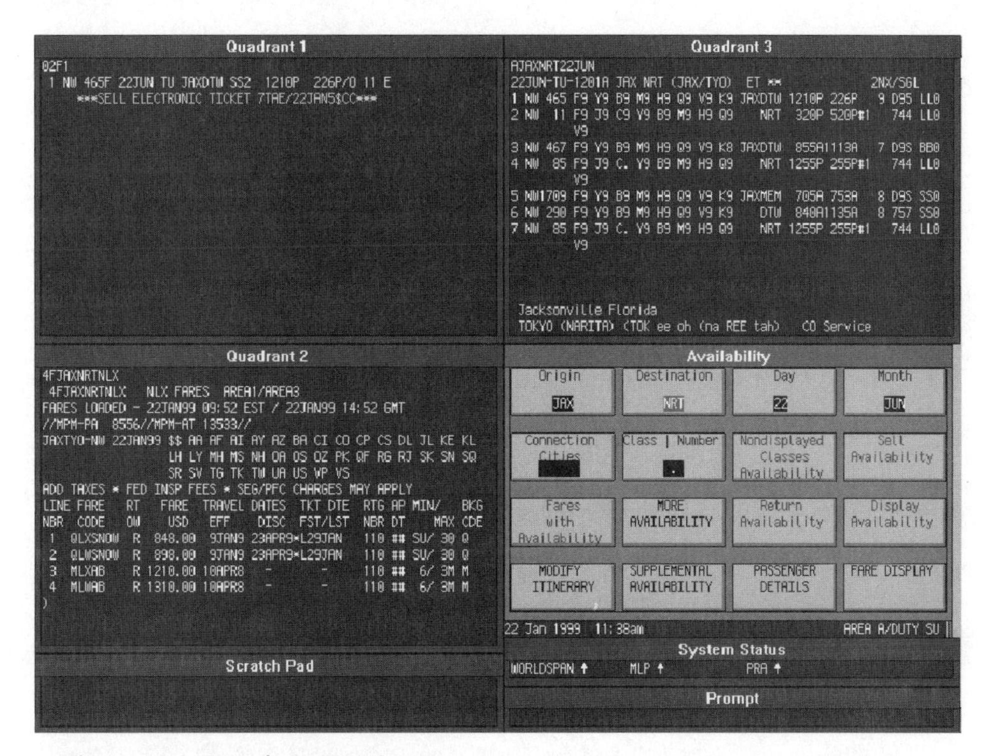

Figure 12-2. Sample ResNet Screen

Figure 12-3 shows the executive summary used in the project plan Peeter and his team created to justify investing in ResNet. Notice that Peeter focused on business issues throughout the executive summary. Northwest's competitors were taking advantage of new technologies to improve their reservations systems and realizing significant cost savings and revenue enhancements. The detailed plan included a table listing major competitors and how much their productivity and sales had increased after upgrading their reservation systems. For example, Peeter showed that Qantas and Canadian Airlines had increased their sales conversion by 11 percent after installing a new reservation system, and American Airlines decreased their sales agents' training time by 22 percent with their new system. Peeter also presented the risk involved in the project as minimal, primarily due to the success of the ResNet Beta project and the fact that most of the money would be spent on off-the-shelf hardware and facilities versus software development. Senior management knew the risks involved in large software development projects, and Peeter's approach was to wait until the vendor's software they planned to use was successfully implemented at two other clients' sites before adopting it for ResNet. All of these items convinced senior management that ResNet was an important project for NWA.

Northwest Reservations offices currently operate with the PARS reservations system using 30 year-old technology. As a result, reservations agents must remember cryptic commands to access vast amounts of quickly changing marketing information. Because of the inflexibility of the current system, the time, and consequently the cost of taking a reservation is difficult to manage, agents are not able to consistently follow procedures, and training costs are high. Most of Northwest's competitors, and now also travel agencies, have been aggressively addressing reservations inefficiencies by installing intelligent reservations terminals. These airlines and agencies have realized significant operating improvement through cost savings and revenue enhancements.

Marketing requests approval to install intelligent sales terminals - to be called ResNet - in the eight domestic Reservations Sales offices, in the new Chisholm WorldPerks office, and in the Sales Action Center. The 5 year NPV of the project is $37.7 M.

There is limited downside risk to the project as it has been proven in a National Sales beta test at NW and similar versions of the product are in service with many domestic and international competitors. As the extensions of the application are built for other parts of Reservations, they will be built to the same standards as the current application. Of the total one-time costs to implement ResNet, the majority (65%) is hardware or facility related.

Furthermore, there are substantial upside benefits possible. The software is a small part of the total cost of the system, but can be modified without change to the hardware to accommodate a wide variety of revenue enhancing and cost reducing add-ons to the base software. Because of the software chosen for this application and the ability of the reservations users to complete many of their own enhancements, rapid deployment of new functionality throughout the project term will be encouraged.

Written by Peeter Kivestu, NWA, October 25, 1994

Figure 12-3. Executive Summary for Justifying ResNet Project

DEVELOPING THE PROJECT CHARTER

Northwest Airlines did not use official project charters in the early 1990s. Receiving formal budget approval seemed to be the main means for chartering projects. They did have internal communications, however, recognizing the need for and existence of the ResNet project.

The October 1994 project plan mentioned earlier provided detailed information on the ResNet project. This document was called a Purchase Request 2 (PR2) and was based on the dollar amount involved. Peeter prepared it by following the guidelines of NWA's Information Services Application Development Methodology. NWA developed this internal methodology for planning and developing their information systems projects. These guidelines include information on creating the following documents (more information related to these planning documents is provided in Chapter 13, *Planning*):

- Project Approach
- Technical Approach
- Development Approach
- Applications Approach
- Operations Approach
- Change Management Approach

The PR2 included a summary of the next steps involved in the ResNet project after completing the Beta ResNet project. Table 12-4 provides information from the PR2 describing an overall approach for the ResNet 1995-1996 implementation process. Notice that the focus was on installing the new reservation system in the various business areas involved in the reservation process.

Table 12-4: Next Steps After ResNet Beta Project

Step After ResNet Beta Project	Date
First facility ready for information systems installation	2/95
First agent training for live calls	3/95
First office completed	5/95
National/International Sales offices completed	2/96
Chisholm WorldPerks office completed	3/96
Specialty Sales/Support Desks completed	12/96

The PR2 was prepared as part of the funding process. When NWA's Finance Committee approved funding for the ResNet 1995 project in November of 1994, that project was officially recognized and funded, and the PR2 served as the initial project charter and plan. Likewise, after the ResNet 1996 project was funded in November of 1995, it was officially recognized. By the late 1990s, NWA had developed project charter forms to formally recognize projects and provide overall guidance for them.

ACTIONS OF THE PROJECT MANAGER AND SENIOR MANAGEMENT IN PROJECT INITIATION

As mentioned throughout this text, strong leadership is an essential part of good project management. Project managers must provide leadership to their project team members, develop good relationships with key project stakeholders, understand the business needs of the project, and prepare realistic project plans. Senior management must participate in the project by providing overall support and direction. Following are some of the actions Peeter and other senior managers took to get the ResNet projects off to a good start:

Quickly assembling a strong project team. As mentioned earlier, Peeter had twenty people working on ResNet before there was even an official budget for the Beta ResNet. There was a lot of work to do, and Peeter saw no need to wait for the annual budget approval before getting started. He assembled a project team for the Beta ResNet system in July and August of 1993, with Fay and other senior managers' support. Peeter convinced Arvid Lee, a crucial asset to the project, that ResNet was a critical project for the company and for the IS Department. After he was convinced of the project's importance and Peeter's ability to successfully lead the project, Arvid helped recruit other Information Services staff and users. Peeter and Arvid's passion for the project drew people to ResNet. This strong start for the project created a momentum that contributed to its success.

Getting key stakeholders involved in the project early. Peeter included stakeholders from all of the areas involved in ResNet, especially top management, marketing, the user community, and the IS Department. He knew from past experience how important it was to have users involved in projects and to have senior management support. Peeter held regular meetings with people in marketing, and Fay was instrumental in getting their support as well as support from the executive management team. Peeter also brought in several people from the reservation call centers. They would be the primary users of ResNet, and they understood the business of dealing with potential customers to sell airline tickets. Peeter had an experienced facilitator lead several focus groups with the sales agents to understand what the call center issues were. Peeter also held several meetings with Information Services people to have them start evaluating different technologies that could be used to improve the reservation system.

Preparing detailed analysis of the business problem and developing project measurement techniques. Peeter knew that senior management would constantly question the value of ResNet, so even before the ResNet Beta project officially started, Peeter developed plans to prove that the system would save the company money. Peeter recognized the need to bring in industrial engineering experts to help document the reservation process at NWA and recommend how to close more sales. These experts found that sales agents had to handle over thirty different types of calls. NWA developed a call strategy based on the detailed analysis done by these engineers. The ResNet system would have to support this call strategy and allow agents to handle any type of call without additional steps. The engineers also developed techniques for measuring the effects of various aspects of the new system on increasing sales and reducing call time. Peeter knew this project would be scrutinized, so he made sure they had evidence that the new technology was indeed increasing productivity.

Using a phased approach. Senior management at NWA decided to divide ResNet into several phases or distinct projects. First they would fund a beta project to prove the concept of the new reservation system. The beta phase would demonstrate that the new technology would work and that the system would be profitable. Originally, Peeter wanted the second project to involve installing ResNet in all of the reservation call centers. Instead, senior management at NWA insisted on continuing a phased approach. They agreed to fund ResNet 1995 after the successful ResNet Beta and made ResNet 1996 contingent on the success of the previous project. This approach would break the work into smaller, more manageable projects, provide motivation for success, and minimize the financial risk for the company. If the first phase did not work, the project could be canceled; the same went for the second phase. As mentioned in Chapter 2, *The Project Management Context and Processes*, this is a very wise approach.

Preparing useful, realistic plans for the project. Having realistic plans is another important factor for project success. Arvid Lee had worked on several information technology projects at NWA and understood what was involved in all phases of ResNet. He was also an experienced planner. Before they did detailed planning, Arvid decided to let implementation dates drive the rest of the plans, and he and Peeter ensured that they never missed any of the critical implementation dates during project execution. Deciding to focus on implementation dates during the initiating process on ResNet helped guide the process groups in later phases of the project.

Project initiation is a critical activity in organizations. Companies must spend their time and money wisely by selecting important projects to work on and by getting them off to a good start. NWA did an excellent job of project initiation with ResNet. They focused on key business objectives, provided top management support, assembled a strong project team, involved key stakeholders, performed detailed analysis, developed measurement techniques for the project, used a phased approach, and prepared an approach for developing useful, realistic project plans.

CASE-WRAP UP

Fay and Peeter were very pleased when Northwest Airlines' Finance Committee agreed to fund the Beta ResNet project and continue to fund ResNet 1995 and ResNet 1996. Their vision and leadership inspired all ResNet stakeholders to ensure project success. Fay understood the business needs of NWA and provided strong user sponsorship. Peeter formed a dynamic project team to successfully initiate ResNet and to later deliver even more than everyone had expected. They proved that a large information technology project could be accomplished successfully by focusing on business needs and using good project management.

CHAPTER SUMMARY

Initiation is the process of recognizing and starting a new project or project phase. Supporting key business objectives is a key reason for funding projects. NWA initiated ResNet to reduce costs from its reservation call centers and implement a new vision of focusing on closing sales, which would increase profits.

As with many information technology projects, the concept of ResNet was discussed in the IS Department for a few years. However, a strong project sponsor outside information technology was needed to provide the vision for the project and convince senior management of its value.

ResNet involved three distinct projects: a ResNet Beta project, ResNet 1995, and ResNet 1996. This phased approach helped break the work into more manageable pieces and minimized financial risks.

Selecting the project manager and forming the core project team were important parts of ResNet's project initiation. Developing strong business justification for the project was also crucial.

Actions in the project initiation phase and during preproject planning helped set the stage for the success of ResNet. These actions included quickly assembling a strong project team, involving key stakeholders early and often in the project, preparing a detailed analysis of the business problem, developing measurement techniques, using a phased approach, and developing an approach for preparing useful, realistic project plans.

Discussion Questions

1. What was Fay's role in initiating ResNet? Why did she succeed when other NWA employees in the IS Department, who had similar ideas years earlier, did not?
2. What role did senior managers, Peeter, and Arvid have in initiating ResNet? Did they make good decisions? Explain your answer.
3. What were some of Peeter's strengths as a project manager? How do these strengths relate to concepts discussed in earlier chapters?
4. Review the executive summary Peeter prepared to help justify investing in ResNet. What points were made to convince the finance committee to support the project?
5. Discuss the major differences between the three ResNet projects. What was the emphasis for each project? How might each be managed differently?
6. Describe a project you have seen initiated. Compare how it was handled to the ResNet project. What were the similarities and differences?

Exercises

1. Review the actions taken by Peeter and other senior managers that helped get ResNet off to a good start. How do these actions compare to what earlier chapters presented as good project management practice? List each action and find specific statements in earlier chapters of this book that support (or do not support) each.
2. Research information about how air travel and the airline reservation system process has changed over the past ten years. How many people currently book flights directly through the airlines? How many people book flights through travel agents? How many people use Web-based systems to book flights? What are the main differences between each approach? Do you see any trends developing? Do you think new services are better for customers? What are the potential disadvantages of increased Web-based bookings?
3. Research at least three different airline reservation systems available on the World Wide Web (nwa.com, other airlines' Web sites, expedia.com, travelocity.com, sidestep.com, and so on). List the basic features of each, then compare them based on the following criteria: ease of use, availability of flights to various locations, costs for specific flights, convenience in travel times, related services such as car rental, hotel information, and payment process. Add additional criteria, if you wish. Which system do you prefer and why?
4. Ask five people who have flown in the past year how they booked their flights. Be sure to talk to people of different ages and backgrounds (for example, your roommate, your parents, your grandparents, your boss, and your teachers). Compare the methods they used and why. What is your personal preference for booking flights?

MINICASE

None of the ResNet projects had a project charter as described in Chapter 4, *Project Scope Management*.

Part 1: Using Table 4-2 as a guide, write a project charter for each of the three ResNet projects.

Part 2: Review ads for project managers on Web sites such as www.monster.com. Then write an ad to advertise for the position of project manager for the ResNet projects. Also develop a list of questions you would ask when interviewing people for the position.

SUGGESTED READINGS

1. Microsoft's Travel Web site, www.expedia.com.

   Microsoft and several other companies have Web sites to help people book flights, reserve rental cars, find lodging, and so on.

2. Northwest Airlines Web site, www.nwa.com.

   A good way to learn more about a company is to visit its Web site. Review important information about Northwest Airlines from their Web site.

3. Sackman, Ralph B. *Achieving the Promise of Information Technology.* Newton Square, PA: Project Management Institute, 1998.

   Sackman's text offers a framework to help companies use the power of information technology to transform their mainstream operations and services. Based on the nature of systems work itself, this book describes initiatives to achieve growth, productivity, cost control, asset utilization, and customer satisfaction objectives.

4. Weiss, Joseph W. and Robert K. Wysocki. *5-Phase Project Management.* New York: Harper-Collins, 1992.

   This book provides practical, step-by-step information on each phase of a complex project, starting with project initiation.

5. WORLDSPAN's Web site, www.worldspan.

   WORLDSPAN is a computer reservation system used by Northwest Airlines and many other airlines and travel agents. Their Web site provides information on the history of WORLDSPAN and more recent products and services.

13

Planning

Objectives

After reading this chapter, you will be able to:

1. *Discuss the project management planning processes and outputs and describe how they were used on ResNet*
2. *Describe how Northwest Airlines organized the scope of work on ResNet using work breakdown structures*
3. *Discuss how Microsoft Project was used to aid in project planning on ResNet*
4. *Review and discuss real-world examples of work breakdown structures, cost estimates, staffing plans, and a project organizational chart for a large information technology project*
5. *Discuss key decisions the project manager and team made in the planning process*
6. *Relate some of the planning events in ResNet to concepts described in previous chapters*
7. *Understand the contribution that good planning makes to project success*

Peeter Kivestu and Arvid Lee took the lead in creating project plans for all the ResNet projects. Peeter had a strong finance background and understood the airline business well. His work in writing the PR2 was instrumental in getting funding for the ResNet 1995 and 1996 projects. Arvid had the most experience working on information systems projects at NWA, and he was good at getting input from the people who would be involved in executing the plans.

However, Arvid was having a difficult time deciding what level of detail to use in planning the project scope and schedule. A friend of his had spent a lot of time planning a large information technology project for his organization, and he had nothing but horror stories to tell. He couldn't keep track of the details, the changes, and what actually occurred on the project versus what was in the plan, even

with the aid of project management software. How should Arvid and the ResNet team plan the work that needed to be completed so they could actually focus on getting the job finished? How should they communicate the plan and keep it updated?

WHAT IS INVOLVED IN PROJECT PLANNING?

Planning is often the most difficult and most unappreciated process in project management. Many people have negative views of planning because the plans created are not used to facilitate action. The main purpose of project plans, however, is to guide project execution. To guide execution, plans must be realistic and useful. To create realistic and useful plans, a fair amount of time and effort must go into the planning process, and people knowledgeable in doing the work need to plan the work.

Unlike initiating, which generally involves only one knowledge area, planning concerns activities in every knowledge area:

- Project integration requires developing the overall project plan.
- Project scope management includes scope planning and scope definition.
- Project time management includes activity definition, sequencing, duration estimating, and project schedule development.
- Project cost management includes resource planning, cost estimating, and cost budgeting.
- Project quality management includes quality planning.
- Project human resource management includes organizational planning and staff acquisition.
- Project communications management includes communications planning.
- Project risk management includes risk management planning, risk identification, qualitative risk analysis, quantitative risk analysis, and risk response planning.
- Project procurement management includes procurement planning and solicitation planning.

Table 13-1 lists the project management knowledge areas, processes, and outputs of project planning. You can see there are many outputs from the planning process group. ResNet documents produced during the planning process, and discussed in this chapter, include a project plan, work breakdown structure, project schedule, cost estimates, and an organizational chart.

Table 13-1: Planning Processes and Outputs

KNOWLEDGE AREA	PROCESS	OUTPUTS
Integration	Project Plan Development	Project Plan
		Supporting Detail
Scope	Scope Planning	Scope Statement
		Supporting Detail
		Scope Management Plan
	Scope Definition	WBS
		Scope Statement Updates
Time	Activity Definition	Activity List
		Supporting Detail
		WBS Updates
	Activity Sequencing	Project Network Diagram
		Activity List Updates
	Activity Duration Estimating	Activity Duration Estimates
		Basis of Estimates
		Activity List Updates
	Schedule Development	Project Schedule
		Supporting Detail
		Schedule Management Plan
		Resource Requirement Updates
Cost	Resource Planning	Resource Requirements
	Cost Estimating	Cost Estimates
		Supporting Detail
		Cost Management Plan
	Cost Budgeting	Cost Baseline
Quality	Quality Planning	Quality Management Plan
		Operational Definitions
		Checklists
		Inputs to Other Processes

Table 13-1: Planning Processes and Outputs (continued)

Knowledge Area	Process	Outputs
Human Resource	Organizational Planning	Role and Responsibility Assignments
		Staffing Management Plan
		Organizational Chart
		Supporting Detail
	Staff Acquisition	Project Staff Assigned
		Project Team Directory
Communications	Communications Planning	Communications Management Plan
Risk	Risk Management Planning	Risk Management Plan
	Risk Identification	Risks
		Triggers
		Inputs to Other Processes
	Qualitative Risk Analysis	Overall Risk Ranking for the Project
		List of Prioritized Risks
		List of Risks for Additional Analysis and Management
		Trends in Qualitative Risk Analysis Results
	Quantitative Risk Analysis	Prioritized List of Quantified Risks
		Probabilistic Analysis of the Project
		Probability of Achieving the Cost and Time Objectives
		Trends in Quantitative Risk Analysis Results
	Risk Response Planning	Risk Response Plan
		Residual risks
		Secondary risks
		Contractual Agreements
		Contingency Reserve Amounts Needed
		Inputs to Other Processes
		Inputs to a Revised Project Plant

Table 13-1: Planning Processes and Outputs (continued)

KNOWLEDGE AREA	PROCESS	OUTPUTS
Procurement	Procurement Planning	Procurement Management Plan
		Statement(s) of Work
	Solicitation Planning	Procurement Documents
		Evaluation Criteria
		Statement of Work Updates

DEVELOPING THE PROJECT PLANS

Recall that ResNet actually involved three related projects: the ResNet Beta project, ResNet 1995, and ResNet 1996. Planning for the first two projects will be discussed briefly, then the emphasis of this and subsequent chapters will be on the ResNet 1996 project. Descriptions of the ResNet 1996 project plans are provided by knowledge area.

ResNet Beta Project Planning

In May of 1993, Peeter asked Arvid Lee to lead development of a project plan for the ResNet Beta project. Recall that Peeter wanted the plan completed in one week. The main purpose of the ResNet Beta project was to prove the concept and business value of moving to a PC-based reservation system. The tasks involved were purchasing new reservations software, developing new software, installing the system on a small group of new PCs, testing the system, and documenting the potential benefits. Peeter insisted that the ResNet Beta project be completed by August of 1994 so they could use the results to justify obtaining funds for a follow-on ResNet project in the next budget cycle.

Arvid was familiar with planning and implementing hardware, software, and networking technologies for NWA and had experience working with the users and vendors involved. He was familiar with PCs and performing beta projects. He also knew whom to contact for help in quickly preparing a good plan. Arvid recruited people from various areas to help him craft and later implement a plan for the ResNet Beta project. He was careful to create a plan that focused on everyone doing his or her part to be successful. A few important decisions that were made in planning the ResNet Beta project included:

- Planning at a fairly high level: Learning from a friend's experience, Arvid was careful not to get carried away by planning in too much detail. He trusted his colleagues to perform their jobs well, so he focused on deliverables and results and not the details of how the work would be accomplished. Since performing any beta project involves a high degree of creativity, Arvid felt it was important to let everyone work together creatively to produce those deliverables and results.

- Getting strong user involvement: Kathy Christenson, an analyst from the marketing and sales division of NWA, had firsthand experience working with sales agents and led the ResNet application development area. Kathy and her team of agents were a critical part of defining the requirements for the software and actually did most of the software customization (*see* Chapter 14, *Executing*). Arvid was open to this innovative approach to ensuring strong user involvement in the project and worked well with non-IS staff.
- Developing a solid measurement tool: Following Peeter's advice, Arvid had several industrial engineers assigned to the project. They planned to do detailed studies of the current reservation system to find areas for improvement (a procedure known as process reengineering) and then measure the impact of the new reservation system. They asked questions such as "Did the new system increase the sales conversion percentage? Did the new system decrease call handle time?" Arvid included plans for developing this measurement tool for ResNet in the project plan for the beta test.

ResNet 1995 Planning

After successfully completing the ResNet Beta project, the ResNet team had to create a plan for installing PCs in various sales offices. The ResNet 1995 plan was to install PCs and software in Baltimore (339 PCs), Tampa (289 PCs), and the international portion of the Minneapolis/St. Paul (82 PCs) reservations offices. The plan also involved developing customized applications for specialty and support desks, two business areas related to the reservations process.

Arvid consulted with key stakeholders to develop these plans. The IS Department had experience installing PCs and networks, so he used that knowledge base to plan the installations. He also had a good idea of the amount of work and time involved in doing more software development and used that knowledge to create plans for customized application development. Arvid consulted with other experts inside the company and outside the company to develop realistic timelines for the work involved.

ResNet 1996 Planning

Because of the ResNet Beta project and the 1995 project, Peeter, Arvid, and their team had a wealth of information available when they prepared the plan for the ResNet 1996 project. Their successful experiences with the ResNet Beta and 1995 projects helped them define scope, time, cost, and other dimensions of the 1996 project. In August of 1995, Peeter prepared another Purchase Request 2 (PR2) for NWA's finance committee to review and then decide on continued funding for ResNet. The PR2 included the following information:

- Executive summary
- Brief justification
- Background
- Need
- Opportunity
- Justification
- Risk
- Other benefits
- Next steps
- Post implementation audit
- NPV schedule
- Attachments

The justification section included information on increased revenue, enhanced quality processes, reduced operating costs, and new revenue generation based on the ResNet 1995 project. The risk section included information about software, change management, and implementation. The attachments included one-time and ongoing expenses, an application development timeline, and a summary of ResNet performance. The summary of ResNet performance included the implementation schedule performance in 1995, the budget versus actual spending performance for the ResNet 1995 project, measurement objectives, performance results in the Baltimore call center, and a qualitative feedback summary of the project. The PR2—for all intents and purposes a project plan—provided a great foundation for convincing the finance committee to fund the ResNet 1996 project. It also provided a great foundation for enabling the project team to execute the project.

What Went Right?

The PR2 for the ResNet 1996 project included performance results from implementing ResNet in Baltimore in 1995. Because the project team developed measurement tools and used them on the ResNet 1995 project, they had a strong business case for justifying further investments in the ResNet 1996 project. The actual results of the call handle time in Baltimore showed a 6.7 percent improvement after implementing ResNet. The percentage of direct ticket sales increased from 3.2 percent to 7.5 percent. This increase in direct sales translated to over $15 million in profits that ResNet generated in 1995 alone.

DETERMINING PROJECT SCOPE AND SCHEDULES

The two primary goals of ResNet 1996 were further software development and seven office implementations. These two efforts were interdependent because application development drove the timing of office implementation. Table 13-2 shows the planned software development and office implementation milestones as presented in the PR2 project plan dated August 29, 1995.

Table 13-2: Planned Software Development and Office Implementation Milestones

Software Development Milestone	Date	Office Implementation Milestone	Date
International sales in production	10/95	Minneapolis/St. Paul (MSP) international desk completed	10/95
WorldPerks in production	3/96	New York City (NYC) and Detroit (DTT) facilities ready	12/95
Specialty and support desk production rollout	8/96-12/96	WorldPerks office rollout	3/96-12/96
Sales Action Center in production	12/96	National/International sales offices completed	11/96
		Specialty sales/support desks completed	12/96

Northwest Airlines used Microsoft Project (one of the first versions available) to aid in planning the scope and schedule for ResNet. They created two separate files for the software development and office implementation efforts. Figure 13-1 shows the actual Gantt chart used to plan ResNet 1996 software application development. The project team also had more detailed schedules based on this Gantt chart; however, they did not link tasks using the software. Instead, they manually entered the dates for each task. The ResNet staff only created Gantt charts to aid them in initial planning. They did not plan to enter actual information or use critical path analysis, so they did not link the tasks in Microsoft Project. They were very careful in setting the milestone dates, and *they missed none of them.*

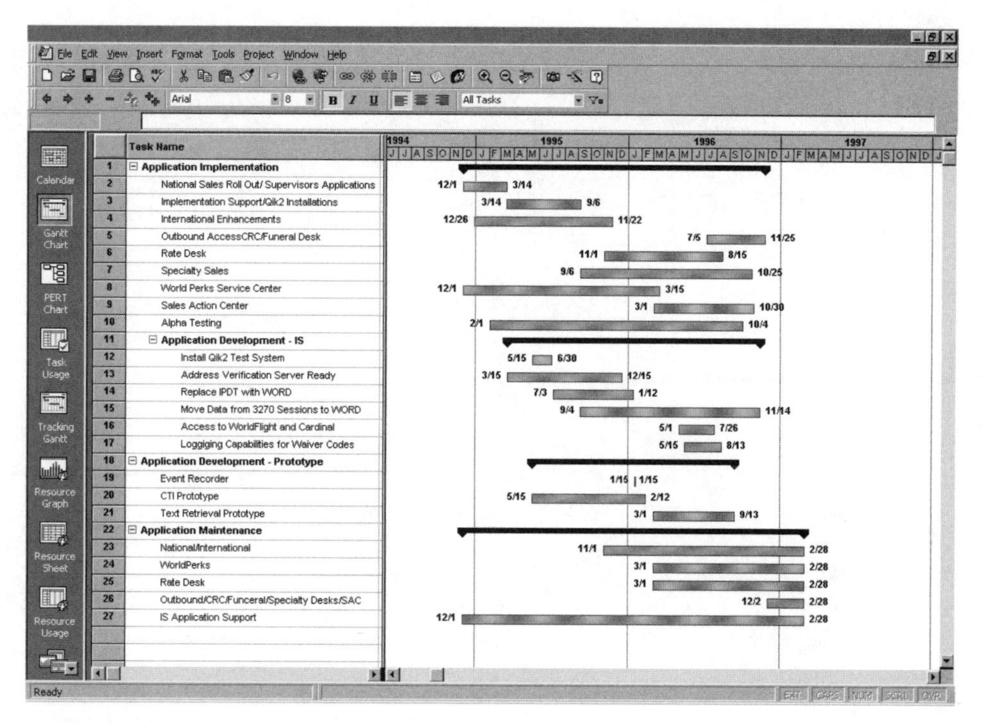

Figure 13-1. ResNet 1996 Application Development Gantt Chart

NWA developed a much more detailed Microsoft Project file for the ResNet 1996 office installations than for the software application development effort. Figure 13-2 shows part of the WBS for the office installations, which NWA staff created using a combination of the phase-oriented and product-oriented approaches for developing a WBS (*see* Chapter 4, *Project Scope Management*). For example, each office was a separate Level 1 WBS item. Figure 13-2 also shows a WBS Level 2 view of the office implementation plan for Office 1, New York City. Each of the seven offices followed the same phase-oriented approach. There were phases for each office for pre-implementation preparation, infrastructure implementation, change management, agent conversion, and achieving office goals.

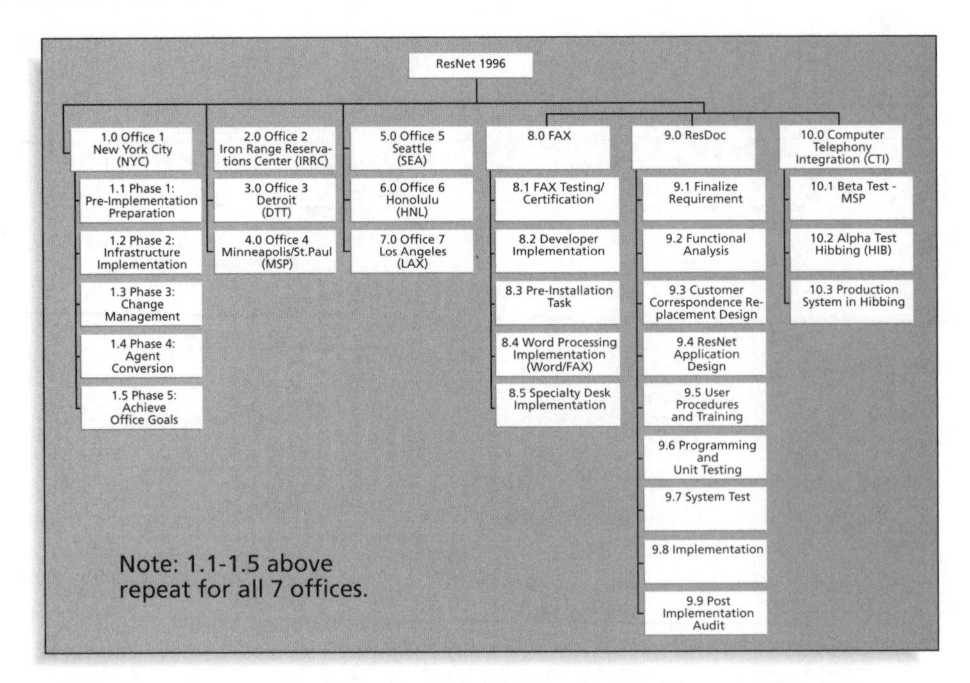

Figure 13-2. ResNet 1996 Level 2 WBS for Office Implementations

The development and implementation of facsimile (FAX) capabilities, the ResDoc customer documentation system, and the Computer Telephony Integration (CTI) system, which could be viewed as major products of the project, were also Level 1 WBS items. The Level 1 WBS item called FAX included the development and implementation of facsimile capability from the ResNet PCs. The ResDoc WBS item refers to the letter-writing system that generates correspondence for members of NWA's frequent flier program called WorldPerks. The CTI WBS item represents the Computer Telephony Integration (CTI) part of the ResNet 1996 project. CTI is a system that connects the telephone system to the ResNet computers. For example, when customers call NWA, they can respond to telephone prompts. Depending on what telephone buttons they press, the ResNet system will direct the call to a particular type of ResNet agent—domestic versus international, for example—and that agent's computer screen will pull up information on the customer's needs, according to the telephone prompt responses.

What Went Wrong?

There was a lot of political pressure related to the Iron Range Reservations Center (IRRC) in Chisholm, Minnesota. The state of Minnesota funded a large portion of this office, which was the central point for handling the WorldPerks system. The WorldPerks implementation date of March 31, 1996, was firm. The ResNet team was a little overconfident in their planning, and several problems developed. WorldPerks involved a new communications protocol, and the ResNet team did not plan enough time up-front to analyze all of the system requirements. They expedited testing, and there were several quality problems that were difficult to resolve because of a lack of precision in planning.

Figure 13-3 shows part of the actual Microsoft Project Gantt chart used for ResNet 1996 office installations. Notice that the WBS follows the structure shown in Figure 13-2. Although the Level 2 WBS items for each office were the same, the Level 3 items beneath them (not shown in this figure) varied to meet the particular needs of each office. There were a total of 821 tasks in the Microsoft Project file for the 1996 office installations. The ResNet project team focused on meeting key milestone dates, so they worked to finish tasks on time. *They missed none of these completion dates even though there were definite challenges in doing so.*

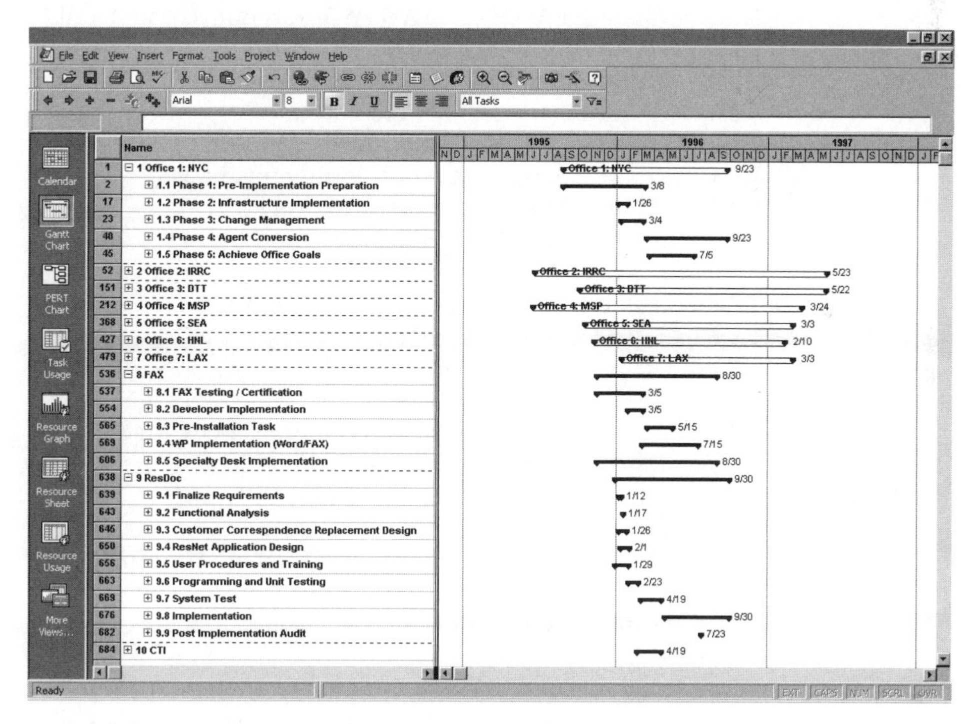

Figure 13-3. ResNet 1996 Office Implementation Gantt Chart

The ResNet 1996 project team broke down the scope of the office installations to WBS Level 3, as follows:

1.1 Phase 1: Pre-Implementation Preparation

1.1.1 Modify Facilities for the Floor

- 1.1.2 Modify Facilities for Communications Room
- 1.1.3 Conduct Pre-Implementation Site Visit
- 1.1.4 Prepare for Network Pre-Installation
- 1.1.5 Order all ResNet Equipment Components
- 1.1.6 Request Communication lines
- 1.1.7 Develop Training Schedule
- 1.1.8 Develop Phased Implementation Plan
- 1.1.9 Order Physical Security Devices for PCs
- 1.1.10 Conduct On-Site Planning Session
- 1.1.11 Conduct Change Management Site Visit
- 1.1.12 Conduct Awareness Programs
- 1.1.13 Validate PC Image
- 1.1.14 Define Electronic Distribution Manager (EDM) services for office

1.2 Phase 2: Infrastructure Implementation
- 1.2.1 Install all components in Communications Room
- 1.2.2 All native WORLDSPAN (WSP) training terminated
- 1.2.3 Install PCs in Training Room
- 1.2.4 Test and Stabilize IS Components
- 1.2.5 IS Certification Completed

1.3 Phase 3: Change Management
- 1.3.1 Train Supervisor Agent Coordinators in MSP
- 1.3.2 Provide 2 Temporary Awareness PCs
- 1.3.3 Install Buffer Area PCs and Demo PCs
- 1.3.4 Train the Trainer (TTT)
- 1.3.5 Install PCs: Trainers/Early Trained Agents
- 1.3.6 Train Supervisors in ResNet
- 1.3.7 Train Managers in ResNet
- 1.3.8 Install PCs for Managers, Supervisors, Podium
- 1.3.9 Conduct Manager Measurement Strategy Meeting
- 1.3.10 Develop Continuous Improvement Plan
- 1.3.11 Train Floor Support Agents
- 1.3.12 Install PCs for Floor Support Agents
- 1.3.13 Conduct Manager/Supervisor Strategy Meeting
- 1.3.14 Train Office Administration and Staff in Windows Applications
- 1.3.15 Install PCs for Office Administrative Staff
- 1.3.16 Conduct Office Kickoff

1.4 Phase 4: Agent Conversion
- 1.4.1 Conduct Agent Conversion Training (ACT) [phased]
- 1.4.2 Install PCs for Conversion Agents [phased]
- 1.4.3 Office Conversion Completed
- 1.4.4 ResNet Agent Measurement Completed

1.5 Phase 5: Achieve Office Goals
   1.5.1 Implement Continuous Improvement Plan
   1.5.2 Analyze Airline Control Protocol (ALC) Line Utilization
   1.5.3 Conduct Recurrent Training
   1.5.4 Complete Test and Balance of Facility
   1.5.5 Conduct Focus Groups
   1.5.6 Conduct Post Implementation Review

The ResNet 1996 project plans were not broken down any further than WBS Level 3 for the offices, but they were broken down to Level 4 for the Fax and CTI areas. For example, samples of Level 4 items for the FAX part of the ResNet 1996 project were:

8 FAX
   8.1 FAX Testing / Certification
      8.1.1 Install Fax Server for Evaluation
         8.1.1.1 Order equipment/software
         8.1.1.2 Set up the PBX
         8.1.1.3 Install T1
         8.1.1.4 Install Fax Server
         8.1.1.5 Install FAX client software

RESNET COST ESTIMATES

Peeter, Arvid, and other members of the ResNet team prepared detailed cost estimates, especially for the ResNet 1996 project. Most of the costs involved purchasing hardware, software, and network equipment, and they used vendor quotes for these items. They also had a good understanding of internal costs for installing hardware, software, and networks. The internal development costs for coding and training were based on estimates of the number of people needed from various departments and the labor cost per hour for those employees.

Table 13-3 shows the maximum departmental headcounts by year for the 1995 and 1996 ResNet projects and for supporting ResNet through 1998. This information was included in the PR2 and provided the basis for the labor cost estimates. The IS Department staff included personnel to perform industrial engineering functions (such as developing and implementing the measurement system), application development, network operations and development, and marketing application management and support. The staff from the Marketing Systems Department included a project analyst, an application development person to assist with the industrial engineering functions, and an administrative assistant. The Reservations Department staff included reservation agents who worked on application development, additional reservation agents for testing, change management analysts, training and awareness staff, applications support/maintenance staff, and reservation agents for the ResNet Hotline (a special help desk for handling agents' questions related to ResNet). The contractor headcount included people from Sabre Decision Technologies who assisted in application design and development and people to assist in ResNet distributed system implementation.

Table 13-3: Maximum Departmental Headcounts by Year

DEPARTMENT	1994	1995	1996	1997	1998	TOTALS
Information Systems	24	31	35	13	13	116
Marketing Systems	3	3	3	3	3	15
Reservations	12	29	33	9	7	90
Contractors	2	3	1	0	0	6
Totals	41	66	72	25	23	227

Peeter and his team had a high degree of confidence in their cost estimates because they had drawn on a large experience base. They knew what it would cost to purchase PCs and other hardware. They knew what their labor costs were. The main risk was in labor costs for application development. Peeter and Fay determined the maximum headcount by year, as shown in Table 13-3. They budgeted for people to work on developing, testing, and enhancing the ResNet software, and they chose to make the functionality of the system fit within the confines of what the people allocated to work on it could accomplish. This choice forced everyone to prioritize his or her work and focus on doing what was most important.

Table 13-4 summarizes the cost estimate for the ResNet 1995 and 1996 projects and on-going and recurring costs for ResNet after the projects would be completed. Peeter and his team provided many detailed calculations to support this estimate in the PR2. Notice the heavy investments in computer equipment, facilities/buildings, and software in 1994, 1995, and 1996. Also notice that costs were projected based on a five-year system life for ResNet, or the system's life cycle cost. The total life cycle cost estimate for ResNet was over $42 million.

Tab e 13-4: Summary of ResNet Tota Life Cyc e Costs

	1994 ($)	1995 ($)	1996 ($)	1997 ($)	1998 ($)	1999 ($)	TOTAL ($)
Computer equipment	325,779	4,337,036	8,487,495	-	-	-	13,150,310
Facilities/ buildings	-	974,372	1,072,969	-	-	-	2,047,341
Software	563,627	2,517,590	1,704,575	-		-	4,785,792
Total one-time capital expenditures	889,406	7,828,998	11,265,039	-	-	-	19,983,443
Capitalized labor one-time costs	88,627	1,589,191	1,704,575				3,382,393
One-time labor costs	38,040	748,267	1,265,323	60,000			2,111,630
Ongoing operating labor	108,675	1,101,809	3,150,414	2,787,984	2,710,982	2,710,981	12,570,845
Total labor and consultant costs	235,342	3,439,267	6,120,312	2,847,984	2,710,982	2,710,981	18,064,868
Recurring computer equipment	-	601,249	559,794	805,549	829,715	854,607	3,650,914
Recurring facilities/ buildings	-	6,100	41,610	42,858	44,144	45,468	180,180
Recurring software	-	-	162,050	206,912	213,119	219,512	801,593
Total recurring expenses	-	607,349	763,454	1,055,319	1,086,978	1,119,587	4,632,687
TOTAL COSTS	**1,124,748**	**11,875,614**	**18,148,805**	**3,903,303**	**3,797,960**	**3,830,568**	**42,680,998**

HUMAN RESOURCE AND COMMUNICATIONS PLANNING

Two of Peeter's strongest assets were his ability to motivate people and to provide strong communications with project stakeholders. He was good at using informal networking to accomplish a great deal of work. For example, recall that Peeter recruited people to work on ResNet before the beta project was officially funded.

Peeter and his team were also good at creating formal human resource and communications plans. Table 13-5 shows the staffing plan Peeter created for the ResNet 1995-1996 PR2. This plan provided detailed estimates of the number of person-months ResNet would require from each department for its five-year system life. Peeter and his team further broke down the staffing needs by the category of people in each department. This detailed staffing plan provided a strong basis for the labor cost estimates and for assigning people to the project.

Table 13-5: ResNet Staffing Plan (in person-months)

DEPARTMENT	1994	1995	1996	1997	1998	TOTALS
Information Systems						
Application Development (IE)	4	12	6	0	0	22
Application Development (IS)	30	135	86	0	0	251
WorldPerks Service Center Application Development (IS)	0	15	5	0	0	20
Network Operations and Development	20	60	10	0	0	90
Network and Systems Support	30	96	96	96	96	414
Marketing Application Management	4	12	12	12	12	52
Application Support	0	6	36	48	48	138
Total Information Systems Department	**88**	**336**	**251**	**156**	**156**	**987**
Marketing Systems						
Project Analyst	4	12	12	12	12	52
Application Development (IE)	4	12	12	12	12	52
Administrative Assistant	4	12	12	12	12	52
Total Marketing Systems Department	**12**	**36**	**36**	**36**	**36**	**156**

Table 13-5: ResNet Staffing Plan (in person-months) (continued)

DEPARTMENT	1994	1995	1996	1997	1998	TOTALS
Reservations Department						
Application Development (Reservations Agents)	20	87	42	0	0	149
Alpha Testers (Reservations Agents)	0	24	12	0	0	36
WorldPerks Service Center Application Development	16	44	0	0	0	60
Sales Action Center Application Development	0	4	18	0	0	22
Change Management Analyst	4	12	12	12	12	52
Training and Awareness (Reservations Agents)	8	24	6	0	0	38
Support/Maintenance (Reservations Agents)	0	27	106	96	72	301
ResNet Hotline (Reservations Agents)	0	60	30	0	0	90
Total Reservations Department	**48**	**282**	**226**	**108**	**84**	**748**
Contractors						
Application Design and Development (Sabre)	3	18	6	0	0	27
ResNet Distributed System Implementation Support	4	4	0	0	0	8
Total Contractors	**7**	**22**	**6**	**0**	**0**	**35**
Grand Total Person-Months (Including Contractors)	**155**	**676**	**519**	**300**	**276**	**1926**

Figure 13-4 shows the organization chart for ResNet as of 4/19/95. Even though most ResNet staff did not directly report to Peeter (noted by the many-dotted line relationships on the chart), he seemed to have no problem managing them. Peeter motivated people by providing them with challenging work and strong leadership. He also made sure that the functional area managers supported ResNet and encouraged their staff to do a good job on this highly visible project. NWA was a strong functional organization, but most of the people assigned to work on ResNet were assigned full-time. Having people assigned full-time also made it easier for Peeter to manage such a large group of people.

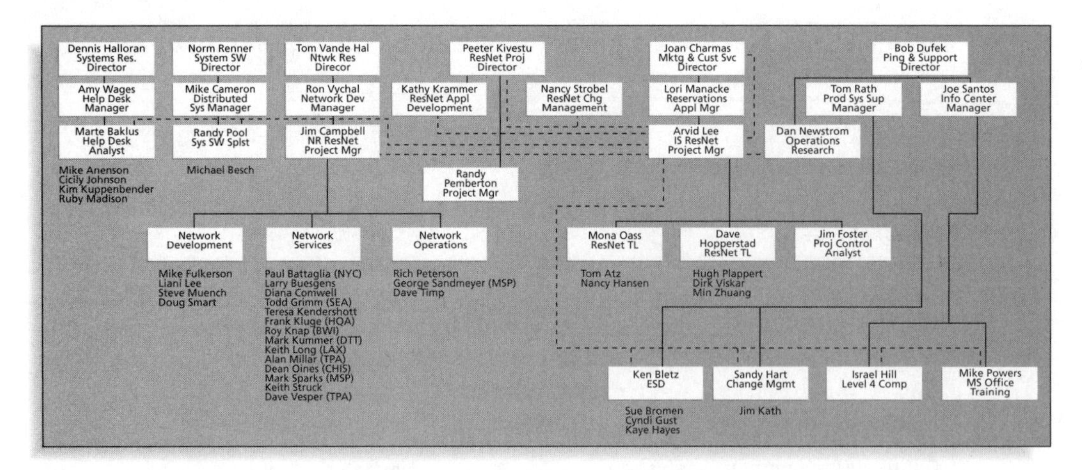

Figure 13-4. ResNet Project Organization Chart

The ResNet team created templates for communicating project information. For example, they had templates for weekly and monthly status reports. They also set up a shared network drive for all project information. Anyone working on the ResNet project could access project files, including the Microsoft Project files. The templates and shared network drive helped the team share information and be more efficient in creating project documents.

The ResNet team also established regular meeting times for the technical project personnel, the management team, and senior management. Peeter, Arvid, and Kathy all had prior project experience, and they knew how important it was to have good communications with all project stakeholders. Establishing regular meeting times facilitated strong communications throughout the project.

QUALITY, RISK, AND PROCUREMENT PLANNING

The PR2 included a few paragraphs related to project quality, risk, and procurement planning, but ResNet did not include formal plans in each of these areas. The ResNet team, however, developed and followed strategies for success in each of these areas. For example, the measurement techniques for tracking call handling time and number of direct sales closed were critical elements in assessing the quality of the system. The PR2 also included plans for which hardware and software vendors ResNet would use based on their technical approach and NWA's prior experience with preferred vendors.

The PR2 included a brief discussion of plans to manage risks, especially for software development. Risk is always a part of software development, and NWA relied on advice from their primary software vendor, Qantas, to mitigate these risks. Qantas' approach to reducing software development risks was to follow the best practices in the airline reservations industry. For example, Qantas recommended that NWA use QIK ACCESS (QIK 1 and later QIK 2) software. QIK ACCESS is interface application software that many other airlines have used for creating interfaces to their reservation systems. NWA chose to purchase software that was a proven tool instead of creating a customized tool. Qantas also recommended that NWA have reservation agents instead of information services people use the QIK ACCESS software to help develop the ResNet interface. By having users be the primary software developers for the ResNet interface, the project team minimized the risk of not understanding user requirements. (The software development process is discussed further in Chapter 14, *Executing*.)

In the area of procurement planning, the ResNet team analyzed several vendors for hardware and software. They developed good relationships with their vendors, primarily Qantas, the company that provided QIK Access. They also hired outside contractors from Sabre Decisions Technology to assist them in the software development process. These contractors had experience in using QIK ACCESS for other reservation systems, and they provided training and advice for using the software. NWA also had good contracts with their vendors. For example, when the communications hardware for the Chisholm office was not working properly, NWA exercised their rights in a contract clause to have the vendor fix the hardware at no charge to NWA.

It is difficult to summarize all of the planning done on such a large, important information technology project. NWA did an extensive job of planning its ResNet projects. The PR2 provided detailed planning information tailored to the needs of the ResNet project. The ResNet project team took planning very seriously, and their plans provided the basis for successful project execution.

CASE-WRAP UP

Arvid decided to focus on Level 3 in most of the project planning to avoid getting lost in the details. Each person responsible for the Level 3 items could do their own, more detailed, planning. Peeter and Arvid found that having project plans focused on key deliverables and at a Level 3 or 4 WBS worked well in managing the project. Following NWA's PR2 guidelines for submitting project plans also aided Peeter in preparing a good project plan. Microsoft Project was useful for project scheduling. NWA liked to keep its overhead low, and this software tool provided a very inexpensive way to prepare several planning documents such as Gantt charts and

schedule-related reports. Arvid also set up a shared drive on their local area network so the ResNet team members could all easily view the Microsoft Project files and other documentation. Arvid and other team members ran several reports from Microsoft Project, such as a list of tasks due by each person each month. The ResNet plans were a critical asset for guiding project execution.

CHAPTER SUMMARY

Project planning involves all nine of the project management knowledge areas. Many processes and outputs are involved in project planning. ResNet planning documents included a project plan, work breakdown structure, project schedule, cost estimates, and project organizational chart.

Factors that contributed to successful planning, and later successful implementation, of ResNet included planning at a WBS Level 3 or 4, getting strong user involvement, and developing a solid measurement tool to track overall project benefits.

Planning for the ResNet 1996 project focused on software development and office implementation milestones. The project team used experience gained during the ResNet Beta and ResNet 1995 projects to create realistic estimates for the time and costs involved. They also followed company guidelines in creating their PR2, NWA's version of a project plan.

The NWA ResNet team used Microsoft Project to assist in project planning, although they did not use all of the Microsoft Project features and tools. They developed Gantt charts and set milestone dates that allowed them to establish realistic schedules and helped them meet their goals.

The project team used vendor quotes to estimate ResNet 1996 costs for purchasing hardware, software, and network equipment. They based internal development costs for coding and training on estimates of per-hour labor costs of people from various departments. Because the team drew on a large experience base, they had a high degree of confidence in their cost estimates.

The ResNet team created staffing plans and documents and procedures to aid in project communications management. The PR2 for the ResNet 1996 project provided detailed staffing information for ResNet throughout the project's life cycle. The team prepared templates for project communications and shared information on a local area network drive.

The PR2 included information related to quality, risk, and procurement planning. Although they did not write separate plans for each of these areas, the ResNet team developed and followed successful strategies for managing these areas.

Discussion Questions

1. Review the list of processes and outputs involved in project planning. Which outputs did NWA emphasize?
2. Discuss Peeter's request for Arvid to create a plan for the ResNet Beta project in one week. How would you react to this request? Is it realistic to take only one week to plan a project that will last more than a year? How do you think Peeter made this request so that Arvid responded so well?
3. What planning decisions were made on the ResNet Beta project that helped ensure successful completion?
4. What made it easier for the ResNet team to develop plans for the 1996 ResNet project? What challenges did they face?
5. Review the Gantt charts for the ResNet 1996 application development and office implementation. Why do you think these were created as two separate Gantt charts? Why do you think the project team did not link activities in Microsoft Project? Do you think linking activities in project management software is done often on real projects? Why or why not?
6. Review the WBSs and cost estimates provided. Discuss any questions you might have about them.

Exercises

1. Review the items NWA includes in their PR2 planning documents. Suggest other items that should be included. Create a new outline for a project plan based on your suggestions. Briefly describe what should be included in each area.
2. Compare the WBSs provided in this chapter with information in Chapter 4 on project scope management. Provide five suggestions for improving the WBSs for ResNet.
3. Develop a list of questions for which you would need the answers in order to link tasks and create a network diagram for ResNet software application development. Do the same for the office implementations.
4. Review the project organizational chart for ResNet. List which groups you think had the main responsibility for each task in the Gantt charts. Create a responsibility assignment matrix as described in Chapter 8, *Project Human Resource Management*.

Minicase

Developing a good project schedule is an important output of the planning process. As stated in this chapter, NWA did an excellent job of developing realistic project schedules and used Microsoft Project to generate Gantt charts, but they did not link tasks in their files.

Part 1: Review the Gantt chart for the ResNet 1996 Application Development. Using Project 2000, re-create this file with some improvements. For example, link tasks that you think should be linked. Also include the milestones for software development listed in Table 13-2 on your Gantt chart.

Part 2: Review the Gantt chart for the ResNet 1996 Office Implementation. Using Project 2000, re-create this file with some improvements. For example, link tasks that you think should be linked. Also include the milestones for office implementation listed in Table 13-2 on your Gantt chart.

Suggested Readings

1. Kouzes, James M. and Barry Z. Posner. *The Leadership Challenge*. San Francisco, CA: Jossey-Bass, 1990.

   Kouzes and Posner show that leadership is not the private preserve of a few charismatic people, but a learnable set of practices that virtually anyone can master. They suggest that leadership involves five basic practices—challenge the process, inspire a shared vision, enable others to act, model the way, and encourage the heart.

2. Levesque, Paul. *Breakaway Planning*. Amacom Books, A Division of AMA, 1998.

   This book walks readers through the entire planning process, from creating a compelling vision to making that vision a reality. Levesque describes eight important questions to address in planning, such as: How do we spread the word internally? How will we make things better for employees? How do we measure success?

3. Lientz, Bennet P. and Kathryn P. Rea. *Breakthrough Technology Project Management*. San Diego, CA: Academic Press, 1999.

   This text describes how to carry out collaborative scheduling and planning for information systems projects. The authors share their experiences by providing over 250 guidelines and lessons learned that address many of the issues project teams face in developing information systems.

4. Maguire, Steve. "Getting Your Team Off on the Right Foot." *Software Development* (May 1997): 37–44.

   A former Microsoft project manager offers advice for planning and managing project teams. He suggests stressing that project team members can create high-quality products, that the work can be done on time, that team members should not be required to put in long hours or seven-day workweeks, and that people should be excited about the work.

14

Executing

Objectives

After reading this chapter, you will be able to:

1. *Understand how important good project execution is to getting work results*
2. *Discuss the executing processes and outputs and how they were used on ResNet*
3. *Describe Peeter's leadership style and how he developed the core team*
4. *Discuss methods used to verify project scope and assure quality on ResNet*
5. *Describe how the ResNet team disseminated information to project stakeholders and managed project procurement*
6. *Explain NWA's rationale for having sales agents write some of the code for the ResNet system*
7. *Relate some of the executing events in ResNet to concepts described in previous chapters*

Over 100 people gathered in the large conference room at the headquarters of Northwest Airlines in Minneapolis, Minnesota. The invited guests included reservation sales agents, their supervisors and office managers, people from the Information Services Department, and people from the Facilities, Purchasing, and Quality departments. They were all looking forward to the ResNet 1996 Kickoff/Strategy Meeting. This project would include developing more custom software and completing the installation of ResNet at the remaining seven reservations and support offices.

Based on the success of the Beta project and the ResNet 1995 project, stakeholders had high expectations of Peeter Kivestu and his ResNet team. Peeter believed in making projects fun, and the ResNet team was especially known for having great project kickoff events. In addition to clarifying project objectives, the ResNet team provided great food, gave away coffee mugs and other items with the ResNet

logo, and made everyone feel welcome and excited about the project. Peeter even hired professional actors to coach his team on delivering engaging presentations and to help them act out some of the skits performed at the kickoff meeting. Now the ResNet team had to deliver what they had promised and make it an enjoyable experience for all stakeholders. Could they install over 2,000 more PCs in seven different offices as scheduled? Given the budget and time constraints, could they develop the necessary software?

WHAT IS INVOLVED IN EXECUTING PROJECTS?

Project execution involves taking the actions necessary to ensure that activities in the project plan are completed. The products of the project are produced during project execution, and it usually takes the most resources to accomplish. Table 14-1 lists the knowledge areas, processes, and outputs of project execution. The most important output is work results, or delivery of products. The ResNet 1996 project involved installing over 2,000 PCs in seven different offices, creating more customized software, training the reservation sales agents, and measuring the benefits of the system. This chapter will focus on ResNet's execution phase activities, such as providing project leadership, developing the core team, verifying project scope, assuring quality, disseminating information to stakeholders, procuring necessary resources, and training users to develop code.

Table 14-1: Executing Processes and Outputs

Knowledge Area	Process	Outputs
Integration	Project Plan Execution	Work Results Change Requests
Quality	Quality Assurance	Quality Improvement
Human Resources	Team Development	Performance Improvements Inputs to Performance Appraisals
Communications	Information Distribution	Project Records
		Project Reports
		Project Presentations
Procurement	Solicitation Source Selection Contract Administration	Proposals Contracts Correspondence Contract Changes Payment Requests

PROVIDING PROJECT LEADERSHIP

Peeter Kivestu was an experienced project manager, having led several project teams before becoming ResNet project manager. When asked what factors contributed to his success as a project manager, he mentioned the importance of *having clear goals, making the work fun, and sticking to schedules.*

Peeter's belief is that it is human nature to want to achieve goals. To achieve goals, however, people must have a very clear understanding of what those goals are. Peeter cares deeply about the business, and he wanted to show everyone involved in the ResNet project that a large information technology project can be successful. Anyone who has worked with Peeter knows that he likes to talk things out and get people to think all of the time. Peeter is known for having very long meetings to hammer out project goals. He always wants everything to be crystal clear to every member of the team before ending a meeting. He challenges people to use every spare moment to think through ideas—while driving, while waiting in line, and even while taking a shower. His ResNet team would tease him about needing a whiteboard in their showers to write down ideas.

Peeter also believes that work should be fun. He knows most information technology professionals are introverted by nature, and this quality often interferes with their ability to communicate with their users. He wanted his technical staff to get to know the reservation agents and other people involved in ResNet. Peeter also believes that strong beginnings are important. Many project managers underestimate the amount of confusion at the start of projects. To help put people at ease and enhance communications, Peeter made sure that various aspects of ResNet got off to good starts. For example, the ResNet team made special plans for each project kickoff meeting and office installation. They planned special themes for each event to make them fun and memorable, as described in the *What Went Right* section. Whenever possible, they used humor to enhance the project. As mentioned in the opening case, project kickoff meetings were huge and entertaining productions; everyone looked forward to them.

What Went Right?

To make ResNet fun, Peeter's team created videotapes and themes for important project activities. For example, NWA prepared a videotape of the ResNet beta test agents sharing their responses to the new system and another tape of the final project recognition dinner. The ResNet team also planned and executed exciting kickoff meetings. They worked with each ResNet office to develop and follow a theme for each office implementation. For example, the Tampa office used a "Flintstones to the Future" theme for ResNet. Months before the Tampa office would receive their ResNet PCs, the ResNet team put up posters about the new system, alerting the office staff to get ready for the future. At the first meeting to coordinate the Tampa installation, people dressed to look like Star Trek characters—they wore silver collars and put the ResNet logo on their shirts. Peeter, who became a Vice President at

NWA in 1999, still proudly displays his silver ResNet logo in his office. The Minneapolis office used a "Broadway Shows" theme, and the ResNet team and sales agents danced in a chorus line. Each office looked forward to their implementation and had fun creating their own theme.

A project management consultant once told Peeter of the chaos caused by missing dates on projects. Peeter decided to set milestone dates with his team and stick to those dates. Everyone liked Peeter's charismatic and fun approach to work, but they also knew that he did not budge on dates unless absolutely necessary. This strict adherence to schedules helped people focus on accomplishing their work on time.

DEVELOPING THE CORE TEAM

Peeter knew that he had to work well with people on his management team and that they, in turn, had to work well with other project stakeholders. Peeter had several strategy meetings with Arvid Lee, the head of the Information Services people on the ResNet project, and Kathy Christenson, the head of software application development for the ResNet interface. They worked together to determine how to motivate different people involved in the project. Kathy came from the marketing area, and she knew that the sales agents and other people in marketing were very excited by themes, office decorations, gifts, contests, and so on. Arvid knew that he and most other members of the Information Services Department would think those types of things were silly, but he fully supported the approach. Arvid motivated the information services staff by providing them with challenging work and keeping them informed of project progress.

Peeter was a hands-on manager and he felt that every single person involved in ResNet was important. He gave a lot of responsibility to Arvid and Kathy, keeping in constant communication with them and relying heavily on their judgments. Peeter kept their project sponsor, Fay Beauchine, well informed of the project's progress. He went out of his way to talk to lower-level people involved in ResNet. He wanted to know what everyone was doing and how he or she thought things were going. Peeter had an excellent memory and made a point of having focused discussions with all ResNet stakeholders.

Peeter provided resources to people as they were needed. In addition to hiring professional actors to help his team run better meetings, he sent people to technical training classes. He shared his experience and advice with others, and what's more important, he worked with them to develop critical thinking skills. Peeter knew that a key part of his job was supporting and developing his staff.

VERIFYING PROJECT SCOPE

As mentioned in Chapter 13, *Planning*, most of the project plans for ResNet were done at a WBS Level 3 or 4. Peeter believed in focusing first on the broad goals of the project and then narrowing down to the details. He held long meetings to help brainstorm and develop important ideas and issues regarding the scope of various aspects of ResNet. Once everyone shared their thoughts, the team worked together to nail down exactly what could be done to meet the overall project goals.

Peeter believed that many information technology projects fail to focus on the necessary requirements and become sidetracked by scope creep. He was determined to have ResNet meet requirements on time and on budget. He met with project members as scope changes were proposed, and together they worked to do what was best for the project. Peeter always stressed the importance of focusing on business needs and not succumbing to unnecessary requests.

Peeter and Fay planned for incrementally developing the ResNet interface. Even though the system worked after the beta project, they knew it could still be improved. They budgeted for additional people to continue developing enhancements to ResNet after it was implemented to more and more sales agents. For example, ResNet included a feature for sales agents to electronically send in suggestions for improving the system. Handling these requests for enhancements will be discussed further in Chapter 15, *Controlling*.

ASSURING QUALITY

During planning for the ResNet beta project, Peeter decided to have people with industrial engineering experience analyze the reservations process and develop techniques to measure the impact of ResNet. This decision proved to be very wise. The industrial engineers used a systematic approach to document the reservations process workflow. For example, in the PARS system, agents often had to go back and forth between several screens and jot down information on a piece of paper to help a caller. The industrial engineers documented over thirty different scenarios for sales calls. This information helped the software developers design the user interface and flow between screens for ResNet.

Figure 14-1 shows Quadrant 4 of the ResNet screen. Notice the many different cells—16 in this example. The industrial engineers worked with the sales agents to design this quadrant and determine the optimum cell positions based on the workflow of making reservations. Each cell corresponded to a key on a special ResNet keypad. After an agent pressed one of the keys on the keypad for a specific cell, the information in Quadrant 4 would change to show related functions. Spending time on the human engineering aspects of the system helped the team design a high-quality application that streamlined the reservation process.

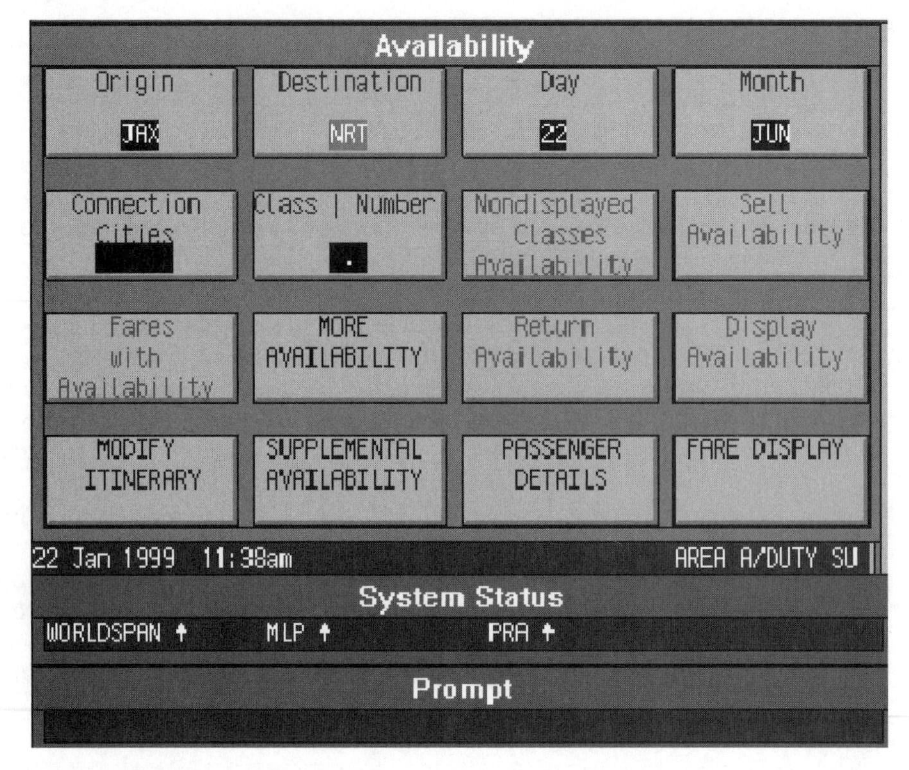

Figure 14-1. ResNet Quadrant 4

Because of Northwest Airlines' poor financial status, Peeter knew that any big project would be scrutinized. He asked the industrial engineers to develop benefits measurement techniques to track performance in increasing direct sales and decreasing call handle time, two of the main business objectives of ResNet. The industrial engineers physically observed what happened when ResNet PCs were installed during the beta project and office installations. They graphed the sales agents' progress on increasing direct sales and reducing call handle times. The industrial engineers were the eyes and ears that provided objective feedback on the business value of the system.

Peeter's team also assured quality by following internal procedures for software development and hardware installation. For example, the software developers created and followed design guidelines for the system. One of the design standards was that users would never be more than two keystrokes away from the next function. They used best practice for determining colors, fonts, and so on. They also developed alternate features to accommodate users who were color-blind. The Information Services staff followed internal testing procedures after installation of new PCs and network hardware. Many experienced professionals were working on ResNet, and they constantly tested their work to make sure it was done well.

DISSEMINATING INFORMATION

Communication was a key factor in ResNet's success. As Kathy Christenson put it, "Fear of the unknown is detrimental to a project." The ResNet team made a point of disseminating project information often and in different ways. Just as people respond to different motivating factors, they also respond to different forms of communication. The ResNet team used meetings, posters, electronic communications, and written reports to disseminate project information.

Figure 14-2 shows one report format that many stakeholders found useful and informative. The report uses a visual representation to show office implementation progress for the training, communications, and network rooms, and for installation of ResNet PCs for sales agents and office managers. Items shaded in gray (they were green in the actual report) indicated areas where the ResNet PCs or other related hardware were installed. Most offices, like the Detroit Reservations Center highlighted in Figure 14-2, had separate rooms for training, communications hardware, network hardware, and managers' offices or specialty desks. The reservation sales agents sat in pods or circular seating areas with six PCs per pod. In addition to coloring the completed areas for each office, this one-page report included at the top a quick count of the number of PCs installed to date.

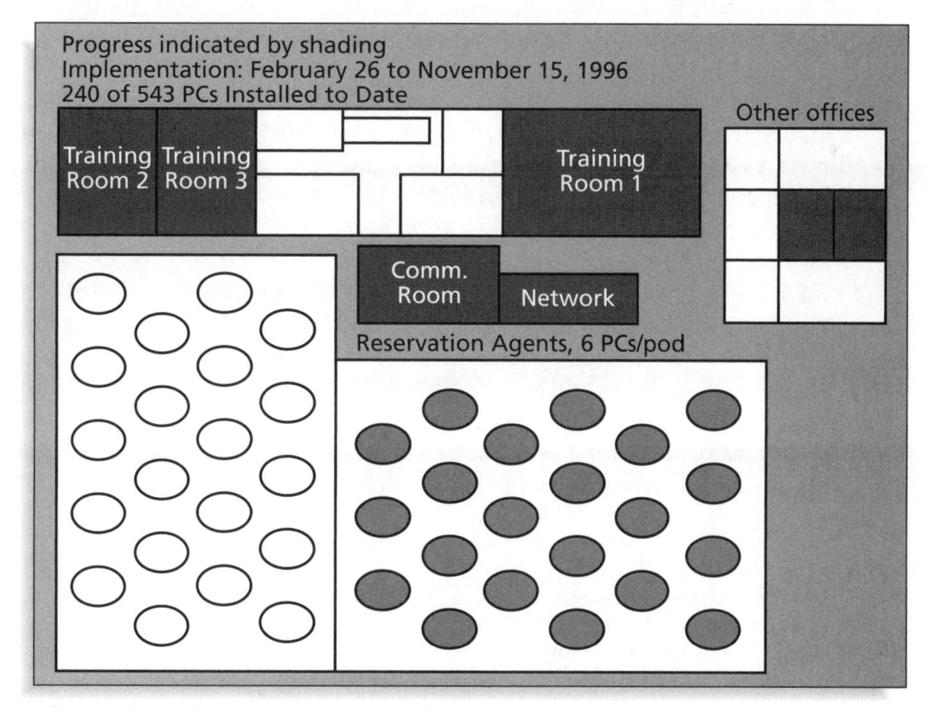

Figure 14-2. Detroit Reservations Center Progress Report

The ResNet team set up change management teams at each office to keep each geographic area informed of specific activities for their offices. For example, the change management team coordinated the delivery and installation of all hardware and scheduled ResNet training for all office personnel. Keeping people well informed helped to ease their fears of using a new system. The training program included brief one-on-one sessions with each reservation sales agent in addition to formal classroom training. Because many sales agents were wary about moving to a new computer system, ResNet management decided it was important to address the unique questions and concerns of each individual sales agent. They knew that many agents would be afraid to ask questions in a large training session. The ResNet team also had a strong informal communications network. At all stages of the ResNet project, Peeter and his team tried to keep everyone informed about the ResNet activities that mattered to them most.

PROCURING NECESSARY RESOURCES

ResNet involved the procurement of many off-the-shelf hardware, software, and networking products. Several vendors provided personal computers, servers, printers, networking hardware, operating systems, communications software, reservations system software, and training.

Figure 14-3 shows a schematic of the ResNet network architecture. Note the many different hardware platforms, communications protocols, and operating systems involved. The main reservation system and database to which ResNet interfaces is Worldspan, which runs on mainframe computers in Atlanta. The Worldspan computer reservation system supplies over 20,000 customers in more than 60 countries with availability, fare, and rate information. In 2001, Worldspan was the market leader in e-commerce for the travel industry, processing more than 50 percent of all online travel agency bookings.[1] Northwest Airlines had its own IBM mainframes running IBM's Multiple Virtual Storage (MVS) and Virtual Machine (VM) operating systems. ResNet also used information from Unisys mainframe computers and several distributed servers using Unix, OS/2, and Windows NT operating systems. NWA's local area network (LAN) included other servers and systems such as Unix, Novell, and Lucent. The ResNet PCs were Pentium 75 MHz machines, state-of-the-art at the time of purchase.

[1]Worldspan's Web site, www.worldspan.com (February 14, 2001).

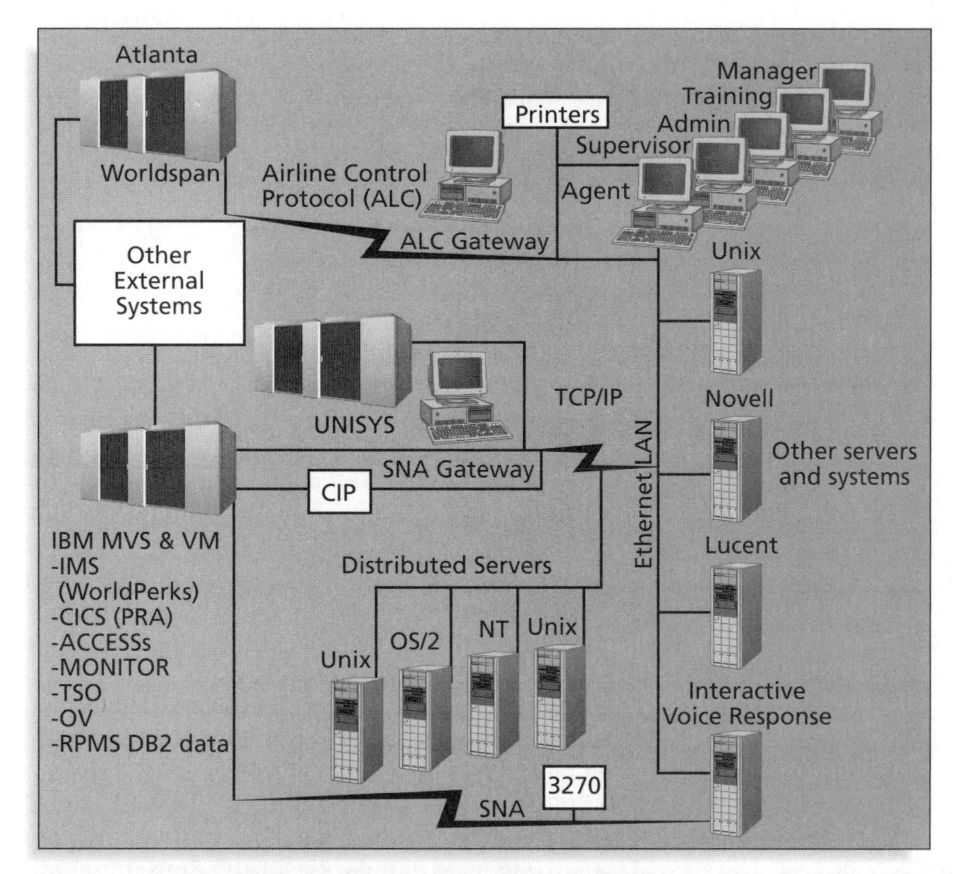

Figure 14-3. ResNet Data Network Overview Schematic

Using a variety of hardware and operating systems is typical on large information technology projects. The main challenge for the Information Services department was getting all of the different computers to communicate. The Information Services staff procured and installed the necessary hardware and wrote the system software for the ResNet infrastructure. As shown in Figure 14-3, Worldspan uses IBM's Transaction Processing Facility (TPF), a high-volume transaction-processing platform used by many airlines. Other communications technologies included Systems Network Architecture (SNA), a set of network protocols developed by IBM, and Transmission Control Protocol/Internet Protocol (TCP/IP). the communications protocol used to connect hosts on the Internet. The ResNet 1996 project was completed before the rapid expansion of Internet applications. ResNet+, the follow-on project to ResNet, uses newer technologies and is described briefly in Chapter 16, *Closing*.

Arvid Lee was instrumental in working with various vendors to procure hardware, software, and technical training. He used his twenty-five years of experience in information technology to the company's advantage. He researched the best information technology products and vendors and consulted with other information services professionals to make purchasing decisions. Much of the hardware and software needed was available off-the-shelf, so the ResNet team was not pushing the state of the art. Arvid also had a good relationship with people at Qantas, the vendor that provided the QIK-ACCESS software as well as the training to use it for developing the graphical user interface for ResNet.

TRAINING USERS TO DEVELOP CODE

A critical decision concerning the ResNet software was who would develop the graphical user interface the sales agents would use. The ResNet team decided that QIK-ACCESS, an off-the-shelf software package from Qantas, was the best tool available. Software development for the ResNet user interface involved using the QIK-ACCESS scripting language to develop code that would generate the user interface to meet NWA's unique needs for ResNet. Qantas recommended that the users—the reservation sales agents—be involved in developing the software and writing the code. Qantas had experience training non-information technology people to customize their software, so they provided the training as well as the QIK-ACCESS software.

Having users—sales agents—actively participate in software development was a new concept for NWA. Arvid researched the advantages and disadvantages of having users do the coding for the user interface. (Information Services staff did all of the systems programming, which is much more technical.) The sales agents understood the reservation process, so understanding user requirements would be less of an issue if they wrote most of the code for the ResNet user interface. On the other hand, coding would go faster if experienced programmers did most of the actual programming. In the mid-1990s, the majority of sales agents had college degrees, but none had coding experience. Training the agents to use the scripting language would take time, and, once trained, the agents might not work as quickly as someone with years of programming experience.

Arvid learned that another major airline was having a difficult time implementing their new reservation system, which also used Qantas' software. Their Information Services department was doing all of the coding. Arvid consulted with Kathy Christenson, and they decided that having users do as much of the coding as they could for the ResNet interface made the most sense. However, they also decided to have Information Services staff assist them.

Peeter agreed to this combined approach. He felt it was good to take a more conservative, business-oriented approach and have the users develop the code that was absolutely necessary for them to do their jobs. Information Services Department members might generate code more quickly, but they could easily have trouble interpreting what users needed. Peeter interviewed several sales agents who volunteered to help develop the software. There were over 4,000 sales agents, so Peeter had a large pool of people from which to choose. The sales agents were union employees, so Peeter had to work with union management to be fair to all union workers. Peeter budgeted for six sales agents and three Information Services people to write the ResNet interface for the beta project. After that, Peeter budgeted for even more people to help test and enhance the software, as described in Chapter 13, *Planning*.

As soon as funds were approved for the ResNet beta project, the six sales agents and three Information Services application developers flew to Texas for a special training course provided by Qantas. Table 14-2 provides a sample of the scripting code they used to write a subroutine that compares agent statistics between reservation offices. As you can see, it would take time for anyone to become proficient in using this text-based, procedural programming language. The team of sales agents and Information Services staff worked throughout the beta test and the ResNet 1995 and ResNet 1996 projects to develop the graphical user interface for ResNet.

Table 14-2. Sample Scripting Language for ResNet User Interface

```
;*****************************************************************************
;* Subroutine: B_A_OFCCOMPARE_SUB
;* Called By:
AR_MTD_DISP_SUB,AR_MTD_DISP_GLD_SUB,AR_MTD_DISP_INL_SUB,AR_MTD_DISP
_OTH_SUB
;* Parameters:      # 1 = column to display data
;*                  # 2 = d.i. for city code
;*                  # 3 = d.i. for dt
;*                  # 4 = d.i. for dt2
;*                  # 5 = d.i. for netrev
;*                  # 6 = d.i. for cph
;*                  # 7 = d.i. for ctt
;*                  # 8 = d.i. for acw
;*                  # 9 = d.i. for occ
;*                  # 10 = d.i. for sit
;* Function:        Builds office comparison of agent statistics
;* Date/Init:       17NOV97 MMR
;* Updates:
;* SPLIT COMMENTS 01SEP98 TLF Renamed applicable objects for application split
;*****************************************************************************
set ar_temp_col"
display #2 ns_cur_window '5' #1 ns_bright_yellow ns_blue
display #3 ns_cur_window '6' #1 ns_bright_white ns_blue
display #4 ns_cur_window '7' #1 ns_bright_white ns_blue
calculate ar_temp_col = #1 - '2'
display #5 ns_cur_window '8' ar_temp_col ns_bright_white ns_blue
calculate ar_temp_col = #1 - '1'
display #6 ns_cur_window '9' ar_temp_col ns_bright_white ns_blue
display #7 ns_cur_window '10' #1 ns_bright_white ns_blue
display #8 ns_cur_window '11' #1 ns_bright_white ns_blue
calculate ar_temp_col = #1 + '1'
display #9 ns_cur_window '12' ar_temp_col ns_bright_white ns_blue
;*** office comparison for sit displayed if DRS gmmores3 DISP COL SIT: IS NOT 'N'
if ar_disp_col_sit <> 'N'
  display #10 ns_cur_window '13' #1 ns_bright_white ns_blue
end_if
```

Based on the workflow of the reservations process, the beta software developers decided to have four quadrants on the ResNet screen at one time. The Quadrant 4 section of the screen, as shown in Figure 14-1, took the most time to develop. The other quadrants were similar to the old reservation system, but the agents could now see important data all on one screen. The system included prompts to help sales agents pronounce unfamiliar words or read prepared promotion scripts. For example, after developing a partnership with KLM, sales agents often handled flights to Japanese cities whose names they did not know how to pronounce. ResNet screens included prompts that explained how to pronounce these cities phonetically, such as "TOK ee oh" for Tokyo. Agents were also prompted to ask for the direct ticket sale by saying, "What type of credit card would you like for me to hold that with?"

What is most important, the majority of sales agents, over 4,000 of them, liked the new system. ResNet sales agents made the following comments:[2]

- "This computer is just wonderful. The colors are great. It's easy to look at. It's a lot of fun."
- "It makes the customer feel confident that you know what you're doing."
- "You have time to think about more important things like nonstop service to certain cities that we have that other airlines don't."
- "Not going home at night with headaches. That's my favorite part."
- "It's fun. It's new and it's different. You look at the big picture and what it can save Northwest and the revenue we can bring in. The bottom line is it's a great system—it's a moneymaker. It's quick, it's fun, and it's easy. What more can you want?"

CASE-WRAP UP

Key deliverables and dates were well defined for ResNet. Peeter, Arvid, Kathy, and the rest of the ResNet team worked hard to ensure that work was delivered on time, close to budget, according to specifications, and with high quality. The PCs and other hardware were delivered according to plan, and the sales agents were satisfied with the new software. The ResNet team successfully installed over 3,000 new PCs, developed several software programs, and provided training to all the reservations sales agents.

[2]JUNTUNEN VIDEO, Inc. Northwest Airlines' ResNet videotape, Minneapolis, MN, April 4, 1995.

ResNet met its business objectives by increasing the number of direct sales and reducing the average time sales agents took to handle calls. In 1996 alone, ResNet generated over $15 million in savings, and the sales agents, all union employees, were very happy with the new system. The ResNet team completed work as planned, celebrated each small achievement, and proved that a large information technology project could be successful and improve the company's bottom line.

CHAPTER SUMMARY

Project execution involves taking the actions necessary to ensure that activities in the project plan are completed. The results of the project are produced in this phase, and it is usually when the most resources are needed.

The main knowledge areas involved in execution are integration, scope, quality, human resources, communications, and procurement. Outputs include work results, change requests, quality improvement, and various procurement items such as contracts.

Factors that contributed to successful project execution on ResNet were having clear goals, making the work fun, and sticking to schedules.

Leadership, team development, scope verification, and quality assurance also contributed to the successful execution of the ResNet projects. Peeter, Arvid, and Kathy worked together to motivate the ResNet project team. Peeter and Fay managed potential scope creep by budgeting for people to develop enhancements to ResNet throughout the project, and office managers set priorities for what enhancements were most important. Industrial engineers helped ensure quality by providing detailed analyses of the reservations process and developing benefits measurement techniques.

Other factors that contributed to project success included information dissemination, procurement, and strong user involvement. Because people respond well to different forms of communication, the project team used various forms of communications to meet stakeholder needs. ResNet involved procuring resources from several different hardware and software vendors, and staff at NWA used their past experiences to select the best vendors, create good contracts, and develop good vendor relationships. Strong user involvement, especially in software development of the ResNet user interface, was a critical success factor for the project.

Discussion Questions

1. What are the main knowledge areas, processes, and outputs of project execution?
2. Discuss how executing processes were done on ResNet and what the outputs of each process were. What were some of the unique ways that Peeter and his team handled project execution?
3. Describe Peeter's leadership style. What made him an effective project leader? Would he be an effective project leader in a different organization? Why or why not? What role did organizational culture play in his leadership style?
4. What were the three main success factors for ResNet, according to Peeter? Can these factors be applied to all large information technology projects? Explain your answer.
5. What do you think about having users go to training classes to learn to write code to develop their own systems? Do you think this could or should be done on more information technology projects?
6. How much impact do you think making the project fun had on project success? Do you think that most large information technology projects could copy this approach?

Exercises

1. Review the ResNet Data Network Overview Schematic shown in Figure 14-3. Research the various technologies listed: Worldspan, Airline Control Protocol (ALC), SNA, IBM MVS & VM, WorldPerks, and so on. Pick three of these technologies and write a brief paper describing each. Include the advantages and disadvantages of using each technology.
2. Review research on project success factors, such as studies done by the Standish Group (www.standishgroup.com). Write a paper summarizing the research results, then compare those success factors with the three success factors mentioned by Peeter Kivestu.
3. Find other examples of projects that took a fun approach to performing the work. Analyze these projects: Were they successful? Why or why not? Write a one- to two-page paper summarizing what you learned by examining examples of other projects.
4. Review the sample progress report in Figure 14-2. Why do you think stakeholders liked this format so much? Find an example of another project status report. Try to make the report more visual and clear to users, like this example.
5. Read one of the books mentioned in the Suggested Readings and write a summary of it.

MINICASE

Peeter was an outstanding leader. He was particularly good at communicating project information and motivating the ResNet team.

Part 1: Pretend that you were Peeter at the ResNet 1996 Kickoff/Strategy meeting described in the opening case. Create a short presentation and script that you would use to start the meeting. Limit your presentation to five to ten minutes.

Part 2: Peeter and the ResNet team did their best to make the ResNet projects fun. Review some of the themes described in the *What Went Right?* section of this chapter. Then develop a theme for your class or workgroup to help make the work more fun. Create a poster or other visual aid to help promote your theme.

SUGGESTED READINGS

Peeter, Arvid, and Kathy mentioned several authors that influenced their leadership styles. Following are a few of the books they found helpful:

1. Kanter, Rosabeth Moss. *When Giants Learn to Dance*. New York: Touchstone Books, 1990.

   In this book about new management strategies and techniques, Dr. Kanter shows how the truly innovative companies are leading the way. She describes how corporate "giants" are actually joining this post-entrepreneurial revolution.

2. Miller, John G. *Personal Accountability*. Denver, CO: Denver Press, 1998.

   The first part of this book explains the Question Behind the Question, a tool to help eliminate blame, victim thinking, and procrastination from people's lives. The second part explores the Pillar Principles, which include ideas such as courage, excellence, ownership, trust, and integrity.

3. Oakley, Ed and Doug Krug. *Enlightened Leadership*. New York: Fireside, 1994.

   Being able to change to keep pace with a rapidly changing world is key to business success today. Managers and leaders at all levels can use Oakley and Krug's proven techniques, including planning, communication, and motivational tools, to support their employees in effecting the positive changes that will make the difference in achieving their organizations' bottom-line goals.

4. Peters, Tom. *Thriving on Chaos*. New York: Harper Collins, 1991. (Also, see other books by the same author.)

   Peters, the coauthor of In Search of Excellence *and* A Passion for Excellence, *provides readers with a book that describes fifty specific courses of action essential to corporate survival in today's turbulent world. Tom Peters also wrote a book called* The Project 50, *published in 1999, emphasizing the importance of good project management.*

5. Roberts, Wess. *Leadership Secrets of Attila the Hun*. New York: Warner Books, 1991.

Wess Roberts draws from the imaginary thoughts of one of history's most effective and least-beloved leaders—Atilla the Hun—to discover leadership principles that can apply to modern situations. This book discusses principles for successful morale building, decision making, delegating, and negotiating, and gives advice on overcoming setbacks and achieving goals.

15

Controlling

Objectives

After reading this chapter, you will be able to:

1. *Understand the importance of good project control to keeping things on track*
2. *Discuss the controlling processes and outputs and how they were used on ResNet*
3. *Describe the tools and techniques used for project control on ResNet*
4. *Discuss challenges the ResNet team faced in controlling the project and decisions they made to manage these challenges*
5. *Describe the use of change management on this project*
6. *Relate some of the controlling events in ResNet with concepts described in previous chapters*

Peeter learned that managing change was 50 percent of a project manager's job. He also learned that every stakeholder was different. His team had to develop strategies for dealing with the unique personalities involved in ResNet. The executive management team, in particular, needed special consideration, as did the reservation sales agents. Peeter's team knew that members of the executive team liked to make decisions. They also knew that typical status reports would bore this group. Therefore, at the 1996 ResNet review meetings, instead of reporting on how things were going, each presenter focused on key issues and decisions needed by the executive team. Would this strategy help keep ResNet on track and get the executive management team to make important decisions quickly? The unionized reservation sales agents were also crucial stakeholders in ResNet's success. How should the ResNet team manage the problems inherent in developing and delivering a new information system to these sales agents? How could they overcome potential resistance to change?

WHAT IS INVOLVED IN CONTROLLING PROJECTS?

Controlling is the process of measuring progress towards project objectives, monitoring deviation from the plan, and taking corrective action to match progress with the plan. Controlling cuts across all other phases of the project life cycle. It also involves seven of the nine project management knowledge areas:

- Project integration requires integrated change control. Outputs include updates to the project plan, corrective actions, and lessons learned.
- Project scope management includes scope verification and change control. A key output is scope changes. (Note that some aspects of scope verification for ResNet were described in Chapter 14, *Executing*.)
- Project time management includes schedule control. The output of this process is schedule updates.
- Project cost management involves cost control. Outputs include revised cost estimates, budget updates, and estimates at completion.
- Project quality management includes quality control. Outputs are quality improvements, acceptance decisions, rework, completed checklists, and process adjustments.
- Project communications management includes performance reporting, and outputs of this process are performance reports and change requests.
- Project risk management involves risk monitoring and control. Outputs of the risk monitoring and control process include work-around plans, corrective actions, project change requests, and updates to the risk response plan.

Table 15-1 lists the knowledge areas, processes, and outputs that are generally part of project controlling. This chapter will focus on the ResNet project's key controlling activities such as schedule control, scope change control, quality control, performance and status reporting, and employee change management.

Table 15-1: Controlling Processes and Outputs

Knowledge Area	Process	Outputs
Integration	Integrated Change Control	Project Plan Updates Corrective Actions Lessons Learned
Scope	Scope Verification Scope Change Control	Formal Acceptance Scope Changes Corrective Actions Lessons Learned Adjusted Baseline
Time	Schedule Control	Schedule Updates Corrective Actions Lessons Learned
Cost	Cost Control	Revised Cost Estimates Budget Updates Corrective Actions Estimate At Completion Project Closeout Lessons Learned
Quality	Quality Control	Quality Improvement Acceptance Decisions Rework Completed Checklists Process Adjustments
Communications	Performance Reporting	Performance Reports Change Requests
Risk	Risk Monitoring and Control	Work-around Plans Corrective Actions Project Change Requests Updates to the Risk Response Plan Risk database Updates to Risk Identification Checklists

SCHEDULE CONTROL

Recall that Peeter's main strategy for avoiding schedule-related problems and conflicts was to focus on meeting schedule dates. Peeter worked with his project team to determine important milestones and set realistic dates for their completion. Peeter also focused on meeting business needs. Problems with WorldPerks and the Chisholm office caused the start of the Detroit office installations to slip by about three weeks, but it was in the best interest of the business to allow that slippage to occur.

Peeter admitted that they sacrificed some functionality of the ResNet system to meet deadlines. Recall that Peeter and Fay budgeted for a certain number of people to work on ResNet software enhancements after writing the beta software. These people developed as many enhancements as they could, given their schedule constraints. Peeter felt that making this trade-off was in the best interests of the overall project and business. Overall, the project was completed on time and just slightly over budget.

Recall that part of a project manager's job is making trade-offs related to the triple constraint—scope, time, and cost. Peeter often focused on meeting time constraints to avoid the problems that missed deadlines would have on the rest of the project. This strategy meant that he and his team worked long hours during certain parts of the project.

Arvid recalled several times when he did not think his team could meet a deadline, and Peeter would say, "Failure is not an option." Because the ResNet team were all salaried employees, they did not receive paid overtime. They worked extra hours as needed to help an important and challenging project succeed.

What Went Right?

Peeter had gone through project management training when he worked in Canada. One piece of advice he remembered was to focus on meeting the project schedule goals. Date changes tend to cause chaos throughout a project. Therefore, his goal for ResNet was to establish key dates and vary the scope as needed to stay on schedule. To emphasize the importance of schedule, the ResNet team made the theme for the 1996 ResNet kickoff meeting "What About Wendy?" Wendy was the office manager of the last call center scheduled to get ResNet PCs. The theme stressed the importance of meeting schedule dates so that Wendy's group would not be disappointed. The ResNet team displayed a Gantt chart and explained that missing any deadlines would hurt Wendy. This personal touch helped everyone focus on meeting deadlines.

Throughout the ResNet projects, Peeter stayed firm on dates, and all stakeholders knew that key dates were real and not subject to change unless there were insurmountable obstacles. Peeter used a football analogy to explain his commitment to keeping the schedule: If a team keeps getting first downs, it will score touchdowns. If his ResNet team focused on meeting key incremental dates, they would finish the project successfully. This strategy worked well on ResNet most of the time.

What Went Wrong?

The WorldPerks application development and office rollout in Chisholm, Minnesota, provided one of the toughest challenges for the ResNet team. They had underestimated how long it would take to develop the WorldPerks software, plus they ran into technical and personnel problems. The WorldPerks software had to interface with the IBM 3270 system, and the ResNet team and vendors had to develop special communications software for this interface. It was much more complicated than they anticipated. Also, the state of Minnesota funded the Chisholm office to develop more jobs in the Minnesota Iron Range. All of the office employees were new hires, so they did not understand Northwest Airlines' reservations systems work flow. Many of the new employees had little experience with computers, so the ResNet team had to send in people to work with them and help coordinate technical and training problems with the main office in Minneapolis. The project incurred additional costs, and the extra work for this office delayed the start of the Detroit implementation by about three weeks.

Peeter and Fay wanted to change the sequencing of office installations after the problems that occurred in Chisholm. The Detroit office installation also used the WorldPerks application and could not be completed until the Chisholm office was completed. (Note the finish-to-finish relationship between Office 2 and Office 3 on the Gantt chart in Figure 13-3.) Peeter and Fay did not want the Detroit installation to be late, as this would affect the entire project schedule. They also did not want people waiting around in Detroit because their start had been delayed. Arvid and other ResNet team members banded together to successfully convince Peeter and Fay not to change the sequencing of office installations because it would disrupt many of their plans and cause even more problems. With a lot of extra hours and some extra money for additional temporary staff, the Chisholm and Detroit offices were both completed on time.

SCOPE CHANGE CONTROL

All stakeholders were involved in determining and controlling the scope of ResNet. Sales agents and their managers defined the basic requirements for the software and made suggestions on enhancements. Senior management determined the overall scope, budget, and schedule goals. The 1996 kickoff meeting emphasized the purpose of the ResNet projects and their scopes. Table 15-2 presents some of the statements about purpose and scope that were used at the kickoff meeting.

Table 15-2: Kickoff Meeting Statements About ResNet Purpose and Scope

PROJECT PURPOSE	PROJECT SCOPE
Provide ResNet presentation screens for all reservations personnel	3,000 PCs in 8 reservations offices and Iron Range Reservations Center (IRRC) (Note: These numbers include the completion of the 1995 ResNet project.)
Connect ResNet to an information pipeline and to other distributed systems at Northwest and alliance partners	Four applications: • Sales • Specialty Sales • Support Desks • Integrated WorldPerks
Design for rapid evolution to meet changing business needs	Preproduction prototypes for intelligent call processing and keyword search

Peeter and his team worked hard to get buy-in from all stakeholders of ResNet. Peeter used the following words to describe the purpose of the 1996 ResNet kickoff meeting: "to launch the ResNet implementation and develop a shared responsibility for a successful outcome." By creating a solid sense of unity among all project stakeholders, everyone shared in the responsibility and success of the project.

Peeter held regular meetings with team members to handle potential scope changes. They accepted some small changes in scope, but only if those changes made business sense and would not be detrimental to the whole project. The number of PCs and software applications never changed from the original plan, but the sales agents requested additional capabilities from the ResNet software.

As mentioned in Chapter 14, *Executing*, people were assigned to ResNet to develop enhancements to the software even as it was being presented to the sales agents. The ResNet screen included the capability to send enhancement requests to the software developers. Kathy Christenson was surprised and yet excited about the number of requests they received—over 11,000. She developed a change control process with the managers who sponsored the four main software applications—sales, specialty sales, support desks, and WorldPerks. Each of these managers had to prioritize the software enhancement requests and decided as a group what changes to approve. The developers of the system enhancements then implemented as many items as they could, in priority order, given the time they had. About 38 percent of the 11,000 suggestions were implemented.

QUALITY CONTROL

Because the software development team knew the reservation process and what shortcuts agents might take while using the new system, the team was able to develop user-friendly, foolproof software. For example, some sales agents were concerned about meeting their required performance statistics for the average length of calls and the percentage of direct sales. Some agents dreaded customers who wanted to make rental car reservations because of the increased call time. Therefore, the ResNet software developers streamlined the rental car reservation process. They added scripted prompts to help the sales agents ask about car rentals and make them more comfortable when trying to close sales directly by asking for the customer's credit card information.

The industrial engineers supporting ResNet took several important actions to control quality. In addition to analyzing the reservation process in detail and making suggestions on improvements, they helped select a random sample of sales agents to properly test the impact of ResNet. They also developed several charts for measuring the business benefits of ResNet.

Arvid Lee mentioned that he had never seen as much statistical analysis done on a project as was done on ResNet. As part of the beta project, the industrial engineers used statistical techniques to select a sample of reservation sales agents to test ResNet. At the time, there were over 4,000 reservation sales agents. They wanted to make sure their testing was not biased by factors such as gender, age, location, shift, or efficiency of the agents. They ensured that the agents involved in testing the benefits of the new system were a random sample of the total group of all NWA reservation sales agents. The final project audit mentioned that biasing factors, such as weather, new hires, and the changing sales environment, were isolated using multiple regression analysis to determine the true impact of ResNet.

Many people ignore or discount statistical analysis and quality control reports, but the people supporting ResNet developed several particularly useful reports for measuring the quality and progress of ResNet. Figure 15-1 shows a chart used to track the learning curves for new hires using ResNet versus NWA's native reservations system on TELEX terminals. The shapes of the curves are similar, but the ResNet agents started and ended at a shorter call handle time. This shorter call handle time was believed to be due to the friendlier design of the new interface provided by ResNet.

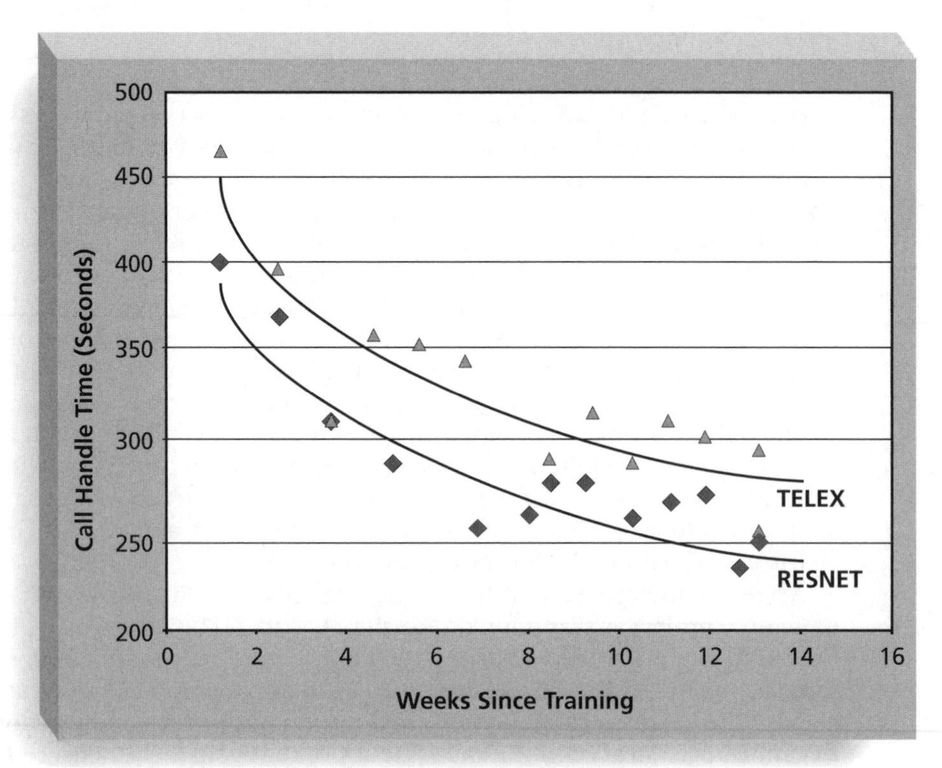

Figure 15-1. New Hire Call Handle Time Learning Curves

The industrial engineers also created charts plotting the predicted versus actual number of direct ticket sales per 100 calls, as shown in Figure 15-2. The forecasting model they developed proved to be 87 percent accurate in forecasting performance on the TELEX system. Because the high accuracy percentage showed that the engineers had very good forecasting capabilities, senior management believed their estimates for the direct ticket sales projections for ResNet. Increasing the number of direct ticket sales would provide great financial savings to NWA.

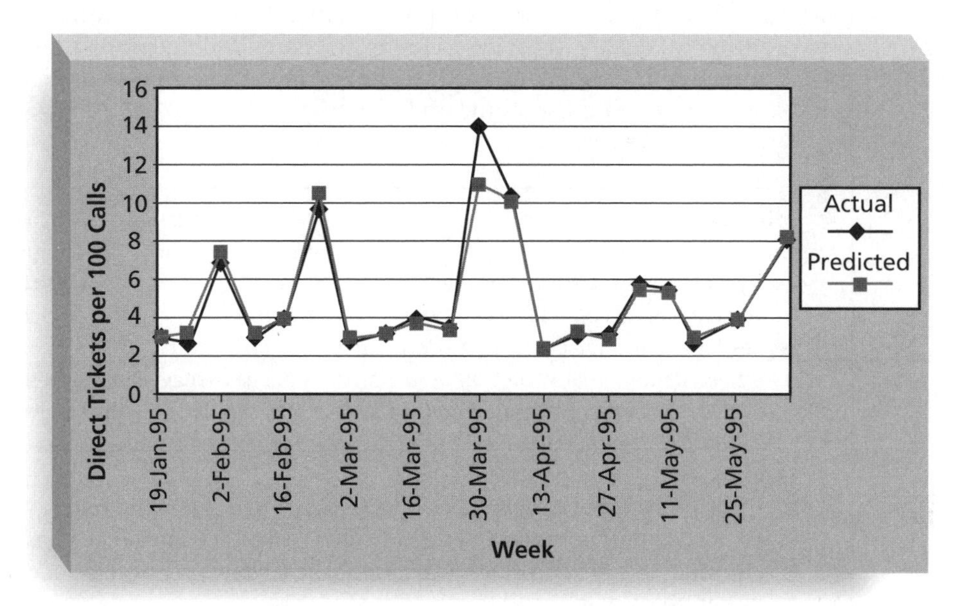

Figure 15-2. Direct Tickets per 100 Calls — Actual vs. Predicted

Having forecasts and tracking their progress against the forecasts helped the ResNet team and sales agents using the new system focus on the project's business goals. The immediate feedback available each week from these charts prompted the team to push forward. Reservation sales agents actually looked forward to seeing these quality control reports.

PERFORMANCE AND STATUS REPORTING

In addition to quality control reports, the ResNet team generated other types of performance and status reports. Emphasis was placed on key issues, decisions that needed to be made, and numerical progress on the project. Important numbers tracked were the number of PCs installed, the average call handle times for sales agents, and the number of calls resulting in direct ticket sales. This emphasis on issues, decisions, and progress helped keep people, especially senior management, interested and involved in reading and listening to performance reports.

Figure 15-3 shows a sample status report created by a ResNet team member. The format of the weekly status report included four sections: milestones, issues, accomplishments, and goals. The first section listed milestones along with planned and actual target dates, whether the item was completed, on schedule, or behind schedule. The second section listed issues requiring discussion and assistance. The majority of status reports included some items in this section, consistent with Peeter's directive to focus on issues and decisions that had to be made, instead of boring people with technical facts. The final two sections listed the last period's accomplishments and the next period's goals.

• **Milestones**	**Plan**	**Actual**	**Status**
— Complete 3270 testing	5 Jun	9 Jun	Completed
— Event Recorder Activated in Tampa	6 Aug		On schedule
— Presentation to NWA Finance Com	6 Sep		On schedule

- **Issues Requiring Discussion/Assistance**
  - — Need definition of WORLDSPAN/CORDA Change Management Process
  - — IMPACT: Need process that notifies ResNet of scheduled system changes
  - — ASSIGNED TO: J. Huss
- **Last Period's Accomplishments**
  - — Completed initial design of Event Recorder analysis and began coding
  - — Researched slow response times in Tampa. Looked at Baltimore response times
  - — Completed minor adjustments to measurement reports per customer request
- **Next Period's Goals**
  - — Conduct formal review session for Address Verification approval
  - — Beta test Event Recorder in the Hotline area in Bldg C

Figure 15-3. Sample Weekly Status Report

The graphical progress report described in Chapter 14 (*see* Figure 14-2) also helped communicate project status. Every few weeks each office received an office layout chart that showed what installations were completed. This report, along with the learning curve charts for new hires' call handle time and the charts plotting direct ticket sales, helped keep the ResNet project on track in meeting performance goals. The staff focused on meeting key business goals of the project, and the report formats kept ResNet stakeholders informed of progress.

MANAGING RESISTANCE TO CHANGE

Arvid and Kathy, the information systems and application development team leaders, did not have the traditional view of change management as the process of handling requests to change project scope, time, or cost. They were used to developing good project plans and executing them accordingly. They viewed change management as dealing with people's natural resistance to change, especially when that change involves new technology. The resulting change management process was a key factor in managing the project and ensuring its success.

NWA formalized the importance of change management in their organizational structure for large information technology projects. The ResNet project included a full-time analyst responsible for ResNet change management and several reservation sales agents responsible for training and awareness. Nancy Strobel was the head of ResNet Change Management and reported directly to Peeter. She worked with a team of people from the reservation offices to help people get involved in, get prepared for, and deal with the changes resulting from a new reservation system. They prepared a change management plan and were instrumental in getting all of the reservation sales agents and their managers to support ResNet. As part of their change management tactics, they helped develop and implement themes for each office, as described in Chapter 14, *Executing*. For example, they helped create posters with a Broadway show theme saying "ResNet—Coming Soon". They also provided special ResNet wrist pads, coffee mugs, and other items to get the agents excited about the new system.

The change management team was also instrumental in identifying people who might be resistant to the new system. Because the reservations sales agents would be most affected by the new reservation system, the change management team decided to create a videotape after the ResNet beta test. Below are the opening remarks from this videotape:

> "Recently we began testing a new reservation system called ResNet with agents from several of our offices. At the end of the test, we sat down with these agents and asked them their candid opinion of ResNet. Now, we know that change can be difficult. But what we found and what you're about to hear will hopefully show you that this is change for the better. Our vision is that ResNet will improve customer service, increase efficiency and revenue, and ultimately make your job easier. Once again, you will be hearing from your peers—agents who have used this new system—giving their honest opinion of ResNet."[1]

The videotape included comments from many different sales agents, representing several sales offices. There were male and female sales agents of various nationalities and age groups in the videotape. It was very effective in getting the sales agents interested in the new system and calming some of their fears. For example, a couple of the sales agents in the videotape mentioned that it was hard to get used to the new system at first, but that it was definitely worth it. Several agents mentioned that they had fewer headaches because of the new screen colors, and their job performance and job satisfaction improved with the new system.

The change management team also decided that every single reservation agent should get at least thirty minutes of one-on-one training on ResNet. This one-on-one training helped personalize the experience and allowed people to feel comfortable asking questions. Providing this individualized training for about 4,000 sales agents showed that the ResNet team cared about the sales agents and was willing to invest the time and money to help them adjust to the new system.

[1]Comments made by NWA Director Crystal Knotek. Northwest Airlines "ResNet," JUNTUNEN VIDEO, Inc. April 4, 1995.

CASE-WRAP UP

Peeter's idea to have status review meetings with upper management focus on discussing issues and making decisions was very successful. The managers were engaged in the meetings and offered good suggestions for keeping ResNet on track. Peeter made sure his team concentrated on using both business and technical terms at these meetings. They also used charts to track the project's numerical progress, such as the number of PCs installed to date. This focus on decision-making and numerical measures of progress worked well to get senior management's support in controlling the project.

The change management team was instrumental in helping sales agents adjust to the new information system. The sales agents involved enjoyed creating the beta test videotape, and other agents enjoyed watching it. The agents also appreciated the one-on-one training on ResNet, and they looked forward to seeing their progress in reducing call handle times and increasing direct sales in the quality control charts. All stakeholders were very pleased to see the financial results of ResNet when the final audit report showed that the increase in direct sales attributable to ResNet greatly surpassed expectations. Instead of increasing direct sales by 5.5 percent as planned, ResNet had increased direct sales by 17.7 percent. This 17.7 percent increase meant an additional $2.3 million in 1996 commission savings.

CHAPTER SUMMARY

Controlling is the process of measuring progress toward project objectives, monitoring deviation from the plan, and taking corrective action to match progress with the plan. Controlling cuts across all phases of the project life cycle.

Peeter made keeping on schedule a priority and focused on achieving milestones one at a time. The ResNet team faced several challenges in meeting project schedules, and they delayed the Detroit office installation to meet more important business needs. They did, however, successfully complete all of the ResNet projects according to schedule.

The ResNet project was designed to accept software enhancement requests, but the software developers could not implement all of the suggested changes. Managers prioritized enhancements, and the software developers completed as many as they could while still meeting schedule deadlines. They implemented about 38 percent of 11,000 suggested enhancements.

Quality control techniques were a factor in communicating progress and pushing staff toward project success. Industrial engineers played an important role in project control.

Performance and status reports focused on issues and tracking key performance numbers.

The ResNet project included a full-time analyst for change management in addition to reservation sales agents responsible for training and awareness. NWA understood the importance of helping people adjust to changes and invested funds in creating a videotape about ResNet for the sales agents and in one-on-one training.

Discussion Questions

1. Recall that the 1995 Standish Group study of information technology projects found that the average schedule overrun was 222 percent. Also recall that schedules are the main source of conflict on projects. Discuss Peeter's decision to focus on meeting schedule milestones. Could this approach be taken on more information technology projects? Justify your answer.
2. The ResNet group adopted a philosophy of giving up some system functionality to meet milestones. What do you think about this approach? Could this approach be taken on more information technology projects? Justify your answer.
3. Discuss the charts created for quality control (Figures 15-1 and 15-2). Do you think most users would understand and appreciate these types of charts? Do you think they influenced the reservation sales agents to work harder because they knew their progress on reducing call handle time and making direct sales was being tracked? Could this approach be taken on more information technology projects?
4. Discuss the format of the sample weekly status report. What do you like/dislike about it? Does emphasizing issues and decisions make sense? Could this approach be taken on more information technology projects?
5. Discuss the use of a change management team on ResNet. Could this approach be taken on more information technology projects?

Exercises

1. Read the article by Elton and Roe listed in the Suggested Readings. They provide and critique project management suggestions made by Eli Goldratt in his books *The Goal* and *The Critical Chain*. Write a paper summarizing the key points of this article and how it relates to the ResNet project. Be sure to address the authors' point that measurements should induce the parts to do what is good for the whole. Also discuss Goldratt's view that the fewer the milestones, the fewer the delays, and how it relates to ResNet.

2. Find two examples of recent status reports for projects. Compare the information they contain with the example in Figure 15-3. Write a one- to two-page paper summarizing and analyzing your findings. Include the two examples you find as attachments. Include suggestions on how to prepare good status reports in your paper.
3. Do an Internet search on "project control." Summarize some of the many vendors that offer their services to help control projects. Try to find any real advice or information on what helps control projects, especially in information technology.
4. Research quality control techniques. Write a two to three-page paper summarizing at least two different techniques and try to find examples of how they helped keep projects under control.
5. Research books and articles on helping people deal with change, especially new technologies. Kanter's book (*see* the Suggested Readings list) is a classic in this area. Summarize information from three good sources on change management and provide your opinion on their recommendations.

MINICASE

Peeter did an excellent job of controlling the ResNet projects, but he faceed several challenges and used unique approaches to managing change.

Part 1: Review the information provided in *What Went Wrong*? and the paragraph following it. Pretend that you are Arvid Lee and that you want to rally other members of the ResNet team to convince Peeter and Fay not to change the sequencing of office installations. Develop a plan for how you would convince your teammates to take a stand. Include in the plan how the team would convince Peeter and Fay not to make the schedule changes.

Part 2: The ResNet team's approach to scope change control was unique. Specific people were assigned to develop enhancements to the software, but these people could not handle all of the enhancement requests. Develop a weighted scoring model that could be used to help prioritize which enhancement requests should be implemented. For example, criteria might be the potential value of the new enhancement, how long it would take to implement, how many agents would use the new enhancement, and so on (*see* Chapter 4, *Project Scope Management*, for information on weighted scoring models). Also develop a form for submitting enhancement requests.

SUGGESTED READINGS

1. DeMarco, Tom. *The Deadline*. New York: Dorset House Publishing, 1997.

   Tom DeMarco has published several articles and books related to managing software development projects. This book is a business novel that illustrates the basic principles that affect the productivity of software development teams.

2. Elton, Jeffrey and Justin Roe. "Bringing Discipline to Project Management." *Harvard Business Review* (March/April 1998).

    This article discusses Eli Goldratt's books on project control and offers suggestions for using a more disciplined approach to project management. Suggestions include having measurements that induce the parts to do what is good for the whole, having strong senior management involvement, and focusing on personal skills.

3. Goldratt, Eliyahu and Jeff Cox. *The Goal: A Process of Ongoing Improvement*, Great Barrington, MA: North River Press, 1994.

    In this book Goldratt uses a fictional story to illustrate the fundamentals of running a business. He describes several problem-solving techniques for managers who lead organizations through change and improvement, which are inevitable in any industry.

4. Kanter, Rosabeth Moss. *Change Masters: Innovation and Entrepreneurship in the American Corporation*. New York: Simon & Schuster, 1985.

    Dr. Kanter's book describes how to be effective in overcoming the kind of "stalled" thinking that inhibits progress in almost all organizations. The author draws strongly from her own research in several different organizations to offer strategies for mastering change and ensuring progress.

16

Closing

Objectives

After reading this chapter, you will be able to:

1. *Understand the importance of formally closing projects*
2. *Discuss closing processes and outputs and how they were used on ResNet*
3. *Describe the tools and techniques used to aid project closing on ResNet*
4. *Explain how NWA measured the business benefits of ResNet*
5. *Describe the methodology and findings in the ResNet final audit report*
6. *Discuss the lessons learned from the ResNet projects*
7. *See how NWA continues to enhance ResNet and develop the discipline of project management in the 21st century*
8. *Relate some of the closing events in ResNet with concepts described in previous chapters*

The ResNet team held a final recognition event as part of closing the ResNet project. As usual, it was a big affair. The software development team wrote and performed a "developers' rap" to share their experiences on the project. In addition to having a recognition event, the project team conducted a formal audit on the project to document its value to Northwest Airlines. The audit was designed to answer the following questions: Did the project deliver what it said it would? Could the ResNet team demonstrate that ResNet was indeed helping sales agents in booking more direct sales and decreasing their call handle times? Should the company take on other large information technology projects like this again?

WHAT IS INVOLVED IN CLOSING PROJECTS?

The closing process involves gaining stakeholder and customer acceptance of the final product and bringing the project, or project phase, to an orderly end. It includes verifying that all of the deliverables have been completed, and it often includes a project audit. Even though many information technology projects are canceled before completion, it is still important to formally close them and reflect on what can be learned to improve future projects. As philosopher George Santayana said, "Those who cannot remember the past are condemned to repeat it."

It is also important to plan for and execute a smooth transition of the project into the normal operations of the company. Most projects produce results that are integrated into the existing organizational structure. For example, ResNet produced a new reservation system interface that replaced the native TELEX terminals. The sales offices adopted the new system, and the Information Services Department provided maintenance for the system. Other projects result in the addition of new organizational structures, such as a new department to manage a new product line.

When senior management cancels projects, they sometimes do so in one of two ways: extinction or starvation. For example, if NWA decided not to fund the 1995 or 1996 ResNet projects, the project would have been extinct. If management had only provided a fraction of the money required to successfully complete any of the ResNet projects, the projects would have terminated by starvation. Project termination by starvation is very hard on people and is usually done when organizations are unwilling to admit project failure. Fortunately for NWA and the ResNet team, the ResNet projects were very well managed, successfully completed, and integrated into the existing organizational structure. Even successful projects, however, must be closed properly.

Table 16-1 lists the knowledge areas, processes, and outputs of project closing. During the final closing of any project, project team members should take the time to communicate project results by documenting the project and sharing lessons learned. If items were procured during the project, the project team must formally complete or close out all contracts. Closing activities on the ResNet project described in this chapter include administrative closure, a final project audit, a final recognition event, personnel transition, and discussions of lessons learned. This chapter also provides a summary of another ResNet project, ResNet+, which addressed new business needs related to the ResNet system, and a description of ResNet+ and project management at NWA in the 21st century.

Table 16-1: Closing Processes and Output

Knowledge Area	Process	Outputs
Communications	Administration Closure	Project Archives Project Closure Lessons Learned
Procurement	Contract Close-out	Contract File Formal Acceptance and Closure

ADMINISTRATIVE CLOSURE

Administrative closure involves verifying and documenting project results to formalize stakeholders' acceptance of the product(s) of the project. It includes collecting project records, ensuring the product(s) meet final specifications, analyzing whether the project was successful and effective, and archiving project information for future use. On the ResNet project, inputs to administrative closure were performance measurement reports (described in Chapters 14 and 15) and the product documentation or specifications used in developing and procuring the hardware and software. Because stakeholders were active participants in ResNet, they knew the project was completed when all the ResNet workstations were installed, all the agents were trained, the system had been thoroughly tested, and the benefits had been measured.

NWA had several reviews of the benefits of ResNet. The beta test documented sales agents' reactions to the new reservations software and measured their performance in handling calls and making more direct sales. In 1996, a formal ResNet benefits review meeting summarized sales agents' performance twenty weeks after ResNet was implemented in the Tampa and Baltimore offices. Table 16-2 shows the results, as documented in a memo from Peeter Kivestu to Fay Beauchine on March 13, 1996. Note that call handle time decreased, direct sales increased, and new hire training took less time. Developing measurement techniques and measuring progress toward meeting business goals helped the ResNet team close the beta and ResNet 1995 projects and receive funding for the 1996 project.

Table 16-2: ResNet Benefits Results from March 1996 Benefits Review

Measurement	Result
Call Handle Time	4.4% Reduction
Direct Ticket Sales/100 Calls	6.0% Increase
New Hire Training	Better Call Handle Time with 25% Less Training

RESNET AUDIT

A formal ResNet audit was completed on December 10, 1996. Project audits are a good technique for formally reviewing project progress and results. The ResNet audit was designed to address two questions: What benefits, in terms of selling and call handle time, have been realized by the Reservations Department as a result of ResNet, and how do those real results compare with projected results in the PR2 plan? The answers to these questions were documented in the audit as follows:

- The Reservations Department has significantly increased selling as a result of management leadership, the market environment, and the opportunities provided by ResNet as a selling tool.
- While increased selling, new hires, E-Ticket sales, and other factors have increased reservations' average call handle time for 1996, statistics support a handle time benefit from ResNet conversion.
- Of the two items audited, the higher than expected increase in selling compensated for the shortfall in call handle time, resulting in a net gain of $0.8 million to Northwest Airlines in 1996 versus the PR2 projections.[1]

The actual call handle time improvement of 5.4 percent, based on the 1996 audit, was slightly lower than the PR2 projections of 6.7 percent. The projected financial benefits of ResNet were based primarily on an improvement in call handle times and an increase in direct sales. The real improvement in call handle times was lower than the projection, resulting in lower financial benefits from that area. However, it did not result in lower overall financial benefits because there was a greater increase in direct sales than what was originally projected.

The audit report included several charts to portray the impact of ResNet on selling, the effect of major factors on average call handle time, the ResNet agents' call handle times after the learning curve effects, and so on. Figure 16-1 shows the chart used in the audit report to display the impact of ResNet on selling.[2] Notice that its simple format helped to communicate one of the significant benefits of this new system—a 17.7 percent increase in direct sales over the old system (referred to as TELEX because of the hardware it ran on). The PR2 goal for increased selling was only 5.5 percent. This increase in direct sales translated into an additional $2.3 million in 1996 commissions savings above the PR2 projection.

[1]Reservations Sales & Services, "ResNet Audit," Northwest Airlines internal document, December 10, 1996.

[2]*Ibid.*

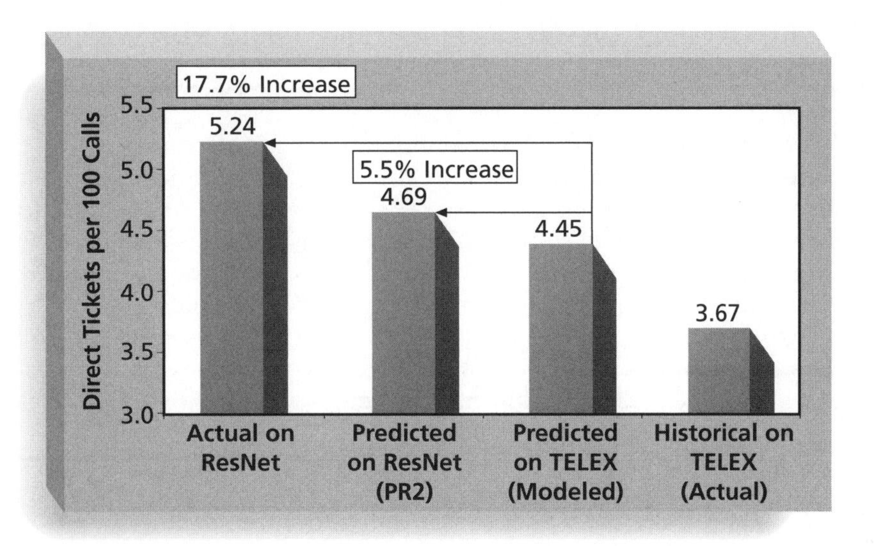

Figure 16-1. Selling Impact of ResNet

What Went Right?

The audit methodology showed the amount of rigor that NWA put into its project audits and benefits measurement methodology. All three ResNet projects made good use of experts who understood the importance of using a sound benefits measurement methodology. These experts analyzed historical information, ensured that data was not biased, and used appropriate statistical techniques to develop and then measure the benefits of ResNet. The following points were included in the audit to describe the selling audit methodology:

- In order to create a stable population for sampling, agents in the survey were required to be full-time, to be working at least 20 hours per week, and to have a seniority date prior to 1/1/95.
- On the basis of four months of historical performance, ongoing TELEX sales data, and statistical modeling, direct ticket sales per 100 calls on the old TELEX system were predicted.
- Actual ResNet selling performance was then compared to the predicted TELEX performance for three national offices and MSP International.
- Biasing factors, such as weather, new hires, and the changing sales environment, were isolated using multiple regression analysis to determine the true impact of ResNet.[3]

[3]*Ibid.*

RESNET FINAL RECOGNITION PARTY AND PERSONNEL TRANSITION

Peeter and his team knew how important strong starts were for ResNet. They decided to also have a strong finish. They held a large luncheon to celebrate ResNet's success and recognize the people who worked so hard on the project. Kathy and her team of software developers wrote a rap song to describe their experiences on the project. You can find the words to the rap in Figure 16-2. Peeter authorized a small amount of funds to hire a professional actor to help the group perform this rap. Many people received awards and gifts to recognize their contributions to the project. The final recognition event was a fun way to share in the celebration of finishing this important and highly successful project.

[click, click, click, click]
Gotta Code, Code [click, click]
Gotta Cooode, Code [click, click]
Gotta Code, Code [click, click]
Gotta Load...
Gotta load to Code.
[click, click]
We developed ResNet, yes siree..
We hadta get it done in big hurry.

We were given a timeline that was set...
the bosses said it must be met.

Started with some charts, listened to some calls...
spent lots of time, climbin the walls.

Hadta come in early, hadta stay real late...
Hadda deadline comin, and it couldn't wait.
[click, click]
Gotta Code, Code [click, click]
Gotta Cooode, Code [click, click]
Gotta Code, Code [click, click]
Gotta Load...
Gotta load to Code.
[click, click, click, click]

Eleven applications make ResNet cool...
But before we coded, we hadta go to school.

Popups, Quadrants, command cells too...
How to make'em fit on a screen of blue?

Config class gave us an over view...
Went to Baltimore to really get a clue.

Alpha test, Beta tests were a must...
If we didn't pass'em, we'd a had a bust.
[click, click]
Gotta Code, Code [click, click]
Gotta Cooode, Code [click, click]
Gotta Code, Code [click, click]
Gotta Load...
Gotta load to Code.
[click, click, click, click]

National, International then Supervisor...
By this time - everyone was wiser.

On came the WorldPerks Application...
That's when we realized the fun begun.
CTI, ResDoc, Iron Range too...
Everything - suddenly, seemed brand new.

TBM and Rates together as one...
Beta in Detroit, Fun, Fun, Fun.
ET's, DT's, and MetroMail...
Calculate those taxes for the next sale.

Specialty Sales 12 desks strong...
Who would've thought they'd all belong.
Consolidators, Groups, conventions and Cruises..
The way we've coded, nobody loses.
[click, click]
Gotta Code, Code [click, click]
Gotta Cooode, Code [click, click]
Gotta Code, Code [click, click]
Got a load...
Gotta load to Code.
[click, click, click, click]

We built an application for CRC...
Meal Dispatch and IMC.
They call their office Jurassic Park...
Now with ResNet they're not in the dark.

Service Support had so much to do
Using ResNet, they're right on Queue.

Sales Action Center, our grand finalee...
built to help the travel agency.

Promote, distribute, activate...
The IS group - kept us straight.
With each new office to define...
we hadta rely on our great hotline.

Challenging, though the task may be
We managed to keep our sanity.

* Written by Nancy Desch, Northwest Airlines

Figure 16-2. ResNet Developers' Rap

Another aspect of project closing is transferring project personnel back into other parts of the organization. Recall that Northwest Airlines had a strong functional organization. People on the project returned to their functional areas and started working on other assignments. A few information services personnel were assigned to maintenance support for ResNet, and others, including Arvid Lee, started working on the next important information technology project for NWA. In 1998 Arvid was selected to be the project manager for the ResNett project, as described later in this chapter. A few people, including Kathy Christenson, transferred from the Reservations and Marketing Departments to the Information Services Department to begin new careers in information technology. Kathy was selected to be the Account Manager for ResNet+, and Peeter Kivestu is now a vice president at NWA.

What Went Wrong?

NWA still struggles with transferring people, roles, and responsibilities between projects and on-going operations. For example, there were problems after completing the Computer Telephony Integration (CTI) work for ResNet 1996. This new technology provided better call handling by having customers respond to telephone prompts before talking to a sales agent. When the ResNet team had completed the work, the functional group that should have taken over support did not understand or willingly accept their new roles and responsibilities. Also, NWA did not budget enough people to support this new technology. This lack of transition planning caused several conflicts after the ResNet project ended. Senior management eventually had to intervene to resolve these conflicts.

LESSONS LEARNED

Although no formal lessons learned were written after completing the ResNet project, former project members continue to share their experiences and take actions based on lessons they learned from the project. Peeter, Kathy, and Arvid highlight the following as lessons they learned from the ResNet project:

- Let workers have fun. Many information technology professionals are introverted, and this characteristic often causes communications problems. Creating a fun working environment helps technical people and other project stakeholders take more interest in information technology projects and promotes buy-in, creativity, and teamwork.
- Beginnings are important. It is easy to underestimate confusion, especially when defining the goals of a project. The project manager needs to get people together early to discuss key project issues. It is also important to have a strong beginning at each phase of a project.

- Top management support is critical. Fay Beauchine was the main sponsor of ResNet, and she provided support throughout the project.
- Managing change is 50 percent of project management. Everyone is different, so different approaches must be used to help people adjust to change. Lots of communication in various forms is crucial.
- Make management reviews interactive. During a review, always include the decision to keep management actively engaged. In addition, use business terms and focus on issues, not just status.
- Set realistic milestone dates, and then stick to the schedule as much as possible. Vary the scope, if necessary, and prioritize user requirements to meet dates. Missing dates often produces chaos. Reach one incremental goal at a time.
- Plan at a workable level. It is easy to get bogged down in too much detail. Focus on getting work accomplished.

RESNET+

Business and market forces usually require most new information systems to change, and ResNet is no exception. In July 1998, NWA launched a new ResNet project called ResNet+. Kathy Christenson was the Account Manager for the ResNet+ project, and Arvid Lee was the project manager for this $12.5 million project.

Using their experiences from previous ResNet and other information technology projects, NWA focused on business needs as the basis for ResNet+. They were negotiating more global alliances with other companies, customers were becoming more computer savvy and accessing flight information on the Web, and new technologies such as computer-based training (CBT) provided opportunities to reduce training costs.

Figure 16-3 shows a chart that Kathy Christenson created to document the ResNet application and system requirements from 1995 to 1999 and beyond. The x-axis represents time, and the y-axis represents ResNet application size in megabytes (MB). ResNet 1995 and 1996 applications are shown using various symbols on the line graph. Important business decisions, such as integrating the WorldPerks operation, improving E-ticketing, and integrating with KLM airlines, are shown above the line graph. The figure shows that additions to the ResNet application and related tools on the ResNet PCs had grown steadily since January 1995. The figure also shows that ResNet had continued to support new business initiatives.

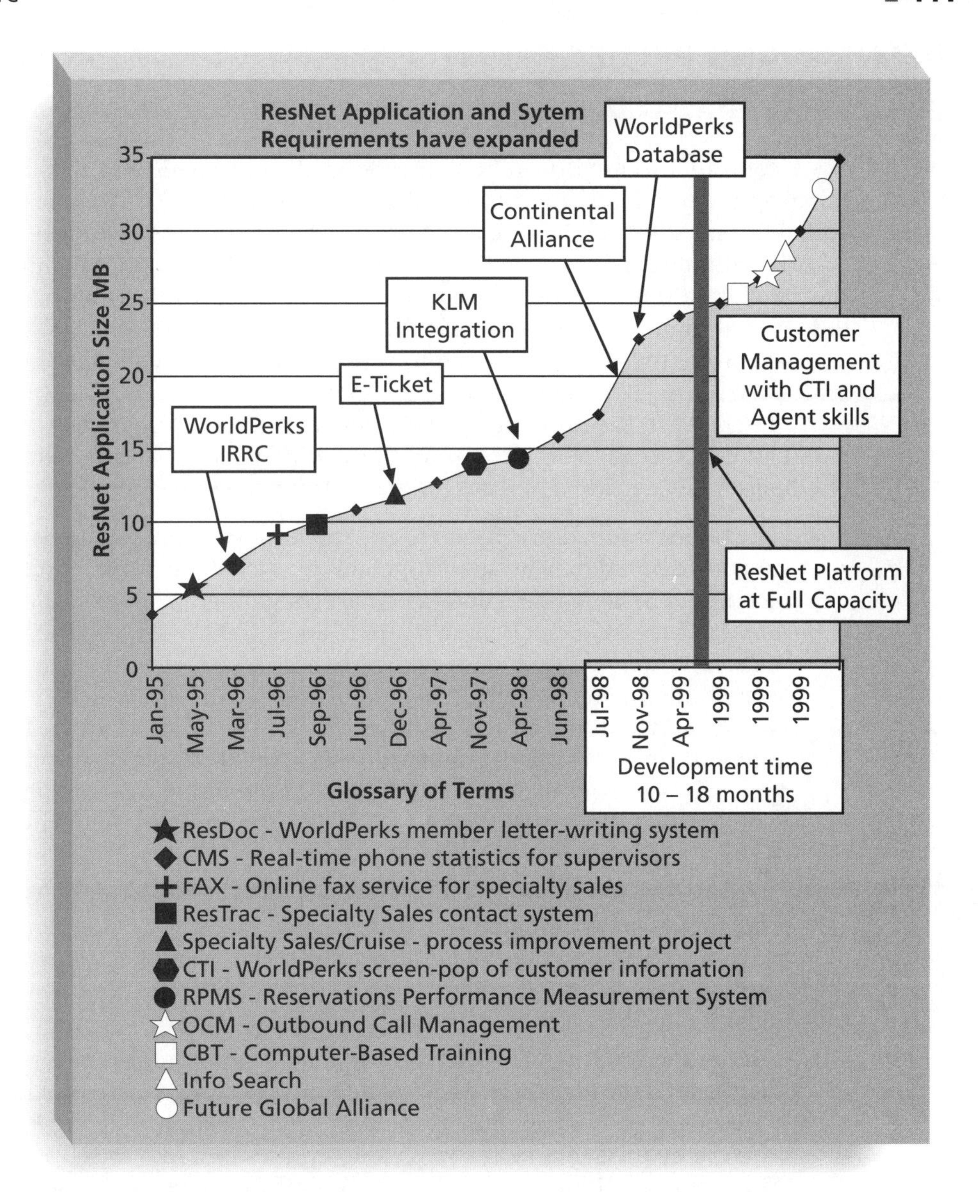

Figure 16-3. Business Need for ResNet+

You can see from the figure that in mid-1999, the ResNet PCs would have reached full capacity. The ResNet+ project upgraded the ResNet PCs to NT workstations and provided new software applications for the NWA reservations sales agents, such as computer-based training and an information search engine.

RESNET+ AND PROJECT MANAGEMENT AT NWA IN THE 21ST CENTURY

NWA continues to modify and enhance ResNet+ to provide additional functionality and to address increases in Web-based ticketing. The ResNet+ project ended in May 2001, delivering all of the planned scope under budget and slightly behind schedule. Arvid says they would have been on schedule, but the supplier of the computer-based training part of the project ran into difficulties and fell slightly behind schedule. The success of the original ResNet projects also prompted NWA to expand its internal work in developing the profession of project management.

As described in the previous section, ResNet+ involved upgrading the original ResNet PCs to NT workstations and providing new software applications. The major new functions of the ResNet+ software include:

- New business functions: The ResNet+ application is regularly updated to support required new business functions. For example, NWA implemented a new frequent flier system and new marketing programs with alliance partners. ResNet+ included new software to address these new business functions. New software was also required to address changes made to Worldspan, the main reservation system and database to which ResNet+ interfaces. Agents are largely insulated from all technical changes, however, as their system presentation and processes remain relatively constant. This approach is a major benefit because training/retraining costs are minimized, learning curves for new functions are reduced, and agent productivity, customer service, and revenue generation are maintained at a high level.
- Computer-based training: New call center agents at NWA can now complete a full training curriculum online. The training includes step-by-step training, simulated telephone calls, full online simulations, "try-it" sessions, and learning assessments. Training updates for new system features and marketing programs are easily delivered to existing call center employees.
- ResQuest: NWA has enhanced the ResNet+ workstation to use ResQuest, an integrated intranet-based online reference system, which provides access to nearly 40,000 documents. These documents cover a wide range of topics required by the agent to answer customer questions and to sell and service customer travel itineraries. The customer may have questions about products or services offered by NWA, its Alliance airline partners, and other business partners for car rentals, hotel reservations, and so on. ResQuest includes a search engine that enables the agent to find information quickly via keyword searches. The search engine utilizes synonyms, "soundex" searches, similar spelling, and so on, to aid the search. Examples of reference information include geographic maps, policies and procedures, aircraft seating schematics, copies of recent promotional materials, sample tickets, and other documents that NWA may have sent to customers.

The number of people making flight reservations and purchasing other travel-related products and services over the Internet has increased dramatically in the last few years. NWA and other travel companies will continue to offer new and improved services from their Web sites. The ResNet+ application fully supports travel itineraries booked directly by customers via the Internet. Customers can call NWA agents to add, change, or cancel those itineraries.

Some people might ask if ResNet+ or similar applications will be totally converted to Web-based applications. Arvid states that converting to a Web-based application is not an automatic "no-brainer" decision, as some people might expect. The current ResNet+ interface has been designed for high productivity. Web-based presentations may be more visually attractive, but often they are not more productive. A major financial investment is often required to convert systems to Web-based applications, and any major financial investment must be supported by a demonstrated financial benefit. The ResNet+ application continues to be a very successful business project and system.

Project management is gaining acceptance and momentum as a recognized discipline at NWA. The initial ResNet project was a strong catalyst for that acceptance because it is an excellent example of the value of effective project management. Most sponsors of major projects at NWA specifically request project managers because they believe this will improve the success of the project. NWA has established a formal project management office (PMO), which in 2001 was staffed by more than twenty-five project managers. Arvid's title is now project manager, and he is part of the PMO. As mentioned earlier, Peeter Kivestu moved on to become a vice president at NWA after completing the ResNet 1996 project. His experience and success as a project manager helped prepare him well for an executive management position.

The project management office encourages NWA project managers to seek PMP certification and is actively developing and promoting a standard project methodology to support infrastructure and application development projects. However, NWA realizes that establishing a project management office, in itself, is not a magic guarantee of project success. The willingness of an organization to embrace the discipline of project management is still a key to successful projects.

CASE-WRAP UP

ResNet was and continues to be a highly successful information technology project for Northwest Airlines. Peeter and his team delivered a new reservation system interface that enabled sales agents to increase direct ticket sales and reduce call handle times. The call centers moved from being a financial drain on the company to being a very profitable part of the company. In 1996, ResNet saved NWA over $15 million, and

savings were over $33 million in 1997. Sales revenues were over $1.2 billion dollars in 1998, compared to $300 million in 1993. The ResNet projects met stakeholder expectations, and everyone involved had the added benefit of enjoying their work on the project. Peeter proved that a large information technology project could be managed well by using good project management, and Kathy Christenson and Arvid Lee are continuing to lead ResNet into the 21st century.

CHAPTER SUMMARY

The closing process group involves gaining acceptance of the end product and bringing the project to an orderly conclusion. It includes verifying that all of the deliverables have been completed and often includes a project audit.

Administrative closure activities on ResNet included creating a final project audit and holding a final recognition event. The purpose of the project audit was to determine the impact of ResNet on the direct ticket sales rate and agent call handle time, compare actual performance improvements to projections, and determine the bottom line impact of actual performance versus the plan.

Results of the audit found that ResNet exceeded expectations by increasing direct ticket sales by 17.7 percent, when the plan projection was 5.5 percent. The call handle time reduction of 5.4 percent was less than the plan estimate of 6.7 percent, but the increase in sales more than made up for the longer call handle time.

The final recognition luncheon provided a fun way to share in the success of the project. Giving people awards and gifts helped provide closure and a shared sense of community for NWA employees.

Lessons learned by members of the ResNet project included making projects fun, having strong beginnings, having strong top management support, actively managing change, focusing on decisions at project review meetings, focusing on meeting schedule deadlines, and planning at a workable level.

NWA initiated a project in 1998, ResNet+, to further enhance ResNet so that the system would continue to support reservations success. NWA continues to modify and enhance ResNet+ in the 21st century to provide additional functionality and to address increases in Web-based ticketing. NWA established a project management office and continues to promote the discipline of project management.

DISCUSSION QUESTIONS

1. Describe the key processes and outputs of the project closing process group. Describe some of the outputs of closing the ResNet project.

2. Review the goals of the ResNet final audit report. What was the focus of the audit? Why was it important to document the methodology assumptions? What other questions could be included in a final project audit?
3. Which lessons learned do you think Peeter, Kathy, and Arvid will most likely use on future projects?
4. Which lessons learned do you think will be hardest to replicate on future projects? Explain your answer.
5. Discuss some of the changes NWA made in ResNet+ and some of the changes in their approach to project management in the 21st century.

EXERCISES

1. Research information on project audits. Write a one- to two-page paper summarizing at least two articles on this topic. Be sure to include items such as purposes and contents of audits and when they should be done.
2. Interview two to three people who worked on the same information technology project. Ask each person what lessons he or she learned from working on the project. Did they have similar answers to this question? How do their lessons learned compare to those described by people who worked on ResNet? Document your findings in a one- to two-page paper.
3. Chapter 9, *Project Communications Management*, suggests that people take time to formally document lessons learned from projects and share those lessons with others. Read at least two articles (from the Internet, magazines, books, or personal interviews) about formal lessons learned. Suggested Reading 4 is one possible source. Do most projects include formal lessons learned at the end? Why or why not? Write a one- to two-page paper summarizing your findings.
4. People assigned to ResNet returned to their functional areas after the project ended. Many projects do not draw people from strong functional organizations, so reallocating people is often an important issue in closing projects. Read at least two articles that address the issue of staffing and what happens to people after projects are completed. Summarize the issues involved in project staffing in a one- to two-page paper.
5. Review the suggested reading by Todd Weiss and other recent articles related to online ticket sales. Changes in the market often affect information systems in terms of their required functionality and their financial benefits. In a one- to two-page paper, summarize the issues involved in online ticket sales as they relate to profitability for airlines and travel agencies.

MINICASE

An important output of the closing process is creation and sharing of lessons learned. As good as they were at project management, the ResNet team did not write any formal lessons learned.

Part 1: Assume that you have been asked by Northwest Airlines to develop a template for all project teams to start documenting their lessons learned. Create a template, then use it to document the lessons learned on ResNet.

Part 2: In addition to documenting lessons learned, it's critical to share them with others. Develop a proposal on how organizations can share lessons learned among people and projects. Include specific policies that organizations could create, and explain how they might use technology to facilitate sharing of lessons learned.

SUGGESTED READINGS

1. Collier, Bonnie, Tom DeMarco, and Peter Fearey. "A Defined Process for Project Postmortem Review." *IEEE Software* (July 1996).

   The authors describe a practical approach for holding project postmortem reviews. A postmortem review is like lessons learned that document what went right and wrong on a project.

2. Greco, Susan, Christopher Caggiano, and Marc Ballon. "I Was Seduced by the New Economy." *Inc* (February 1999): 34–46.

   Inc. magazine's cover story describes the lessons several smart CEOs learned the hard way about doing business in the late 1990s. Several myths they discussed include the need to be a virtual organization, the belief that technology makes life easy, and the belief that businesses must be on the Web in a big way.

3. Ruskin, A. M. and W. E. Estes. "The Project Management Audit: Its Role and Conduct." *Project Management Journal* (August 1985).

   This article describes what project audits entail and offers suggestions on performing a project audit.

4. Segil, Larraine. "Global Work Teams: A Cultural Perspective." *PM Network* (March 1999): 25–29.

   Staffing and creating teams are issues throughout a project's life cycle. This article discusses issues related to global work teams. The author's suggestions also apply to transitioning people on and off projects.

5. Weiss, Todd R. "Travelocity Strikes Back as Northwest and KLM Drop Online Ticket Commissions." *Computerworld* (March 2, 2001).

   This article describes how Internet travel agency Travelocity.com Inc. began charging a $10 fee on all Northwest and KLM tickets for travel in the U.S. and Canada one day after those airlines dropped their commissions on plane tickets sold online. Links are provided to related stories from www.computerworld.com.

6. Whitten, Neal. "Are You Learning From Project to Project?" *PM Network* (March 1999): 16.

   This short article stresses the fact that many projects do not have a mandatory post-project review, and even fewer require project managers on new projects to review other lessons learned.

Appendix A

Guide to Using Microsoft Project 2000

Introduction 448

Overview of Microsoft Project 2000 450
- Starting Project 2000 and the Project Help Window 450
- Main Screen Elements 454
- Project 2000 Views 455
- Project 2000 Filters 458

Project Scope Management 459
- Creating a New Project File 460
- Developing a Work Breakdown Structure 461
- Saving Project Files with or Without a Baseline 465

Project Time Management 466
- Entering Task Durations 466
- Establishing Task Dependencies 470
- Changing Task Dependency Types and Adding Lead or Lag Time 472
- Gantt Charts 474
- Network Diagrams 476
- Critical Path Analysis 477

Project Cost Management 478
- Fixed and Variable Cost Estimates 479
- Assigning Resources to Tasks 481
- Baseline Plan, Actual Costs, and Actual Times 485
- Earned Value Management 488

Project Human Resource Management 489
- Resource Calendars 489
- Resource Histograms 490
- Resource Leveling 493

Project Communications Management 494
- Common Reports and Views 494
- Using Templates and Inserting Hyperlinks and Comments 496
- Saving Files as Web Pages 498

Using Project 2000 in Workgroups 500
Project Central 501
Exercises 502
Exercise A-1: Web Site Development 502
Exercise A-2: Software Training Program 505
Exercise A-3: Project Tracking Database 507
Exercise A-4: Real Project Application 507

INTRODUCTION

This appendix provides a concise guide to using Microsoft Project 2000 to assist in performing project management. Project 2000 can help users manage different aspects of all nine project management knowledge areas. Most users, however, focus on using Project 2000 to assist with scope, time, cost, human resource, and communications management. This appendix uses these project management knowledge areas as the context for learning how to use Project 2000.

As described in Chapter 1, *Introduction to Project Management*, hundreds of project management software products are available today. Microsoft Project is the clear market leader among midrange applications. A 1998 survey of 1,000 project managers showed that Microsoft Project was by far the most widely used computerized project management tool. When asked to list three project management tools they were currently using or had used within the past three years, 48 percent of the 159 survey respondents listed Microsoft Project. Only 13.8 percent of respondents mentioned the second most-listed tool, Primavera Project Planner. The third most-listed tool was Microsoft Excel, the only one of the ten tools listed that is not categorized as a project management software tool. Survey respondents used Microsoft Project for project control and tracking, detailed scheduling, early project planning, communication, reporting, high-level planning, Gantt, CPM, and PERT. Respondents used Excel for budgeting, cost analysis, variance analysis, tracking and reporting, and creating work breakdown structures.[1]

Even with its popularity, however, Microsoft Project is not nearly as widespread in its use as other Microsoft products such as Word or Excel. At Inmark Communications' Project World Conference in December 1998, Microsoft issued a press release stating that they had surpassed the 3-millionth-customer mark for Microsoft Project products.[2] Sales of Word or Excel are much higher—well over 100 million. In January 2000 Microsoft announced that revenues from Microsoft

[1] Fox, Terry L. and J. Wayne Spence, "Tools of the Trade: A Survey of Project Management Tools," Project Management Journal (September 1998): 20–27.

[2] Microsoft Corporation, "Microsoft Project Surpasses 3 Millionth-Customer Mark: Customers Help Make Microsoft Project a Category Leader," San Jose, CA, December 7, 1998.

Project had grown to over $100 million, and Microsoft recognizes the importance of the project management market as it continues to develop and enhance applications.[3] Microsoft also includes Project 2000 among its Office family of products.

Before being able to use Microsoft Project well, you must understand the fundamental concepts of project management. Therefore, to master the proper use of Project 2000, use this guide in conjunction with the main text.

This appendix uses a fictitious information technology (IT) project—the Project Tracking Database—to illustrate how to use the software. The lessons assume that you are familiar with other Windows-based applications. Each section includes several hands-on activities. You will gain the most from this guide if you complete these hands-on activities. As you work through the activities, periodically save your file. In addition, after each major section of the appendix, save your file using a different name, so you can open the appropriate file at the start of each new section. Figures showing Project 2000 screens will help you check your work. You can also check your work against files that are available for download from the Course Technology Web site for this book (go to www.course.com, then search for this text by author or subject). For further practice, exercises based on a variety of information technology projects are provided at the end of this appendix.

OVERVIEW OF MICROSOFT PROJECT 2000

The first step to mastering Project 2000 is to become familiar with the Help facility and online tutorials, major screen elements, views, and filters. This section describes each of these features.

Starting Project 2000 and the Project Help Window

To start Microsoft Project 2000, click Start on the Windows taskbar, select Programs, and then click Microsoft Project. Alternately, a shortcut or icon Proj... might be available on the desktop; in this case, double-click the icon to start the software.

Once Project 2000 has started, depending on how the options have been configured, you might see the Microsoft Project Help window, as shown in Figure A-1. You can also make this window appear by selecting Help from the menu bar, then selecting Contents and Index. The Microsoft Project Help window presents six options:

- What's New: For those who've used previous versions of Microsoft Project, this option summarizes new features, including new scheduling features, network diagram improvements, and so on.

[3] Business Wire, "Pacific Edge Software Appoints Jim Dunnigan as Vice President of Marketing" (1/19/00).

- Quick Preview: This option provides a short animation on how Project 2000 can assist you in building a project plan, managing a project, and communicating project information.
- Tutorial: The online tutorial provides more detailed information on how to use Project 2000.
- Project Map: The project map describes common project management activities such as defining a project, planning project activities, planning for and procuring resources, and so on. This map can serve as a general guide and provide detailed steps on using Project 2000 for setting up, managing, and completing your projects.
- Office Assistant: As in other Microsoft applications, Office Assistant allows you to type a question and search for related information.
- Reference: The Reference option opens a window with links to information on Project 2000 features, reference content in Help, and additional resources.

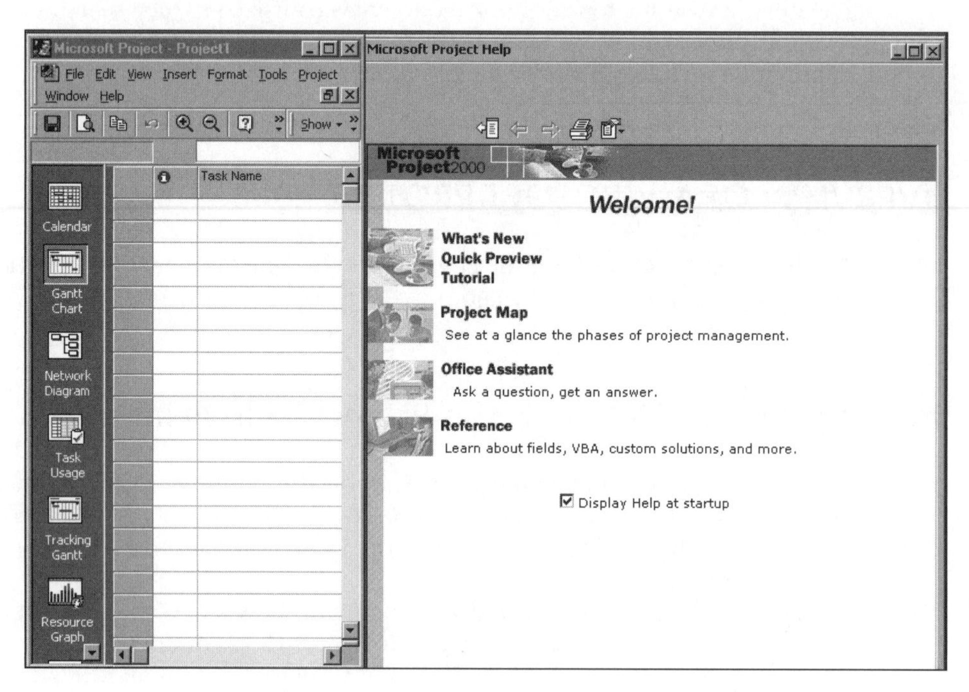

Figure A-1. Microsoft Project Help Dialog Box

Table A-1 provides some highlights of what's new in Project 2000. Several new features make it easier to enter, track, and share project information. For example, to help create a good WBS for your projects, you can now use the toolbar to quickly see different levels of tasks in your WBS. You can also enter months as a unit for task duration, and Project 2000 will determine the appropriate number of days for that month. Project 2000 includes several improvements to scheduling and network diagrams, such as a new Network Diagram view that includes filtering capabilities. You can also enter durations as estimates, allowing you to see where you need to go back and spend more time in developing better estimates. Project 2000 also includes new administration and programming features. For example, you can install a language pack to display menus and dialog boxes in various languages. Project 2000 is also an OLE-DB provider, making it easier to integrate project data across the enterprise.

Table A-1: What's New in Project 2000

NEW FEATURES

- WBS outline levels can now be expanded and collapsed more efficiently. You can now directly select the outline level for displaying your tasks.
- Months are now supported as a unit of duration. Typing "3mon" in the duration field is recognized as three months.
- Projects can now be based on templates; several detailed templates for different types of projects are included.
- Options allowing you to clear the baseline or to clear an interim plan on a project are included.
- Fill handles make fill up or down operations easier, and they are now included on the Gantt Chart view.
- Microsoft Project Central is available as a companion product to Project 2000, which builds on the Web-based workgroup features that were available in Microsoft Project 98. Enhancements include personal Gantt charts, filtering, sorting, grouping, task delegation, reports and a reporting capability, the ability to work offline, the ability to set message rules, and many other features.

SCHEDULING AND NETWORK DIAGRAM IMPROVEMENTS

- Calendars, including working time, for tasks can now be set. Task calendars allow you to create schedules that affect only selected tasks.
- Master projects can now calculate the latest finish date across all subprojects. This allows you to see one critical path across the master project.
- You can now set a deadline date that allows Microsoft Project 2000 to alert you if a task is scheduled to finish after its deadline.
- A task's duration may now be entered as an estimated duration. Estimated durations allow you to quickly find tasks with durations that may not be firm, using the new Tasks With Estimated Durations filter.
- The Network Diagram view in Microsoft Project is new, and completely replaces the PERT chart from previous versions. You can now apply a filter to the network diagram.
- In the Network Diagram view, you can use outlining symbols to hide or display the subtasks of a summary task in much the same way that you use outlining in the Gantt Chart view.

Table A-1: What's New in Project 2000 (continued)

ADMINISTRATION AND PROGRAMMABILITY
■ Just like the Office 2000 applications, Project uses the Microsoft Installer technology. Only the files you use are installed on your system. If a critical file is missing, the installer technology reinstalls the missing piece.
■ A language pack can be installed, so that your version of Microsoft Project can display menus and dialog boxes in another language.
■ Files can be saved to the Microsoft Project 98 MPP file format. This format allows you to easily exchange projects with users who have not upgraded to Microsoft Project 2000.
■ Microsoft Project 2000 is an OLE-DB provider. OLE-DB is a specification for a set of data access interfaces that enables a multitude of data stores in an enterprise to work seamlessly together. This makes it possible for other applications to easily access Microsoft Project data as well as its scheduling capabilities, making it much easier to integrate project data in the enterprise.

Microsoft Project 2000 Help, What's New Dialog Box

You can access the Quick Preview, Tutorial, and Project Map features from the Help menu under Getting Started. The Help menu also includes options for accessing Microsoft Project Help, Contents and Index, Showing or Hiding the Office Assistant, What's This?, Office on the Web, Detect and Repair, and About Microsoft Project. All of these features help you to use this powerful project management software.

Microsoft realizes that Project 2000 can take some time to learn and provides a number of resources on its Web site. Select Office on the Web from the Project 2000 Help menu to get to Microsoft's Web site for Office products. Microsoft's Web site for Project 2000 (www.microsoft.com/project) provides files for users to download, case studies, articles, and other useful materials.

Within Project 2000, Office Assistant helps answer questions as they arise. You can access Office Assistant at any time by selecting Microsoft Project Help from the Help menu. You can also press F1 or click the Office Assistant icon on the Standard toolbar. As in other Microsoft applications, Office Assistant allows you to type a question and search for related information.

Main Screen Elements

Figure A-2 shows the main screen you see after starting Project 2000. The Project 2000 default main screen, called the Gantt Chart view, has three parts: a Gantt chart, an Entry table, and the View bar. At the top of the main screen, the menu bar and toolbar are similar to those in other Windows applications. Note that the order and appearance of icons on your toolbar may vary from those shown in Figure A-2, depending on the features you are using. You can display the toolbar in one row or two, and icons often change according to which ones you use. Several commonly used icons such as the Link tasks icon,

Zoom in and out icons, and Indent icon are also identified in Figure A-2. Below the toolbar is the Entry bar, which displays entries you make in the Entry table, located right below the Entry text box. The Gantt chart appears on the right of the split bar, which separates the Entry table and the Gantt chart. Use the Entry table to enter task information, such as task names and durations. The column to the left of the Task Name column is the Indicators column. The Indicators column displays indicators or symbols related to items associated with each task, such as task notes or hyperlinks to other files.

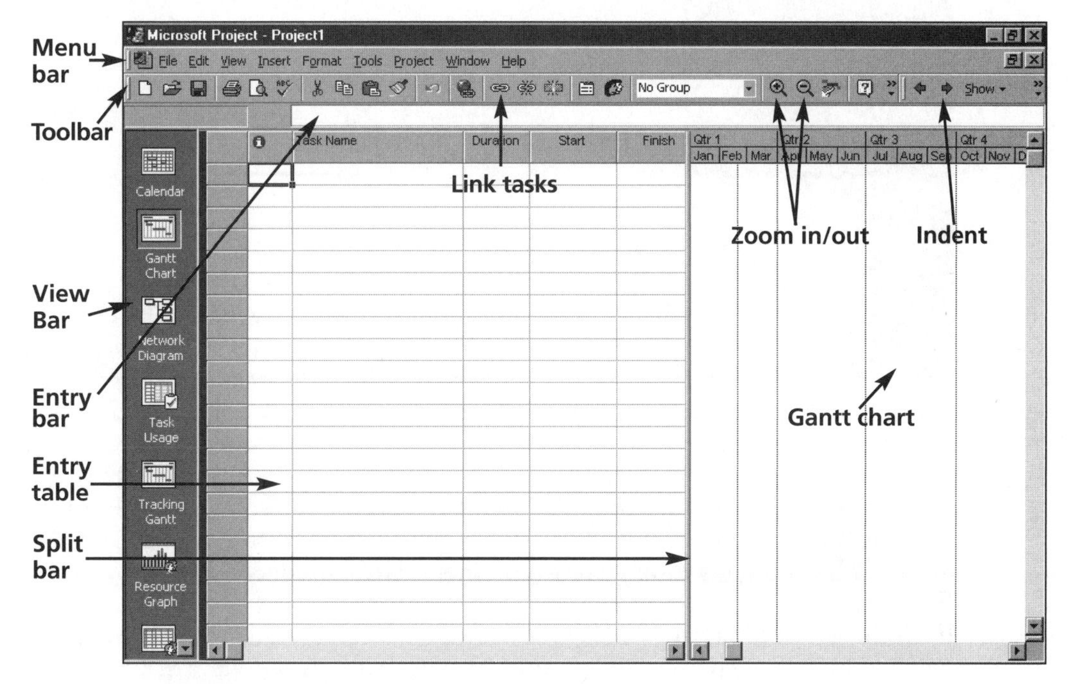

Figure A-2. Project 2000 Main Screen

Many features in Project 2000 are similar to features in other Windows applications. For example, to view different toolbars, select Toolbars from the View menu. To access shortcut items, right-click in either the Entry table area or the Gantt chart. Many of the Entry table operations in Project 2000 are very similar to operations in Excel. For example, to adjust a column width, click and drag or double-click between the column heading titles.

If you select another view and want to return to the Gantt Chart view, select Gantt Chart from the expanded View bar on the left of the screen or select View from the menu bar, and click Gantt Chart, as shown in Figure A-3. If the table on the left appears to be different, select View and click Table: Entry to return to the default table entry settings.

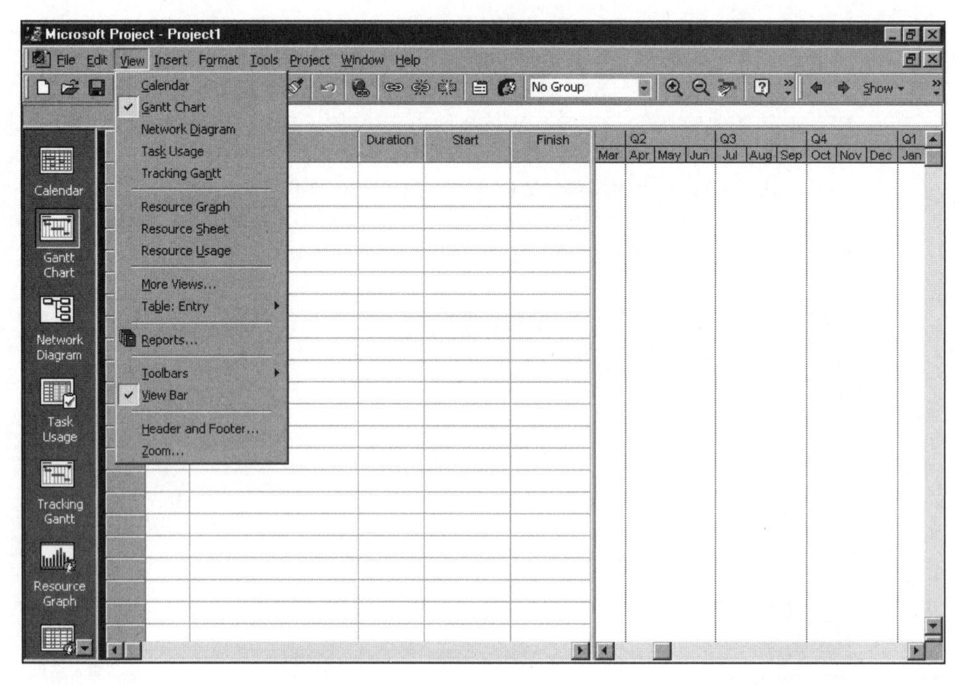

Figure A-3. Project 2000 View Menu Options

Notice the split bar, which separates the Entry table from the Gantt chart. When you run the mouse over the split bar you see the resize pointer ◂||▸. Clicking and dragging the split bar to the right reveals other task information in the Entry table, including start and finish dates, predecessors, and resource names.

To the left of the Entry table is the View bar. Instead of using commands on the View menu to change views, you can click icons on the View bar. To save screen space, you can hide the View bar by selecting View from the menu bar and deselecting View Bar. When the View bar is not visible, a blue line appears to the far left of the main screen. When you right-click this blue line, a shortcut menu appears, which gives you quick access to the other views.

Project 2000 Views

In Project 2000, there are many different ways to display project information. These displays are called views. Several views discussed in the main text are on the default View bar: Gantt Chart, Network Diagram, Tracking Gantt Chart, and Resource Graph. Other views include Calendar, Task Usage, Resource Sheet, Resource Usage, and an option to display More Views. These different views allow you to examine project information in different ways, which helps you analyze and understand what is happening on your project.

The View menu also provides access to different tables and reports that display information in a variety of ways. Some tables that you can access from the View menu include Schedule, Cost, Tracking, and Earned Value. Some reports you can access from the View menu include Overview, Current Activities, Costs, Assignments, and Workload.

Some Project 2000 views, such as the chart views, present a broad look at the entire project, whereas others, such as the form views, focus on specific pieces of information about each task. Three main categories of views are available:

- Graphical: A chart or graphical representation of data using bars, boxes, lines, and images
- Task Sheet or Table: A spreadsheet-like representation of data in which each task appears as a new row, and each piece of information about the task is represented by a column. Different tables are applied to a sheet to display different kinds of information.
- Form: A specific view of information for one task. Forms are used to focus on the details of one task.

Table A-2 describes some of the predesigned views within each category that Project 2000 provides to help you display the project or task information that you need.

Next, you will use sample files to access and explore some of the views available in Project 2000. Project 2000 comes with several sample files that should be loaded onto your hard drive. You can open these files by selecting New from the File menu and then selecting the Project Templates tab. Additional templates are available from the Office 97 tab. Reviewing various template files can give you ideas on how to create your own project files. As mentioned earlier, the toolbars and icons can be configured in different ways, as with other Office 2000 products. When you first start Project 2000, the menus and toolbars display basic commands and buttons. As you work with Project, the commands and buttons that you use most often are stored as personalized settings and displayed on menus and toolbars. Your toolbar might be displayed on two rows instead of one. To make the toolbar appear on one line, as shown in this text, click the left side of the second toolbar row and drag it to the first toolbar row. You can also select Customize from the Tools menu, then Toolbars, and deselect the option for the Standard and Formatting toolbars to share one row.

Table A-2. Common Project 2000 Views

CATEGORY OF VIEW	VIEW NAME	DESCRIPTION OF VIEW
Graphical	Gantt Chart	Standard format for displaying project schedule information that lists project tasks and their corresponding start and finish dates in a calendar format. Shows each task as a horizontal bar with the length and position corresponding to the time scale at the top of the Gantt chart.
	Network Diagram	Schematic display of the logical relationships or sequencing of project activities. Shows each task as a box with linking lines between tasks to show sequencing. Critical tasks appear in red.
Task Sheet or Table	Entry Table	Default table view showing columns for the Task Name and Duration. By revealing more of the Entry table, you can enter start and end dates, predecessors, and resource names.
	Schedule Table	Displays columns for Task Name, Start, Finish, Late Start, Late Finish, Free Slack, and Total Slack
	Cost Table	Displays columns for Task Name, Fixed Cost, Fixed Cost Accrual, Total Cost, Baseline, Variance, Actual, and Remaining
	Tracking Table	Displays columns for Task Name, Actual Start, Actual Finish, % Complete, Actual Duration, Remaining Duration, Actual Cost, and Actual Work
	Earned Value	Displays columns for Task Name, BCWS, BCWP, ACWP, SV, CV, EAC, BAC, and VAC
Form	Task Details Form	Displays detailed information about a single task in one window
	Task Name Form	Displays columns for the Task Name, Resources, and Predecessors for a single task

To practice accessing different views and to explore those views, start Project 2000, then do the following:

1. *Open the file called Infrastructure Deployment.* Click **File** on the menu bar, then click **New**. Click the **Project Templates** tab, then double-click the **Infrastructure Deployment** template icon. The Project Information dialog box opens. Type **6/01/02** in the highlighted Start Date: text box, then click **OK** or press **Enter** to close the Project Information dialog box.

If the Infrastructure Deployment file is not available on your hard drive, you can download it from the Course Technology Web site for this text.

2. *Change the Gantt chart time scale.* Click **Zoom out** on the toolbar three times, or as needed, until the Gantt chart time scale displays quarters. You should see that the entire project starts in June 2002 and ends in December 2002. The Zoom in icon on the toolbar makes the time scale smaller—months instead of quarters, for example.
3. *Explore different views.* Click different views using the View bar. Alternately, select different views from the View menu. For example, click the Calendar icon on the View bar to see project information in a calendar format, click the Network Diagram icon to see the project's network diagram, and click the Task Usage icon to see how hours are assigned for tasks.
4. *Examine columns in the Entry table.* Return to the Gantt Chart view. Move the split bar by moving your mouse between the Entry table and Gantt chart until you see the resize pointer. Click and drag your mouse to the right to reveal other columns in the Entry table. To make all the text display in the Task Name column, move the mouse over the right-column gridline in the Task Name header until you see the resize pointer, then double-click the **left mouse** button to make the column width automatically resize to show all the text.
5. *Examine the Table:Schedule view.* Click **View** on the menu bar, move your mouse to Table:Entry, then click **Schedule** in the cascading menu to the right. Notice that the columns to the left of the Gantt chart now display more detailed schedule information. Experiment with other table views, then return to the Table:Entry view.
6. *Explore the Reports feature.* Click **View** on the menu bar, click **Reports**, then double-click **Overview** from the Reports dialog box. Double-click **Project Summary** in the Overview Reports dialog box. Notice that the insertion point now resembles a magnifying glass. Click inside the report to zoom in or out. Click the **Close** button to close this report, and then experiment with viewing other reports.

Project 2000 Filters

Project 2000 uses an underlying relational database to filter, sort, store, and display information. This database is compliant with Microsoft Access, and it is relatively easy to import information from and export information to other applications. Project 2000 is an OLE-DB provider, making it possible for other applications to easily access Microsoft Project data as well as its scheduling capabilities.

You can filter information by clicking the Filter text box list arrow on the toolbar. Figure A-4 shows the resulting list of filters. You can access more filter options by using the scroll bar.

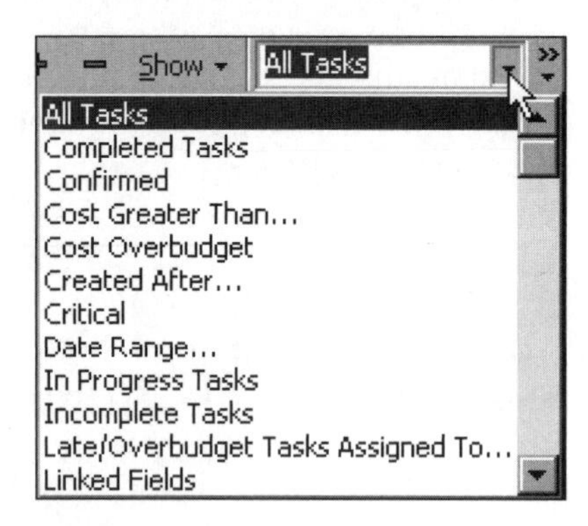

Figure A-4. Filter List Options

Filtering project information is very useful. For example, if a project includes hundreds of tasks, you might want to view only summary or milestone tasks to get an overview of the project. To get this type of overview of a project, select the Milestones or Summary Tasks filter from the Filter list. If you are concerned about the schedule, you can select a filter that shows only tasks on the critical path. Other filters include Completed Tasks, Late/Overbudget Tasks, and Date Range, which displays tasks on the basis of dates you provide. You can also click the Show list arrow on the toolbar to quickly display different levels in your WBS. For example, Outline Level 1 shows the highest-level items in your WBS, Outline Level 2 shows the next level of detail in your WBS, and so on.

To explore Project 2000 filters:

1. *Filter to show specific tasks.* Using the Infrastructure Deployment file, apply a filter to see only summary tasks. If the Filter text box [All Tasks] does not appear on your toolbar, click the **More buttons** on the toolbar to find it. Click the **filter list arrow**, scroll down until you see Summary Tasks, and click **Summary Tasks**. Click **Zoom out** on the toolbar three times, or as needed, so the time scale displays quarters. Your screen should resemble Figure A-5.

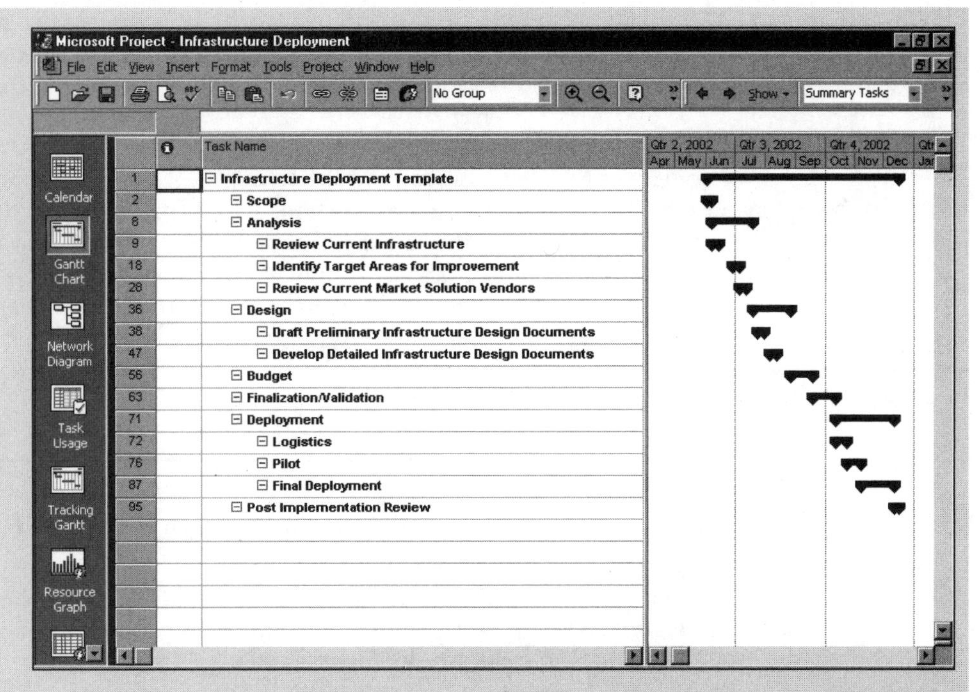

Figure A-5. Summary Tasks Filter for Infrastructure Deployment File

2. *Show outline levels.* Select **All tasks** from the Filter list box to reveal all the tasks in your WBS again. Click the **Show list arrow**, then click **Outline Level 2**. Only level 2 items and above appear in your WBS. Experiment with other outline levels. Note: The Project 2000 template files show the entire project as level 1 instead of level 0, which is the more common syntax, as explained in Chapter 4 of this text.
3. *Examine other template files.* When you are finished reviewing the Infrastructure Deployment file, close it. After you select **Close** from the File menu, a save changes dialog box will appear. Click **No**. Open and examine other template files as you desire.

PROJECT SCOPE MANAGEMENT

Project scope management involves defining the work to be done to carry out the project. To use Project 2000, you must first determine the scope of the project. To begin determining the project's scope, create a new file with the project name and start date. Develop a list of tasks that need to be done to carry out the project. This list of tasks is called a work breakdown structure (WBS; *see*

Chapter 4). If you intend to track actual project information against the initial plan, you must set a baseline. This section explains how to create a new project file, develop a WBS, and set a baseline.

In this section you will go through several steps to create a Project 2000 file named scope. If you wish to download the Project 2000 file to check your work or continue to the next section, a copy of the Project 2000 file named scope is available on the Course Technology Web site for this text or from your instructor. You should try to complete an entire section of this appendix (project scope management, project time management, and so on) in one sitting to create the complete file.

Creating a New Project File

To create a new project file:

1. *Create a blank project.* Click **File** on the menu bar, then click **New**. The New dialog box appears with the General tab active. Double-click the **Blank Project** icon to display the Project Information dialog box, as shown in Figure A-6. The Project Information dialog box enables you to set dates for the project, select the calendar to be used, and view project statistics. By default, the project start date is the date you create the file. The default filenames are Project1, Project2, and so on.

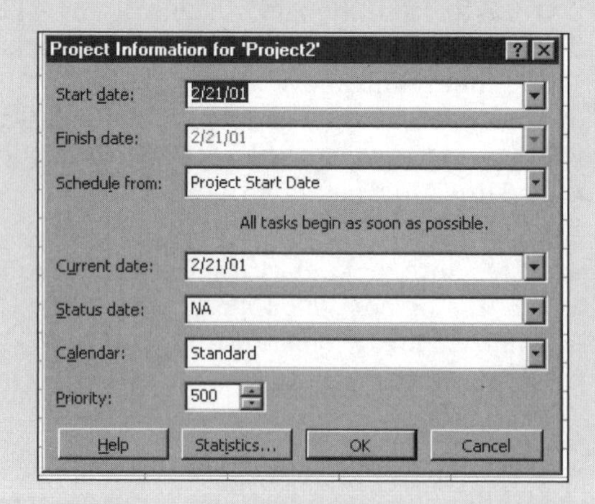

Figure A-6. Project Information Dialog Box

2. *Enter the project start date.* In the Start date text box, enter **6/03/02**. Setting your project start date to 6/03/02 will ensure that your work matches the results that appear in this appendix. Leave the finish date and current date at the default settings. Click **OK**.

3. *Enter project properties.* Click **File** on the menu bar, then click **Properties**. You may need to wait for Properties to appear, or click the arrows at the bottom of the menu to display the expanded menu. Any command that you click in the expanded menu is added immediately to the personalized (short) version of the menu. If you stop using a command for a while, Project 2000 stops showing it on the short version of the menu. The Project Properties dialog box opens, as shown in Figure A-7. If necessary, click the Summary tab to make it active. This dialog box allows you to enter the project title, subject, author, manager, company, and so on. Type **Project Tracking Database** in the Title text box, and type your name in the Author text box. Click **OK**.

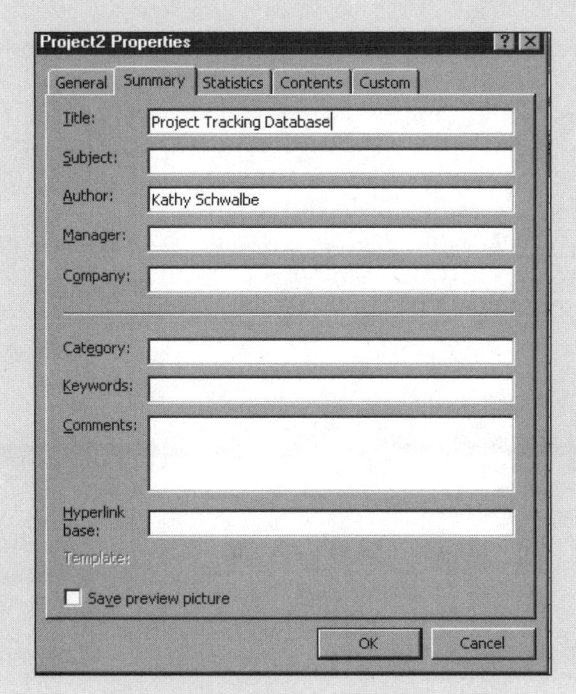

Figure A-7. Project Properties Dialog Box

As with other Office 2000 applications, you can change the default toolbar and menu options. Select Tools, Customize, Toolbars from the menu bar. Under the Options tab, you can personalize menus and toolbars. To display the Project 2000 toolbars on one row, check the first check box for Standard and Formatting Toolbars share one row. To always display the full set of menu options, make sure the check box for the second item, Menus show recently used commands first, is unchecked.

Developing a Work Breakdown Structure

Before using Project 2000, you must develop a work breakdown structure (WBS) for your project. Developing a good WBS takes time, and it will make entering tasks into the Entry table easier if you develop the WBS first. It is also a good idea to establish milestones before entering tasks in Project 2000. See Chapter 4, *Project Scope Management,* for more information on creating a WBS and determining milestones. You will use the information in Table A-2 to enter tasks for the Project Tracking Database exercise. Be aware that this example is much shorter and simpler than most WBSs.

To develop a work breakdown structure and enter milestones for the Project Tracking Database:

1. *Enter task names.* Enter the twenty tasks in Table A-3 into the Task Name column in the order shown. Don't worry about durations or any other information at this time. Type the name of each task into the Task Name column of the Entry table, beginning with the first row. Press [**Enter**] or the **down arrow** key to move to the next row. If you accidentally skip a task, highlight the task row and select **Insert** from the menu bar, then select **New Task**, to get a blank row. To edit a task entry, click the text for that task, click the entry bar under the formatting toolbar, and either type over the old text or edit the existing text.

Entering tasks into Project 2000 and editing the information is similar to entering and editing data in an Excel spreadsheet.

2. *Adjust the Task Name column width.* To make all the text display in the Task Name column, move the mouse over the right-column gridline in the Task Name header until you see the resize pointer ✥, then click the **left mouse** button and drag the line to the right to make the column wider.

Table A-3: Project Tracking Database Tasks

ORDER	TASKS	ORDER	TASKS
1	Initiating	11	Design
2	Kickoff meeting	12	Implementation
3	Develop project charter	13	System implemented
4	Charter signed	14	Controlling
5	Planning	15	Report performance
6	Develop project plans	16	Control changes
7	Review project plans	17	Closing
8	Project plans approved	18	Prepare final project report
9	Executing	19	Present final project
10	Analysis	20	Project completed

Notice that this WBS separates tasks according to the project process groups of initiating, planning, executing, controlling, and closing. These tasks will be the level 1 items in the WBS for this project. It is a good idea to include all of these process group tasks, rather than focus only on the executing tasks of a project, as many people do. Recall that the WBS should include all of the work required for the project. In the Project Tracking Database WBS, the tasks will purposefully be left at a high WBS level (level 2). You will create these levels, or the WBS hierarchy, next when you create summary tasks. For a real project, you would break the WBS into even more levels to provide more details to describe all the work involved in the project. For example, analysis tasks for a database project might be broken down further to include preparing entity relationship diagrams for the database and developing guidelines for the user interface. Design tasks might be broken down to include preparing prototypes, incorporating user feedback, entering data, and testing the database. Implementation tasks might include more levels such as installing new hardware or software, training the users, fully documenting the system, and so on.

Creating Summary Tasks

After entering the WBS tasks in Table A-3 into the Entry table, the next step is to show the WBS levels by creating summary tasks. The summary tasks in this example are Tasks 1 (initiating), 5 (planning), 9 (executing), 14 (controlling), and 17 (closing). You create summary tasks by highlighting and indenting their respective subtasks.

To create the summary tasks:

1. *Select lower-level or subtasks.* Highlight **Tasks 2** through **4** by left-clicking the text for Task 2 and dragging the mouse through the text for Task 4.
2. *Indent subtasks.* Click the **Indent** icon on the Formatting toolbar; after the subtasks (Tasks 2 through 4) are indented, notice that Task 1 automatically becomes boldface, which indicates that it is a summary task. A minus sign appears to the left of the new summary task name. Clicking the minus sign will collapse the summary task and hide the subtasks beneath it. When subtasks are hidden, a plus sign appears to the left of the summary task name; click the plus sign to expand the summary task. Also notice that the symbol for the summary task on the Gantt chart has changed from a blue line to a black line with arrows indicating the start and end dates.
3. *Create other summary tasks and subtasks.* Create subtasks and summary tasks for Tasks 5, 9, 14, and 17 by following the same steps. Indent **Tasks 6** through **8** to make Task 5 a summary task. Indent **Tasks 10** through **13** to make Task 9 a summary task. Indent **Tasks 15** through **16** to make Task 14 a summary task. Indent **Tasks 18** through **20** to make Task 17 a summary task.

To change a task from a subtask to a summary task or to change its level in the WBS, you can "outdent" it. Highlight the task or tasks you wish to change, and click the Outdent icon on the Formatting toolbar. Remember that the tasks in Project 2000 should be entered in a good WBS format with several levels in the hierarchy.

Numbering Tasks

Depending on how Project 2000 is set up on your computer, you may or may not see numbers associated with tasks as you enter and indent them.

To display automatic numbering of tasks, using the standard tabular numbering system for a WBS:

1. *Display the Options dialog box.* Click **Tools** on the menu bar, then click **Options**. The Options dialog box opens.
2. *Show outline numbers.* In the Options dialog box, click the **View** tab, if necessary. In the Outline options section of the View options for 'Project 2' section (at the bottom of the dialog box), click **Show outline number** so that a check mark appears in the check box. Click **OK**.

Saving Project Files with or Without a Baseline

An important part of project management is tracking performance against a baseline. When you first save a file in Project 2000, you are prompted to enter a filename and then save the file with or without a baseline. The default is to save without a baseline. It is important to wait until you are ready to save your file with a baseline, because Project 2000 will show changes against a baseline. Because you are still developing your project file for the tracking database project, you will save the file without a baseline. Later in this appendix you will save with a baseline and then enter actual information to compare planned and actual performance data. Project 2000 also allows you to clear a baseline by selecting Tools, Tracking, Clear Baseline.

To save a file without a baseline:

1. *Save your file.* Save your file by clicking **File** on the menu bar and then clicking **Save**, or by clicking the **Save** icon.
2. *Enter a filename.* In the Save as dialog box, type **scope** in the File name text box. Save the file in your personal folder, on a floppy disk, or in the C:\temp folder, depending on how your computer is set up.
3. *Save without a baseline.* The Planning Wizard dialog box displays two options: saving the file without a baseline (the default value), or saving with a baseline. Be sure the first option to save without a baseline is selected, then click **OK**.

When you have finished entering all twenty tasks, created the summary tasks and subtasks, set the options to show the standard WBS tabular numbering system, and saved your file, your Project 2000 file should look like Figure A-8. (*Note*: You can adjust the time scale to show months by clicking Zoom in or out.)

If you wish to download the Project 2000 file to check your work or continue to the next section, a copy of the Project 2000 file named scope is available on the Course Technology Web site for this text or from your instructor.

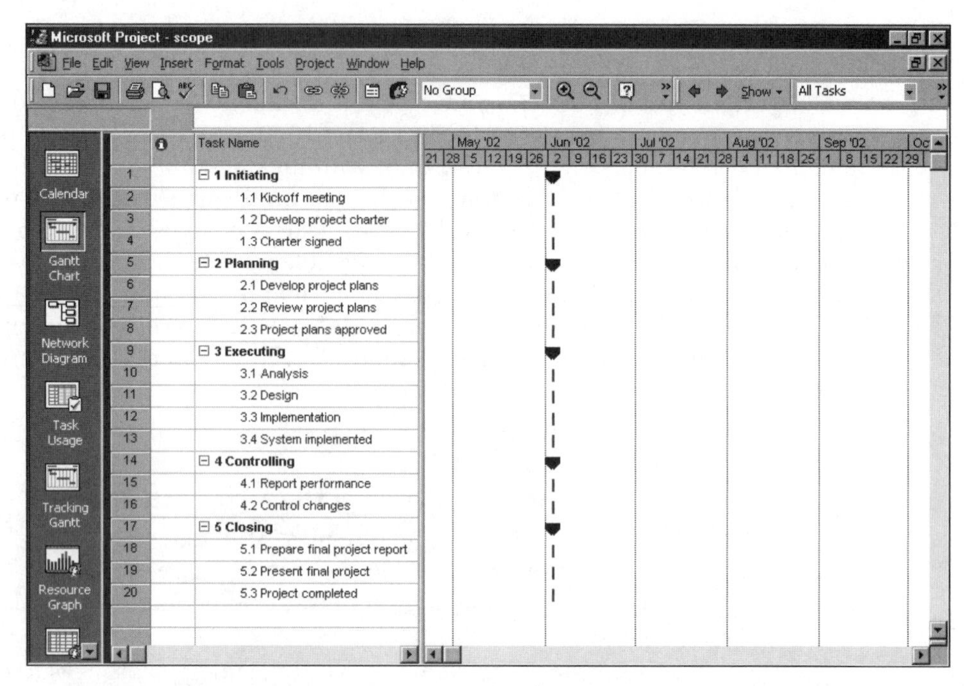

Figure A-8. Project 2000 Screen with All Tasks Entered, Summary Tasks, and Standard Numbering System

PROJECT TIME MANAGEMENT

Many people use Project 2000 for its time management features. The first step in using these features, after entering the WBS for the project, is to enter durations for tasks or specific dates when tasks will occur. Entering durations or specific dates will automatically update the Gantt chart. If you plan to do critical path analysis, you must also enter task dependencies. After entering durations and task dependencies, you can view the network diagram and critical path information. This section describes how to use each of these time management features.

Entering Task Durations

When you enter a task, Project 2000 automatically assigns to it a default duration of one day, followed by a question mark. To change the default duration, type a task's estimated duration in the Duration column. If you are unsure of an estimate and want to review it again later, enter a question mark after it. For example, you could enter 5d? for a task with an estimated duration of five days that you want to review later. One new feature of Project 2000 is the Tasks With Estimated Durations filter. This filter allows you to quickly see the tasks for which you need to review duration estimates, which you indicated by entering a question mark after the duration.

To indicate the length of a task's duration, you must type both a number and an appropriate duration symbol. If you type only a number, Project 2000 automatically enters days as the duration unit. Duration unit symbols include:

- d = days
- w = weeks
- m = minutes
- h = hours
- mon = months

For example, to enter one week for a task duration, type 1w in the Duration column. To enter two days for a task duration, type 2d in the Duration column. The default unit is days, so if you enter 2 for the duration, it will be entered as 2 days. Months is a new duration unit in Project 2000. You can also enter elapsed times in the Duration column. For example, led means one elapsed day, and lew means one elapsed week. You would use an elapsed duration for a task like "Allow paint to dry." The paint will dry in exactly the same amount of time regardless of whether it is a workday, a weekend, or a holiday.

If the Duration column is not visible, move the mouse over the vertical split bar until the resize pointer is visible. Click and drag the split bar to the right until the Duration column is in view.

Entering time estimates or durations might seem like a straightforward process. However, there are a few important procedures you must follow:

- Do not enter durations for summary tasks. Summary task durations are calculated automatically, based on the subtasks. If you enter a duration for a task and then make it a summary task, its duration will automatically change to match the durations of its subtasks. Project 2000 will not allow you to enter or change the duration of a summary task.
- To mark a task as a milestone, enter 0 for the duration.
- To enter recurring tasks, such as weekly meetings or monthly status reports, select Recurring Task from the Insert menu. Enter the task name, the duration, and when the task occurs. Project 2000 will automatically insert appropriate subtasks based on the length of the project and the number of tasks required for the recurring task. For example, if you enter a recurring task for monthly review meetings that occur on the first day of every month for a twelve-month project, Project 2000 will enter a summary task for monthly review meetings and twelve subtasks—one meeting for each month.

- You can enter the exact start and finish dates for activities, instead of entering durations. To enter start and finish dates, move the split bar to the right to reveal the start and finish columns. Enter start and finish dates only when those dates are certain. If you want task dates to adjust according to any other task dates, do not enter exact start and finish dates. Instead, you should enter a duration and then establish a dependency to related tasks. *The real scheduling power of Project 2000 comes from setting up dependencies or relationships among tasks, as described in the next section.*
- Project 2000 uses a default calendar with standard workdays and hours. Duration estimates will vary according to the calendar used. For example, entering 5d in the standard calendar may result in more than 5 days on the Gantt chart if the time period included Saturday or Sunday. You can change specific working and nonworking days or the entire project calendar by selecting Change Working Time from the Tools menu.
- You often need to adjust the time scale on the Gantt chart to see your project's schedule in different time frames, such as weeks, months, quarters, or years. You expand the time scale by clicking the Zoom out icon . The Zoom in icon collapses the time scale.

Next, you will set task durations in the Project Tracking Database file called scope, created in the previous section. If you did not create this file, you can download it from the Course Technology Web site. You will create a new recurring task and enter its duration, and then you will enter other task durations. First, create a new recurring task called Status Reports above Task 15, Report performance.

To create a new recurring task:

1. *Insert a recurring task above Task 15, Report performance.* Using the scope file, click **Report performance** (Task 15) in the Task Name column to select that task. Click **Insert** on the menu bar, then click **Recurring task**. The Recurring Task Information dialog box opens, as shown in Figure A-9. The new recurring task will appear above Task 15, Report performance.
2. *Enter task and duration information for the recurring task.* Type **Status Reports** as the task title in the Name text box. Type **1h** in the Duration text box. Select **Weekly** from the Recurrence pattern radio buttons. Select **every** from the week on: drop-down list. On the list of days, check **Wednesday**. In the Range of recurrence section, type **6/12/02** in the Start: text box, click the **End by**: radio button to select it, and type **11/13/02** in the End by: text box. Alternately, select the drop-down list in the Start: and End by: text boxes and scroll through

the calendar that appears to find and then select the desired month and day. Figure A-9 shows this calendar in the Range of recurrence section. You can also enter a number of occurrences instead of a To date for a recurring task.

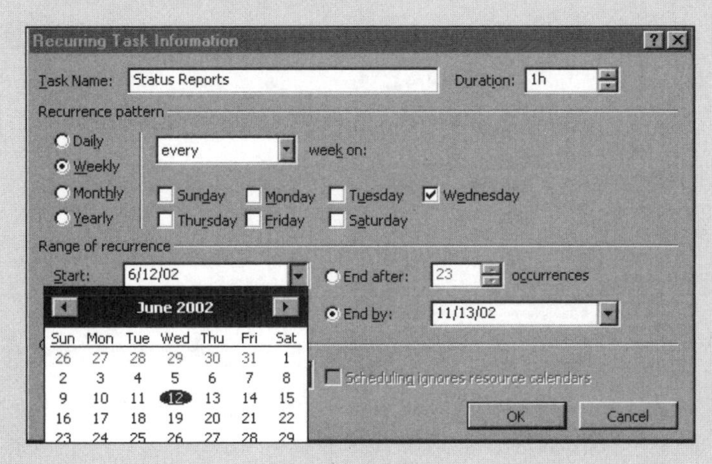

Figure A-9. Recurring Tasks

3. View the new summary task and its subtasks. Click OK. Project 2000 inserts a new Status Reports subtask in the Task Name column. Note that this new subtask is boldface and has a plus sign [+] to the left of the task name. Expand the new subtask, by clicking [+] to the left of the words Status Reports. To collapse the recurring task, click the minus sign [–].
4. *Adjust the Duration column width and Gantt chart time scale.* Notice the # signs that appear in the Duration column for the Status Reports task. Like Excel, Project 2000 uses these symbols to denote that the column width needs to be increased. Increase the Duration column's width by moving your mouse to the right of the Duration column heading until you see a resize pointer. Double-click to have the column width adjust automatically to display the information. Adjust the time scale of the Gantt chart, if necessary, to display quarters, using the Zoom out or Zoom in icons on the toolbar. Your screen should resemble Figure A-10. Notice that the recurring task appears on the appropriate dates on the Gantt chart.

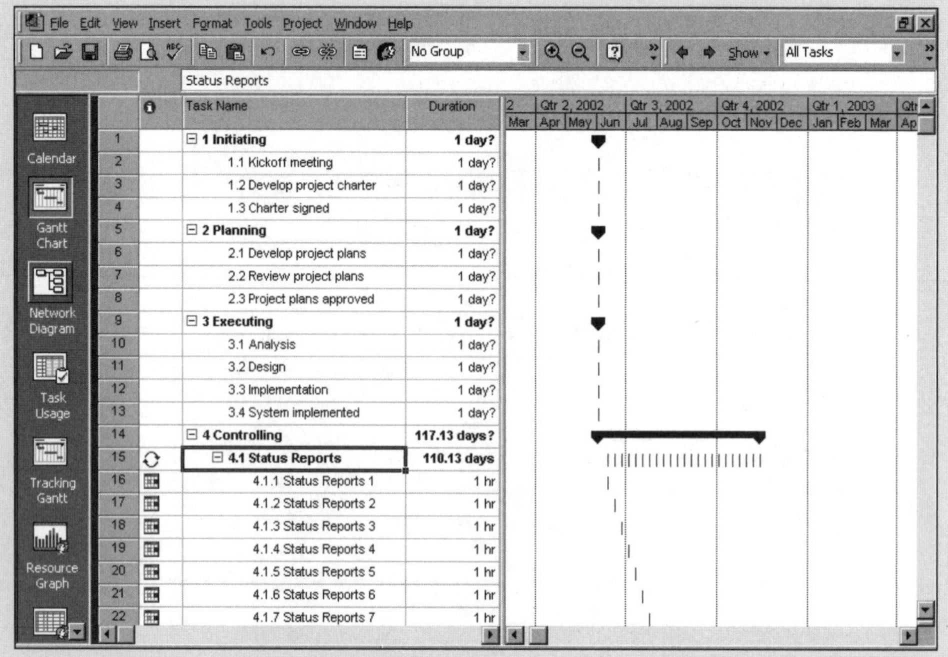

Figure A-10. Expanding a Recurring Task

Use the information in Table A-4 to enter durations for the other tasks for the Project Tracking Database project. The Project 2000 row number is shown to the left of each task name in the table. Remember that you already entered a duration for the recurring task. Also remember that you should not enter durations for summary tasks. Durations for summary tasks are automatically calculated to match the durations and dependencies of subtasks, as described further in the next section, "Establishing Task Dependencies."

Table A-4: Durations for Project Tracking Database Tasks

Task Number/Row	Task Name	Duration
2	Kickoff meeting	2h
3	Develop project charter	10d
4	Charter signed	0d
6	Develop project plans	3w
7	Review project plans	4mon
8	Project plans approved	0d
10	Analysis	1mon
11	Design	2mon
12	Implementation	1mon
13	System implemented	0d
39	Report performance	5mon
40	Control changes	5mon
42	Prepare final project report	2w
43	Present final project	1w
44	Project completed	0d

To enter task durations for the other tasks:

1. *Enter the duration for Task 2.* Click the **Duration** column for row 2, Kickoff meeting, type **2h**, then press **[Enter]**.
2. *Enter the duration for Task 3.* In the **Duration** column for row 3, Develop project charter, type **10d**, then press **[Enter]**.
3. *Enter remaining task durations.* Continue to enter the durations, using information in Table A-4.
4. *Save your file and name it.* Click **File** on the menu bar, then click **Save As**. Enter **time** for the filename, and save without a baseline.

Establishing Task Dependencies

To use Project 2000 to adjust schedules automatically and to do critical path analysis, you *must* determine the dependencies or relationships among tasks (*see* Chapter 5, *Project Time Management*, for a full discussion of task dependencies). Project 2000 provides three methods for creating task dependencies: using the Link tasks icon, using the Predecessors column of the Entry table, or clicking and dragging on the Gantt chart symbols for tasks with dependencies.

To create dependencies using the Link tasks icon, highlight tasks that are related and click the Link tasks icon on the toolbar. For example, to create a finish-to-start dependency between Task 1 and Task 2, click any cell in row 1, drag down to row 2, and then click the Link tasks icon. The default type of link is finish-to-start. In the Project Tracking Database example, all the tasks use this default relationship. Other types of dependencies will be explained later.

Selecting tasks is similar to selecting cells in Microsoft Excel. To select adjacent tasks, you can click and drag the mouse. You can also click the first task, hold down the Shift key, and then click the last task. To select nonadjacent tasks, hold down the Control key as you click tasks in order of their dependencies.

When you use the Predecessors column of the data entry table to create dependencies, you must manually enter the information. To manually create dependencies, type the task row number of the preceding task in the Predecessors column of the Entry table. For example, Task 3 in Table A-4 has Task 2 as a predecessor, as entered in the Predecessors column, meaning that Task 3 cannot start until Task 2 is finished. To see the Predecessors column of the Entry table, move the split bar to the right.

You can also create task dependencies by clicking the Gantt chart symbol for a task and then dragging to the Gantt chart symbol for a task that succeeds it. For example, you could click the milestone symbol ◆ for Task 4, hold down the left mouse button, and drag to the task bar symbol for Task 6 to create a dependency, as shown in the Gantt chart in Figure A-11. Note the Finish-to-Start Link dialog box that appears when you use this method.

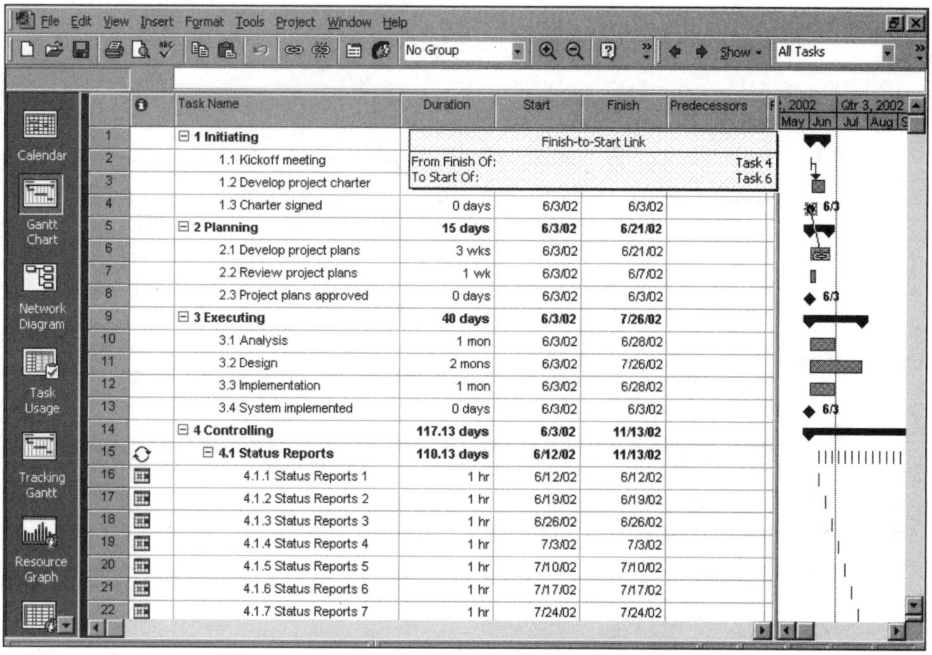

Figure A-11. Creating a Task Dependency Using Gantt Chart Symbols

Next, you will use information from Figure A-12 to enter the predecessors for tasks as indicated. You will create some dependencies by manually typing the predecessors in the Predecessors column, some by using the Link tasks icon, some by using the Gantt chart symbols, and the remaining dependencies by using whichever method you prefer.

To link tasks or establish dependencies for the Project Tracking Database:

1. *Display the Predecessors column in the Entry table.* Move the split bar to the right to reveal the Predecessors column in the file named time that you created earlier.
2. *Highlight the cell where you wish to enter a predecessor, then type the task number for the preceding task.* Click the **Predecessors cell** for Task 3, type **2**, and press **[Enter]**.
3. *Enter predecessors for Task 4.* Click the **Predecessors cell** for Task 4, type **3**, and press **[Enter]**. Notice that as you enter task dependencies, the Gantt chart changes to reflect the new schedule.
4. *Establish dependencies using the Link tasks icon.* To link Tasks 10 through 13, click the task name for Task 10 in the Task Name column and drag down through Task 13. Then, click the **Link tasks** icon on the toolbar.

5. *Create a dependency using Gantt chart symbols.* Click the **milestone** symbol ◆ for Task 4 on the Gantt chart, hold down the **left mouse** button, and drag to the **task bar** symbol for Task 6. (See Figure A-11.)
6. *Enter remaining dependencies.* Link the other tasks either by manually entering the predecessors into the Predecessors column, using the Link tasks icon, or clicking and dragging the Gantt chart symbols. If you have entered all data correctly, the project should end on 11/18/02. (*See* Figure A-12 to double-check the duration and predecessor information.) When you finish, your screen should look like Figure A-12.

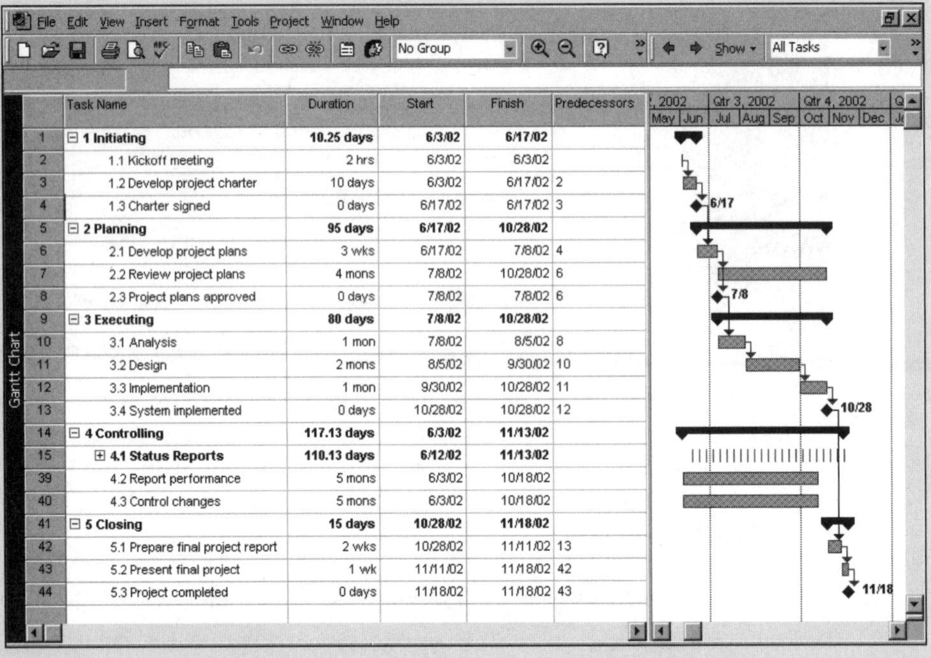

	Task Name	Duration	Start	Finish	Predecessors
1	**1 Initiating**	**10.25 days**	**6/3/02**	**6/17/02**	
2	1.1 Kickoff meeting	2 hrs	6/3/02	6/3/02	
3	1.2 Develop project charter	10 days	6/3/02	6/17/02	2
4	1.3 Charter signed	0 days	6/17/02	6/17/02	3
5	**2 Planning**	**95 days**	**6/17/02**	**10/28/02**	
6	2.1 Develop project plans	3 wks	6/17/02	7/8/02	4
7	2.2 Review project plans	4 mons	7/8/02	10/28/02	6
8	2.3 Project plans approved	0 days	7/8/02	7/8/02	6
9	**3 Executing**	**80 days**	**7/8/02**	**10/28/02**	
10	3.1 Analysis	1 mon	7/8/02	8/5/02	8
11	3.2 Design	2 mons	8/5/02	9/30/02	10
12	3.3 Implementation	1 mon	9/30/02	10/28/02	11
13	3.4 System implemented	0 days	10/28/02	10/28/02	12
14	**4 Controlling**	**117.13 days**	**6/3/02**	**11/13/02**	
15	**4.1 Status Reports**	**110.13 days**	**6/12/02**	**11/13/02**	
39	4.2 Report performance	5 mons	6/3/02	10/18/02	
40	4.3 Control changes	5 mons	6/3/02	10/18/02	
41	**5 Closing**	**15 days**	**10/28/02**	**11/18/02**	
42	5.1 Prepare final project report	2 wks	10/28/02	11/11/02	13
43	5.2 Present final project	1 wk	11/11/02	11/18/02	42
44	5.3 Project completed	0 days	11/18/02	11/18/02	43

Figure A-12. Project Database File with Durations and Relationships

7. *Preview and save your file.* Select **Print Preview** from the File menu or click the **Print Preview** icon on the toolbar. If your previewed file does not resemble Figure A-12, you may need to adjust the location of the split bar between the Entry table and the Gantt chart. When you are finished, save your file again without a baseline by clicking the **Save** icon. If you desire, print your file by clicking the Print icon or by selecting Print from the File menu.

If information in a column on the Entry table does not appear in the Print Preview mode, close the Print Preview mode and move the split bar to fully reveal the column, then select Print Preview again. You should not print from Project 2000 until the Print Preview displays the desired information.

Changing Task Dependency Types and Adding Lead or Lag Time

A task dependency or relationship describes how a task is related to the start or finish of another task. Project 2000 allows for four task dependencies: finish-to-start (FS), start-to-start (SS), finish-to-finish (FF), and start-to-finish (SF). You can find detailed descriptions of these dependencies in Chapter 5 and in the Project 2000 Help topic "Dependencies." By using these dependencies effectively, you can modify the critical path and shorten your project schedule. (See Chapter 5 for a detailed discussion of critical path.) The most common type of dependency is finish-to-start (FS). All of the dependencies in the Project Tracking Database example are FS dependencies. However, sometimes you need to establish other types of dependencies. This section describes how to change task dependency types. It also explains how to add lead or lag times between tasks. You will shorten the duration of the Project Tracking Database project by adding lead time between some tasks.

To change a dependency type, you would open the Task Information dialog box for that task by double-clicking the task name. On the Predecessors tab of the Task Information dialog box, you can select a new dependency type from the Type column list arrow.

The Predecessor tab also allows you to add lead or lag time to a dependency. You can enter both lead and lag time using the Lag column on the Predecessor tab. Lead time reflects an overlap between tasks that have a dependency. For example, if Task B can start when its predecessor, Task A, is half-finished, you can specify a finish-to-start dependency with a lead time of 50 percent for the successor task. You enter lead times as negative numbers. In this example, you would enter -50% in the first cell of the Lag column. Adding lead times is also called fast tracking and is one way to compress a project's schedule (*see* Chapter 5).

Lag time is the opposite of lead time—it is a time gap between tasks that have a dependency. If you need a 2-day delay between the finish of Task C and the start of Task D, establish a finish-to-start dependency between Tasks C and D and specify a 2-day lag time. Enter lag time as a positive value. In this example, you would type 2d in the Lag column.

In the Project Tracking Database example, notice that work on design tasks does not begin until all the work on the analysis tasks has been completed (*see* Rows 10 and 11 on the WBS), and work on implementation tasks does not begin until all the work on the design tasks has been completed (*see* Rows 11 and 12 on the WBS). In reality, it is rare to wait until all of the analysis work has been completed before starting any design work, or to wait for all of the design work to be completed before starting any implementation work. It is also a good idea to add some additional time, or a buffer, before crucial milestones, such as a system being implemented. To create a more realistic schedule, add lead times to the design and implementation tasks and lag time before the system implemented milestone.

To add lead and lag times:

1. *Open the Task Information dialog box for Task 11, Design.* In the Task Name column of the time file, double-click the text for Task 11, **Design**. The Task Information dialog box opens. Click the **Predecessors** tab to make it active.
2. *Enter lead time for Task 11.* Type **-10%** in the Lag column, as shown in Figure A-13. Click **OK**. You could also type a value like -5d to indicate a 5-day overlap. In the resulting Gantt chart, notice that the bar for this task has moved slightly to the left. Also notice that the project completion date has moved from 11/18 to 11/14.

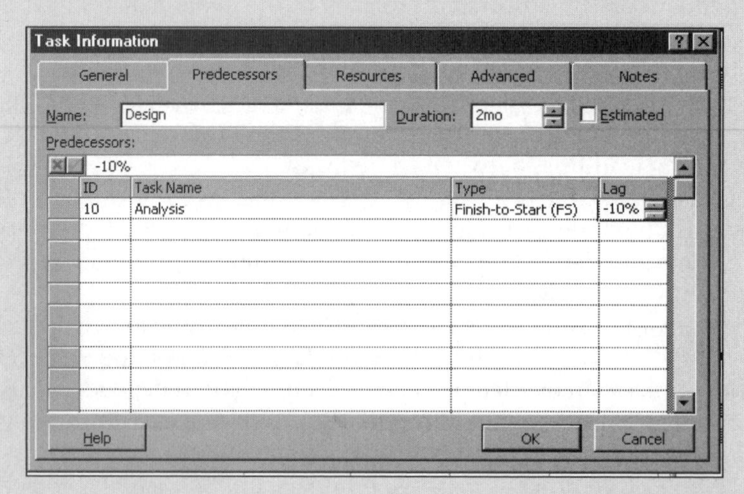

Figure A-13. Adding Lead or Lag Time to Task Dependencies

3. *Enter lead time for Task 12.* Double-click the text for Task 12, **Implementation**, enter **-3d** in the Lag column, and click **OK**. Notice that the project now ends on 11/11 instead of 11/14.
4. *Enter lag time for Task 13.* Double-click the text for Task 13, **System implemented**, and type **5d** in the Lag column for this task. Move the split bar to the right, if necessary, to reveal the Predecessor column.

When you are finished, your screen should resemble Figure A-14. Notice the overlap in the taskbars for Tasks 10 and 11 and the short gap between the taskbars for Tasks 12 and 13. The project completion date should be 11/18.

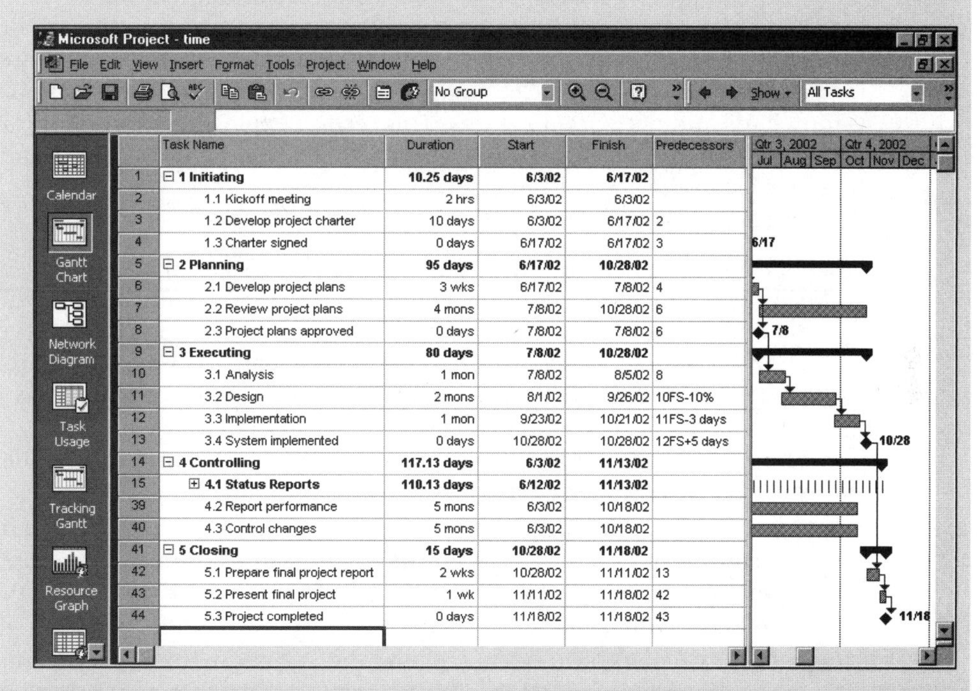

Figure A-14. Schedule for Project Tracking Database with Lead and Lag Times

5. *Save your file again without a baseline.* Click **File** on the menu bar, then Save or click the **Save** icon on the toolbar. Save the file named time without a baseline.

You can enter or modify lead or lag times directly in the Predecessors column of the Entry table. Notice how the predecessors appear for Tasks 11, 12, and 13. For example, the Predecessor column for Task 11 shows 10FS-10%. This notation means that Task 11 has a finish-to-start (FS) dependency with Task 10 and a 10 percent lead. You can enter lead or lag times to task dependencies directly into the Predecessors column of the Entry table by using this same format: task row number, followed by type of dependency, followed by the amount of lead or lag.

Gantt Charts

Project 2000 shows a Gantt chart in its default view along with the Entry table. Gantt charts show the time scale for a project and all of its activities. In Project 2000, dependencies between tasks are shown on the Gantt chart by the arrows between tasks. Many Gantt charts, however, do not show any dependencies. Instead, as you might recall, project network diagrams or PERT charts are used to show task dependencies. See Chapter 5 for more detailed information about Gantt charts and network diagrams. This section explains important information about Gantt charts and describes how to make critical path information more visible in the Gantt Chart view.

There are a few important things to know about using Gantt charts:

- To adjust the time scale, click the Zoom out icon or the Zoom In icon. Clicking these icons automatically makes the dates on the Gantt chart show more or less information. For example, if the time scale for the Gantt chart is showing months and you click the Zoom out icon, the time scale will adjust to show quarters. Clicking Zoom out again will display the time scale in years. Likewise, each time you click the Zoom in icon, the time scale changes to display more detailed time information—from years to quarters, quarters to months, and months to weeks.
- You can also adjust the time scale and access more formatting options by selecting Timescale from the Format menu. Adjusting the time scale can enable you to see the entire Gantt chart on one screen and in the time increments you desire.
- A Gantt Chart Wizard is available on the Format menu. This wizard helps you adjust the format of the Gantt chart. For example, you can select an option to display critical path information on the Gantt chart, and those items on the critical path will automatically be displayed using a red bar.
- You can view a tracking Gantt chart by setting a baseline project plan and entering actual durations for tasks. The tracking Gantt view displays two taskbars, one above the other, for each task. One bar shows planned or baseline start and finish dates, and the other bar shows actual start and finish dates. You will find a sample tracking Gantt chart later in this appendix, after we enter actual information for the Project Tracking Database example.

Because you have already created task dependencies, you can now find the critical path for the Project Tracking Database project. You can view the critical tasks by changing the color of those items on the critical path in the Task Name column. You can also change the color of bars on the Gantt chart. Tasks on the critical path will automatically be red in the Network Diagram view, as described in the following section.

also shows information such as the start and finish dates, task ID, and duration. The dashed line at the far left of Figure A-17 represents a page break. You often need to change some of the default settings for the Network Diagram view before printing it.

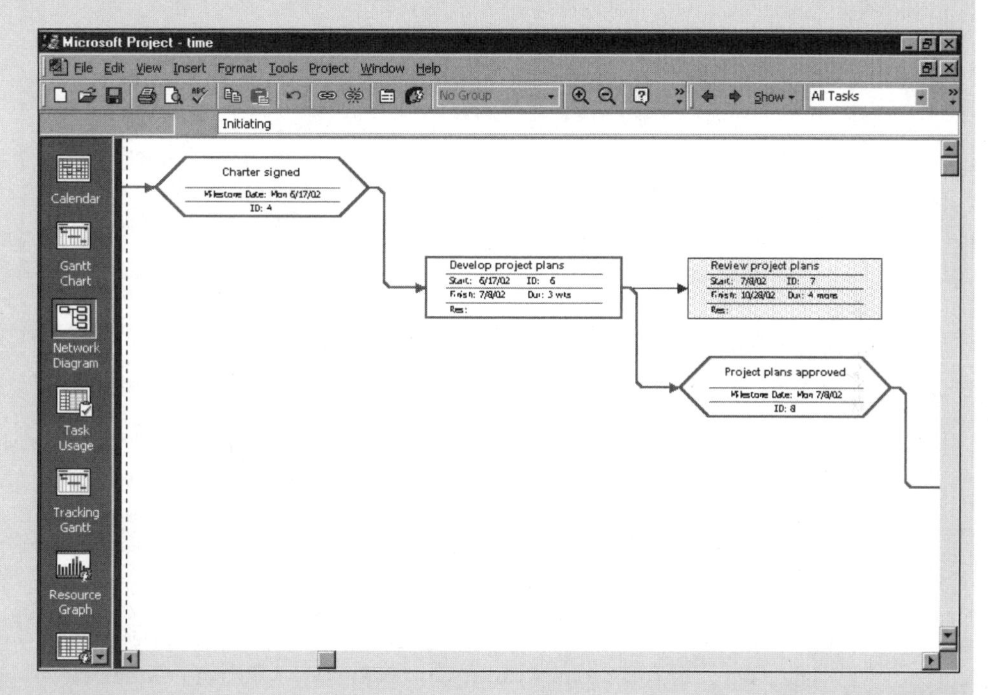

Figure A-17. Network Diagram View

3. *View the Help topic on rearranging network diagrams.* Click the **Microsoft Project Help** icon on the toolbar to display the Office Assistant. You can also press **F1** to display the Office Assistant. Type "**How do I rearrange the network diagram?**" in the text box. Then click the resulting Help topic called "**Change the arrangement of boxes in the Network Diagram view**." Figure A-18 shows the resulting Project 2000 Help screen. Note that you can change several layout options for the network diagram, hide fields, and manually position boxes. Close the Help window.

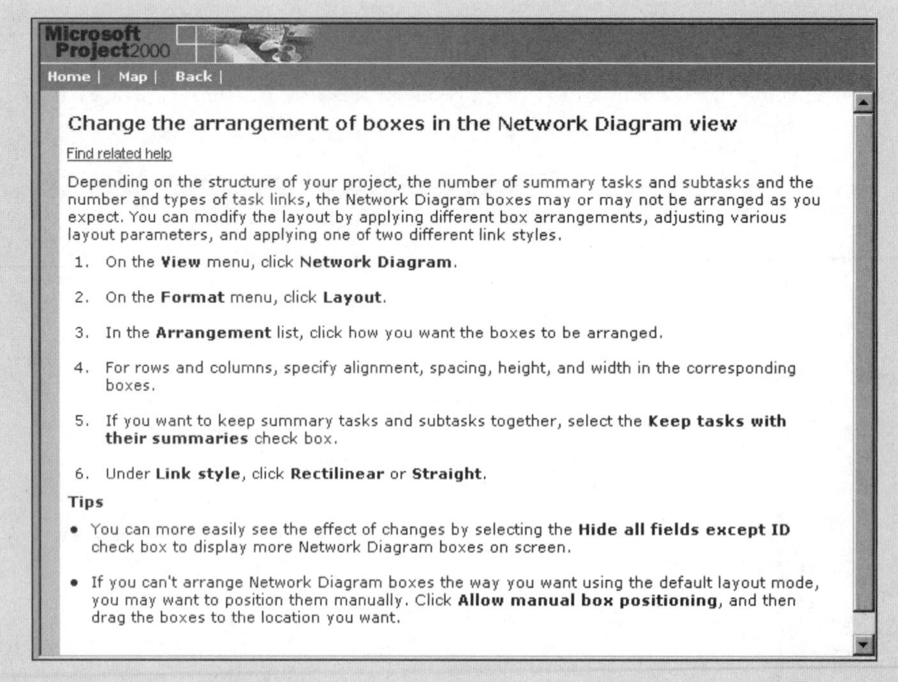

Figure A-18. Project 2000 Help Screen for Rearranging the Network Diagram View

4. *Return to Gantt Chart view.* Return to the Gantt Chart view by clicking the **Gantt chart** icon in the View bar or by selecting **Gantt Chart** from the View menu on the main menu bar.

Some people create new Project 2000 files or add tasks to existing files in the Network Diagram view instead of the Gantt Chart view. To add a new task or node in the Network Diagram view, select New Task from the Insert menu, or press the Insert key on your keyboard. Double-click the new node to add a task name and other information. Create finish-to-start dependencies between tasks in Network Diagram view by clicking the preceding node and dragging to the succeeding node. To modify the type of dependency and add lead or lag time, double-click the arrow between the dependent nodes.

Critical Path Analysis

Recall that the critical path is the path through the network diagram with no slack; it represents the shortest possible time to complete the project (*see* Chapter 5). If a task on the critical path takes longer than planned, the project schedule will slip unless time is reduced on a task later on the critical path. Sometimes you can shift resources between tasks to help keep a project on schedule. Project 2000 has several views and reports that help you analyze critical path information.

Two particularly useful features are the Schedule Table view and the Critical Tasks report. The Schedule Table view shows the early and late start dates for each task, the early and late finish dates for each task, and the free and total slack for each task. This information shows you how flexible the schedule is and helps in making schedule compression decisions. The Critical Tasks report lists only tasks that are on the critical path for the project. If meeting schedule deadlines is essential for a project, project managers want to closely monitor tasks on the critical path.

To access the Schedule Table view and view the Critical Tasks report for a file:

1. *View the Schedule table.* Click **View** on the menu bar, point to **Table: Entry**, then click **Schedule**. The Schedule table replaces the Entry table to the left of the Gantt chart.
2. *Reveal all columns in the Schedule table.* Move the split bar to the right until you see the entire Schedule table. Your screen should resemble Figure A-19. This view shows the start and finish (meaning the early start and early finish) and late start and late finish dates for each task, as well as free and total slack. Click **View** on the menu bar, then point to **Table: Schedule**, and click **Entry** to return to the Entry table view.

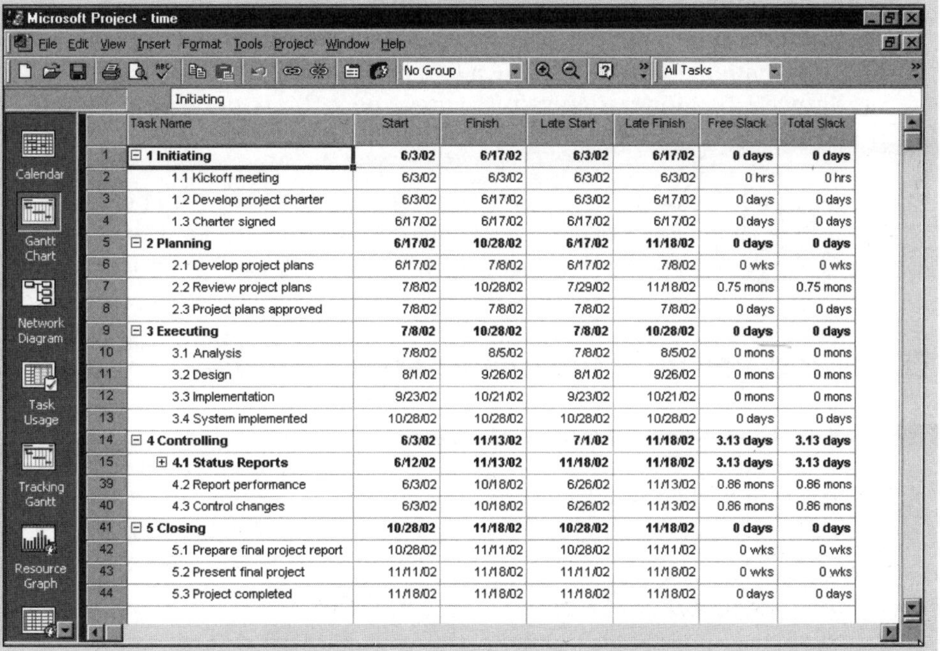

Microsoft Project - time

File Edit View Insert Format Tools Project Window Help

No Group | All Tasks

Initiating

Calendar | Gantt Chart | Network Diagram | Task Usage | Tracking Gantt | Resource Graph

	Task Name	Start	Finish	Late Start	Late Finish	Free Slack	Total Slack
1	**1 Initiating**	**6/3/02**	**6/17/02**	**6/3/02**	**6/17/02**	**0 days**	**0 days**
2	1.1 Kickoff meeting	6/3/02	6/3/02	6/3/02	6/3/02	0 hrs	0 hrs
3	1.2 Develop project charter	6/3/02	6/17/02	6/3/02	6/17/02	0 days	0 days
4	1.3 Charter signed	6/17/02	6/17/02	6/17/02	6/17/02	0 days	0 days
5	**2 Planning**	**6/17/02**	**10/28/02**	**6/17/02**	**11/18/02**	**0 days**	**0 days**
6	2.1 Develop project plans	6/17/02	7/8/02	6/17/02	7/8/02	0 wks	0 wks
7	2.2 Review project plans	7/8/02	10/28/02	7/29/02	11/18/02	0.75 mons	0.75 mons
8	2.3 Project plans approved	7/8/02	7/8/02	7/8/02	7/8/02	0 days	0 days
9	**3 Executing**	**7/8/02**	**10/28/02**	**7/8/02**	**10/28/02**	**0 days**	**0 days**
10	3.1 Analysis	7/8/02	8/5/02	7/8/02	8/5/02	0 mons	0 mons
11	3.2 Design	8/1/02	9/26/02	8/1/02	9/26/02	0 mons	0 mons
12	3.3 Implementation	9/23/02	10/21/02	9/23/02	10/21/02	0 mons	0 mons
13	3.4 System implemented	10/28/02	10/28/02	10/28/02	10/28/02	0 days	0 days
14	**4 Controlling**	**6/3/02**	**11/13/02**	**7/1/02**	**11/18/02**	**3.13 days**	**3.13 days**
15	**4.1 Status Reports**	**6/12/02**	**11/13/02**	**11/18/02**	**11/18/02**	**3.13 days**	**3.13 days**
39	4.2 Report performance	6/3/02	10/18/02	6/26/02	11/13/02	0.86 mons	0.86 mons
40	4.3 Control changes	6/3/02	10/18/02	6/26/02	11/13/02	0.86 mons	0.86 mons
41	**5 Closing**	**10/28/02**	**11/18/02**	**10/28/02**	**11/18/02**	**0 days**	**0 days**
42	5.1 Prepare final project report	10/28/02	11/11/02	10/28/02	11/11/02	0 wks	0 wks
43	5.2 Present final project	11/11/02	11/18/02	11/11/02	11/18/02	0 wks	0 wks
44	5.3 Project completed	11/18/02	11/18/02	11/18/02	11/18/02	0 days	0 days

Figure A-19. Schedule Table View

3. *Open the Reports dialog box.* Click **View** on the menu bar, then click **Reports**. Double-click **Overview** to open the Overview Reports dialog box.
4. *Display the Critical Tasks report.* Double-click **Critical tasks**, and a Critical Tasks report as of today's date appears.
5. *Close the report and save your file.* When you are finished examining the Critical Tasks report, click **Close**. Click **Close** on the Reports dialog box.
6. Click the **Save** icon on the toolbar to save your time file, again without a baseline.

If you wish to download the Project 2000 file to check your work or continue to the next section, a copy of the time file is available on the Course Technology Web site for this text or from your instructor.

PROJECT COST MANAGEMENT

Many people do not use Project 2000 for cost management. Most organizations have more established cost management software products and procedures in place. Using the cost features of Project 2000, however, makes it possible to integrate total project information more easily. This section offers brief instructions for entering fixed and variable cost estimates and actual cost and time information after establishing a baseline plan. It also explains how to use Project 2000 for earned value management. More details on these features are available in Project 2000 Help, online tutorials, or other books.

To complete the hands-on steps in this section, you need to download the Project 2000 file named resource from the Course Technology Web site for this book and save it on your Data Disk or in your data files folder on your hard drive.

Fixed and Variable Cost Estimates

The first step to using the cost features of Project 2000 is entering cost-related information. You enter costs as fixed or variable based on per use material costs, or variable based on the type and amount of resources used. Costs related to personnel are often a significant part of project costs.

Entering Fixed Costs in the Cost Table

The Cost table enables you to enter fixed costs related to each task. To access the Cost table, select Table: Cost from the View menu. Figure A-20 shows the

Adjusting Resource Costs

To make a resource cost adjustment, such as a raise, for a particular resource, double-click the person's name in the Resource Name column, select the Costs tab in the Resource Information dialog box, and then enter the effective date and raise percentage. You can also adjust other resource cost information, such as standard and overtime rates.

To give the project manager a 10 percent raise starting 9/02/02:

1. *Open the Resource Information dialog box.* From the Resource Sheet view, double-click **Kathy** in the Resource Name column. The Resource Information dialog box opens.
2. *Enter an effective date for a raise.* Select the **Costs** tab to make it active, and select **tab A** to make it active, if needed. Type **9/2/02** in the second cell in the Effective Date column. Alternately, click the **list arrow** in the second cell and use the calendar that appears to enter the effective date.
3. *Enter the raise percentage.* Type **10%** in the standard rate column, then press **[Enter]**. The Resource Information screen should resemble the screen shown in Figure A-21. Click **OK**.

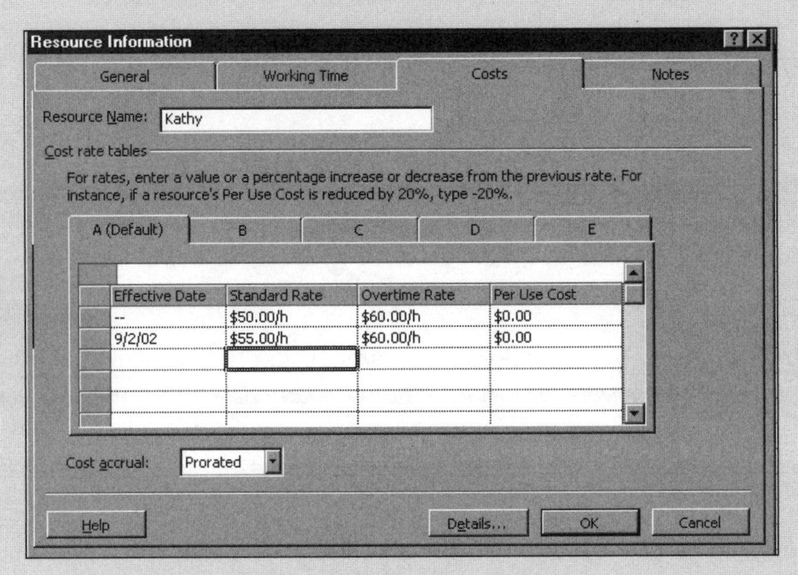

Figure A-21. Adjusting Resource Costs

If you know that some people will be available for a project only part-time, enter their percentage of availability in the Max. Units column of the Resource Sheet. Project 2000 will then automatically assign those people on the basis of their maximum units. For example, if someone can work only 25 percent of their time on a project throughout most of the project, enter 25% in the Max. Units column for that person. When you enter that person as a resource for a task, his or her default number of hours will be 25 percent of a standard 8-hour workday, or 2 hours per day.

Assigning Resources to Tasks

For Project 2000 to calculate resource costs, you must assign the appropriate resources to each task in your WBS. There are several methods for assigning resources. As mentioned earlier, in the Entry table there is a column called Resources for selecting a resource using a drop-down list. However, you can use this method to assign only one resource to a task full-time. You often need to use other methods to assign resources, such as the Assign Resources icon on the toolbar or the Resource Cost window, to ensure that Project 2000 captures resource assignments as you intend them. Next you will use these three methods for assigning resources to the Tracking Database Project.

Assigning Resources Using the Entry Table

To assign resources using the Entry table:

1. *Select the task to which you want to assign resources.* Click the **Gantt chart** icon on the View bar to return to the Gantt Chart view. Click the task name for Task 2, **Kickoff meeting**, in the second row of the Task Name column.
2. *Reveal the Resource Names column of the Entry table.* Move the split bar to the right to reveal the entire Resource Names column in the Entry table.
3. *Select a resource from the Resource Names column.* Click the **drop-down arrow** in the Resource Names column, then click **Kathy** to assign her to Task 2. Notice that the resource choices are based on information that was entered in the Resource Sheet. If you had not entered any resources, you would not have a drop-down arrow or any choices to select.
4. *Try to select another resource for Task 2.* Again click the **drop-down arrow** in the Resource Names column for Task 2, then click **John** and press **Enter**. Notice that only John's name appears in the Resource column. You can assign only one resource to a task using this method.
5. *Clear the resource assignment.* Click the Resource Names column for Task 2, click **Edit** on the menu bar, then select **Clear** and **Contents** to remove the resource assignments.

Do not press the Delete key to try to clear a resource assignment. If you press the Delete key, the entire task will be deleted, not just the resource assignment.

Assigning Resources Using the Toolbar

To assign resources using the toolbar:

1. *Select the task to which you want to assign resources.* Click the **Gantt chart** icon on the View bar to return to the Gantt Chart view. Click the task **Kickoff meeting** in the second row of the Task Name column.
2. *Open the Assign Resource dialog box.* Click the **Assign Resources** icon on the toolbar. The Assign Resources dialog box, which lists the names of the people assigned to the project, appears, as shown in Figure A-22. This dialog box remains open while you move from task to task to assign resources.

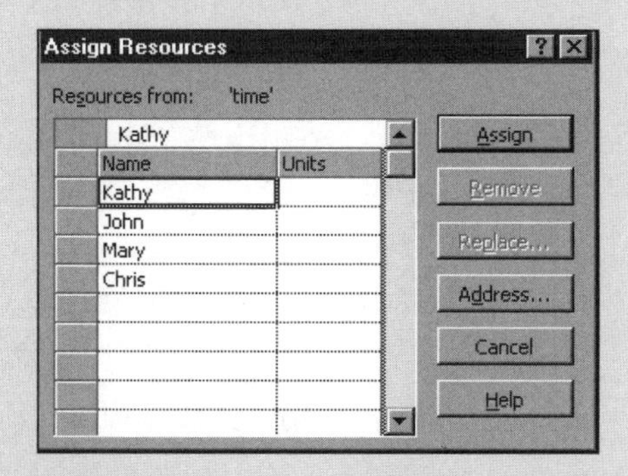

Figure A-22. Assign Resources Dialog Box for Time File

3. *Assign Kathy to Task 2.* With Kathy selected, click the **Assign** button. Notice that the duration estimate for Task 2 remains at 2 hours, and Kathy's name appears on the Gantt chart by the bar for Task 2.
4. *Assign John to Task 2.* Click John's name in the Assign Resources dialog box, then click the **Assign** button. Note that the duration for Task 2 has changed to 1 hour, *but you do not want this change to occur*. Click **Close** in the Assign Resources dialog box.
5. *Clear resource assignments.* Click the **Resource Names** column for Task 2, click **Edit** on the menu bar, then select **Clear** and **Contents** to remove the resource assignments.
6. *Reenter the duration for Task 2.* Type **2h** in the duration column for Task 2 and press **[Enter]**.

As you can see, you must be careful when making resource assignments. Project 2000 assumes that the durations of tasks are not fixed, but are effort-driven, and this assumption can create some problems when you are assigning resources. For example, when you assigned two full-time people, Kathy and John, to a two-hour task, Project 2000 automatically adjusted the duration of that task to half its original duration, or one hour. In the Project Tracking Database example, however, several tasks have durations that will not change no matter how many resources you assign to them.

Assigning Resources Using the Split Window and Resource Cost View

Even though using the Assign Resources icon seems simpler, it is often better to use a split window and the Resource Cost view when assigning resources. When you assign resources using the split view, the durations of tasks will not automatically change when more resources are assigned, and you have more control over how information is entered. Project 2000 online Help has more details on different options for resource assignment, under the topic Assign Resources.

To assign both Kathy and John to attend the two-hour kickoff meeting:

1. *Split the window to reveal more information.* Select **Window, Split** from the menu bar. The screen displays the Gantt Chart view at the top and a resource information table at the bottom.
2. *Open the Resource Cost window.* Right-click in the lower window and select the **Resource Cost** option.
3. *Assign Kathy to Task 2.* With Task 2 selected, click under **Resource Name** in the Resource Cost window. Click the **down arrow** and select **Kathy**.
4. *Assign John to Task 2.* Click in the cell below Kathy, and select **John**.
5. *Enter the resource assignment and review the Gantt chart.* Click **OK** in the lower window. Your screen should resemble Figure A-23. Notice that the duration for Task 2 is still 2 hours, and both Kathy and John are assigned to that task 100%. You can also enter a different percentage under the Units column if resources will not work 100% of their available time on a task.
6. *Open the Resource Schedule window.* Right-click in the lower window and select the **Resource Schedule** option. Notice that Kathy and John are assigned to task 2 for 2 hours, as you intended. You can also enter resource information using the Resource Schedule option.
7. *Close the file and do not save it.* Close the file, but do not save the changes you made. Other resource information has been entered for you in the Project 2000 file named resource, available on the Course Technology Web site.

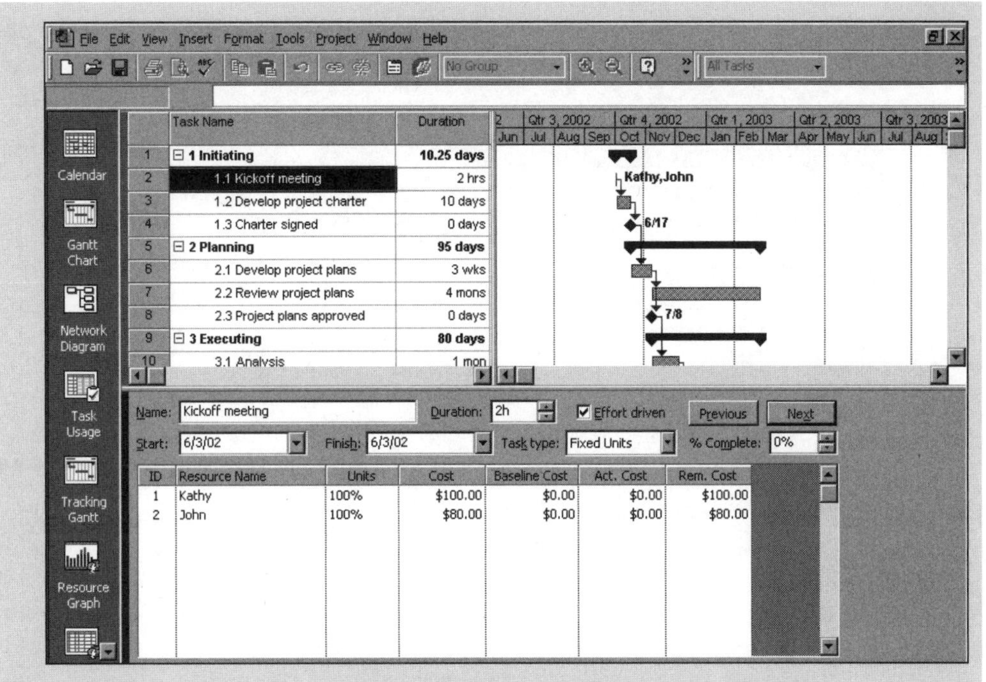

Figure A-23. Split Screen View for Entering Resource Information

A copy of the resource file is available on the Course Technology Web site for this text or from your instructor. You must use this file to continue the steps in the next section.

Viewing Project Cost Information

Once you enter resource information, resource costs are automatically calculated for the project. There are several ways to view project cost information. You can view the Cost table to see cost information, or you can run various cost reports. Next you will view cost information for the Project Tracking Database.

To view cost information:

1. *View the resource file.* After downloading the resource file from the Course Technology Web site or receiving it from your instructor, open it by clicking **Open** on the toolbar or by clicking File on the menu bar, then clicking **Open** and selecting the file named **resource**.
2. *Open the Cost table.* Click **View** on the menu bar, then point to **Table: Entry**, and click **Cost**. The Cost table displays cost information along with the Gantt chart. Move the split bar to the right to reveal all of the columns. Your screen should resemble Figure A-24.

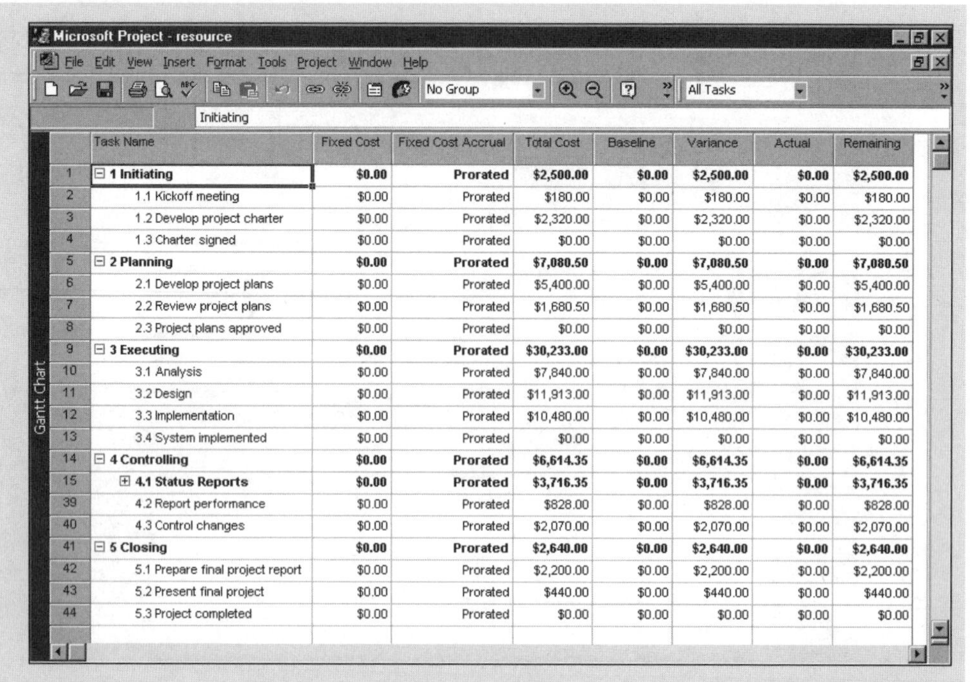

	Task Name	Fixed Cost	Fixed Cost Accrual	Total Cost	Baseline	Variance	Actual	Remaining
1	1 Initiating	$0.00	Prorated	$2,500.00	$0.00	$2,500.00	$0.00	$2,500.00
2	1.1 Kickoff meeting	$0.00	Prorated	$180.00	$0.00	$180.00	$0.00	$180.00
3	1.2 Develop project charter	$0.00	Prorated	$2,320.00	$0.00	$2,320.00	$0.00	$2,320.00
4	1.3 Charter signed	$0.00	Prorated	$0.00	$0.00	$0.00	$0.00	$0.00
5	2 Planning	$0.00	Prorated	$7,080.50	$0.00	$7,080.50	$0.00	$7,080.50
6	2.1 Develop project plans	$0.00	Prorated	$5,400.00	$0.00	$5,400.00	$0.00	$5,400.00
7	2.2 Review project plans	$0.00	Prorated	$1,680.50	$0.00	$1,680.50	$0.00	$1,680.50
8	2.3 Project plans approved	$0.00	Prorated	$0.00	$0.00	$0.00	$0.00	$0.00
9	3 Executing	$0.00	Prorated	$30,233.00	$0.00	$30,233.00	$0.00	$30,233.00
10	3.1 Analysis	$0.00	Prorated	$7,840.00	$0.00	$7,840.00	$0.00	$7,840.00
11	3.2 Design	$0.00	Prorated	$11,913.00	$0.00	$11,913.00	$0.00	$11,913.00
12	3.3 Implementation	$0.00	Prorated	$10,480.00	$0.00	$10,480.00	$0.00	$10,480.00
13	3.4 System implemented	$0.00	Prorated	$0.00	$0.00	$0.00	$0.00	$0.00
14	4 Controlling	$0.00	Prorated	$6,614.35	$0.00	$6,614.35	$0.00	$6,614.35
15	4.1 Status Reports	$0.00	Prorated	$3,716.35	$0.00	$3,716.35	$0.00	$3,716.35
39	4.2 Report performance	$0.00	Prorated	$828.00	$0.00	$828.00	$0.00	$828.00
40	4.3 Control changes	$0.00	Prorated	$2,070.00	$0.00	$2,070.00	$0.00	$2,070.00
41	5 Closing	$0.00	Prorated	$2,640.00	$0.00	$2,640.00	$0.00	$2,640.00
42	5.1 Prepare final project report	$0.00	Prorated	$2,200.00	$0.00	$2,200.00	$0.00	$2,200.00
43	5.2 Present final project	$0.00	Prorated	$440.00	$0.00	$440.00	$0.00	$440.00
44	5.3 Project completed	$0.00	Prorated	$0.00	$0.00	$0.00	$0.00	$0.00

Figure A-24. Cost Table for the Resource File

3. *Open the Cost Reports dialog box.* Click **View** on the menu bar, then click **Reports.** Double-click **Costs** to display the Cost Reports dialog box.
4. *Set the time units for the report.* Click **Cash Flow**, then click the **Edit** button. The Crosstab Report dialog box opens. Click the **drop-down arrow** and select **Months** instead of weeks in the Crosstab Report dialog box, if needed, as shown in Figure A-25. Click **OK**.

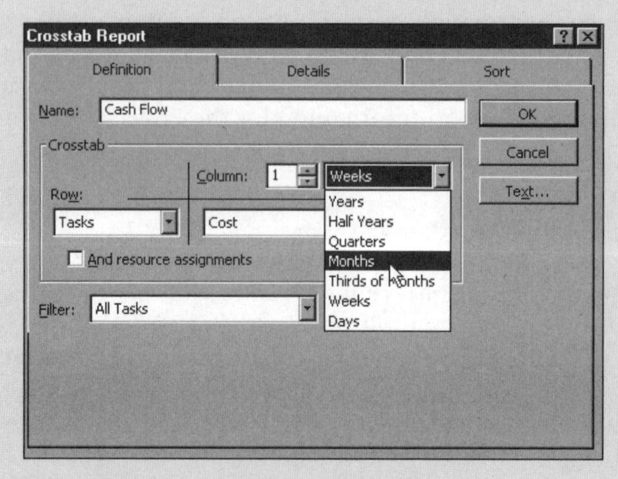

Figure A-25. Crosstab Report Dialog Box

5. *View the Cash Flow report.* Click the **Select** button in the Cost Reports dialog box. A Cash Flow report for the Project Tracking Database appears, as shown in Figure A-26. Click **Close**.

	June	July	August	September	October	November	Total
Initiating							
Kickoff meeting	$180.00						$180.00
Develop project charter	$2,320.00						$2,320.00
Charter signed							
Planning							
Develop project plans	$3,510.00	$1,890.00					$5,400.00
Review project plans		$355.00	$440.00	$462.00	$423.50		$1,680.50
Project plans approved							
Executing							
Analysis		$6,958.00	$882.00				$7,840.00
Design			$6,438.00	$5,475.00			$11,913.00
Implementation				$3,013.00	$7,467.00		$10,480.00
System implemented							
Controlling							
Status Reports	$426.40	$754.40	$721.60	$730.80	$800.40	$282.75	$3,716.35
Report performance	$160.00	$184.00	$176.00	$184.80	$123.20		$828.00
Control changes	$400.00	$460.00	$440.00	$462.00	$308.00		$2,070.00
Closing							
Prepare final project report					$825.00	$1,375.00	$2,200.00
Present final project						$440.00	$440.00
Project completed							
Total	$6,996.40	$10,601.40	$9,097.60	$10,327.60	$9,947.10	$2,097.75	$49,067.85

Figure A-26. Cash Flow Report

6. View the Project Summary report. In the Reports dialog box, double-click **Overview**, then double-click **Project Summary**. A Project Summary report for the Project Tracking Database appears, listing information such as project baseline start and finish dates; actual start and finish dates; summaries of duration, work hours, and costs; as well as variance information. Since you have not yet saved the file as a baseline, much of this information is blank in the report. The Project Summary report provides a good big-picture view of a project. Close the report after reviewing it. Close the Reports dialog box.

You can edit many of the report formats in Project 2000. Instead of double-clicking the report, highlight the desired report, then click Edit.

The total estimated cost for this project, based on the information entered, should be $49,067.85, as shown in the Cash Flow and Project Summary reports. In the next section you will save this data as a baseline plan and enter actual information.

Baseline Plan, Actual Costs, and Actual Times

Once you complete the initial process of creating a plan—entering tasks, establishing dependencies, assigning costs, and so on—you are ready to set a baseline plan.

By comparing the information in your baseline plan to an updated plan during the course of the project, you can identify and solve problems. After the project ends, you can use the baseline and actual information to plan similar, future projects more accurately. To use Project 2000 to help control projects, you must establish a baseline plan, enter actual costs, and enter actual durations.

Establishing a Baseline Plan

An important part of project scope management is setting a baseline plan. If you plan to compare actual information such as durations and costs, you must first save the Project 2000 file as a baseline. The procedure involves selecting Tracking, Save Baseline from the Tools menu. Before setting a baseline, you must complete the baseline plan by entering time, cost, and human resources information. Be careful not to save a baseline until you have completed the baseline plan. In case you do save a baseline before completing the plan, Project 2000 has added a new feature to clear the baseline.

Even though you can clear a baseline plan, it is a good idea to save a separate, backup file of the baseline for your project. Enter actuals and save that information in the main file, but always keep a backup baseline file without actuals. Keeping separate baseline and actual files will allow you to return to the original file in case you ever need to use it again.

To rename the resource file and then save it as a baseline plan in Project 2000:

1. *Save the resource file as a new file named baseline.* With the resource file open, click **File** on the menu bar, then click **Save as**. Type **baseline** in the File name: text box, then click the **Save** button. In the Planning Wizard dialog box, leave the first option, **Save baseline without a baseline**, selected, and click **OK**.
2. *Open the Save Baseline dialog box.* Click **Tools** on the menu bar, select **Tracking**, and then click **Save Baseline**.
3. *Review the Save Baseline options, then save the entire project as a baseline.* Notice that there are several options in the Save Baseline dialog box. You can save the file as an interim plan if you anticipate several versions of the plan. You can also select the entire project or selected tasks when saving a baseline or interim plan. If necessary, click the **Save baseline** radio button and the **Entire project** radio button. These options should be the default settings in Project 2000. Click **OK**.

Entering Actual Costs and Times

After you set the baseline plan, it is possible to track information on each of the tasks as the project progresses. You can also adjust planned task information for tasks still in the future. The Tracking table displays tracking information, and the

Tracking toolbar helps you enter this information. Figure A-27 describes each button on the Tracking toolbar. The 100% icon and the Update as Schedule icon are the most commonly used icons for entering actual information.

Figure A-27: Buttons on the Tracking Toolbar

BUTTON	NAME	DESCRIPTION
	Project Statistics	Provides summary information about the project baseline, the actual project start and finish dates, and overall project duration, costs, and work
	Update as Scheduled	Updates the selected tasks to indicate that actual dates, costs, and work match the scheduled dates, costs, and work
	Reschedule Work	Schedules the remaining duration for a task that is behind schedule so that it will continue from the status date
	Add Progress Line	Displays a progress line on the Gantt chart from a date that you select on the time scale
0%	0% Complete	Marks the selected tasks as 0% complete as of the status date. (Actual date, work, and duration data is updated.)
25%	25% Complete	Marks the selected tasks as 25% complete as of the status date. (Actual date, work, and duration data is updated.)
50%	50% Complete	Marks the selected tasks as 50% complete as of the status date. (Actual date, work, and duration data is updated.)
75%	75% Complete	Marks the selected tasks as 75% complete as of the status date. (Actual date, work, and duration data is updated.)
100%	100% Complete	Marks the selected tasks as 100% complete as of the status date. (Actual date, work, and duration data is updated.)
	Update Tasks	Displays the Update Tasks dialog box for the selected tasks so that you can enter their percentages completed, actual durations, remaining durations, or actual start or finish dates
	Workgroup Toolbar	Toggles on and off the Workgroup Toolbar, which displays buttons that enable you to share parts of the project with others

Friedrichsen, Lisa and Rachel Biheller Bunin, *Microsoft Project 2000, New Perspectives Series Introductory*, Course Technology (2000) (p. 232).

To practice entering actual information, enter just a few changes to the baseline. Assume that Tasks 1 through 7 were completed as planned, but that Task 10 took longer than planned.

To enter actual information for tasks that were completed as planned:

1. *Display the Tracking toolbar.* Click **View** on the menu bar, then select **Toolbars**. Click **Tracking** from the Toolbars submenu to display the Tracking toolbar. Move the toolbar as desired.

2. *Display the Tracking table.* Click **View** on the menu bar, then point to **Table**: **Cost**, and click **Tracking** to see more information as you enter actual data. Move the split bar to reveal all the columns in the Tracking table.
3. *Mark Tasks 1 though 8 as 100% complete.* Click the Task Name for Task 1, **Initiating**, and drag down through Task 8 to highlight the first seven tasks. Click the **100 percent icon** on the Tracking toolbar. The columns with dates, durations, and cost information should now contain data instead of the default values, such as NA or 0. 100% should be displayed in the % Comp. column. Adjust column widths if needed. Your screen should resemble Figure A-28.

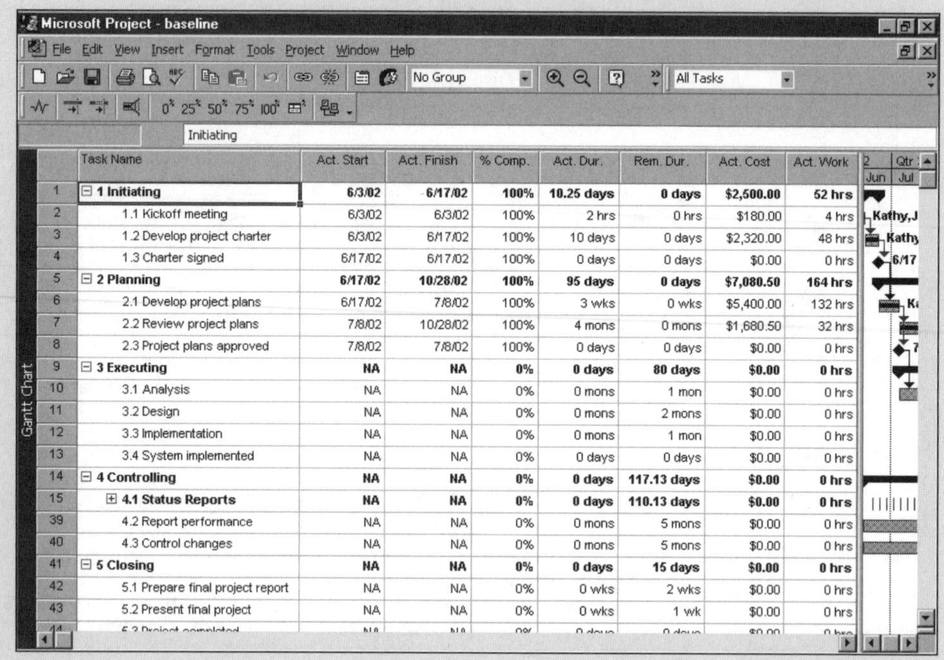

	Task Name	Act. Start	Act. Finish	% Comp.	Act. Dur.	Rem. Dur.	Act. Cost	Act. Work
1	**1 Initiating**	**6/3/02**	**6/17/02**	**100%**	**10.25 days**	**0 days**	**$2,500.00**	**52 hrs**
2	1.1 Kickoff meeting	6/3/02	6/3/02	100%	2 hrs	0 hrs	$180.00	4 hrs
3	1.2 Develop project charter	6/3/02	6/17/02	100%	10 days	0 days	$2,320.00	48 hrs
4	1.3 Charter signed	6/17/02	6/17/02	100%	0 days	0 days	$0.00	0 hrs
5	**2 Planning**	**6/17/02**	**10/28/02**	**100%**	**95 days**	**0 days**	**$7,080.50**	**164 hrs**
6	2.1 Develop project plans	6/17/02	7/8/02	100%	3 wks	0 wks	$5,400.00	132 hrs
7	2.2 Review project plans	7/8/02	10/28/02	100%	4 mons	0 mons	$1,680.50	32 hrs
8	2.3 Project plans approved	7/8/02	7/8/02	100%	0 days	0 days	$0.00	0 hrs
9	**3 Executing**	**NA**	**NA**	**0%**	**0 days**	**80 days**	**$0.00**	**0 hrs**
10	3.1 Analysis	NA	NA	0%	0 mons	1 mon	$0.00	0 hrs
11	3.2 Design	NA	NA	0%	0 mons	2 mons	$0.00	0 hrs
12	3.3 Implementation	NA	NA	0%	0 mons	1 mon	$0.00	0 hrs
13	3.4 System implemented	NA	NA	0%	0 days	0 days	$0.00	0 hrs
14	**4 Controlling**	**NA**	**NA**	**0%**	**0 days**	**117.13 days**	**$0.00**	**0 hrs**
15	**4.1 Status Reports**	**NA**	**NA**	**0%**	**0 days**	**110.13 days**	**$0.00**	**0 hrs**
39	4.2 Report performance	NA	NA	0%	0 mons	5 mons	$0.00	0 hrs
40	4.3 Control changes	NA	NA	0%	0 mons	5 mons	$0.00	0 hrs
41	**5 Closing**	**NA**	**NA**	**0%**	**0 days**	**15 days**	**$0.00**	**0 hrs**
42	5.1 Prepare final project report	NA	NA	0%	0 wks	2 wks	$0.00	0 hrs
43	5.2 Present final project	NA	NA	0%	0 wks	1 wk	$0.00	0 hrs

Figure A-28. Tracking Table Information

4. *Enter actual completion dates for Task 10.* Click the Task Name for Task 10, **Analysis**, then click the **Update Tasks** icon on the Tracking toolbar. The Update Tasks dialog box opens. For Task 10, enter the Actual Start date as **7/10/02** and the Actual Finish date as **8/15/02** in the Update Tasks dialog box, as shown in Figure A-29. Click **OK**.

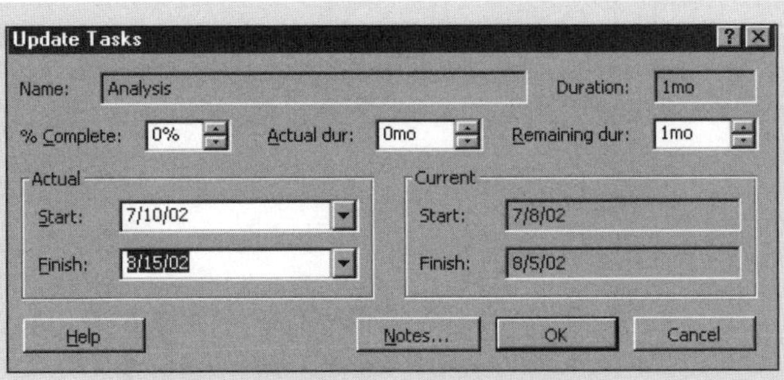

Figure A-29. Update Tasks Dialog Box

5. *Display the Indicators column.* Click **Insert** on the menu bar and click **Column**, click the **Field name** list arrow, then select **Indicators**, and click **OK**. The Indicators column now appears, showing a check mark by completed tasks.
6. *Review changes in the Gantt chart.* Move the split bar back to the left to reveal more of the Gantt chart. Notice that the Gantt chart bars for the completed tasks have changed. Gantt chart bars of completed tasks appear with a black line drawn through the middle.

You can hide the Indicators column by dragging the right border of the column heading to the left.

You can also view actual and baseline schedule information more clearly with the Tracking Gantt Chart view. See Chapter 5 for descriptions of the symbols on a Tracking Gantt chart.

To display the Tracking Gantt chart:

1. *View the Tracking Gantt chart.* Click **View** on the menu bar, then click **Tracking Gantt**. If your View bar is displayed, you can also click the Tracking Gantt icon. Adjust the horizontal scroll bar at the lower-right of the Gantt chart window, if necessary, to see symbols on the Gantt chart.
2. *Display Gantt chart information in months.* Click **Zoom out** to display information in months. Your screen should resemble Figure A-30.

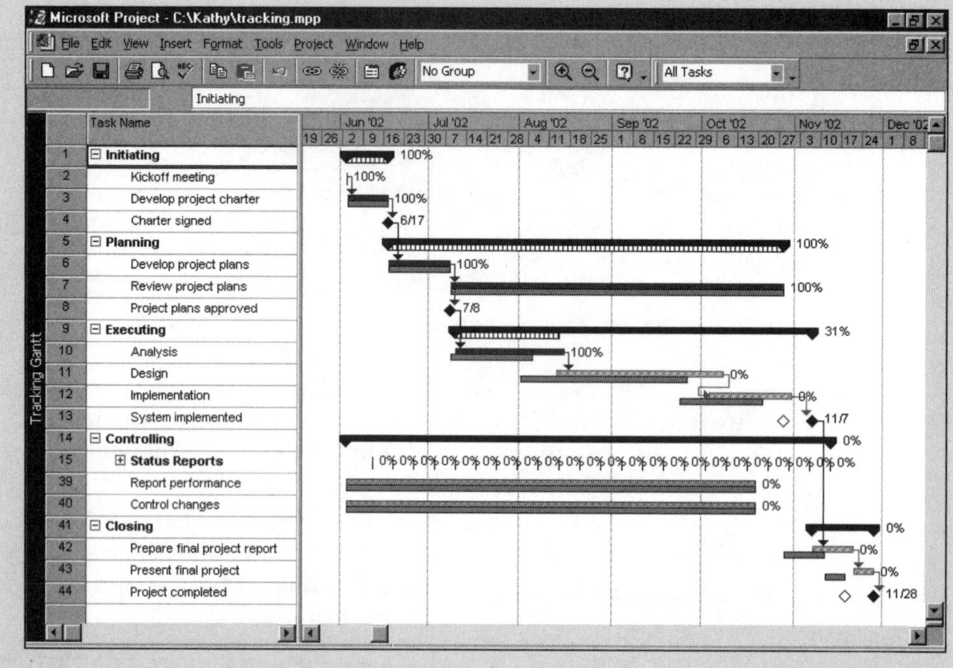

Figure A-30. Tracking Gantt Chart View

3. *Save your file as a new file named tracking.* Click **File** on the menu bar, then click **Save as** and name this file **tracking**. Accept the default option to save without a baseline.

Notice the additional information available on the Tracking Gantt chart. Completed tasks have 100% by their symbols on the Tracking Gantt chart. Tasks that have not started yet display 0%. Tasks in progress, such as Task 9, show the percent of the work completed, 31% in this example. Also note that the project completion date has moved to 11/28, since several tasks depended on completion of Task 10, which took longer than planned. Viewing the Tracking Gantt chart allows you to easily see your schedule progress against the baseline plan.

After you have entered some actuals, you can review earned value information for the initiating tasks of this project.

Earned Value Management

Earned value management (*see* Chapter 6, *Project Cost Management*) is an important project management technique for measuring project performance. Because you have entered actual information for the initiating tasks in the Project Tracking Database project, you can now view earned value information in Project 2000.

To view earned value information:

1. *View the Earned Value table.* With the Project 2000 file named tracking open, click **View** on the menu bar, then point to **Table: Entry**. Select **More Tables** to open the More Tables dialog box, then double-click **Earned Value**.
2. *Display all the Earned Value table columns.* Move the split bar to the right to reveal all of the columns. Note that the Earned Value table includes columns for each earned value term described in Chapter 6.

Project 2000 uses earned value terms from the 1996 PMBOK Guide. Recall that the 2000 PMBOK Guide uses simpler acronyms for earned value terms. See Chapter 6 of this text for explanations of the 1996 and 2000 terms.

3. Close the file without saving it. Click **File** on the menu bar, then click **Close**. Do not save the file.

If you wish to check your work or continue to the next section, copies of the files baseline and tracking are available on the Course Technology Web site for this text or from your instructor.

Further information on earned value is available from Project 2000 Help or in Chapter 6, *Project Cost Management*. To create an earned value chart, you must export data from Project 2000 into Excel. Consult Project 2000 Help for more details.

PROJECT HUMAN RESOURCE MANAGEMENT

In the cost section, you learned how to enter resource information into Project 2000 and how to assign resources to tasks. Two other helpful human resource features include resource calendars and histograms. In addition, it is important to know how to use Project 2000 to assist in resource leveling.

Resource Calendars

When you created the Project Tracking Database file, you used the standard Project 2000 calendar. This calendar assumes that standard working hours are Monday through Friday, from 8:00 a.m. to 5:00 p.m., with an hour for lunch starting at noon. Rather than using this standard calendar, you can create a different calendar that takes into account each project's unique requirements.

To create a new base calendar:

- Click Tools on the menu bar, then click Change Working Time. The Change Working Time dialog box opens.
- In the Change Working Time dialog box, click New. The Create New Base Calendar dialog box opens. Click the Create New Base Calendar radio button, type a name for the new calendar in the Name text box, and click OK. Make adjustments to create your own base calendar. Click OK.

You can use this new calendar for the whole project, or you can assign it to specific resources on the project.

To assign the new calendar to the whole project:

- Click Project on the menu bar, then click Project Information. The Project Information dialog box opens.
- Click the Calendar list arrow to display a list of available calendars. Select your new calendar from this list, and click OK.

To assign a specific calendar to a specific resource:

- Click View on the menu bar, then click Resource Sheet.
- Select the resource name to which you want to assign a new calendar.
- Click the Base Calendar cell for that resource name. Using the left mouse button, click the list arrow that appears, drag to select the appropriate calendar, and release the mouse button.
- Double-click a resource name to display the Resource Information dialog box, then click the Working Time tab. You can block off vacation time for people by selecting the appropriate days on the calendar and marking them as nonworking days, as shown in Figure A-31.

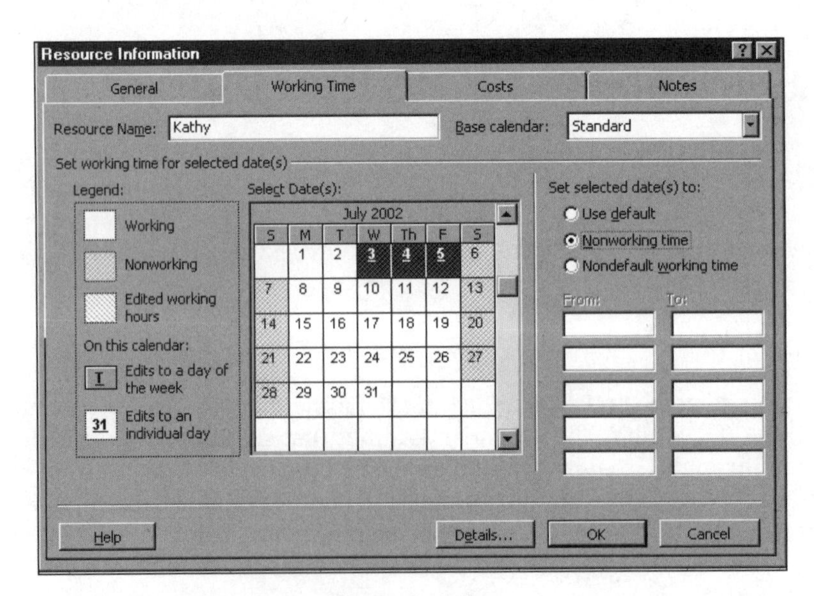

Figure A-31. Blocking off Vacation Time

Resource Histograms

A resource histogram is a type of chart that shows the number of resources assigned to a project over time (*see* Chapter 8, *Project Human Resource Management*). A histogram by individual shows whether a person is over- or underallocated during certain time periods. To view histograms in Project 2000, select Resource Graph from the View bar or select Resource Graph from the View menu. The Resource Graph helps you to see which resources are overallocated, by how much, and when. It also shows you the percentage of capacity each resource is scheduled to work, so you can reallocate resources, if necessary, to meet the needs of the project.

To view resource histograms for the Project Tracking Database project:

1. *Open the Project 2000 file named baseline.* Click **Open** on the toolbar or Click **File** on the menu bar, then click **Open**, and open the Project 2000 file named **baseline**.
2. *View the Resource Graph.* Click the **Resource Graph** icon on the View bar. If you cannot see the Resource Graph icon, you may need to click the up or down arrow on the View bar. Alternately, you can click View on the menu bar, then Resource Graph. A histogram for Kathy will appear first, as shown in Figure A-32. Notice that the screen is divided into two sections. The left pane displays a person's name, and the right pane displays a resource histogram for that person.

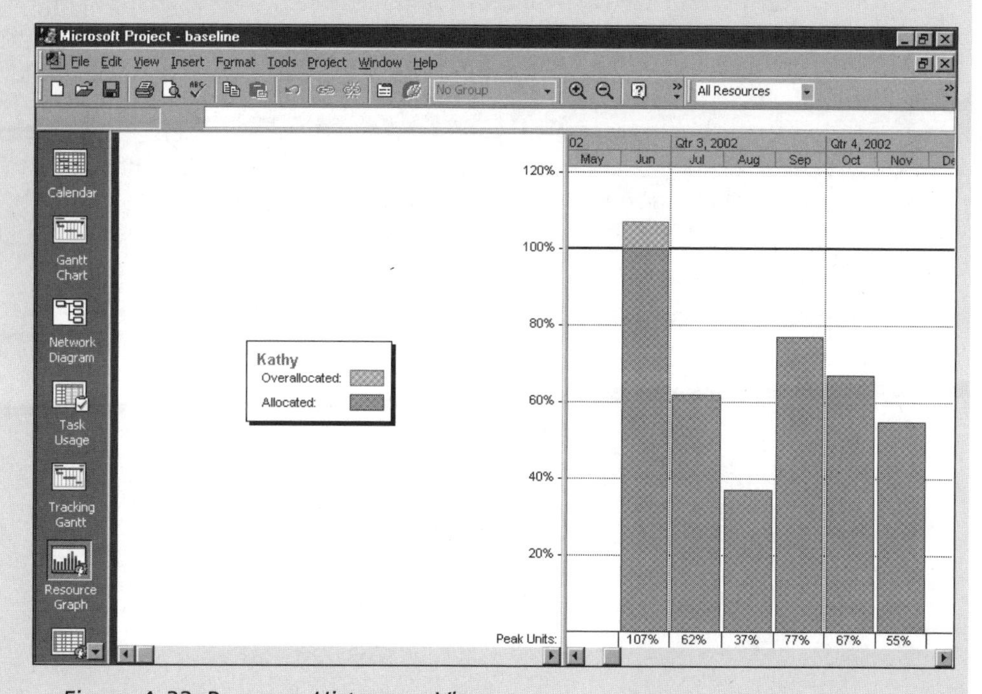

Figure A-32. Resource Histogram View

3. *Adjust the histogram's time scale.* Click Zoom out and Zoom in, if necessary, to adjust the time scale for the histogram so that it appears in quarters and then months.
4. *View the next resource's histogram.* Click the **right scroll arrow** at the bottom of the resource name pane. The resource histogram for the next person appears.

Notice that Kathy's histogram has a partially red bar in June 2002. This red portion of the bar means that she has been overallocated during that month; she has been scheduled to work more than 8 hours a day during June. The percentages at the bottom of each bar show the percentage each resource is assigned to work. For example, Kathy is scheduled to work 107% of her available time in June, 62% of her available time in July, and so on. Project 2000 has two tools that enable you to see more details about resource overallocation: the Resource Usage view and the Resource Management toolbar.

To see more details about an overallocated resource using the Resource Usage view:

1. *Open the Resource Usage view.* Click **Resource Usage** from the View bar or click **View** on the menu bar, then click **Resource Usage**.
2. *Adjust the information displayed.* On the right side of the screen, click the **right scroll arrow** to display the hours Kathy is assigned to work each day starting in June. You may also need to click the scroll down arrow to see all of Kathy's hours. If you need to adjust the time scale to display weeks, click Zoom out or Zoom in. When you are finished, your screen should resemble Figure A-33.
3. *Examine overallocation information.* Notice that Kathy's name appears in red, as does the value 5.57h in the column for Monday. This means that Kathy is scheduled to work more than 8 hours that day.

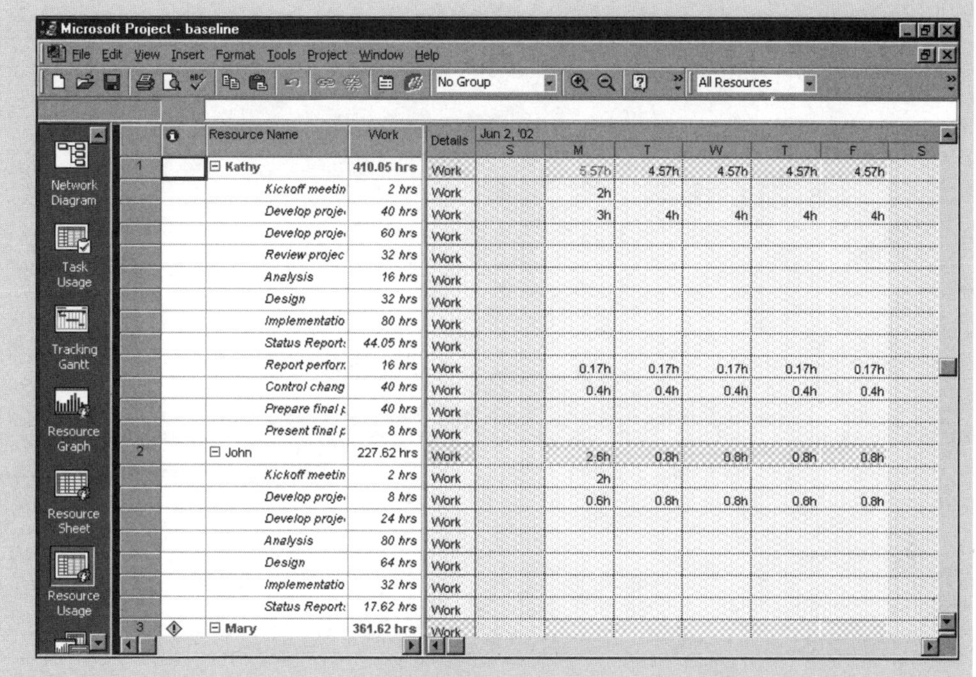

Figure A-33. Resource Usage View

To learn more about an overallocated resource, you can also use a special toolbar for resource allocation.

To see more resource allocation information:

1. *View the Resource Management toolbar.* Click **View** on the menu bar, select **Toolbars**, and then click **Resource Management**. The Resource Management toolbar appears beneath the Formatting toolbar.
2. *Select the Resource Allocation view.* Click **Resource Allocation View** on the Resource Management toolbar. The Resource Allocation view appears, showing the Resource Usage view at the top of the screen and the Gantt Chart view at the bottom part of the screen. Figure A-34 shows this view with Kathy's information highlighted.
3. *Close the file, but do not save any changes.* Click **File** on the menu bar, then click **Close**. Click the **No** button when you are prompted to save changes.

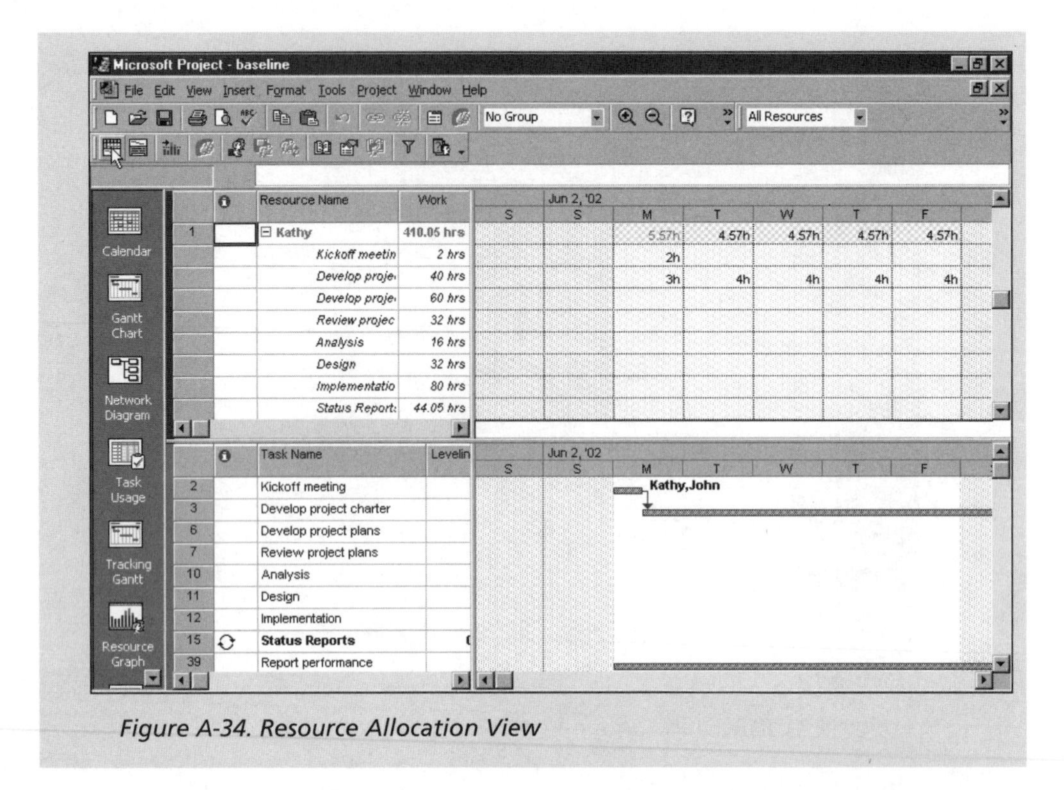

Figure A-34. Resource Allocation View

The Resource Allocation view can help you identify the source of a resource overallocation. It is fairly obvious that Kathy has a slight overallocation the first day of the project because she is assigned to three different tasks, for a total of 8.57 hours. You could fix this problem by reducing Kathy's hours on one of the tasks or by allowing her to work a small amount of overtime. By using the scroll bar, you can view information for other resources. If you scroll down to look at Mary's resource allocation information and scroll to the right to reveal the months of July and August, you can see that overlapping the design and implementation tasks may have caused another overallocation problem. To fix the problem, you can have Mary work overtime, assign another resource to help, or reschedule the implementation task to reduce the overlap. You can also see if resource leveling, as described in the next section, will help solve the problem.

Resource Leveling

Resource leveling is a technique for resolving resource conflicts by delaying tasks. Resource leveling also creates a smoother distribution of resource usage. You can find detailed information on resource leveling in Chapter 8, *Project Human Resources Management,* and in Project 2000 online Help.

To use resource leveling:

1. *Reopen the baseline file.* Click **Open** and select the baseline file. You should see the file in Gantt Chart view again. Notice that the project completion date is 11/18/02.
2. *Open the Resource Leveling dialog box.* Click **Tools** on the menu bar, then click **Resource Leveling**. The Resource Leveling dialog box opens, as shown in Figure A-35. Note the default settings, but do not change them.

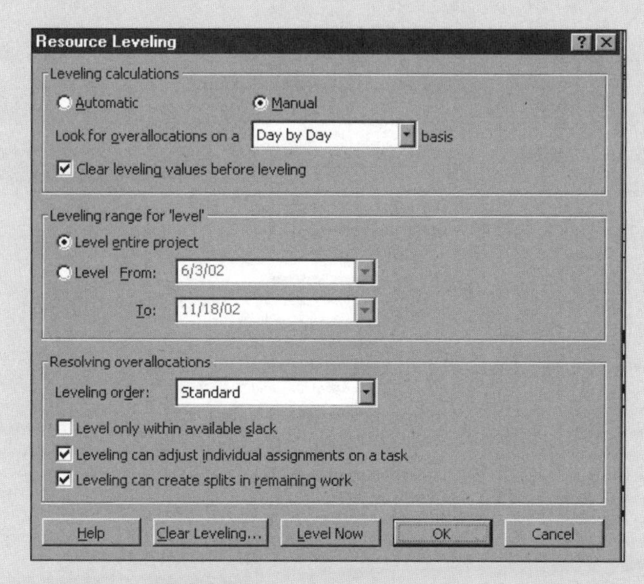

Figure A-35. Resource Leveling Dialog Box

3. *Level the file and review the date changes.* Click the **Level Now** button. Click **OK**. In this example, resource leveling moved the project completion date from 11/18 to 11/21, as shown on the Gantt chart milestone symbol for Task 44, Project completed.
4. *View Kathy's Resource Graph again.* Click the **Resource Graph** icon on the View bar. Notice that Kathy is still overallocated in June.
5. *Level using the Hour by Hour option.* Click **Tools** on the menu bar, then click **Resource Leveling**. Change the basis for leveling from Day by Day to **Hour by Hour**, then click the **Level Now** button. Notice that the project completion date on the Gantt chart changes to 11/25. If a Level Now dialog box appears, accept the default settings, and click **OK**.
6. *View Resource Graphs again.* Click the **Resource Graph** icon on the View bar. Notice that Kathy is no longer overallocated. Use the scroll arrows to reveal information about Mary. Remember that the left

pane displays different resources, and the right pane reveals the resource histogram over time. Mary's overallocations are also gone, as a result of this resource leveling.

7. *View the Leveling Gantt chart.* Click **View** on the menu bar, click **More Views**, and double-click **Leveling Gantt**. Click Zoom out, if necessary, to adjust the time scale to show quarters and months, and use the horizontal scroll bar to reveal all the symbols on the Leveling Gantt chart, until your screen resembles Figure A-36. Project 2000 adds a green bar to the Gantt chart to show leveled tasks.

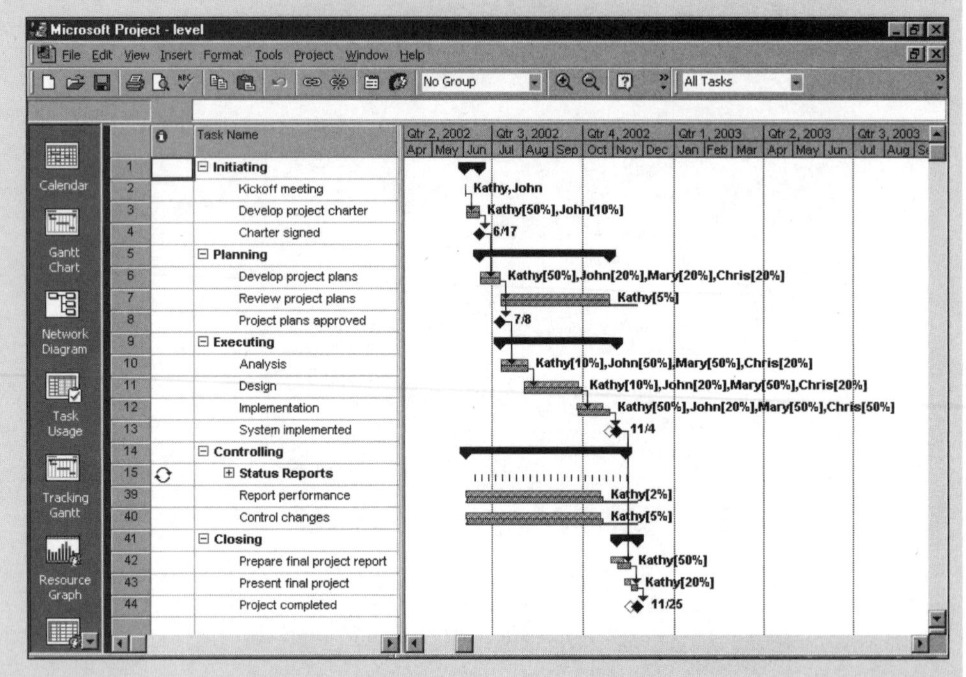

Figure A-36. Leveling Gantt Chart View

8. *Save the file and call it level.* Click **File** on the menu bar, click **Save as**, and name the file **level**. Save without a baseline.

If you want to undo the leveling, immediately select Undo from the Standard toolbar. Alternately, return to the Resource Leveling dialog box, and click the Clear Leveling button.

If you wish to check your work, a copy of the level file is available on the Course Technology Web site for this text or from your instructor.

Consult the Project 2000 Help topic called "Leveling" for more information on resource leveling. Also, be careful when setting options for this feature, so that the software adjusts resources only when it should.

PROJECT COMMUNICATIONS MANAGEMENT

Project 2000 can help you generate, collect, disseminate, store, and report project information. There are many different tables, views, and reports to aid in project communications, as you have seen in the previous sections. This section highlights some common reports and views. It also describes how to use templates and insert hyperlinks from Project 2000 into other project documents, how to save Project 2000 files as Web pages, and how to use Project 2000 in a workgroup setting.

Common Reports and Views

To use Project 2000 to enhance project communications, it is important to know when to use the many different ways to collect, view, and display project information. Table A-6 provides a brief summary of Project 2000 features and their functions, which will help you to understand when to use which feature. Examples of most of these features are provided as figures in this appendix.

You can see from Table A-6 that many different reports are available in Project 2000. The overview reports provide summary information that senior management might wish to see, such as a project summary or report of milestone tasks. Current activities reports help project managers stay abreast of and control project activities. The reports of unstarted tasks and slipping tasks alert project managers to areas having problems. Cost reports provide information related to the cash flow for the project, budget information, over-budget items, and earned value management. Assignment reports help the entire project team by providing different views of who is doing what on a project. You can see who is overallocated by running the overallocated resources report or by viewing the two workload reports. You can also create your own custom reports based on any project information you have entered into Project 2000.

Table A-6: Functions of Project 2000 Features

FEATURE	FUNCTION
Gantt Chart view, Entry table	Enter basic task information
Network Diagram view	View task dependencies and critical path graphically
Schedule table	View schedule information in a tabular form
Cost table	Enter fixed costs or view cost information
Resource Sheet view	Enter resource information
Resource Information and Gantt Chart split view	Assign resources to tasks
Save Baseline	Save project baseline plan
Tracking toolbar	Enter actual information
Earned Value table	View earned value information
Resource Graph	View resource allocation
Resource Usage	View detailed resource usage
Resource Management toolbar	View resource usage and Gantt chart to find overallocation problems
Resource Leveling	Level resources
Overview Reports	Project summary, Top-Level Tasks, Critical Tasks, Milestones Tasks, Working Days
Current Activities Reports	Unstarted Tasks, Tasks Starting Soon, Tasks In Progress, Completed Tasks, Should Have Started Tasks, Slipping Tasks
Cost Reports	Cash Flow, Budget, Overbudget Tasks, Overbudget Resources, Earned Value
Assignments Reports	Who Does What, Who Does What When, To-Do List, Overallocated Resources
Workload Reports	Task Usage, Resource Usage
Custom Reports	Allows customization of each type of report
Save as Web page	Save files as HTML documents
Insert Hyperlink	Inserts hyperlinks to other files or Web sites

Using Templates and Inserting Hyperlinks and Comments

Chapter 9, *Project Communications Management*, provides several templates that you can use to improve project communications. It is difficult to create good project plans, and several organizations, such as Northwest Airlines, Microsoft, and various government agencies, keep a repository of sample Project 2000 files. Project 2000 includes several templates.

To access the Project 2000 templates:

1. *Open the New dialog box.* Click **File** on the menu bar, click **New**, and then click the tab for **Project templates**. Your screen should resemble Figure A-37.

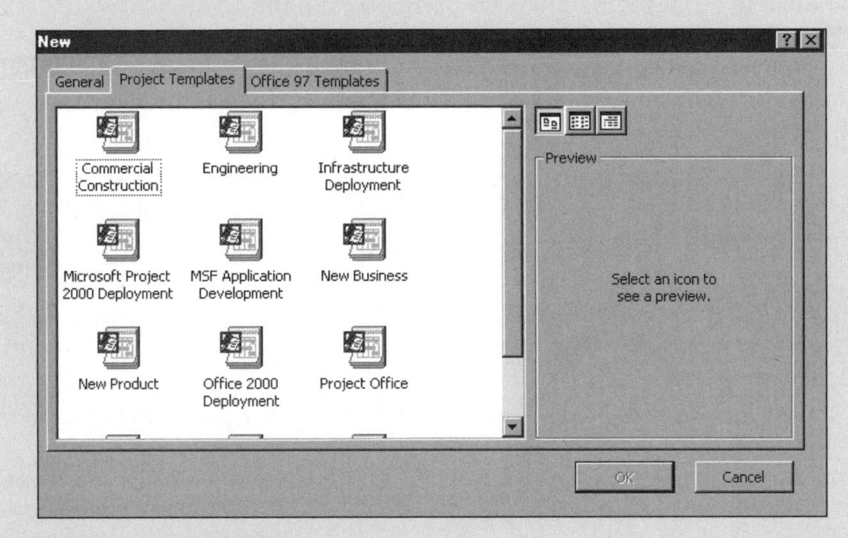

Figure A-37. Project Templates

2. *Open the Project Office template.* Double-click the **Project Office** icon to open that template file.
3. *Enter a start date for the file.* Type **10/01/02** in the Start date text box, then click **OK**.

Project 2000 includes templates for the following types of projects:

- Commercial Construction
- Engineering
- Infrastructure Deployment
- Microsoft Project 2000 Deployment
- Microsoft Solutions Framework (MSF) Application Development
- New Business
- New Product
- Office 2000 Deployment
- Project Office
- Residential Construction

- Software Development
- Windows 2000 Deployment

Using templates can help you prepare your own project files, but be very careful to address the unique needs of your project and organization. For example, although many people do home construction projects, every schedule is different. Even though the Residential Construction file can provide guidance in preparing a Project 2000 file for a specific home construction project, you need to tailor the file for your particular situation.

In addition to using templates for Project 2000 files, it is helpful to use templates for other project documents and insert hyperlinks to them from Project 2000. For example, your organization might have templates for meeting agendas, project charters, status reports, and project plans.

To insert a hyperlink within a Project 2000 file:

1. *Select the task where you want to insert a hyperlink.* With the baseline file open, click the Task Name for Task 2, **Kickoff meeting**.
2. *Open the Insert Hyperlink dialog box.* Click **Insert** on the menu bar, then click **Hyperlink**. The Insert Hyperlink dialog box opens, as shown in Figure A-38.
3. *Enter the filename of the hyperlink file.* Type **kickoffmeeting.doc** in the Type the file or Web page name: text box. For a real project, you would be careful to enter the path to the file or Web page you want to link to. You can also select Browse to browse for files. Click **OK**.
4. *Reveal the Indicators column.* Click **Insert** on the menu bar, then click **Column**. Click the **Field name** drop-down list, click **Indicators**, then click **OK**. A hyperlink icon should appear in the Indicators column to the left of the Task Name for Task 2.

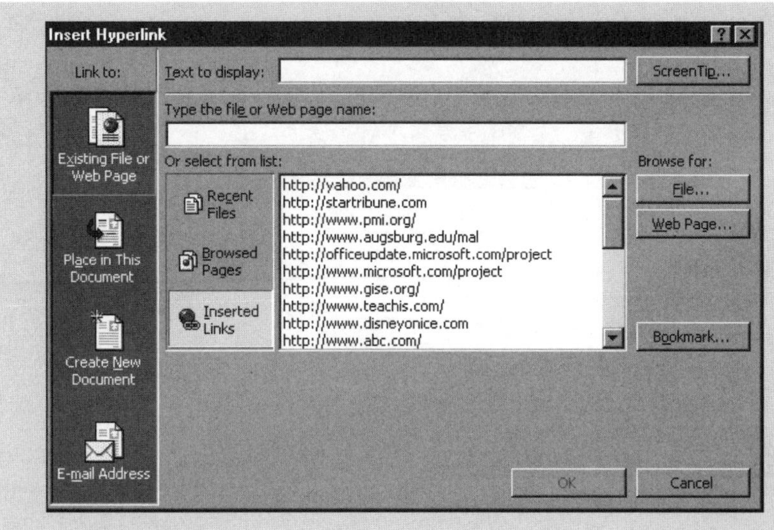

Figure A-38. Insert Hyperlink Dialog Box

Clicking the Hyperlink icon in the Indicators column will automatically open the hyperlinked file. Using hyperlinks is a good way to keep all project documents organized.

It is also a good idea to insert notes or comments into Project 2000 files to provide more information on specific tasks.

To insert a note for Task 3:

1. *Open the Task Information dialog box for Task 3.* With the baseline file open, double-click the Task Name for Task 3, **Develop project charter**, then click the **Notes** tab.
2. *Enter your note text.* Type **We have several sample charters.** in the Notes: text box, as shown in Figure A-39.
3. *See the resulting Notes icon.* Click **OK** to enter the note, and notice the Notes icon in the Indicators column next to Task 3.
4. *Open the note.* Double-click the **Notes** icon in the Indicators column for Task 3 to view the note, then click **OK**.
5. *Close the file without saving it.* Click **File** on the menu bar, then click **Close**. Click **No** when you are prompted to save the file.

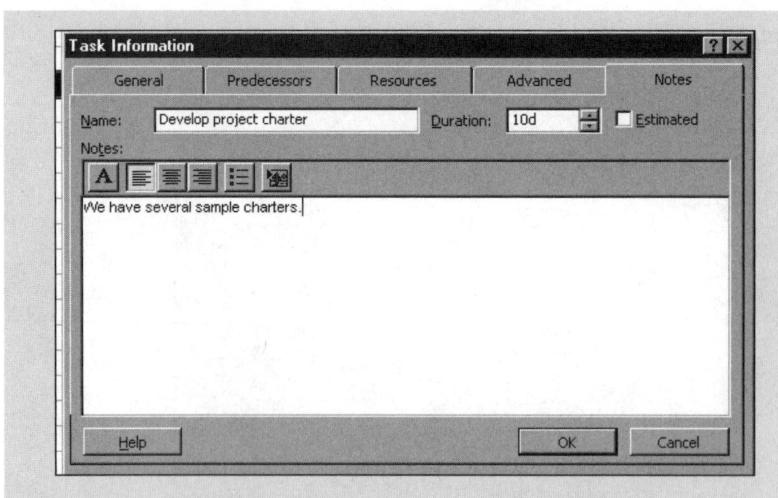

Figure A-39. Task Information Dialog Box Notes Tab

Saving Files as Web Pages

Many organizations use the Web to disseminate information to employees and to communicate with customers. Project 2000 includes a feature that enables you to save project information as a Web page. As with most Microsoft applications, this feature in Project 2000 is available by selecting Save as Web Page from the File menu. You can create Web pages in several standard formats, including Compare to Baseline, Cost data by task, Earned value information, Top Level Tasks list, a "Who Does What" report, and more. You can also design your own formats for information to be saved as Web pages.

To save project information as a Web page:

1. *Open the Save as dialog box for Web pages.* With the baseline file open, click **File** on the menu bar, then click **Save as Web Page**.
2. *Name the Web page file.* Type **taskinfo** in the File name text box as the new filename, and save the file in your personal folder, on a floppy disk, or in the C:\My Documents folder. The Export Mapping dialog box opens, as shown in Figure A-40.
3. *Select the Import/export map.* Click **Default task information** from the Import/export map list, then click **Save**. Project 2000 saves an HTML version of task information from the baseline file. The name of the new web page is taskinfo.html.

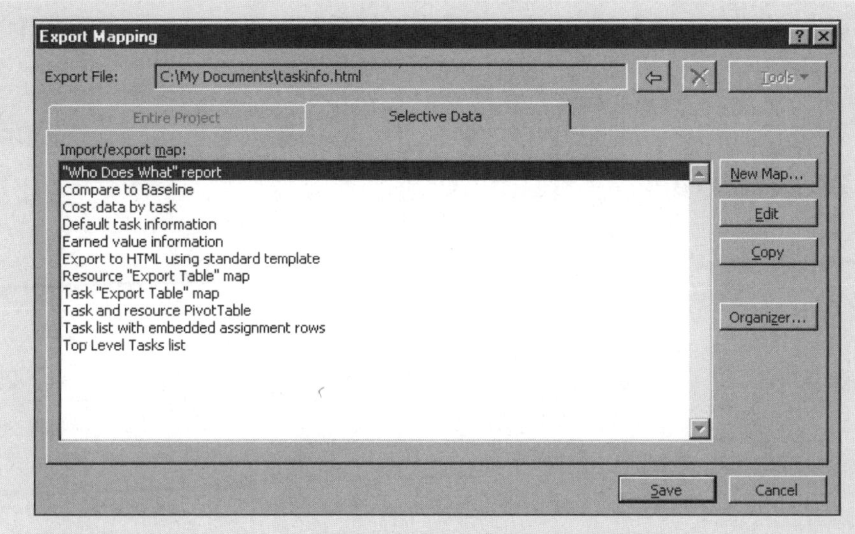

Figure A-40. Export Mapping Dialog Box to Save as Web Page

To open the taskinfo.html file:

1. *Open Windows Explorer.* Click the Windows **Start** button on the lower-left of your desktop, then click **Windows Explorer**. Navigate to the folder in which you saved the taskinfo.html file. You can also use other methods to open Windows Explorer or use My Computer to find the desired file. Alternately, start your Web browser, and use the Open Page feature from the File menu to open the file.
2. *Open the taskinfo.html file in your Web browser.* Double-click **taskinfo.html** to open it in your Web browser. The file extension for the file may not appear, depending on your settings. Figure A-41 shows part of the taskinfo.html file in Microsoft Explorer.
3. *Close the taskinfo.html file.* Click **File**, then click **Close**.

There are several ways to change or customize the standard HTML documents in Project 2000. The Export Format dialog box includes an edit feature that allows you to customize the standard HTML formats, using the Define Import/Export Map dialog box. You can also create your own format or "map" by clicking the New Map button in the Export Format dialog box, which also provides access to the Define Import/Export Map dialog box. Consult the Project 2000 Help topic "HTML" to find more information on HTML features of Project 2000.

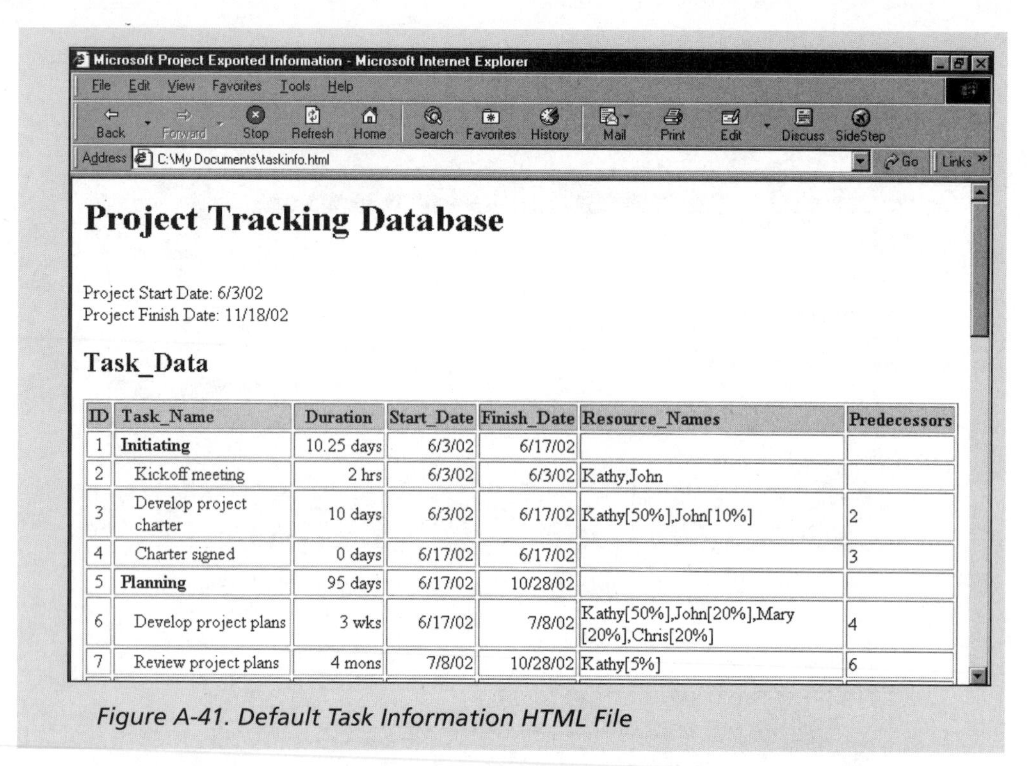

Project Tracking Database

Project Start Date: 6/3/02
Project Finish Date: 11/18/02

Task_Data

ID	Task_Name	Duration	Start_Date	Finish_Date	Resource_Names	Predecessors
1	**Initiating**	10.25 days	6/3/02	6/17/02		
2	Kickoff meeting	2 hrs	6/3/02	6/3/02	Kathy,John	
3	Develop project charter	10 days	6/3/02	6/17/02	Kathy[50%],John[10%]	2
4	Charter signed	0 days	6/17/02	6/17/02		3
5	**Planning**	95 days	6/17/02	10/28/02		
6	Develop project plans	3 wks	6/17/02	7/8/02	Kathy[50%],John[20%],Mary[20%],Chris[20%]	4
7	Review project plans	4 mons	7/8/02	10/28/02	Kathy[5%]	6

Figure A-41. Default Task Information HTML File

Saving project information as Web pages includes text and numerical information, but not graphic information. You can create Web displays of graphic views, such as Gantt charts or network diagrams, but the process is complicated and requires knowledge of how to include images in Web pages. For detailed information on working with graphic views, see the Project 2000 online Help topic "Copy a Microsoft Project picture into another program or a Web page." If Web users have Project 2000 loaded on their local PCs, you can share Project 2000 files over the Web by creating a hyperlink to the Project 2000 file stored on a Web server. The following section describes other ways to share Project 2000 information in workgroups.

Using Project 2000 in Workgroups

Project 2000 has many workgroup features to help you manage and work with other people on a project team. To use these workgroup features, first designate a workgroup manager—the person who builds and maintains the project schedule and makes task assignments.

To set up the workgroup features of Project 2000:

- Click Tools on the menu bar in Project 2000, then click Options.
- Click the Workgroup tab to select it. Figure A-42 shows the resulting dialog box.

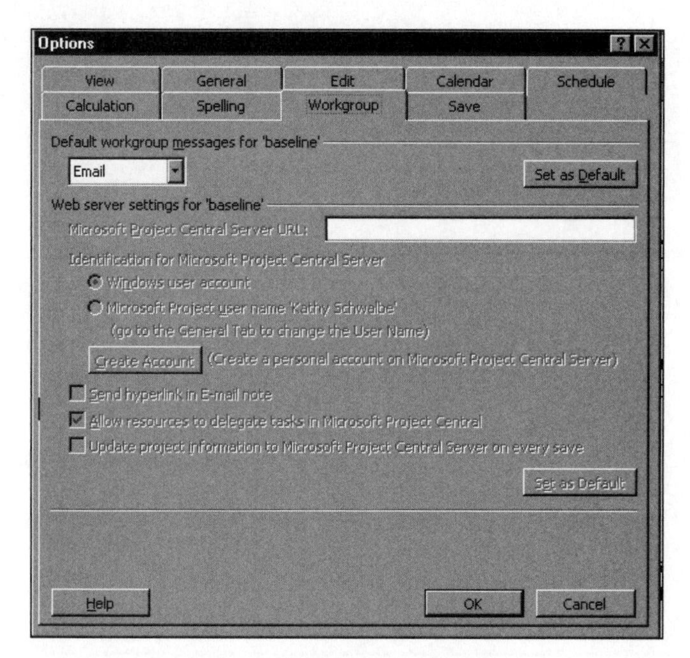

Figure A-42. Options Dialog Box for Workgroup Settings

- Select the default workgroup messaging method from the Default workgroup messages for 'baseline' drop-down list. In this example, Email is selected. You can also choose Web for a Web server setup or Email and Web for sending messages both ways.
- In the Microsoft Project Central Server URL field, enter the URL for the Web server that people will use to access workgroup messages. This Internet address points to your organization's Web server.
- Use the remaining options to set up Project 2000 to enable the sending of hyperlinks in e-mail, allow resources to delegate tasks in Microsoft Project Central, and update project information to Microsoft Project Central Server on every save.
- Click OK to save your workgroup settings.

Once you establish a workgroup, you can use the workgroup communication features. You can manage messages with the TeamInbox in Project 2000. To access the TeamInbox and other workgroup features, you can view the Workgroup toolbar by selecting Toolbars from the View menu, then selecting Workgroup. You can use the Workgroup toolbar to set reminders for team members, send e-mail messages to your team members, assign people to tasks, update tasks, and submit status information. More information on using Workgroup features is available in Project 2000 online Help under "Workgroup." Figure A-43 shows a Help screen about workgroup messaging.

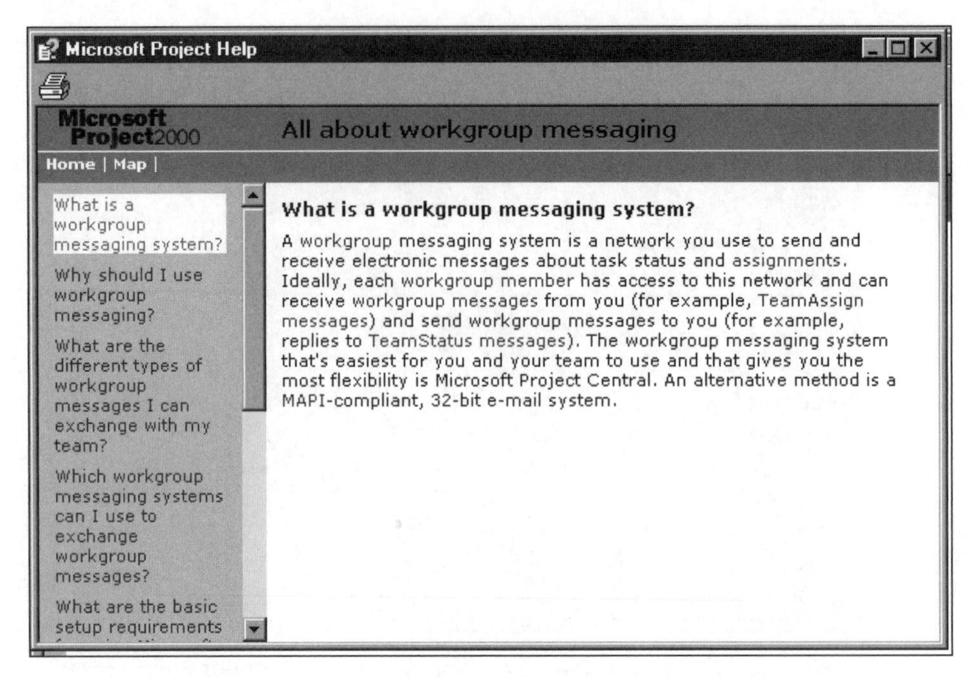

Figure A-43. Project 2000 Help Screen on Workgroup Messaging

Project Central

Project Central is a Microsoft Project companion product that enables collaborative planning among workgroup members, project managers, and other stakeholders. The ability to work with Project Central is one of the most significant new features of Project 2000. Microsoft developed Project Central to allow users to share project details and update a single project by working collaboratively with multiple people across a network.

Microsoft Project Central offers more flexibility and additional benefits over an e-mail workgroup system. With Project Central, workgroup members can:

- View tasks for all of their projects at once, see their tasks in a Gantt chart, and group, sort, and filter their tasks
- View the latest information for the entire project, not just their assigned tasks, at the Microsoft Project Central administrator's discretion
- Create new tasks and send them to the project administrator for incorporation into the project file
- Delegate tasks to other workgroup members, so that the project manager can send tasks to a team leader to be reassigned to individual resources

Project managers can:

- Request and receive status reports in the format they specify and easily consolidate individual status reports into one project status report

- Establish message rules to automatically accept updates from workgroup members. They can specify that updates from all members or only certain members be automatically accepted. And, they can specify that certain updates require the project manager's review, such as actual work in excess of planned work.

Figure A-44 shows a Project Central screen displaying the Personal Gantt view. The Personal Gantt view presents a graphical display of all project information available to a user. By default, resources see all tasks to which they have been assigned, for all projects on which they have been assigned a task. Users can choose from predefined views and chart styles made available by the project administrator, or customize their view by reordering columns, filtering for specific tasks, and sorting or grouping tasks by project name, start date, or other task characteristics. Users can also modify this view to include activities from their Microsoft Outlook Tasks list.[4] Microsoft's Web site provides several white papers describing Project Central.

As you can see, Project 2000 is a very powerful tool. If used properly, it can greatly help users successfully manage projects.

EXERCISES

The more you practice using the different features of Project 2000, the more quickly you will master the application and be able to use it to manage projects. This section includes exercises based on three examples of information technology projects that could benefit from using Project 2000 to assist in project scope, time, cost, human resource, and communications management. It also includes an exercise you can use to apply Project 2000 on a real project.

Exercise A-1: Web Site Development

A nonprofit organization would like you to lead a Web site development project for them. The organization has Internet access that includes space on a Web server, but has no experience developing Web pages or Web sites. In addition to creating their Web site, they would like you to train two people on their staff to do simple Web page updates. The organization wants their Web site to include the following basic information, as a minimum: description of the organization (mission, history, and recent events), list of services, and contact information. They want the Web site to include graphics (photographs and other images) and have an attractive, easy-to-use layout.

[4]Microsoft, Microsoft Project Central White Paper, www.microsoft.com/office/project/ProjCen.htm (November 1999).

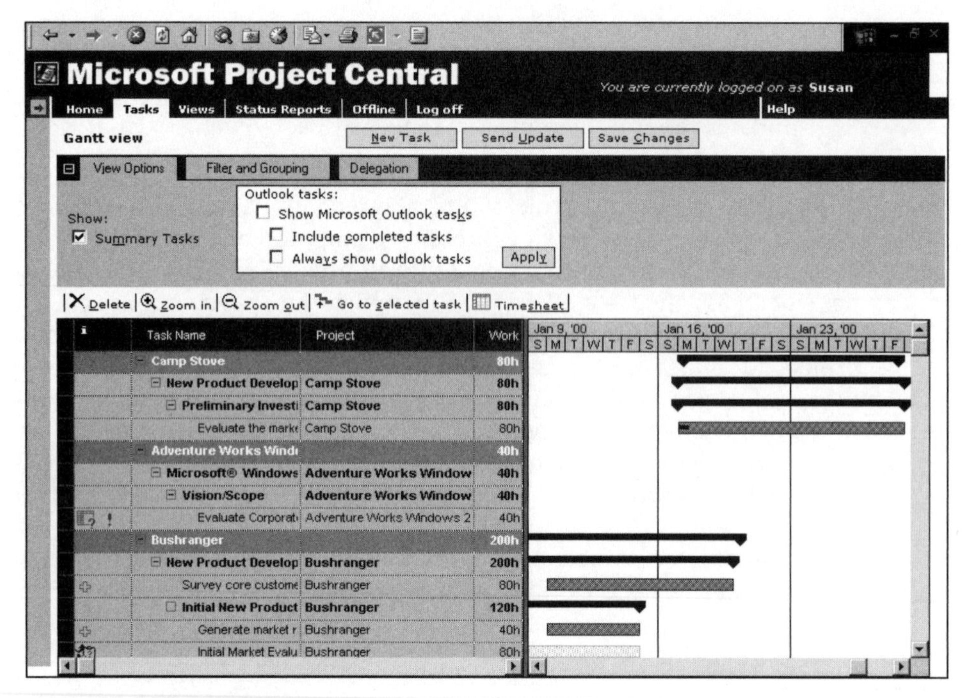

Figure A-44. Project Central Screen

Microsoft, Microsoft Project Central White Paper (November 1999) (www.microsoft.com/office/project/ProjCen.htm)

1. Project Scope Management

   Create a WBS for this project and enter the tasks in Project 2000. Create milestones and summary tasks. Assume that some of the project management tasks you need to do are similar to tasks from the Project Tracking Database example. Some of the specific analysis, design, and implementation tasks will be to:

- Collect information on the organization in hard-copy and digital form (brochures, reports, organization charts, photographs, and so on)
- Research Web sites of similar organizations
- Collect detailed information about the customer's design preferences and access to space on a Web server
- Develop a template for the customer to review (background color for all pages, position of navigation buttons, layout of text and images, typography, including basic text font and display type, and so on)
- Create a site map or hierarchy chart showing the flow of Web pages
- Digitize the photographs and find other images for the Web pages; digitize hard-copy text
- Create the individual Web pages for the site

- Test the pages and the site
- Implement the Web site on the customer's Web server
- Get customer feedback
- Incorporate changes
- Create training materials for the customer on how to update the Web pages
- Train the customer's staff on updating the Web pages

2. Project Time Management

- Enter realistic durations for each task, then link the tasks as appropriate. Be sure that all tasks are linked (in some fashion) to the start and end of the project. Assume that you have four months to complete the entire project. (*Hint*: Use the Project Tracking Database as an example.)
- Print the Gantt chart and Network Diagram view for the project.
- Print the Schedule table to see key dates and slack times for each task.

3. Project Cost Management

- Assume that you have three people working on the project, and each of them would charge $20 per hour. Enter this information in the Resource Sheet.
- Estimate that each person will spend about five hours per week on average for the four-month period. Assign resources to the tasks, and try to make the final cost in line with this estimate.
- Print the budget report for your project.

4. Project Human Resource Management

- Assume that one project team member will be unavailable (due to vacation) for two weeks in the middle of the project. Make adjustments to accommodate this vacation so that the schedule does not slip and the costs do not change. Document the changes from the original plan and the new plan.
- Use the Resource Usage view to see each person's work each month. Print a copy of the Resource Usage view.

5. Project Communications Management

- Print a Gantt chart for this project. Use a time scale that enables the chart to fit on one page.
- Print a "To-do List" report for each team member.
- Create a "Who Does What Report" HTML document. Print it from your Web browser.

Exercise A-2: Software Training Program

ABC Company has 50,000 employees in its headquarters. The company wants to increase employee productivity by setting up an internal software applications training program for its employees. The training program will teach employees how to use software programs such as Windows, Word, Excel, PowerPoint, Access, and Project 2000. Courses will be offered in the evenings and on Saturdays and will be taught by qualified volunteer employees. The instructors will be paid $40 per hour. In the past, various departments sent employees to courses offered by local vendors during company time. In contrast to local vendors' programs, this internal training program should save the company money on training as well as make its people more productive. The Human Resources Department will manage the program, and any employee can take the courses. Employees will receive a certificate for completing courses, and a copy of the certificate will be put in their personnel files. The company has decided to use off-the-shelf training materials, but is not sure which vendor's materials to use. It needs to set up a training classroom, survey employees on what courses they want to take, find qualified volunteer instructors, and start offering courses. The company wants to offer the first courses within six months. One person from Human Resources is assigned full-time to manage this project, and top management has pledged its support of the project.

1. Project Scope Management

   Create a WBS for this project and enter the tasks in Project 2000. Create milestones and summary tasks. Assume that some of the project management tasks you need to do are similar to tasks from the Project Tracking Database example. Some of the tasks specific to this project will be to:

- Review off-the-shelf training materials from three major vendors and decide which materials to use
- Negotiate a contract with the selected vendor for their materials
- Develop communications information about this new training program. Disseminate the information via department meetings, e-mail, the company's intranet, and fliers to all employees.
- Create a survey to determine the number and type of courses needed and employees' preferred times for taking courses
- Administer the survey
- Solicit qualified volunteers to teach the courses
- Review resumes, interview candidates for teaching the courses, and develop a list of preferred instructors
- Coordinate with the facilities department to build two classrooms with twenty personal computers in each, a teacher station, and an overhead projection system (assume that the facilities department will manage this part of the project)

- Schedule courses
- Develop a fair system for signing up for classes
- Develop a course evaluation form to assess the usefulness of each course and the instructor's teaching ability
- Offer classes

2. Project Time Management

- Enter realistic durations for each task and then link appropriate tasks. Be sure that all tasks are linked in some fashion to the start and end of the project. Use the Project Tracking Database as an example. Assume that you have six months to complete the entire project.
- Print the Gantt chart and Network Diagram view for the project.
- Print the Schedule table to see key dates and slack times for each task.

3. Project Cost Management

- Assume that you have four people from various departments available part-time to support the full-time Human Resources person, Terry, on the project. Assume that Terry's hourly rate is $40. Two people from the Information Technology Department will each spend up to 25 percent of their time supporting the project. Their hourly rate is $50. One person from the Marketing Department is available 25 percent of the time at $40 per hour, and one person from Corporate is available 30 percent of the time at $35 per hour. Enter this information about time and hourly wages into the Resource Sheet. Assume that the cost to build the two classrooms will be $100,000, and enter it as a fixed cost.
- Using your best judgment, assign resources to the tasks.
- View the Resource Graphs for each person. If anyone is overallocated, make adjustments.
- Print the budget report for the project.

4. Project Human Resource Management

- Assume that the marketing person will be unavailable for one week, two months into the project, and for another week, four months into the project. Make adjustments to accommodate this unavailability, so the schedule does not slip and costs do not change. Document the changes from the original plan and the new plan.
- Add to each resource a 5 percent raise that starts three months into the project. Print a new budget report.
- Use the Resource Usage view to see each person's work each month. Print a copy.

5. Project Communications Management

- Print a Gantt chart for this project. Use a time scale that enables the chart to fit on one page.

- Print a "To-do List" report for each team member.
- Create an HTML document using the "Default task information" map. Print it from your Web browser.

Exercise A-3: Project Tracking Database

Expand the Project Tracking Database example. Assume that XYZ Company wants to create a history of project information, and the Project Tracking Database is the best format for this type of history. The company wants to track information on twenty past and current projects and wants the database to be able to handle 100 projects total. It wants to track the following project information:

- Project name
- Sponsor name
- Sponsor department
- Type of project
- Project description
- Project manager
- Team members
- Date project was proposed
- Date project was approved or denied
- Initial cost estimate
- Initial time estimate
- Dates of milestones (for example, project approval, funding approval, project completion)
- Actual cost
- Actual time
- Location of project files

1. Project Scope Management

   Open the scope file and add more detail to Executing tasks, using the information in Table A-7.

2. Project Time Management

- Enter realistic durations for each task, and then link appropriate tasks. Make the additional tasks fit the time estimate: 20 days for analysis tasks, 30 days for design tasks, and 20 days for implementation tasks. Assume that all of the executing tasks have a total duration of 70 days. Do not overlap analysis, design, and implementation tasks.
- Print the Gantt chart and Network Diagram view for the project.
- Print the Schedule table to see key dates and slack times for each task.

Table A-7: Company XYZ Project Tracking Database Executing Tasks

ANALYSIS TASKS	DESIGN TASKS	IMPLEMENTATION TASKS
Collect list of 20 projects	Gather detailed requirements for desired outputs from the database	Enter project data
Gather information on projects	Create fully attributed, normalized data model	Test database
Define draft requirements for fields, queries, and reports about projects	Create list of edit rules, default values, and format masks for data	Make adjustments, as needed
Create entity relationship diagram for database	Develop list of queries required for database	Conduct user testing
Create sample entry screen	Review design information with customer	Make adjustments based on user testing
Create sample report	Make adjustments to design based on customer feedback	Create online Help, user manual, and other documentation
Develop simple prototype of database	Create full table structure in database prototype	Train users on the system
Review prototype with customer	Create data input screens	
Make adjustments based on customer feedback	Write queries	
	Create reports	
	Create main screen	
	Review new prototype with customer	

3. Project Cost Management

- Use the resource and cost information provided in the resource file available on the Course Technology web site for this text or from your instructor.
- Assign resources to the new tasks. Try to make the final cost about the same as shown in the Project Tracking Database example about $50,000.
- Print the budget report for the project.

4. Project Human Resources Management

- As of 10/1/02, give everyone on the team a 10 percent raise. Document the increase in costs that these raises cause.
- Use the Resource Usage view to see each person's work each month. Print a copy of the Resource Usage view.

5. Project Communications Management

- Print a Gantt chart for this project. Use a time scale that enables the chart to fit on one page.
- Print a "Top Level Tasks" report.
- Create a "Cost Data by Task" HTML document. Print it from your Web browser.

Exercise A-4: Real Project Application

If you are doing a group project as part of your class or for a project at work, use Project 2000 to create a detailed file describing the work you plan to do. Enter a good WBS, estimate task durations, link tasks, enter resources and costs, assign resources, and so on. Then save your file as a baseline and track your progress. View earned value information when you are about halfway done with the project or course. Export the earned value information into Excel and create an earned value chart. Continue tracking your progress until the project or course is finished. Print your Gantt chart, Resource Sheet, Project Summary report, and relevant information. Also write a two- to three-page paper describing your experience. What did you learn about Project 2000 from this exercise? What did you learn about managing a project? How do you think Project 2000 helps in managing a project? You may also want to interview people who use Project 2000 and ask them about their experiences and suggestions.

Appendix B

Advice for the PMP Exam and Related Certifications

This appendix provides information on Project Management Professional (PMP) and related certifications and offers some advice for taking the PMP exam. It briefly describes PMP and related certifications, requirements for earning and maintaining PMP certification, and the structure and content of the PMP exam. It also provides suggestions on preparing for the PMP exam, tips for taking the exam, sample questions, and final advice on PMP certification and project management in general.

WHAT IS PMP CERTIFICATION?

The Project Management Institute (PMI) offers certification as a Project Management Professional (PMP). A major milestone occurred in 1999 when PMI's Certification Program Department became the first professional certification program department in the world to attain ISO 9001 recognition. Detailed information about PMP certification, including an application, is available from PMI in a brochure called "Certification & Standards." Basic information is also available on the PMI Web site (www.pmi.org), and you can download the brochure and application forms. The following information is quoted from PMI's Web site:

> PMI conducts a certification program in project management. PMI's Project Management Professional (PMP) credential is the project management profession's most globally recognized and respected certification credential. To obtain PMP certification an individual must satisfy education and experience requirements, agree to and adhere to a Code of Professional Conduct, and pass the PMP Certification Examination.
>
> Worldwide there are over 27,000 PMPs who provide project management services in 26 countries. Many corporations require that, for individual advancement within the corporation or for employment, the individual have the PMP credential.[1]

[1]Project Management Institute, www.pmi.org/certification (March 2001).

WHAT ARE THE REQUIREMENTS FOR EARNING AND MAINTAINING PMP CERTIFICATION?

There are four requirements for earning PMP certification. You must:

1. Have experience working in the field of project management. PMP certification requires all applicants to have 4,500 hours of project management experience if they have a bachelor's degree. Applicants without a bachelor's degree are required to have 7,500 hours of experience. This experience must have been gained within a period of three to eight years prior to the PMP application date. You must submit a resume and fill out the Project Management Experience Verification Form (provided in PMI's Certification & Standards brochure) describing your experience working on projects. PMI staff will review your qualifications and send you a letter within a few weeks to let you know if you are qualified to take the PMP exam. You cannot take the exam without this initial qualification.
2. Sign a PMP Certificant and Candidate Agreement and Release form. This form certifies that application information is accurate and complete and that candidates will conduct themselves in accordance with the PMP Code of Professional Conduct, professional development requirements, and other PMI certification program policies and procedures.
3. Complete the PMP Certification Exam Application and pay a certification fee of $555 for nonmembers and $405 for members.
4. Pass the PMP exam. After completing steps one through three and receiving notification from PMI that you are eligible to take the PMP exam, you have three months to take the exam. The PMP exam consists of 200 multiple-choice questions. The questions on each test are randomly selected from a large test bank, so each person taking the exam receives different questions. Test takers have 4.5 hours to take this computerized exam, and a passing score is 68.5%, or at least 137 correct answers out of 200. PMI reviews and revises the examination annually.[2]

You cannot take the PMP exam before you complete the first three steps. If you do not have the minimum amount of work experience, it is best to wait until you do before submitting certification information to PMI. Also realize that if you do not pass the PMP exam the first time you take it, you must pay the full certification fee to sit for the exam again.

Starting in January 1999, PMI instituted a new professional development program for maintaining the PMP certification. To maintain your PMP status, you must earn at least sixty Professional Development Units within a three-year cycle and agree to continue to adhere to PMI's PMP Code of Professional Conduct. PMI will mail you their Professional Development Program Handbook after you earn PMP certification. This handbook provides more details on maintaining your PMP status.

[2]ibid.

WHAT OTHER CERTIFICATION EXAMS RELATED TO INFORMATION TECHNOLOGY PROJECT MANAGEMENT ARE AVAILABLE?

In recent years, PMI and other organizations have been developing more certifications related to information technology project management. Below are brief descriptions of other existing certifications and some currently in development.

- PMI's Certificates of Added Qualification: The PMP exam is not industry-specific, so you will not be tested on information unique to a particular industry, such as information technology. PMI has developed additional exams called Certificates of Added Qualification (CAQs) to test industry-specific knowledge. PMPs with experience in certain fields can take CAQ exams. The first PMI CAQ examination, given in August 2000, was the Automotive Product Development (APD) CAQ. PMI is working on a CAQ for people in the information technology field. Current plans are to have a CAQ for systems development in early 2002. PMI's Web site (www.pmi.org) is the best source of information on CAQ exams. The PMI Information Systems Specific Interest Group (ISSIG) also provides its members with updates about certifications, in their quarterly newsletters and on their Web site (www.pmi-issig.org).
- CompTIA's IT Project+ Certification: In February 2001, CompTIA, the Computing Technology Industry Association, announced plans to launch certifications in information technology project management and e-Business fundamentals. The new certification programs, recently purchased from Prometric-Thomson Learning, and recognized as the Gartner Institute Certification Program, will be rebranded and launched by the end of 2001. In April 2001 CompTIA started offering their IT Project+ certification. The IT Project+ exam is designed for information technology professionals with one year of experience in leading, managing, and directing small information technology projects. The exam covers the business, interpersonal, and technical skills required to successfully manage information technology projects. As of April 2001, the cost for taking this 90-minute, multiple-choice exam was $190 for non-CompTIA members. More than 250,000 individuals worldwide have earned CompTIA certifications in PC service, networking, document imaging, Internet, and PC Server technologies. Detailed information about the IT Project+ and other CompTIA exams is available on ComTIA's Web site (www.comptia.org). Check the Course Technology Web site for information about study materials for this upcoming certification exam.[3]
- Certified IT Project Manager: In 1998 the Singapore Computer Society collaborated with the Infocomm Development Authority of Singapore to establish an Information Technology Project Management Certification

[3]CompTIA Web site, www.comptia.org (February 2000).

Program. By the fall of 2000, about 400 participants had joined the program, with 248 professionals currently having the title of Certified IT Project Manager (CITPM). PMI recently signed a Memorandum of Understanding with the Singapore Computer Society to support and advance the global credentialing of project management and information technology expertise.[4]

WHAT IS THE STRUCTURE AND CONTENT OF THE PMP CERTIFICATION EXAM?

The PMP exam is based on information from the entire project management body of knowledge (PMBOK). It is essential to review concepts and terminology in PMI's PMBOK Guide, and texts like this one will help to reinforce your understanding of key topics in project management. Starting in January 2002, the PMP exam will include questions from the PMBOK Guide 2000. This textbook reflects the PMBOK Guide 2000. Table B-1 shows the approximate breakdown of questions on the PMP exam by process groups as of early 2001. Beginning in July 2001 the PMP exam will change to include questions related to Professional Responsibility. Candidates should review PMI's Code of Professional conduct and updated exam information on PMI's Web site to make sure they are prepared for their PMP exams.

Table B-1: Breakdown of Questions on the PMP by Process Groups

PROCESS GROUP	PERCENT OF QUESTIONS ON PMP EXAM
Initiating	4
Planning	35.5
Executing	25
Controlling	27.5
Closing	8

It is a good idea to study Table 2-5 to understand the relationships among project management process groups, activities, and knowledge areas. The table briefly outlines which activities are done during each of the project management process groups and what is involved in each of the knowledge areas. It is also important to understand what each of the project management activities involves. Several questions on the certification exam require an understanding of this framework for project management, and many questions require an understanding of the various tools and techniques described in the PMBOK and this text.

[4]Schwalbe, Kathy, "Certification News," *ISSIGreview* (Fourth Quarter 2000).

The PMP exam includes conceptual questions, application questions, and evaluative questions. Conceptual questions test understanding of key terms and concepts in project management. For example, you should know basic definitions such as what a project is, what project management is, and what key activities are part of scope management. Application questions test your ability to apply techniques to specific problems. For example, a question might provide information for constructing a network diagram and ask you to find the critical path. You might be given cost and schedule information and be asked to find the schedule or cost performance index. Evaluative questions provide situations for you to analyze, and your response will indicate how you would handle them. For example, a project might have a lot of problems. You might be asked what you would do in that situation, given the information provided. Remember that all questions are multiple choice, so you must select the best answer from the options provided.

HOW SHOULD YOU PREPARE FOR THE PMP EXAM?

To prepare for the PMP exam, it is important to understand your own learning and testing style, and to use whatever resources and study techniques work best for you. Several companies offer sample tests, CD-ROMs, and courses designed to help people pass the exam. Several chapters of PMI offer PMP exam review courses. Many people form study groups to provide support in preparing for the exam. Having some "peer pressure" might help to ensure that you actually take and pass the exam. If you do not pass the exam, you have to pay to take it again, so you want to make sure you are ready the first time.

Even if you think you know about project management, studying the material in the PMBOK and this book before taking the exam will help you. If you are comfortable with the concepts in the PMBOK and this text and have some experience working on projects, you should do fine.

Recall the triple constraint of project management—scope, time, and cost management. Focus on these three knowledge areas when studying. Be familiar with project charters, WBSs, network diagrams, critical path analysis, cost estimates, earned value, and so on. Many questions on the exam are related to project time management, so study that area closely. Also be sure to memorize earned value formulas.

To pass the exam, you need to answer only 68.5 percent of the questions correctly. Your PMP certificate will not display your final score, so it does not matter if you get 70, 80, 90, or even 100 percent correct.

TEN TIPS FOR TAKING THE PMP EXAM

1. The PMP exam is computer-based and begins with a short tutorial on how to use the testing software. The software makes it easy to mark questions you want to review later, so learning how to mark questions is helpful. Using this feature can give you a feel for how well you are doing on the test. It is a good idea to go through every question fairly quickly and mark those that you want to spend more time on. If you mark fewer then 60 out of 200 questions before you finish, odds are that you will get at least 68.5 percent correct and pass the exam.
2. The time allotted for the exam is 4.5 hours, and each multiple-choice question has four alternatives. You should have plenty of time to complete the exam. Use the first two to three hours to go through all of the questions. Do not spend too much time on any one question. As you work, mark each question that you would like to return to for further consideration. Then use the remaining time to check the questions you are not sure of.
3. Some people believe it is better to change answers you were originally unsure of. If you think that a different answer is better, after reading the question again, then change your answer. If you are still unsure, leave the question marked so you can see a final total of unmarked questions.
4. Do not try to read more into the questions than what is stated. There are no trick questions. Most of the questions are fairly short, and there are only four alternatives from which to choose the answer. Few of the alternatives include "all of the above" or "none of the above." Some questions might include the alternative "There is not enough information."
5. To increase your chances of getting the right answer, first eliminate obviously wrong alternatives, then choose among the remaining alternatives.
6. Some questions require doing calculations such as earned value management. It is worthwhile memorizing the earned value formulas to make answering these questions easier. You may use a non-programmable calculator while taking the exam, so be sure to bring one to make performing calculations easier.
7. You will be given a couple of pieces of blank paper to use during the exam. Before starting the test, you might want to write down important equations so that you do not have to rely on your memory. When you come to a question involving calculations, write the calculations down so you can check your work for errors.
8. Read all questions carefully. A few sections of the test require that you answer three to four questions about a scenario. These questions can be difficult; it can seem as if two of the alternatives could be correct, although you can choose only one. Read the directions for these types of questions several times to be sure you know exactly what you are supposed to do.

9. If you do not know an answer and need to guess, be wary of alternatives that include words such as *always, never, only, must,* and *completely*. These extreme words usually indicate wrong alternatives because there are many exceptions to rules.
10. After about two hours, take a short break to clear your mind. You might want to bring a snack to have during your break.

SAMPLE QUESTIONS

A few sample questions similar to those you will find on the PMP exam are provided below. Correct answers are marked with an asterisk.

1. A document that formally recognizes the existence of a project is a
   A. Gantt chart
   B. WBS
   C. Project charter*
   D. Scope statement
2. Decomposition is used in developing
   A. the management plan
   B. the communications plan
   C. the earned value
   D. the WBS*
3. The critical path on a project represents
   A. the shortest path through a network diagram
   B. the longest path through a network diagram*
   C. the most important tasks on a project
   D. the highest-risk tasks on a project
4. If the EV for a project is $30,000, the AC is $33,000, and the PV is $25,000, what is the cost variance?
   A. $3,000*
   B. -$3,000
   C. $5,000
   D. -$5,000
5. If the EV for a project is $30,000, the AC is $33,000, and the PV is $25,000, how is the project performing?
   A. The project is over budget and ahead of schedule*
   B. The project is over budget and behind schedule
   C. The project is under budget and ahead of schedule
   D. The project is under budget and behind schedule

6. Some organizations are now using 6 sigma for quality control, compared to the usual
   A. 2 sigma
   B. 3 sigma*
   C. 4 sigma
   D. 5 sigma
7. Project human resource management includes which of the following processes?
   A. organizational planning
   B. staff acquisition
   C. team development
   D. all of the above*
8. Determining the information and communications needs of project stakeholders is part of
   A. communications planning*
   B. scope planning
   C. human resource planning
   D. quality planning
9. Reducing the impact of a risk event by reducing the probability of its occurrence is
   A. risk avoidance
   B. risk acceptance
   C. risk mitigation*
   D. contingency planning
10. Which type of contract provides the least amount of risk for the buyer?
    A. firm fixed price*
    B. fixed price incentive
    C. cost plus incentive fee
    D. cost plus fixed fee
11. Suppose you have a project with four tasks as follows:

- Task 1 can start immediately and has an estimated duration of 1.
- Task 2 can start after Task 1 is completed and has an estimated duration of 4.
- Task 3 can start after Task 2 is completed and has an estimated duration of 5.
- Task 4 can start after Task 1 is completed and must be completed when Task 3 is completed. Its estimated duration is 8.

What is the length of the critical path for this project?

A. 9
B. 10*
C. 11
D. 12

12. In which of the following project management process groups are the most time and money typically spent?
    A. initiating
    B. planning
    C. executing*
    D. controlling
13. Measuring the probability and consequences of risks is part of
    A. risk management planning
    B. risk identification
    C. qualitative risk analysis
    D. quantitative risk analysis*
14. A technique that modifies the project schedule to account for limited resources is
    A. critical path analysis
    B. PERT
    C. critical chain*
    D. earned value

The Project Management Institute started providing sample PMP exam questions on their Web site in 1999. You can download forty PMP sample examination questions from www.pmi.org/certification. This Web site also provides answers to frequently asked questions about the PMP certification application and PMI's professional development program. PMI also provides information on this site to help you prepare the PMP examination application.

FINAL ADVICE ON PMP CERTIFICATION AND PROJECT MANAGEMENT IN GENERAL

Now that you have read this text and discussed project management with others, I hope that you have grown to see project management as a valuable skill, especially in the information technology field. Project management certification can be your first step toward advancing your career in the fast-paced world of technology, for no matter how much things change, the need for projects and project managers is a constant.

The knowledge and experience I have gained working on and managing projects continues to help me in my career and my personal life. I was fortunate to step into a project management role very early in my career as an Air Force officer. My first real job at the age of twenty-two was as a project manager. I have held several job titles since then—systems analyst, engineer, technical specialist, information technology management consultant, independent consultant, college professor, and now author. All of these jobs included working on or managing projects. As a mother of three, I can also attest to the fact that project management skills help in planning and executing social activities (weddings, birthday parties, fundraisers, and so on) and in dealing with the challenges of everyday life.

Glossary

A

Acceptance decisions — decisions that determine if the products or services produced as part of the project will be accepted or rejected

Activity — an element of work, normally found on the WBS, that has an expected duration, cost, and resource requirements; also called **task**

Activity definition — identifying the specific activities that the project team members and stakeholders must perform to produce the project deliverables

Activity duration estimating — estimating the number of work periods that are needed to complete individual activities

Activity sequencing — identifying and documenting the relationships between project activities

Activity-on-arrow (AOA) or arrow diagramming method (ADM) — a network diagramming technique in which activities are represented by arrows and connected at points called nodes to illustrate the sequence of activities

Actual cost (AC) — the total of direct and indirect costs incurred in accomplishing work on an activity during a given period, formerly called the **actual cost of work performed (ACWP)**

Administrative closure — generating, gathering, and disseminating information to formalize phase or project completion

Analogous estimates — a cost estimating technique that uses the actual cost of a previous, similar project as the basis for estimating the cost of the current project, also called **top-down estimates**

Analogy approach — creating a WBS by using a similar project's WBS as a starting point

Appraisal cost — the cost of evaluating processes and their outputs to ensure that a project is error-free or within an acceptable error range

B

Backward pass — a project network diagramming technique that determines the late start and late finish dates for each activity

Baseline — the original project plan plus approved changes

Benchmarking — a technique used to generate ideas for quality improvements by comparing specific project practices or product characteristics to those of other projects or products within or outside the performing organization

Bottom-up approach — creating a WBS by having team members identify as many specific tasks related to the project as possible and then grouping them into higher level items

Bottom-up estimates — a cost estimating technique based on estimating individual work items and summing them to get a project total

Brainstorming — a technique by which a group attempts to generate ideas or find a solution for a specific problem by amassing ideas spontaneously and without judgment

Budget at Completion (BAC) — the original total budget for a project

Budgetary estimate — a cost estimate used to allocate money into an organization's budget

Buffer — additional time to complete a task, added to an estimate to account for various factors

Burst — when a single node is followed by two or more activities on a project network diagram

C

Capability Maturity Model (CMM) — a five-level model laying out a generic path to process improvement for software development in organizations

Cash flow analysis — a method for determining the estimated *annual* costs and benefits for a project

Change control board (CCB) — a formal group of people responsible for approving or rejecting changes on a project

Change control system — a formal, documented process that describes when and how official project documents may be changed

Closing processes — formalizing acceptance of the project or phase and bringing it to an orderly end

COCOMO II — a newer, computerized cost-estimating tool based on Boehm's original model that allows one to estimate the cost, effort, and schedule when planning a new software development activity

Coercive power — using punishment, threats, or other negative approaches to get people to do things they do not want to do

Communications infrastructure — a set of tools, techniques, and principles that provide a foundation for the effective transfer of information among people

Communications management plan — a document that guides project communications

Communications planning — determining the information and communications needs of the stakeholders: who needs what information, when will they need it, and how will the information be given to them

Compromise mode — using a give-and-take approach to resolving conflicts. Bargaining and searching for solutions that bring some degree of satisfaction to all the parties in a dispute

Computerized tools — cost-estimating tools that use computer software, such as spreadsheets and project management software

Configuration management — a process that ensures that the descriptions of the project's products are correct and complete

Conformance — delivering products that meet requirements and fitness for use

Conformance to requirements — the project processes and products meet written specifications

Confrontation mode — directly facing a conflict using a problem-solving approach that allows affected parties to work through their disagreements

Constructive change orders — oral or written acts or omissions by someone with actual or apparent authority that can be construed to have the same effect as a written change order

Constructive Cost Model (COCOMO) — a parametric model developed by Barry Boehm for estimating software development costs

Contingency plans — predefined actions that the project team will take if an identified risk event occurs

Contingency reserves — dollars included in a cost estimate to allow for future situations that may be partially planned for (sometimes called **known unknowns**) and are included in the project cost baseline

Contract — a mutually binding agreement that obligates the supplier to provide the specified products or services, and obligates the buyer to pay for them

Contract administration — managing the relationship with the supplier

Contract close-out — completion and settlement of the contract, including resolution of any open items

Control chart — a graphic display of data that illustrates the results of a process over time

Controlling processes — actions to ensure that project objectives are met

Cost budgeting — allocating the overall cost estimate to individual work items to establish a baseline for measuring performance

Cost control — controlling changes to the project budget

Cost estimating — developing an approximation or estimate of the costs of the resources needed to complete the project

Cost management plan — a document that describes how cost variances will be managed on the project

Cost of nonconformance — taking responsibility for failures or not meeting quality expectations

Cost of quality — the cost of conformance plus the cost of nonconformance

Cost performance index (CPI) — the ratio of earned value to actual cost; can be used to estimate the projected cost to complete the project

Cost plus fixed fee (CPFF) contract — a contract in which the buyer pays the supplier for allowable performance costs plus a fixed fee payment usually based on a percentage of estimated costs

Cost plus incentive fee (CPIF) contract — a contract in which the buyer pays the supplier for allowable performance costs along with a predetermined fee and an incentive bonus

Cost plus percentage of costs (CPPC) contract — a contract in which the buyer pays the supplier for allowable performance costs along with a predetermined percentage based on total costs

Cost reimbursable contracts — contracts involving payment to the supplier for direct and indirect actual costs

Cost variance (CV) — the earned value minus the actual cost

Crashing — a technique for making cost and schedule tradeoffs to obtain the greatest amount of schedule compression for the least incremental cost

Critical chain — a method of scheduling that takes limited resources into account when creating a project schedule and includes buffers to protect the project completion date

Critical path — the series of activities in a project network diagram that determines the earliest completion of the project. It is the longest path through the network diagram and has the least amount of slack or float

Critical path method (CPM) or critical path analysis — a project network analysis technique used to predict total project duration

D

Definitive estimate — a cost estimate that provides an accurate estimate of project costs

Deliverable — a product, such as a report or segment of software code, produced as part of a project

Delphi Technique — an approach used to derive a consensus among a panel of experts, to make predictions about future developments

Dependency — the sequencing of project activities or tasks; also called a **relationship**

Deputy project managers — people who fill in for project managers in their absence and assist them as needed, similar to the role of a vice president

Design of experiments — a quality technique that helps identify which variables have the most influence on the overall outcome of a process

Direct costs — costs that are related to a project and can be traced back in a cost-effective way

Directives — new requirements imposed by management, government, or some external influence

Discount factor — a multiplier for each year based on the discount rate and year

Discount rate — the minimum acceptable rate of return on an investment; also called the required rate of return, hurdle rate, or opportunity cost of capital

Discretionary dependencies — sequencing of project activities or tasks defined by the project team and used with care since they may limit later scheduling

Dummy activities — activities with no duration and no resources used to show a logical relationship between two activities in the arrow diagramming method of project network diagrams

Duration — the actual amount of time worked on an activity *plus* elapsed time

E

Early finish date — the earliest possible time an activity can finish based on the project network logic

Early start date — the earliest possible time an activity can start based on the project network logic

Earned value (EV) — the percentage of work actually completed multiplied by the planned cost, formerly called the **budgeted cost of work performed (BCWP)**

Earned value management (EVM) — a project performance measurement technique that integrates scope, time, and cost data

Empathic listening — listening with the intent to understand

Estimate at completion (EAC) — an estimate of what it will cost to complete the project based on performance to date

Executing processes — coordinating people and other resources to carry out the project plans and produce the products or deliverables of the project

Expected monetary value (EMV) — the product of the risk event probability and the risk event's monetary value

Expert power — using one's personal knowledge and expertise to get people to change their behavior

External dependencies — sequencing of project activities or tasks that involve relationships between project and non-project activities

External failure cost — a cost related to all errors not detected and corrected before delivery to the customer

F

Fallback plans — plans developed for risks that have a high impact on meeting project objectives, to be implemented if attempts to reduce the risk are not effective

Fast tracking — a schedule compression technique in which you do activities in parallel that you would normally do in sequence

Features — the special characteristics that appeal to users

Feeding buffers — additional time added before tasks on the critical path that are preceded by non-critical-path tasks

Finish-to-finish dependency — a relationship on a project network diagram where the "from" activity must finish before the "to" activity can finish

Finish-to-start dependency — a relationship on a project network diagram where the "from" activity must finish before the "to" activity can start

Fishbone diagrams — diagrams that trace complaints about quality problems back to the responsible production operations; sometimes called Ishikawa diagrams

Fitness for use — a product can be used as it was intended

Fixed price or lump sum contracts — contracts with a fixed total price for a well-defined product or service

Flowcharts — diagrams that show how various elements of a system relate to each other

Forcing mode — using a win-lose approach to conflict resolution to get one's way

Formal acceptance — documentation that the project's sponsor or customer signs to show they have accepted the products of the project

Forward pass — a project network diagramming technique that determines the early start and early finish dates for each activity

Free slack — the amount of time an activity can be delayed without delaying the early start of any immediately following activities; also called free float

Function points — technology-independent assessments of the functions involved in developing a system

Functional organizational structure — an organizational structure that groups people by functional areas such as information technology, manufacturing, engineering, and accounting.

Functionality — the degree to which a system performs its intended function

G

Gantt chart — a standard format for displaying project schedule information by listing project activities and their corresponding start and finish dates in a calendar format

Groupthink — conformance to the values or ethical standards of a group

H

Hierarchy of needs — a pyramid structure illustrating Maslow's theory that people's behaviors are guided or motivated by a sequence of needs

Human resources frame — focuses on producing harmony between the needs of the organization and the needs of people

I

Indirect costs — costs that are related to the project but cannot be traced back in a cost-effective way

Influence diagrams — diagrams that represent decision problems by displaying essential elements, including decisions, uncertainties, and objectives, and how they influence each other

Information distribution — making needed information available to project stakeholders in a timely manner

Initiating processes — actions to commit to begin or end projects and project phases

Initiation — committing the organization to begin a project or continue to the next phase of a project

Intangible costs or benefits — costs or benefits that are difficult to measure in monetary terms

Integrated change control — coordinating changes across the entire project

Integration testing — testing that occurs between unit and system testing to test functionally grouped components to ensure a subset(s) of the entire system works together

Interface management — identifying and managing the points of interaction between various elements of a project

Internal failure cost — a cost incurred to correct an identified defect before the customer receives the product

Internal rate of return (IRR) — the discount rate that makes the net present value equal to zero, also called **time-adjusted rate of return**

Interviewing — a fact-finding technique that is normally done face-to-face or via telephone

ISO 9000 — a quality system standard developed by the International Organization for Standardization (ISO) that includes a three-part, continuous cycle of planning, controlling, and documenting quality in an organization

J

Joint Application Design (JAD) — using highly organized and intensive workshops to bring together project stakeholders—the sponsor, users, business analysts, programmers, and so on—to jointly define and design information systems

K

Kickoff meeting — a meeting held at the beginning of a project or project phase where all major project stakeholders discuss project objectives, plans, and so on

L

Late finish date — the latest possible time an activity can be completed without delaying the project finish date

Late start date — the latest possible time an activity may begin without delaying the project finish date

Learning curve theory — a theory that states that when many items are produced repetitively, the unit cost of those items normally decreases in a regular pattern as more units are produced

Legacy systems — older information systems that usually ran on an old mainframe computer

Legitimate power — getting people to do things based on a position of authority

Lessons learned — reflective statements written by project managers and their team members

Life cycle costing — considers the total cost of ownership, or development plus support costs, for a project

M

Maintainability — the ease of performing maintenance on a product

Make-or-buy decision — when an organization decides if it is in their best interests to make certain products or perform certain services inside the organization, or if it is better to buy them from an outside organization

Malcolm Baldrige Award — an award started in 1987 to recognize companies that have achieved a level of world-class competition through quality management

Management reserves — dollars included in a cost estimate to allow for future situations that are unpredictable (sometimes called **unknown unknowns**)

Mandatory dependencies — sequencing of project activities or tasks that are inherent in the nature of the work being done on the project

Matrix organizational structure — an organizational structure in which employees are assigned to both functional and project managers

Maturity model — a framework for helping organizations improve their processes and systems

Mean — the average value of a population

Measurement and test equipment cost — the capital cost of equipment used to perform prevention and appraisal activities

Merge — when two or more nodes precede a single node on a project network diagram

Milestone — a significant event on a project with zero duration

Mirroring — the matching of certain behaviors of the other person

Monte Carlo analysis — a risk quantification technique that simulates a model's outcome many times, to provide a statistical distribution of the calculated results

Multitasking — when a resource works on more than one task at a time

Murphy's Law — if something can go wrong, it will

Myers-Briggs Type Indicator (MBTI) — a popular tool for determining personality preferences

N

Net present value (NPV) analysis — a method of calculating the expected net monetary gain or loss from a project by discounting all expected future cash inflows and outflows to the present point in time

Node — the starting and ending point of an activity on an activity-on-arrow diagram

Normal distribution — a bell-shaped curve that is symmetrical about the mean of the population

O

Opportunities — chances to improve the organization

Organization planning — identifying, assigning, and documenting project roles, responsibilities, and reporting relationships

Organizational breakdown structure (OBS) — a specific type of organizational chart that shows which organizational units are responsible for which work items

Overallocation — when more resources than are available are assigned to perform work at a given time

P

Parametric modeling — a cost-estimating technique that uses project characteristics (parameters) in a mathematical model to estimate project costs

Pareto analysis — identifying the vital few contributors that account for most quality problems in a system

Pareto diagrams — histograms that help identify and prioritize problem areas

Parkinson's Law — work expands to fill the time allowed

Payback period — the amount of time it will take to recoup, in the form of net cash inflows, the net dollars invested in a project

Performance — how well a product or service performs the customer's intended use

Performance reporting — collecting and disseminating performance information, which includes status reports, progress measurement, and forecasting

PERT weighted average = (optimistic time + 4 × most likely time + pessimistic time) ÷ 6

Phase exit or **kill point** — management review that should occur after each project phase to determine if projects should be continued, redirected, or terminated

Planned value (PV) — that portion of the approved total cost estimate planned to be spent on an activity during a given period, formerly called the **budgeted cost of work scheduled (BCWS)**

Planning processes — devising and maintaining a workable scheme to accomplish the business need that the project was undertaken to address

Political frame — addresses organizational and personal politics

Politics — competition between groups or individuals for power and leadership

Power — the potential ability to influence behavior to get people to do things they would not otherwise do

Precedence Diagramming Method (PDM) — a network diagramming technique in which boxes represent activities

Prevention cost — the cost of planning and executing a project so that it is error-free or within an acceptable error range

Probabilistic time estimates — duration estimates based on using optimistic, most likely, and pessimistic estimates of activity durations instead of using one specific or discrete estimate

Problems — undesirable situations that prevent the organization from achieving its goals

Process adjustments — adjustments made to correct or prevent further quality problems based on quality control measurements

Process — a series of actions directed toward a particular result

Procurement — acquiring goods and/or services from an outside source

Procurement planning — determining what to procure and when

Profit margin — the ratio between revenues and profits

Profits — revenues minus expenses

Program — a group of projects managed in a coordinated way

Program Evaluation and Review Technique (PERT) — a project network analysis technique used to estimate project duration when there is a high degree of uncertainty with the individual activity duration estimates

Progress reports — reports that describe what the project team has accomplished during a certain period of time

Project — a temporary endeavor undertaken to accomplish a unique purpose

Project acquisition — a common reference to the last two project phases—implementation and close-out

Project archives — a complete set of organized project records that provide an accurate history of the project

Project buffer — additional time added before the project's due date

Project charter — a document that formally recognizes the existence of a project and provides direction on the project's objectives and management

Project communications management — the processes required to ensure timely and appropriate generation, collection, dissemination, storage, and disposition of project information

Project cost management — the processes required to ensure that the project is completed within the approved budget

Project feasibility — a common reference to the first two project phases—concept and development

Project forecasting — predicting future project status and progress based on past information and trends

Project human resource management — the processes required to make the most effective use of the people involved with a project

Project integration management — includes the processes involved in coordinating all of the other project management knowledge areas throughout a project's life cycle

Project life cycle — the collection of project phases—concept, development, implementation, and close-out

Project management — the application of knowledge, skills, tools, and techniques to project activities in order to meet project requirements

Project Management Institute (PMI) — international professional society for project managers

Project management knowledge areas — project integration management, scope, time, cost, quality, human resource, communications, risk, and procurement management

Project management office or **center of excellence** — an organizational entity created to assist project managers in achieving project goals

Project management process groups — the progression of project activities from initiation to planning, executing, controlling, and closing

Project Management Professional (PMP) — certification provided by PMI that requires documenting project experience, agreeing to follow the PMI code of ethics, and passing a comprehensive examination

Project management software — software specifically designed for project management

Project management tools and techniques — methods available to assist project managers and their teams

Project network diagram — a schematic display of the logical relationships or sequencing of project activities

Project organizational structure — an organizational structure that groups people by major projects, such as specific aircraft programs

Project plan — a document used to coordinate all project planning documents and guide project execution and control

Project plan development — taking the results of other planning processes and putting them into a consistent, coherent document—the project plan

Project plan execution — carrying out the project plan by performing the activities it includes

Project procurement management — the processes required to acquire goods and services for a project from outside the performing organization

Project quality management — the processes required to ensure that the project will satisfy the needs for which it was undertaken

Project risk management — the art and science of identifying, analyzing, and responding to risk throughout the life of a project and in the best interests of meeting project objectives

Project scope management — the processes involved in defining and controlling what is or is not included in a project

Project time management — the processes required to ensure timely completion of a project

Prototyping — developing a working replica of the system or some aspect of the system to help define user requirements

Q

Qualitative risk analysis — qualitatively analyzing risks and prioritizing their effects on project objectives

Quality — the totality of characteristics of an entity that bear on its ability to satisfy stated or implied needs

Quality assurance — periodically evaluating overall project performance to ensure the project will satisfy the relevant quality standards

Quality audits — structured reviews of specific quality management activities that help identify lessons learned and can improve performance on current or future projects

Quality circles — groups of nonsupervisors and work leaders in a single company department who volunteer to conduct group studies on how to improve the effectiveness of work in their department

Quality control — monitoring specific project results to ensure that they comply with the relevant quality standards and identifying ways to improve overall quality

Quality planning — identifying which quality standards are relevant to the project and how to satisfy them

Quantitative risk analysis — measuring the probability and consequences of risks and estimating their effects on project objectives

R

Rapport — a relation of harmony, conformity, accord, or affinity

Referent power — getting people to do things based on an individual's personal charisma

Reliability — the ability of a product or service to perform as expected under normal conditions without unacceptable failures

Request for Proposal (RFP) — a document used to solicit proposals from prospective suppliers

Request for Quote (RFQ) — a document used to solicit quotes or bids from prospective suppliers

Required rate of return — the minimum acceptable rate of return on an investment

Reserves — dollars included in a cost estimate to mitigate cost risk by allowing for future situations that are difficult to predict

Residual risks — risks that remain after all of the response strategies have been implemented

Resource histogram — a column chart that shows the number of resources assigned to a project over time

Resource leveling — a technique for resolving resource conflicts by delaying tasks

Resource loading — the amount of individual resources an existing schedule requires during specific time periods

Resource planning — determining what resources (people, equipment, and materials) and what quantities of each resource should be used to perform project activities

Responsibility assignment matrix (RAM) — a matrix that maps the work of the project as described in the WBS to the people responsible for performing the work as described in the OBS

Return on investment (ROI) — income divided by investment

Reward power — using incentives to induce people to do things

Rework — action taken to bring rejected items into compliance with product requirements or specifications or other stakeholder expectations

Risk — the possibility of loss or injury

Risk acceptance — accepting the consequences should a risk occur

Risk avoidance — eliminating a specific threat or risk, usually by eliminating its causes

Risk events — specific circumstances that may occur to the detriment of the project

Risk factors — numbers that represent overall risk of specific events, given their probability of occurring and the consequence to the project if they do

Risk identification — determining which risks are likely to affect a project and documenting the characteristics of each

Risk management plan — a plan that documents the procedures for managing risk throughout the project

Risk management planning — deciding how to approach and plan the risk management activities for a project, by reviewing the project charter, WBS, roles and responsibilities, stakeholder risk tolerances, and the organization's risk management policies and plan templates

Risk mitigation — reducing the impact of a risk event by reducing the probability of its occurrence

Risk monitoring and control — monitoring known risks, identifying new risks, reducing risks, and evaluating the effectiveness of risk reduction throughout the life of the project

Risk response planning — taking steps to enhance opportunities and reduce threats to meeting project objectives

Risk symptoms or triggers — indications for actual risk events

Risk transference — shifting the consequence of a risk and responsibility for its management to a third party

Risk utility or risk tolerance — the amount of satisfaction or pleasure received from a potential payoff

Risk-averse — having a low tolerance for risk

Risk-neutral — a balance between risk and payoff

Risk-seeking — having a high tolerance for risk

Robust design methods — methods that focus on eliminating defects by substituting scientific inquiry for trial-and-error methods

Rough order of magnitude (ROM) estimate — a cost estimate prepared very early in the life of a project to provide a rough idea of what a project will cost

Runaway projects — projects that have significant cost or schedule overruns

S

Schedule control — controlling and managing changes to the project schedule

Schedule development — analyzing activity sequences, activity duration estimates, and resource requirements to create the project schedule

Schedule performance index (SPI) — the ratio of earned value to planned value; can be used to estimate the projected time to complete a project

Schedule variance (SV) — the earned value minus the planned value

Scope — all the work involved in creating the products of the project and the processes used to create them

Scope change control — controlling changes to project scope

Scope creep — the tendency for project scope to keep getting bigger and bigger

Scope definition — subdividing the major project deliverables into smaller, more manageable components

Scope planning — developing documents to provide the basis for future project decisions, including the criteria for determining if a project or phase has been completed successfully

Scope statement — a document used to develop and confirm a common understanding of the project scope

Scope verification — formalizing acceptance of the project scope

Secondary risks — risks that are a direct result of implementing a risk response

Seven run rule — if seven data points in a row on a quality control chart are all below the mean, above the mean, or are all increasing or decreasing, then the process needs to be examined for nonrandom problems

Slack — the amount of time a project activity may be delayed without delaying a succeeding activity or the project finish date; also called **float**

Slipped milestone — a milestone activity that is completed later than planned

SMART criteria — guidelines to help define milestones that are specific, measurable, assignable, realistic, and time-framed

Smoothing mode — deemphasizing or avoiding areas of differences and emphasizing areas of agreements

Solicitation — obtaining quotations, bids, offers, or proposals as appropriate

Solicitation planning — documenting product requirements and identifying potential sources

Source selection — choosing from among potential suppliers

Staff acquisition — getting the needed personnel assigned to and working on the project

Staffing management plan — a document that describes when and how people will be added to and taken off the project team

Stakeholder analysis — an analysis of information such as key stakeholders' names and organizations, their roles on the project, unique facts about each stakeholder, their level of interest in the project, their influence on the project, and suggestions for managing relationships with each stakeholder

Stakeholders — people involved in or affected by project activities

Standard deviation — a measure of how much variation exists in a distribution of data

Start-to-finish dependency — a relationship on a project network diagram where the "from" activity cannot start before the "to" activity can finish

Start-to-start dependency — a relationship in which the "from" activity cannot start until the "to" activity starts

Statement of work (SOW) — a description of the work required for the procurement

Statistical sampling — choosing part of a population of interest for inspection

Status reports — reports that describe where the project stands at a specific point in time

Status review meetings — regularly scheduled meetings used to exchange project information

Strategic planning — determining long-term objectives by analyzing the strengths and weaknesses of an organization, studying opportunities and threats in the business environment, predicting future trends, and projecting the need for new products and services

Structural frame — deals with how the organization is structured (usually depicted in an organizational chart) and focuses on different groups' roles and responsibilities to meet the goals and policies set by top management

Subproject managers — people responsible for managing the subprojects that a large project might be broken into

Sunk cost — money that has been spent in the past

Symbolic frame — focuses on the symbols, meanings, and culture of an organization

Synergy — an approach where the whole is greater than the sum of the parts

System outputs — the screens and reports the system generates

System testing — testing the entire system as one entity to ensure it is working properly

Systems — sets of interacting components working within an environment to fulfill some purpose

Systems analysis — a problem-solving approach that requires defining the scope of the system to be studied, and then dividing it into its component parts for identifying and evaluating its problems, opportunities, constraints, and needs

Systems approach — a holistic and analytical approach to solving complex problems that includes using a systems philosophy, systems analysis, and systems management

Systems development life cycle (SDLC) — a framework for describing the phases involved in developing and maintaining information systems

Systems management — addressing the business, technological, and organizational issues associated with making a change to a system

Systems philosophy — an overall model for thinking about things as systems

Systems thinking — taking a holistic view of an organization to effectively handle complex situations

T

Tangible costs or **benefits** — costs or benefits that can be easily measured in dollars

Team development — building individual and group skills to enhance project performance

Termination clause — a contract clause that allows the buyer or supplier to end the contract

Theory of constraints (TOC) — a management philosophy that states that any complex system at any point in time often has only one aspect or constraint that is limiting its ability to achieve more of its goal

Time and material contracts — a hybrid of both fixed price and cost reimbursable contracts

Top 10 Risk Item Tracking — a qualitative risk analysis tool for identifying risks and maintaining an awareness of risks throughout the life of a project

Top-down approach — creating a WBS by starting with the largest items of the project and breaking them into their subordinate items

Total slack — the amount of time an activity may be delayed from its early start without delaying the planned project finish date; also called **total float**

Tracking Gantt chart — a Gantt chart that compares planned and actual project schedule information

Triple constraint — balancing scope, time, and cost goals

U

Unit price contract — a contract where the buyer pays the supplier a predetermined amount per unit of service, and the total value of the contract is a function of the quantities needed to complete the work

Unit test — a test of each individual component (often a program) to ensure it is as defect-free as possible

Use case modeling — a process for identifying and modeling business events, who initiated them, and how the system should respond to them

User acceptance testing — an independent test performed by end users prior to accepting the delivered system

W

Weighted scoring model — a technique that provides a systematic process for basing project selection on numerous criteria

Withdrawal mode — retreating or withdrawing from an actual or potential disagreement

Work authorization system — a method for ensuring that qualified people do the work at the right time and in the proper sequence

Work breakdown structure (WBS) — an outcome-oriented analysis of the work involved in a project that defines the total scope of the project

Work package — a deliverable or product at the lowest level of the WBS

Index

A

abbreviated maturity level, 219
acceptance decisions, 204
ACM-W (Association for Computing Machinery Committee on Women in Computing), 230
activities. *See* task(s)
activity definition, 121, 122–123
activity duration estimating, 121, 127–128
activity-on-arrow (AOA) method, 124
activity sequencing, 121, 123–127
  project network diagrams, 124–127
activity sequencing relationships, 123–124
  importance of establishing, 146
  project network diagrams, 125–126
actual cost(s) (ACs), 177
  entering using Project 2000, 494–496
actual cost of work performed (ACWP), 177
actual times, entering using Project 2000, 494–496
ACWP (actual cost of work performed), 177
Adamany, Hank, 353
adaptive maturity level, 219
Addeman, Frank, 133
ad-hoc maturity level, 219
ADM (arrow diagraming method), 124
administrative closure, 270, 279–280
  Northwest Airlines example, 435
Air Force, 105, 291, 314–315
Amazon.com, 274
American Airlines, 215
"amiables," 255
AMS REALTIME, 18
analogous estimates, 167
analogy approach for developing WBSs, 105
"analyticals," 255
Anderson, David, 310
Antarctic Support Associates, 140
AOA (activity-on-arrow) method, 124
Apple Computer, 194
appraisal cost, 216
Ariba, 353
arrow diagraming method (ADM), 124
Artemis, 13
Arthur Andersen Worldwide, 350–351
assignment as influence base, 236
Association for Computing Machinery Committee on Women in Computing (ACM-W), 230
audits
  project, Northwest Airlines example, 436–437
  quality, 203
authority as influence base, 236

B

Baan, 353
BAC (budget at completion), 181
backward passes, 135
Bank of America, 144
baseline(s)
  costs, 175
  EVM, 176
  saving files with and without, 465–466
baseline dates, 131
baseline plans, Project 2000, 494
BCWP (budgeted cost of work performed), 178
BCWS (budgeted cost of work scheduled), 177
Beauchine, Fay. *See* Northwest Airlines ResNet project
benchmarking, 203
Blue Cross Blue Shield of Michigan, 48–50
Borg, Anita, 230
bottom-up approach for developing WBSs, 106
bottom-up estimating, 167
brainstorming, 311
Briggs, Katherine C., 253
budget(s), 177
  influence base, 236
budgetary estimates, 166
budget at completion (BAC), 181

budgeted cost of work performed (BCWP), 178
budgeted cost of work scheduled (BCWS), 177
budgeting, cost, 160, 173–175
budget section of project plan, 66
buffers, 139
bursts, 125
business area analysis, 86
business justification, Northwest Airlines example, 368–372
business strategy, 86–87

C

Cabanis-Brewin, Jeannette, 229–230
California, state of, 159
California State University, Monterey Bay, 282
Capability Maturity Model (CMM), 218
careers in project management, 14–15
Carter, Marshall, 40
case modeling to improve requirements process, 110
cash flow analysis, 162
categorizing information technology projects, 89
CCBs (change control boards), 74–75
center of excellence, 40
Certificates of Added Qualification, 529
certification
  Certified IT Project Manager, 529–530
  CompTIA IT Project+, 529
  PMP. *See* Project Management Professional (PMP) certification
Certified IT Project Manager (CITPM), 529–530
change, managing resistance to, 426–427
change control
  contract administration, 351–352
  schedule changes, 142–144
  scope. *See* scope change control
change control boards (CCBs), 74–75
change control systems, 74–76
change orders, constructive, 352
Chemical Bank, 194
Chief Project Officer (CPO), 40
Choiniere, Ray, 254–255
Christenson, Kathy. *See* Northwest Airlines ResNet project
CITPM (Certified IT Project Manager), 529–530
classical systems theory, 235
close-out phase of projects, 29
closing processes, 44, 46–47
  Northwest Airlines example, 433–444
  scope, 434–435
CMM (Capability Maturity Model), 218
Coca-Cola, 240
COCOMO (Constructive Cost Model), 168
COCOMO II, 168
code of ethics, 16
coercive power, 237
comments, inserting in Project 2000 files, 511–512
Commerce One, 353
communications
  change control, 75
  determining complexity, 275–278
communications infrastructure, 291
communication skills
  conflict management using, 281–282
  improving, 282–283
communications management, 9. *See also* project communications management
communications management plans, 270
communications planning, 269, 270–272
compromise mode, 281
CompTIA IT Project+ certification, 529
computerized tools, estimating costs, 168
concept phase of projects, 27
Concur Technologies, 353
configuration management, 75
conflict management, 281–282
conformance, 215
conformance to requirements, 195
confrontation mode, 281
constructive change orders, 352

Constructive Cost Model (COCOMO), 168
contingency plans, 306
contingency reserves (allowances), 164, 306–307
contract(s), 337. *See also* project procurement management
  termination clauses, 344
  types, 341–344
contract administration, 338, 351–352
contract close-out, 338, 352
control charts, 208–209
controlling processes, 44, 46–47
  Northwest Airlines example, 417–429
  scope, 418–419
cost(s)
  actual, 177
  direct, 163
  indirect, 163
  sunk, 163
  triple constraint, 5–7
  viewing project cost information using Project 2000, 491–493
cost baselines, 175
cost budgeting, 160, 173–175
cost control, 160, 175–183
cost estimating, 160, 165–173
  Northwest Airlines example, 391–393
  Project 2000, 484–488
  sample cost estimates, 170–173
  techniques and tools, 167–168
  types of cost estimates, 165–167
  typical problems, 169
cost management, 8. *See also* project cost management
cost management plans, 167
cost of nonconformance, 215
cost of quality, 215–216
cost performance index (CPI), 179
cost plus fixed fee (CPFF) contracts, 342
cost plus incentive fee (CPIF) contracts, 342
cost plus percentage of costs (CPPC) contracts, 342
cost reimbursable contracts, 341–342
Cost Table view, Project 2000, 456
cost variance (CV), 179
Covey, Stephen, 238
CPFF (cost plus fixed fee) contracts, 342
CPI (cost performance index), 179
CPIF (cost plus incentive fee) contracts, 342
CPM. *See* critical path method (CPM)
CPO (Chief Project Officer), 40
CPPC (cost plus percentage of costs) contracts, 342
crashing, 136
critical chain scheduling, 138–140
critical path(s), 132
critical path method (CPM), 132–137
  Project 2000, 482–484
  schedule trade-offs, 134–135
  shortening project schedules, 135–316
  updating data, 137
Crosby, Philip B., 198
Crosstable Report dialog box, 492
culture, organizational, 242
customers, 4–5
CV (cost variance), 179

D

dates
  baseline, 131
  finish. *See* finish dates
  start. *See* start dates
Dayton Tire Co., 308
decision trees, 318–319
Defense Systems Management College (DSMC), 313–314
defined maturity level, 218
definitive estimates, 166
deliverables, 27
Delphi Technique, 312
DeMarco, Tom, 169, 216–217
Deming, W. Edwards, 196–197
Department 56, Inc., 350–351
Department of Defense. *See* U.S. Department of Defense (DOD)
dependencies. *See* relationships
deputy project managers, 242
design of experiments, 201
development phase of projects, 28
diagraming techniques for risk management, 313
direct costs, 163
directives, 89
discount factor, 91–92

discount rate, 91
discretionary dependencies, 123
DOD. *See* U.S. Department of Defense (DOD)
"drivers," 255
DSMC (Defense Systems Management College), 313–314
dummy activities, 127
duration of activities, estimating, 127–128

E

EAC (estimate at completion), 180–181
early finish dates, 134–135
early start dates, 134, 135
earned value (EV), 178
earned value management (EVM), 175–183
  Project 2000, 498–499
Earned Value view, Project 2000, 456
Edison, Thomas, 32
effectiveness of people, improving, 238–241
empathic listening, 239
EMV (expected monetary value), 318–319
enterprise project management software, 18
enterprise resource planning (ERP) systems, 353
Entry table, assigning resources, 488–489
Entry Table view, Project 2000, 456
ERP (enterprise resource planning) systems, 353
estimate at completion (EAC), 180–181
estimating costs. *See* cost estimating
EV (earned value), 178
EVM. *See* earned value management (EVM)
exams, certification. *See* Project Management Professional (PMP) certification
executing processes, 44, 45–46–47
  Northwest Airlines example, 401–413
expected monetary value (EMV), 318–319
experiments, design, 201
expertise as influence base, 236
expert judgment for qualitative risk analysis, 317
expert power, 237
"expressives," 255
external dependencies, 123
external failure cost, 216

F

facilitating areas, 8–9
fallback plans, 306
Farmland Industries Inc., 111
fast tracking, 136
features, 201–202
Federal Express, 86
feeding buffers, 139
Feigenbaum, Armand V., 200
FFP (firm-fixed price) contracts, 341
files
  creating in Project 2000, 460–461
  saving as Web pages, 512–514
  saving with and without baselines, 465–466
filters, Project 2000, 457–459
financial risk, 309
finish dates
  early, 134–135
  late, 135
finish-to-finish relationships, 126
finish-to-start relationships, 126
Finney, Jeff, 231
firm-fixed price (FFP) contracts, 341
fishbone diagrams, 199
fitness for use, 195
fixed costs, entering in Cost table, 484–485
fixed price contracts, 341
fixed price incentive (FPI) contracts, 341
float, 132
  free, 134
forcing mode, 281
Ford Motor Company, 165, 200
formal acceptance, 279, 280
form views, Project 2000, 455, 456
FPI (fixed price incentive) contracts, 341

free slack (float), 134
friendship as influence base, 236
functionality, 201–202
functional structure, 34, 35, 36
function points, 168

G

Gantt, Henry, 11, 129
Gantt charts, 11–12, 71
  creating with software, 145
  Project 2000, 478–480
  tracking, 131
Gantt Chart view, Project 2000, 456
Gates, Bill, 192, 255, 291
GE, 209, 240
General Motors, 192–193
Goodyear, 200
graphical views, Project 2000, 455, 456
groupthink, 282
Gurer, Denise, 230

H

Harry, Mikel, 209
Hawkins, Bill, 371
Help dialog box, 450
Help window, Project 2000, 449–452
Herzberg, Frederick, 235
Hewlett-Packard, 200
hierarchy of needs, 233–234
Higgins, Chris, 144
history of project management, 11–13
"How's it going?," 17
human resource costs, entering in Cost table, 485–486
human resource management, 8, 228–232. *See also* project human resource management
  current state, 228–230
  future, 231–232
human resources frame, 33
hygiene factors, 235
hyperlinks, inserting in Project 2000 files, 510–511

I

Ibbs, William, 219, 302
IEEE Standard 1058.1, 66
implementation phase of projects, 28–29
improving information technology project quality, 213–219
  cost of quality, 215–216
  leadership, 213–215
  maturity models, 217–219
  organizational influences and workplace factors, 216–217
improving project communications, 280–291
  communications skills to manage conflict, 281–282
  developing better communication skills, 282–283
  developing communications infrastructure, 219
  running effective meetings, 283–285
  templates, 285–291
incremental release model of system development life cycle, 29
indirect costs, 163
influence, 236–237
influence diagrams, 313
information distribution, 269, 272–278
  determining complexity of communications, 275–278
  formal and informal methods, 273–275
  technology to enhance, 273
information sources, project risk management, 307–310
information technology projects, integrated change control, 73–74
initial maturity level, 218
initiating processes, 44, 45, 46–47
  Northwest Airlines example, 361–376
  scope, 362
INPUT, 347
Insert Hyperlink dialog box, 511
intangible costs (benefits), 163
integrated change control, 59, 71–76
  change control system, 74–76
  information technology projects, 73–74
  integrated. *See* integrated change control
  objectives, 71
integration management. *See* project integration management
integration testing, 211
interface management, 61–62
internal failure cost, 216

internal rate of return (IRR), 162

Internal Revenue Service (IRS), 159

International Organization for Standardization (ISO)
  definition of quality, 195
  ISO 9000 standard, 200

interviewing for risk management, 312

introduction section, of project plan, 63–64

IRR (internal rate of return), 162

IRS (Internal Revenue Service), 159

Ishikawa, Kaoru, 198–199

ISO. *See* International Organization for Standardization (ISO)

ISO 9000 standard, 200

J

JAD (Joint Application Design), 110

Joint Application Design (JAD), 110

Juran, Joseph M., 197–198, 213

Juran Trilogy, 197

K

Kaman Science Center, 310

Keirsey, David, 254–255

Kennedy, John F., 238

kickoff meetings, 283

kill points, 31

King, Martin Luther, Jr., 238

Kivestu, Peeter. *See* Northwest Airlines ResNet project

knowledge areas, 8
  process groups related to, 46–47

known unknowns, 164

Kodak, 339

Kwak, Young H., 219, 302

L

lag time, 475, 476

late finish dates, 135

late start dates, 135

leadership. *See also* top management
  improving information technology project quality, 213–215
  Northwest Airlines example, 403–404

lead time, 475, 476

learning curve theory, 163

Lee, Arvid. *See* Northwest Airlines ResNet project

legacy systems, 180

legitimate power, 237

lessons learned, 279, 280
  Northwest Airlines example, 439–440

life cycle costing, 161

listening, empathic, 239

Lister, Timothy, 216–217

L.L. Bean, 240

Lucent Technology, 140

lump sum contracts, 341

M

McDonnell Aircraft Company, 105

McDonnell Douglas Corporation, 168

McFarlan, F. W., 308–309

McGregor, Douglas, 235

McNeally, Scott, 337

main screen, Project 2000, 452–454

maintainability, 202

make-or-buy analysis, 340–341

make-or-buy decisions, 338

Malcolm Baldridge Award, 200

managed maturity level, 218

managed project management, 219

management objectives in project plan, 64

management reserves, 164

management skills, 41

management/technical approach section of project plan, 64–65

mandatory dependencies, 123

market risk, 309

Maslow, Abraham, 233–234

matrix organizations, 34, 35, 36

maturity models, improving information technology project quality, 217–219

MBTI (Myers-Briggs Type Indicator), 253–255

mean, 207

measurement and test equipment costs, 216

meetings, effective, 283–285

"Melissa" virus, 194

merges, 125

Merrill, David, 255

methodology, developing, 48–50

Micro-Frame Technologies, Inc., 219

Microsoft, 192–193, 274
Microsoft Project 2000. *See* Project 2000
Microsoft Project Central, 17–18, 294, 516–518
milestones, 129, 130–131
  slipped, 132
Milestones Simplicity, 17
mirroring, 240
money as influence base, 236
Monte Carlo analysis, 320
motivation theories, 233–236
Motorola, Inc., 214
multitasking, 138
Murphy's Law, 139
Myers, Isabel B., 253
Myers-Briggs Type Indicator (MBTI), 253–255

N

NASP (National Aerospace Plan) project, 321
National Aerospace Plan (NASP) project, 321
National Insurance Recording System (Nirs2), 137
needs hierarchy, 233–234
net present value (NPV) analysis, 90–92
network diagrams, 12
  Project 2000, 480–482
Network Diagram view, Project 2000, 456
Nirs2 (National Insurance Recording System), 137
nodes, 124
normal distribution, 207
Northwest Airlines, 109, 203
Northwest Airlines ResNet project
  controlling processes example, 417–429
  executing processes example, 401–413
  initiating processes example, 361–376
  lessons learned, 439–440
  personnel transition, 438–439
  planning processes example, 379–398
  project audits, 436–437
  recognition, 438–439
notes, inserting in Project 2000 files, 511–512
NPV (net present value) analysis, 90–92
numbering tasks using Project 2000, 464

O

OBS (organizational breakdown structure), 244
OPM3 (Organizational Project Management Maturity Model) Standard, 218–219
opportunities, 89
optimizing maturity level, 218
organization(s), 32–40
  frames, 32–34
  improving information technology project quality, 216–217
  structures, 34–36
organizational breakdown structure (OBS), 244
organizational charts in project plans, 64
organizational culture, 242
organizational needs, project selection basis, 88
organizational planning, 232, 241–246
Organizational Project Management Maturity Model (OPM3) Standard, 218–219
organizational skills, 41
organizational standards, 40
organized maturity level, 219
Ouchi, William, 235–236
outsourcing. *See* procurement
over-allocation, 249
overview section of project plan, 63–64
Owens-Corning, 86–87

P

PanEnergy Corp., 110
parametric modeling, 168
Pareto analysis, 204–205
Pareto diagrams, 205
Parkinson's Law, 139
payback period, 93
PDM (precedence diagraming method), 125, 126–127
penalty as influence base, 236
performance, 202
performance indexes
  costs, 179
  schedule, 179–180
performance reporting, 270, 278–279
  Northwest Airlines example, 425–426
perks, 231
personnel transition, Northwest Airlines example, 438–439

PERT (Program Evaluation and Review and Technique), 141
phase exits, 31
plan(s)
  baseline, Project 2000, 494
  communications management, 270
  contingency, 306
  cost management, 167
  fallback, 306
  project. *See* project plans
  risk management, 306
  staffing management, 246
planned value (PV), 177
planning
  communications, 269, 270–272
  ERP systems, 353
  organizational, 232, 241–246
  procurement. *See* procurement planning
  project management. *See* project management planning
  quality, 195–196, 201–203
  resource, 160, 164–165
  risk management, 305–307
  risk response, 305, 321–323
  scope, 85, 98–99
  solicitation, 338, 345–348
  strategic, 85
planning processes, 44, 45, 46–47
  Northwest Airlines example, 379–398
PMI. *See* Project Management Institute (PMI)
PMP. *See* Project Management Professional (PMP) certification; Project Management Professionals (PMPs)
political frame, 33–34
politics, 33
Porter, Michael, 86
power, 237–238
precedence diagraming method (PDM), 125, 126–127
prevention cost, 215–216
probabilistic time estimates, 141
problems, 89
process(es), definition, 44
process adjustments, 204
process flow charts, 313
process groups. *See* project management process groups
procurement
  definition, 336
  purposes, 336–337
procurement management, 9. *See also* project procurement management
procurement planning, 339–345
  contract types, 341–344
  statement of work, 344–345
  tools and techniques, 340–341
product life cycles, 29–31
  project life cycle versus, 30
profit(s), 161
profit margin, 161
program(s), 5
Program Evaluation and Review and Technique (PERT), 141
progress reports, 278
project(s)
  attributes, 4–5
  definition, 4–7
project acquisition, 27
project archives, 279–280
project audits, Northwest Airlines example, 436–437
project buffers, 139
Project Central, 17–18, 294, 516–518
Project+ certification, 529
project charters, 84, 96–98
  Northwest Airlines example, 373–374
Project Commander, 18
project communications management, 267–296
  administrative closure, 270, 279–280
  communications planning, 269, 270–272
  importance, 268–270
  improving. *See* improving project communications
  information distribution. *See* information distribution
  Northwest Airlines example, 394–396, 407–408
  performance reporting, 270, 278–279
  software, 292–295, 507–518
project controls in project plan, 64
project cost management, 157–185
  basic principles, 160–164
  cost budgeting, 160, 173–175
  cost control, 160, 175–183
  cost estimating. *See* cost estimating

cost management
plans, 167
importance, 158–160
resource planning, 160, 164–165
software, 183, 484–499

project feasibility, 27

project forecasting, 279

project human resource management, 227–261
definition, 232
importance, 228–232
improving effectiveness, 238–241
influence and power, 236–238
motivation theories, 233–236
Northwest Airlines example, 394–396
organizational planning, 232, 241–246
resource loading and leveling, 249–252
software, 257–259, 499–507
staff acquisition, 232, 247–248
team development. *See* team development

Project Information dialog box, 460

project initiation, 84, 85–98
identifying potential projects, 85–87
project selection methods. *See* selecting projects

project integration management, 58–77
definition, 59–62
integrated change control. *See* integrated change control
project plan. *See* project plans

project life cycle, 27–32
importance of phases, 31–32
phases, 27–29
product life cycle versus, 30

project management
advantages, 3
definition, 7–10
history, 11–13
relationship to other disciplines, 10–11

Project Management Institute (PMI), 14
Certificates of Added Qualification, 529
certification. *See* Project Management Professional (PMP) certification
code of ethics, 16
Standards Development Program, 218–219

project management office, 40

project management planning
cost estimates, 391–393
human resource and communications planning, 394–396
Northwest Airlines example, 379–398
plan development, 383–385
project scope and schedules, 386–391
quality, risk, and procurement planning, 396–397
scope, 380–383

project management process groups, 43–47
knowledge areas related to, 46–47

project management profession, 13–18
careers, 14–15
certification, 15–16
code of ethics, 16
software, 16–18

Project Management Professional (PMP) certification, 15–16, 527–528, 530–536
advice, 535–536
exam structure and content, 530–531
preparing for exam, 531
requirements for earning and maintaining, 528
sample exam questions, 533–535
tips for taking exam, 532–533

Project Management Professionals (PMPs), 15–16

project management software. *See* software

Project Management Technologies, Inc., 219

project managers
actions, Northwest Airlines example, 374–375
selection, Northwest Airlines example, 368
skills, 41–43

project matrix, 34–35, 36

project network diagrams, 124–127

project organization section of project plan, 64

project phases. *See* project life cycle

project plans
development, 59, 62–67
execution, 59, 68–71, 69
format guidelines, 66
sections, 63–66
stakeholder analysis, 67

project procurement management, 335–356
contract administration, 338, 351–352
contract close-out, 338, 352
definition, 337–338
importance, 336–339
Northwest Airlines example, 396–398, 408–410
procurement planning. *See* procurement planning
software, 352–354
solicitation, 338, 348
solicitation planning, 338, 345–348
source selection, 338, 348–350

Project Properties dialog box, 461

project quality management, 191–221
development of modern quality management, 196–200
definition, 195–196
improving project quality. *See* improving information technology project quality
Northwest Airlines example, 396–398
Pareto analysis, 204–205
processes, 195–196
quality assurance, 203
quality control, 204
quality control charts, 208–209
quality of information technology projects, 192–194
quality planning, 201–203
seven run rule, 211
six sigma, 209–210
statistical sampling and standard deviation, 206–208
testing, 211–213

project responsibilities in project plans, 64

project risk management, 301–329
importance, 302–305
information sources, 307–310
Northwest Airlines example, 396–398
processes, 305
results, 326–327
risk analysis. *See* qualitative risk analysis; quantitative risk analysis
risk identification, 305, 310–313
risk management planning, 305–307
risk monitoring and control, 305, 323
risk response planning, 305, 321–323

project scope management, 83–112
definition, 84–85
processes, 84–85
project initiation. *See* project initiation; selecting projects
scope change control, 108–111
scope creep, 107
scope planning, 98
scope statement, 98–99
scope verification, 108
software, 459–466
work breakdown structures. *See* work breakdown structures (WBSs)

project time management, 119–149
activity definition, 121, 122–123
activity duration estimating, 121, 127–128
activity sequencing, 121, 123–127
definition, 121
importance of schedules, 120–122
processes involved, 121
project network diagrams, 124–127
schedule control, 121, 142–144
schedule development. *See* schedule development
software, 145–147, 466–484

Project 2000, 17, 447–518
creating new project files, 460–461
developing work breakdown structures, 462–464
features, 508
filters, 457–459
Help window, 449–452
main screen elements, 452–454
new features, 451–452
project communications management, 507–518
project cost management, 484–499
project human resource management, 499–507
project scope management, 459–466
project time management, 466–484
starting, 449
Tracking toolbar, 495
views, 454–457
workgroups, 515–516

promotion as influence base, 236

prototyping release model of system development life cycle, 29

prototyping to improve requirements process, 110

PV (planned value), 177

software, 16–18, 71. *See also* Project Central; Project 2000
  cost management, 183
  high-end tools, 18
  human resource management, 257–259
  low-end tools, 17
  midrange tools, 17–18
  preventing misuse, 146–147
  project communications, 292–295
  project procurement management, 352–354
  project risk management, 323–326
  time management, 145–147
  Web sites, 353

Software Engineering Institute (SEI), 218

Software Quality Function Deployment (SQFD) model, 217

solicitation, 338, 348

solicitation planning, 338, 345–348

source selection, 348–350
  project procurement management, 338, 348–350

SOW (statement of work), 344–345

SPI (schedule performance index), 179–180

spiral model of system development life cycle, 29, 30

split window, assigning resources, 490–491

sponsors, 4–5

SQFD (Software Quality Function Deployment) model, 217

staff acquisition, 232, 247–248

staffing, in project plan, 65

staffing management plans, 246

stakeholder(s), 8

stakeholder analysis, 67
  project communications, 271

stakeholder management, 37–38

standard deviation, 207–208

start dates
  early, 134, 135
  late, 135

starting Project 2000, 449

start-to-finish relationships, 126

start-to-start relationships, 126

statement of work (SOW), 344–345

State Street Bank and Trust Company, 40

statistical sampling, 206–207

status reporting, Northwest Airlines example, 425–426

status reports, 278

status review meetings, 70

strategic planning, 85

Strobel, Nancy, 427

structural frame, 32–33

subproject managers, 242

summary task(s), creating using Project 2000, 463–464

Summary Task Information dialog box, 512

sunk costs, 163

supporting details, cost estimates, 167, 170–173

SV (schedule variance), 179

SWOT analysis, 85, 312

symbolic frame, 34

Synergis Technologies Group, 140

synergy, 238

system(s), definition, 25

system flow charts, 313

system outputs, 202

systems analysis, 25

systems approach, 25

systems development life cycle (SDLC), 29–30

systems management, 25–27

systems philosophy, 25

systems thinking, 25

system testing, 211

T

table views, Project 2000, 455, 456

Taco Bell, 240

Taguchi, Genichi, 199–200

tangible costs (benefits), 162–163

task(s). *See also* activity *entries*
  assigning resources, 488–492
  definition, 121
  dummy, 127
  numbering using Project 2000, 464
  recurring, creating in Project 2000, 468–470
  summary, creating using Project 2000, 463–464

task dependencies. *See* relationships

Task Details Form view, Project 2000, 456

task durations, entering using Project 2000, 466–471

Task Name Form view, Project 2000, 456

task sheet views, Project 2000, 455, 456

Taylor, William C., 248

team development, 232, 252–257
- general advice, 2560257
- reward and recognition systems, 256
- team-building activities, 253–256
- training, 252–253

technical processes in project plan, 65

technological skills, 41

technology risk, 309, 310

templates
- Project 2000, 509–510
- project communications, 285–291
- software, proper use, 147

termination clauses, 344

testing, 211–213

Texas Instruments, 240

Thamhain, H. J., 236–237

Theory of Constraints, 138

Theory X, 235

Theory Y, 235

Theory Z, 235–236

3M, 208

time, triple constraint, 5–7

time-adjusted rate of return, 162

time and material contracts, 342

time management, 8. *See also* project time management

toolbar, assigning resources, 489–490

tools and techniques, 9

top-down approach for developing WBSs, 105–106

top-down estimates, 167

top management
- improving information technology project quality, 213–215
- Northwest Airlines example, 374–375

top management commitment, 38–40
- need for, 39–40
- organizational standards, 40

Top 10 Risk Item Tracking, 316

tracking Gantt charts, 131

Tracking Table view, Project 2000, 456

Tracking toolbar, Project 2000, 495

training
- team development, 252–253
- users, 410–412

triggers, 313

triple constraint, 5–7

U

uncertainty, 5

United Airlines, 215

U.S. Air Force, 105, 291, 314–315

U.S. Department of Defense (DOD)
- project plan format guidelines, 66
- WBS development guidelines, 104–105

unit price contracts, 343

unit tests, 211

unknown unknowns, 164

Update Tasks dialog box, 496–497

updating CPM data, 137

user acceptance testing, 211

users, training, 410–412

V

value
- earned, planned, 178
- planned, 177

variances
- cost, 179
- schedule, 179

view(s), Project 2000, 454–457

viewing project cost information, 491–493

W

Wal-Mart, 86

Walt Disney Imagineering, 133

waterfall model of system development life cycle, 29

WBSs. *See* work breakdown structures (WBSs)

Webcritical Technologies, 350
Web pages, saving files as, 512–514
Web sites, 290, 353
  software, 293
weighted averages, 141
weighted scoring model, 93–95
Whirlpool, 209
Wilemon, D. L., 236–237
withdrawal mode, 281
workarounds, 323
work authorization systems, 70
work breakdown structures (WBSs), 27, 99–107
  analogy approach for developing, 105
  developing using Project 2000, 462–464
  guidelines for developing, 104–105
  principles for creating, 106–107
  top-down and bottom-up approaches for developing, 105–106
work challenge as influence base, 236
work description section of project plan, 65
workgroups, Project 2000, 515–516
work packages, 103
workplace factors, improving information technology project quality, 216–217

X

Xerox, 200

Y

Yourdon, Ed, 142